This book reviews one-dimensional reactions, dynamics, diffusion, and adsorption.

In studies of complex systems in biology, chemistry, and physics, understanding can be gained by analytical and numerical analyses of simple models. This book presents integrated reviews at an advanced research level, describing results for one-dimensional models of dynamical processes such as chemical reactions and catalysis, kinetic Ising models, phase separation and cluster growth, monolayer and multilayer adsorption with added relaxation, surface and hard-core, particle dynamics, diffusional transport, random systems. It also covers experimental results for systems ranging from chemical reactions to adsorption and reactions on polymer chains, steps on crystalline surfaces, and DNA. All chapters are written by leading scientists in the field. They present a self-contained review of this subject that will guide a reader from basic concepts, ideas, methods and models to the forefront of research.

This book will be of interest to researchers and graduate students in statistical and theoretical physics, in theoretical chemistry, and in applied mathematics.

NONEQUILIBRIUM STATISTICAL MECHANICS IN ONE DIMENSION

Nonequilibrium Statistical Mechanics in One Dimension

Edited by

VLADIMIR PRIVMAN

Clarkson University

CAMBRIDGE
UNIVERSITY PRESS

CAMBRIDGE UNIVERSITY PRESS
Cambridge, New York, Melbourne, Madrid, Cape Town, Singapore,
São Paulo, Delhi, Dubai, Tokyo, Mexico City

Cambridge University Press
The Edinburgh Building, Cambridge CB2 8RU, UK

Published in the United States of America by Cambridge University Press, New York

www.cambridge.org
Information on this title: www.cambridge.org/9780521559744

© Cambridge University Press 1997

This publication is in copyright. Subject to statutory exception
and to the provisions of relevant collective licensing agreements,
no reproduction of any part may take place without the written
permission of Cambridge University Press.

First published 1997

A catalogue record for this publication is available from the British Library

Library of Congress Cataloguing in Publication data
Nonequilibrium statistical mechanics in one dimension/edited
by Vladimir Privman.
p. cm.
Includes bibliographical references and index.
ISBN 0 521 55974 X (hc)
1. Statistical mechanics. I. Privman, V. (Vladimir), 1955-
QC174.8.N67 1997
530.1'3-dc20 96-45965 CIP

ISBN 978-0-521-55974-4 Hardback
ISBN 978-0-521-01834-0 Paperback

Cambridge University Press has no responsibility for the persistence or
accuracy of URLs for external or third-party internet websites referred to in
this publication, and does not guarantee that any content on such websites is,
or will remain, accurate or appropriate. Information regarding prices, travel
timetables, and other factual information given in this work are correct at
the time of first printing but Cambridge University Press does not guarantee
the accuracy of such information thereafter.

Contents

Contributors

Mustansir Barma
Tata Institute of Fundamental Research
Homi Bhabha Road, Bombay 400 005, India

Daniel ben-Avraham
Clarkson Institute for Statistical Physics and Department of Physics
Clarkson University, Potsdam, NY 13699-5820, USA

Stefan H. Bossmann
Lehrstuhl für Umweltmesstechnik am Engler-Bunte-Institut
Universität Karlsruhe, Richard Willstätter Allee 5
76128 Karlsruhe, Germany

Alan J. Bray
Department of Physics and Astronomy
University of Manchester, Manchester M13 9PL, UK

Stephen J. Cornell
Department of Mathematics and Statistics, University of Guelph
Guelph, Ontario N1G 2W1, Canada
Present address: Department of Physics and Astronomy
University of Manchester, Manchester M13 9PL, UK

Bernard Derrida
Laboratoire de Physique Statistique de l'Ecole Normale Supérieure
24 rue Lhomond, F–75231 Paris 05 Cedex, France
Service de Physique Théorique, C. E. Saclay
F–91191 Gif–sur–Yvette Cedex, France

Ronald Dickman
Department of Physics and Astronomy, Lehman College
City University of New York, Bronx, NY 10468, USA

Charles R. Doering
Center for Nonlinear Studies and Theoretical Division
Los Alamos National Laboratory, Los Alamos, NM 87545, USA
Present address: Department of Mathematics, University of Michigan
Ann Arbor, MI 48109-1109, USA

Timothy C. Elston
Center for Nonlinear Studies and Theoretical Division
Los Alamos National Laboratory, Los Alamos, NM 87545, USA

James W. Evans
Ames Laboratory and Department of Mathematics
Iowa State University, Ames, IA 50011, USA

Martin R. Evans
Theoretical Physics, University of Oxford
1 Keble Road, Oxford OX1 3NP, UK
Present address: Department of Physics and Astronomy
University of Edinburgh, Mayfield Road, Edinburgh EH9 3JZ, Scotland

Harry L. Frisch
Department of Chemistry, State University of New York at Albany
Albany, NY 12222, USA

Nobuyasu Ito
Department of Applied Physics, Faculty of Engineering
The University of Tokyo, Tokyo 113, Japan

Steven A. Janowsky
Department of Mathematics, University of Texas
Austin, TX 78712, USA

Raoul Kopelman
Department of Chemistry, University of Michigan
Ann Arbor, MI 48109-1055, USA

Ron Kroon
Van der Waals-Zeeman Instituut, University of Amsterdam
Valckenierstraat 65-67, 1018 XE Amsterdam, The Netherlands
Present address: Philips Research Laboratories
Prof. Holstlaan 4, 5656 AA Eindhoven, The Netherlands

Joachim Krug
IFF, Forschungszentrum Jülich, D-52425 Jülich, Germany
Present address: Fachbereich Physik
Universität GH Essen, D-45117 Essen, Germany

Joel L. Lebowitz
Departments of Mathematics and Physics, Rutgers University
New Brunswick, NJ 08903, USA

Anna L. Lin
Department of Chemistry, University of Michigan
Ann Arbor, MI 48109-1055, USA

Peter Nielaba
Institut für Physik, Universität Mainz, D-55099 Mainz, Germany

Vladimir Privman
Department of Physics, Clarkson University
Potsdam, NY 13699-5820, USA

Zoltán Rácz
Institute for Theoretical Physics, Eötvös University
Puskin u. 5-7, Budapest 1088, Hungary

Sidney Redner
Center for Polymer Studies and Department of Physics
Boston University, Boston, MA 02215, USA

Lawrence S. Schulman
Department of Physics, Clarkson University
Potsdam, NY 13699-5820, USA

Rudolf Sprik
Van der Waals-Zeeman Instituut, University of Amsterdam
Valckenierstraat 65-67, 1018 XE Amsterdam, The Netherlands

Hideki Takayasu
Graduate School of Information Sciences
Tohoku University, Aoba-ku, Sendai 980-70, Japan

Misako Takayasu
Research Institute for Fracture Technology, Department of Engineering
Tohoku University, Aoba-ku, Sendai 980-70, Japan

Julia M. Yeomans
Theoretical Physics, University of Oxford
1 Keble Road, Oxford OX1 3NP, UK

Klaus Ziegler
Institut für Theorie der Kondensierten Materie
Universität Karlsruhe, D-76128 Karlsruhe, Germany

Preface

A challenge of modern science has been to understand complex, highly correlated systems, from many-body problems in physics to living organisms in biology. Such systems are studied by all the classical sciences, and in fact the boundaries between scientific disciplines have been disappearing; 'interdisciplinary' has become synonymous with 'timely'. Many general theoretical advances have been made, for instance the renormalization group theory of correlated many-body systems [1]. However, in complex situations the value of analytical results obtained for simple, usually one-dimensional (1D) or effectively infinite-dimensional (mean-field), models has grown in importance. Indeed, exact and analytical calculations deepen understanding, provide a guide to the general behavior, and can be used to test the accuracy of numerical procedures.

A generation of physicists have enjoyed the book *Mathematical Physics in One Dimension . . .*, edited by Lieb and Mattis [2], which has recently been re-edited [3]. But what about mathematical chemistry or mathematical biology in 1D? Since statistical mechanics plays a key role in complex, many-body systems, it is natural to use it to define topical coverage spanning diverse disciplines. Of course, there is already literature devoted to 1D models in selected fields, for instance, [4], or to analytically tractable models in statistical mechanics, e.g., [5]. However, in recent years there has been a tremendous surge of research activity in 1D reactions, dynamics, diffusion, and adsorption. These developments are reviewed in this book.

There are several reasons for the flourishing of studies of 1D many-body systems with stochastic time evolution. Not only do such systems offer simplicity and, frequently, exact solvability, but they also correspond to the strong-fluctuation limit; fluctuations in many-body systems often become dominant in lower dimensions. There are also *experimental* systems where 1D dynamics has been observed. Interestingly, examples of such systems, although much less numerous than ordinary (three-dimensional space) or surface (two-dimensional) stochastic-dynamics systems, cover all the classical sciences and range from 1D chemical reactions to certain dynamical effects associated with adsorption and reactions on polymer chains, steps on crystalline surfaces, and DNA—the natural 1D geometries.

This book has been written by leading scientists working in the field. The objective has been to provide an introduction to each topic covered, followed by a technical example of methods and ideas, as well as a survey of related literature. Each chapter brings the reader to the forefront of ongoing research yet also offers a sufficient introductory component to be suitable for readers with only a basic knowledge of statistical mechanics. Furthermore, as with many rapidly growing fields, methods and terminology have developed independently for several topics. This book will serve as an introduction and reference to related ideas for researchers and students in each particular topic.

Part I of the book covers chemical reactions and catalysis, Part II is devoted to kinetic Ising models, and Part III reviews results on phase separation and cluster growth. The methods used in all three of these areas are frequently similar, although they describe different systems. They are also related to the techniques used to study monolayer and multilayer adsorption models, with added relaxation processes—a topic of Part IV. Surface and hard-core particle systems whose dynamics is described by variants of nonlinear diffusion equations are taken up in Part V, while Part VI surveys selected recent results on diffusional transport, including a chapter on random systems. Finally, Part VII presents several experimental systems where 1D dynamical behavior has been measured. Each Part consists of about three chapters prefaced by an editorial note. The latter describes the contents of the chapters, mentions, where applicable, topics not covered in the present monograph, and provides literature citations.

Vladimir Privman

[1] C. Domb and M. S. Green, eds., *Phase Transitions and Critical Phenomena*, Vol. **6** (Academic Press, London, 1976).

[2] E. H. Lieb and D. C. Mattis, eds., *Mathematical Physics in One Dimension. Exactly Solvable Models of Interacting Particles* (Academic Press, New York, 1966).

[3] D. C. Mattis, ed., *The Many-Body Problem. An Encyclopedia of Exactly Solved Models in One Dimension* (World Scientific, Singapore, 1993).

[4] Z. Ha, *Quantum Many-Body System in One Dimension* (World Scientific, Singapore, 1996).

[5] J. L. Lebowitz, ed., *Simple Models of Equilibrium and Nonequilibrium Phenomena* (Elsevier, Amsterdam, 1987).

Part I: Reaction-Diffusion Systems and Models of Catalysis

Editor's note

The first three chapters of the book cover topics in reactions and catalysis. Chemical reactions comprise a vast field of study. The recent interest in models in low dimension has been due to the importance of two-dimensional surface geometry, appropriate, for instance, in heterogeneous catalysis. In addition, several experimental systems realize 1D reactions (Part VII).

The classical theory of chemical reactions, based on rate equations and, for nonuniform densities, diffusion-like differential equations, frequently breaks down in low dimension. Recent advances have included the elucidation of this effect in terms of fluctuation-dominated dynamics. Numerous models have been developed and modern methods in the theory of critical phenomena applied. The techniques employed range from exact solutions to renormalization-group, numerical, and scaling methods.

Models of reactions in 1D are also interrelated with many other 1D systems ranging from kinetic Ising models (Part II) and deposition (Part IV) to nucleation (Part III). Chapter 1 reviews the scaling theory of basic reactions and summarizes numerous results. One of the methods of obtaining exact solutions in 1D, the interparticle-distribution approach, is reviewed in Ch. 2. Other methods for deriving exact results in 1D are not considered in this Part. Instead, closely related systems and solution techniques based on kinetic Ising models and cellular automata are presented in Chs. 4, 6, 8. Coagulation models in Ch. 9 employ methods that have also been applied to reactions [1].

More complicated models of catalysis, directed percolation, and kinetic phase transitions, are treated in Ch. 3. The emphasis is on scaling theories, universality, and a summary of results for various model systems and processes. Further details, and a presentation not biased towards 1D, can be found in a forthcoming book by this publisher [2].

Finally, we cite [3] recent results in reaction-diffusion systems and related 1D models, obtained by converting the problem to that of a 1D quantum-chain Hamiltonian with Euclidean time. This approach provides relations between various models in a systematic way, as well as reproducing some exact solutions.

[1] See, e.g., V. Privman, *Phys. Rev.* E**50**, 50 (1994).
[2] J. Marro and R. Dickman, *Nonequilibrium Phase Transitions and Critical Phenomena* (Cambridge University Press, Cambridge), in preparation.
[3] F. C. Alcaraz, M. Droz, M. Henkel and V. Rittenberg, *Ann. Phys. (NY)* **230**, 250 (1994); G. M. Schütz, *J. Stat. Phys.* **79**, 243 (1995).

1

Scaling theories of diffusion-controlled and ballistically controlled bimolecular reactions

Sidney Redner

Basic features of the kinetics of diffusion-controlled two-species annihilation, $A + B \to 0$, as well as that of single-species annihilation, $A + A \to 0$, and coalescence, $A + A \to A$, under diffusion-controlled and ballistically controlled conditions, are reviewed in this chapter. For two-species annihilation, the basic mechanism that leads to the formation of a coarsening mosaic of A- and B-domains is described. Implications for the distribution of reactants are also discussed. For single-species annihilation, intriguing phenomena arise for 'heterogeneous' systems, where the mobilities (in the diffusion-controlled case) or the velocities (in the ballistically controlled case) of each 'species' are drawn from a distribution. For such systems, the concentrations of the different 'species' decay with time at different power-law rates. Scaling approaches account for many aspects of the kinetics. New phenomena associated with discrete initial velocity distributions and with mixed ballistic and diffusive reactant motion are discussed. A scaling approach is outlined to describe the kinetics of a ballistic coalescence process which models traffic on a single-lane road with no passing allowed.

1.1 Introduction

There are a number of interesting kinetic and geometric features associated with diffusion-controlled two-species annihilation, $A + B \to 0$, and with single-species reactions, $A + A \to 0$ and $A + A \to A$, under diffusion-controlled and ballistically controlled conditions.

In two-species annihilation, there is a spontaneous symmetry breaking in which large-scale single-species heterogeneities form when the initial concentrations of the two species are equal and spatially uniform. Underlying this domain formation is an effective repulsion between A and B that favors segregation of particles into single-species domains. In low spatial dimension, this effective repulsion dominates over the mixing due to diffusion. The resulting spatial organization invalidates the mean-field approximation and its corresponding predictions. Scaling approaches provide an understanding for the origin of this spatial organization and some of its consequences. Related ideas can be applied to the situation where the reactants move by driven diffusive motion, the particles hopping in only one direction. Counter to naive intuition based on Galilean invariance, the reaction kinetics with driven diffusion is qualitatively different from that which occurs with isotropic diffusion and hard-core repulsion.

While single-species reactions are fundamentally simpler than two-species annihilation, there is a wide range of phenomenology that is not fully explored. For example, the 'heterogeneous' single-species reaction, where each reactant moves at a different rate, naturally raises new questions regarding the relation between the initial mobility distribution and the decay rate of different mobility 'species'. These issues are more central when particles move ballistically, so that the initial condition is the only source of stochasticity in the system. In spite of this simplicity, rich and unanticipated phenomena occur, especially for a discrete initial velocity distribution. The case of $A + A \rightarrow 0$ with combined diffusive and ballistic motion is particularly surprising because the concentration decay in this composite process is faster than that of the reaction with only ballistic motion or only diffusion. This intriguing behavior can be understood through dimensional analysis. Finally, in ballistically driven aggregation, $A_i + A_j \rightarrow A_{i+j}$, scaling arguments can be advantageously combined with analytic methods to give a comprehensive account of the kinetics in a momentum-conserving process and in a model that mimics traffic flow on a single-lane road.

In Sec. 1.2, the primary features that characterize the kinetics and spatial organization in two-species annihilation are outlined. The emphasis is on qualitative approaches that should be applicable to many nonequilibrium phenomena. This section closes with a (necessarily incomplete) outline of recent and not fully understood results for $A + B \rightarrow 0$ with driven diffusive reactant motion. Section 1.3 is devoted to single-species reactions. In spite of the wide diversity of phenomena, scaling analyses for time-dependent population distributions provide unifying and comprehensive descriptions of the kinetics. A number of case studies are presented, with mention of some open questions in addition to discussions of known results. A brief summary is given in Sec. 1.4.

1.2 Diffusion-controlled two-species annihilation

In diffusion-limited two-species annihilation, an intriguing aspect of the kinetics is that the density decays more slowly than the rate equation (mean-field) prediction of $1/t$, for a random initial condition of reactants with equal densities for the two species. This fundamental observation, first made by Zel'dovich and coworkers [1] and independently rediscovered by Toussaint and Wilczek [2], stimulated considerable interest and continues to foster current research [3-11]. To appreciate the basic issues, a mean-field description of the decay kinetics in single species reactions is outlined in the next subsection. A heuristic approach for the kinetics of two-species annihilation, which is based on the existence of large-scale spatial heterogeneity, is then presented.

1.2.1 Preliminary: mean-field theory for single-species reactions

In the single-species reaction, irreversible annihilation occurs whenever two particles approach within a reaction radius R. To determine the decay of the concentration within a mean-field approximation, note that in a time of order $1/c$, where $c \equiv c(t)$ is the concentration, each particle will typically encounter another particle. Consequently, in a time $\Delta t \propto 1/kc$, where k is the reaction rate, the concentration decrement, Δc, will be of order c. Combining these gives the mean-field rate equation,

$$\dot{c} \cong \Delta c/\Delta t \propto -kc^2, \qquad (1.1)$$

with solution $c(t) = c(0)/[1 + c(0)\,kt] \sim (kt)^{-1}$. Thus the exponent of the power-law decay is -1, and the time scale is set by k. As discussed below, this exponent value is correct only for spatial dimension $d \geq 4$, the regime of validity of the mean-field approach.

It is instructive to determine the scaling of the reaction rate k [2]. By dimensional analysis of (1.1), k has units $[\ell]^d/[t]$. Furthermore, k is a function only of the diffusion coefficient, D, and the particle radius, R. The only combination of these quantities that possesses the correct units is

$$k \propto DR^{d-2}. \qquad (1.2)$$

For $d > 2$, this ansatz agrees with the Smoluchowski theory [12], in which the reaction rate is given by the steady-state flux towards an absorbing test particle due to the remainder of the particles in the system. However, for $d < 2$, (1.2) leads to the nonsensical conclusion that the reaction rate decreases with increasing particle radius. To determine the appropriate behavior for $d < 2$, one can still apply the Smoluchowski theory, but its interpretation must be modified. Because the incident flux is time dependent, even as $t \to \infty$, the reaction rate now acquires a time dependence of the form $k \propto D^{d/2}/t^{1-d/2}$ for $d < 2$ and $k \propto D/\ln Dt$ for $d = 2$. The lack of dependence on the particle radius is a manifestation of the recurrence of random walks [13] for $d \leq 2$. Therefore, with respect to diffusion, the collision radius is effectively infinite and thus drops out of the system. Employing these reaction rates in the rate equation gives the asymptotic behaviors appropriate for $d \leq 2$:

$$c(t) \propto (Dt)^{-d/2}, \quad d < 2; \qquad c(t) \propto (\ln Dt)/Dt, \quad d = 2. \qquad (1.3)$$

These results can also be obtained from a more microscopic, but equally nonrigorous approach. Because of the recurrence of random walks for $d \leq 2$, the time for a particular reaction to occur should be of order $\Delta t \propto \ell^2/D$, where $\ell \propto c^{-1/d}$ is the typical interparticle spacing. In this formulation, the reaction rate does not enter in the collision time because random walk trajectories are compact. Therefore if two reactants collide once, they will collide an infinite number of times and the reaction rate rescales to a large value. Consequently, for $d < 2$ the rate equation becomes

$$\dot{c} \cong \Delta c/\Delta t \propto -\left(c/D^{-1}c^{-2/d}\right) = -Dc^{1+2/d}, \qquad (1.4)$$

with solution $c(t) \sim (Dt)^{-d/2}$. A straightforward adaptation of this approach also gives a logarithmic correction for the case $d = 2$. Notice that for $d = 1$ the above mean-field approaches give either $\dot{c} \propto -c^2/\sqrt{t}$ or $\dot{c} \propto -c^3$ for the 'effective' rate equation. Both give the correct decay of the concentration, but neither can be rigorously justified. In fact, from the exact solution in one dimension, it may be appreciated that a polynomial rate equation is inadequate in many respects [14], even though it can be engineered to reproduce the correct exponents.

1.2.2 Fluctuation-driven kinetics in two-species annihilation

For $d \leq 4$, the above embellishments of the rate equation are inadequate because single-species domains form, thus invalidating the homogeneity assumption implicit in the mean-field approach. The long-time behavior of $c(t)$ can be understood from the following account of local density fluctuations. Roughly speaking, the difference in the number of A's and B's in a finite volume of linear dimension L remains nearly constant during the time for a particle to traverse the volume by diffusion, $t_L \sim L^2/D$. At $t = 0$, this difference is of the order of the square root of the initial particle number: $N_A - N_B \approx \pm\sqrt{c(0)}\, L^{d/2}$. After a time t_L has elapsed, only the local majority species remains in the domain, whose number $N_>(t_L)$ is of order $\sqrt{c(0)}\,L^{d/2}$. Elimination of L in favor of t gives

$$c(t) \approx N_>(t)/L^d \sim \sqrt{c(0)}\,(Dt)^{-d/4}, \qquad d \le 4. \qquad (1.5)$$

Thus a homogeneous system evolves into a continuously growing domain mosaic whose individual identities are determined by the local majority species in the initial state. At time t, these domains will be of typical linear dimension $\sqrt{Dt}$, within which a single species of concentration $\sqrt{c(0)}\,(Dt)^{-d/4}$ remains.

However, for $d > 4$ the domains are unstable and mean-field theory applies. To justify this consider, for example, the fate of an A particle inside a B domain of linear dimension L and local concentration $\sim L^{-d/2}$. The impurity needs L^2 time steps to exit the domain, during which L^2 distinct sites will have been visited. At each site, the A particle will react with a probability of the order of the B concentration, $L^{-d/2}$. Therefore the probability that an A particle is unsuccessful in exiting a B domain is of order $L^{(4-d)/2}$. Since this vanishes as $L \to \infty$ if $d > 4$, a growing domain mosaic is eventually unstable to diffusive homogenization for $d > 4$.

1.2.3 Multiple microscopic lengths

The above arguments indicate that two lengths characterize the reactant distribution in 1D: the average domain size $L \propto (Dt)^{1/2}$ and the typical interparticle spacing, which scales as $c(t)^{-1} \propto t^{1/4}$. A surprising feature, which reveals a richer structure for the reactant distribution, is that the typical distances between AA and AB closest-neighbor pairs, ℓ_{AA} and ℓ_{AB}, grow with different powers of time for $d < 3$ [15]. The latter characterizes the interdomain 'gap' that separates adjacent domains (Fig. 1.1). This gap controls the kinetics, since each reaction event involves diffusion of an AB pair across a gap. The different scalings of ℓ_{AA} and ℓ_{AB} indicate that nontrivial modulation exists in the reactant concentration over the extent of a domain.

To determine the evolution of ℓ_{AB} in 1D, consider the time dependence of the concentration of closest-neighbor AB pairs, c_{AB}. Typical AB pairs react in a time $\Delta t \sim \ell_{AB}^2/D$. Since the number of reactions per unit length is of order c_{AB}, the time rate of change of the overall concentration is

$$\Delta c/\Delta t \approx -c_{AB}\Big/\big(\ell_{AB}^2/D\big). \qquad (1.6)$$

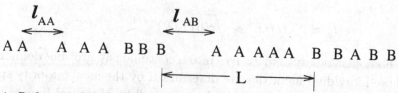

Fig. 1.1. Definition of fundamental interparticle distances in 1D: the typical distance between closest-neighbor same species particles, ℓ_{AA}, the distance between closest-neighbor unlike species, ℓ_{AB}, i.e., the gap between domains, and the typical domain length, L.

The l.h.s. is known from $c(t)$ itself, while in 1D $c_{AB} \propto (Dt)^{-1/2}$, since there is one AB pair per domain of typical size $(Dt)^{1/2}$. These give

$$\ell_{AB} \propto c(0)^{-1/4} (Dt)^{3/8}. \tag{1.7}$$

Thus at least three lengths characterize the reactant distribution: in addition to the average domain size and the typical interparticle spacing, there is the interdomain gap $\ell_{AB} \propto t^{3/8}$. The inequality $\ell_{AB} \gg \ell_{AA}$ is a manifestation of the effective repulsion between opposite species.

The above results can be generalized to spatial dimension $1 \leq d \leq 2$. The time dependence of ℓ_{AB} still follows by applying (1.6), since it holds whenever random walks are compact. Under the assumption of a smooth domain perimeter of linear dimension $t^{(d-1)/2}$ and particles in the perimeter zone separated by a distance of the order of ℓ_{AB}, irrespective of identity, it is straightforward to obtain

$$\ell_{AB} \propto t^{\frac{(d+2)}{4(d+1)}}, \qquad c_{AB}(t) \propto t^{-\frac{d(d+3)}{4(d+1)}}, \tag{1.8}$$

which gives $\ell_{AB} \sim t^{1/3}$ and $c_{AB}(t) \sim t^{-5/6}$ for $d = 2$. For $d > 2$, the transience of random walks implies that two opposite-species particles within a region of linear dimension ℓ_{AB} will react in a time of order ℓ_{AB}^d (rather than ℓ_{AB}^2). Consequently, (1.6) should be replaced by $\Delta c/\Delta t \approx -c_{AB}/\ell_{AB}^d$. This relation, together with the assumption of a smooth interfacial region between domains, gives, for $d > 2$,

$$\ell_{AB} \approx t^{\frac{d+2}{4(2d-1)}}, \qquad c_{AB} \approx t^{-\frac{d^2+5d-4}{4(2d-1)}}. \tag{1.9}$$

These coincide with (1.8) at $d = 2$, but yield $c_{AB} \approx t^{-1}$ and $\ell_{AB} \approx t^{1/4}$ for $d = 3$. The latter represents the limit at which ℓ_{AB} is of the same order as ℓ_{AA}. Thus the nontrivial scaling of interparticle distances disappears in three dimensions and above.

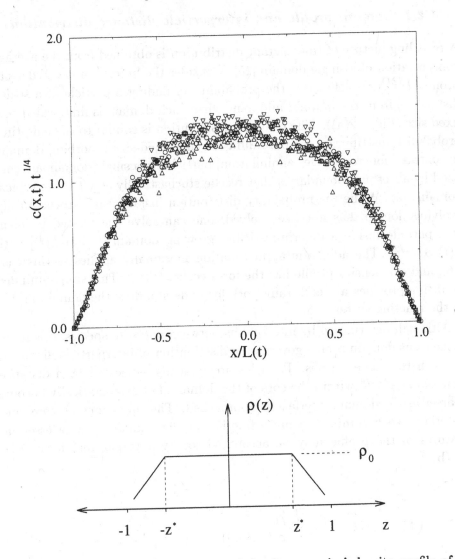

Fig. 1.2. (a) Simulation data for the scaled microcanonical density profile of a single domain, for $A + B \rightarrow 0$ with isotropic diffusion, at $t = 1.5^{13} \cong 194$ ($\triangle$), $t = 1.5^{18} \cong 1477$ ($\triangledown$), and $t = 1.5^{23} \cong 11222$ ($\circ$). Plotted is the scaled local concentration, $\rho(z) \equiv c(x,t)\,t^{1/4}$ (in arbitrary units) vs. $z \equiv x/L(t)$. Here $\pm L(t)$ defines the extent of the domain, and x is a local coordinate with respect to the center of the domain. (b) Idealized trapezoidal form. In the domain core the density is relatively constant, while the density vanishes linearly as a function of the distance to the domain edge.

1.2.4 Domain profile and interparticle distance distribution

A revealing picture of the reactant distribution is obtained from the average concentration of a single domain [15]. Consider the 'microcanonical' density profile, $P^{(M)}(x)$, defined as the probability of finding a particle at a scaled distance x from the domain midpoint when each domain is first scaled to a fixed size (Fig. 1.2(a)). The resulting distribution is similar to the long-time probability distribution for pure diffusion in a fixed-size absorbing domain. In contrast, for two-species annihilation, particles in a single domain are confined by absorbing boundaries that recede stochastically as $\sqrt{t}$ —the typical domain size. While the probability distribution inside such a stochastically evolving domain has not been solved, one can solve the related problem of a particle inside a deterministically growing domain $[-L(t), L(t)]$ with $L(t) \propto t^{1/2}$. The adiabatic approximation marginally applies in this case [16], and the density profile has the form $\cos[\pi x/L(t)]$. This simple-minded modeling provides a useful framework for understanding the domain profile in the reacting system.

Although determined by interactions between *opposite* species, this inhomogeneous domain profile governs the distribution of interparticle distances between the *same* species. Particles are typically separated by a distance that grows as $t^{1/4}$ within the core of the domain, but systematically become sparser as the domain interface is approached. The subregions of 'core' and 'interface' each comprise a finite fraction of the domain. These essential features of the profile may be accounted for by a trapezoidal form, Fig. 1.2(b),

$$\rho(z) \equiv c(x,t)\, t^{1/4} = \begin{cases} \rho_0, & |z| \le z^*; \\ \rho_0(1 - |z|), & z^* < |z| < 1 - \epsilon. \end{cases} \qquad (1.10)$$

Here $z \equiv x/L(t)$ is the scaled spatial coordinate, where $x \in [-L(t), L(t)]$, and ρ_0 and $z^* \lesssim 1$ are constants. The upper limit for $|z|$ in the second line of (1.10) reflects the fact that there are no particles within a scaled distance of $\epsilon \equiv \ell_{AB}/L(t) \sim t^{-1/8}$ from the domain edge. The linear decay of the concentration near the domain edge arises from the finite flux of reactants that leave the domain. Thus, the local nearest-neighbor distance is $\rho(z)^{-1}$, where $\rho(z) = \rho_0$ in the core ($|z| \le z^*$), and $\rho(z) = \rho_0(1 - |z|)$ near the boundary; the time dependence of the reduced moments of the AA distance distribution are then

$$M_n \equiv \langle \ell_{AA}^n \rangle^{1/n} = \left(\int_0^\infty x^n \, P_{AA}(x,t) \, dx \right)^{1/n}$$

$$\approx t^{1/4} \left(2 \int_0^{z*} \frac{dz}{\rho_0^n} + 2 \int_{z*}^{1-\epsilon} \frac{dz}{\rho_0^n (1-z)^n} \right)^{1/n}$$

$$\sim \begin{cases} t^{1/4}, & n < 1; \\ t^{1/4} \ln t, & n = 1; \\ t^{(3n-1)/8n}, & n > 1. \end{cases} \tag{1.11}$$

For $n < 1$, the dominant contribution to M_n originates from the ρ_0^{-n} term in the parentheses, while for $n \geq 1$ the term involving $\rho_0^{-n}(1-z)^{-n}$ dominates, the second term giving a logarithmic singularity at the upper limit for $n = 1$. Thus the large-scale modulation in the domain profile leads to moments $M_n(t)$ that are governed by both the gap lengths, ℓ_{AB} and ℓ_{AA}. As $n \to \infty$, the reduced moment is dominated by the contribution from the sparsely populated region near the domain periphery where nearest-neighbor particles are separated by a distance of order $t^{3/8}$.

1.2.5 Driven diffusive motion

A recent surprising development has been the discovery by Janowsky [17] that the concentration decays as $t^{-1/3}$ when particles move by driven diffusion. In this mechanism, each particle attempts to move only to the right and actually moves only if the target site is at that instant unoccupied by a particle of the same species. If the incident particle lands on a site already occupied by an opposite species particle, annihilation occurs. By Galilean invariance, one might anticipate that $c(t) \propto t^{-1/4}$, as in $A + B \to 0$ with isotropic diffusion and with the same hard-core exclusion.

A rough argument for the $t^{-1/3}$ decay with driven diffusion follows from a description of the dynamics of a single AB interface [18] using the continuum inviscid Burgers' equation. Consider the 'separated' initial condition with $c_A(x, t = 0) = \bar{c}_A \Theta(-x)$ and $c_B(x, t = 0) = \bar{c}_B \Theta(x)$, where $\Theta(x)$ is the Heaviside step function. By applying mass balance as the interface moves, one finds that the interfacial velocity is the following function of the initial densities:

$$v_{AB} = \begin{cases} 1 - 2\bar{c}_B(\sqrt{2} - 1), & \bar{c}_A \geq \bar{c}_B(\sqrt{2} - 1); \\ 1 - (\bar{c}_A^2 + \bar{c}_B^2) / (\bar{c}_A + \bar{c}_B), & \bar{c}_A < \bar{c}_B(\sqrt{2} - 1). \end{cases} \tag{1.12}$$

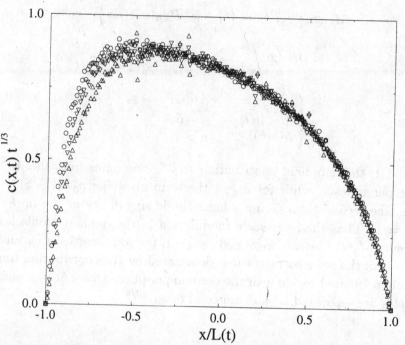

Fig. 1.3. Simulation data for the scaled microcanonical density profiles for $A+B \to 0$ with driven diffusive motion at $t = 1.5^{17} \cong 985$ ($\triangle$), $t = 1.5^{20} \cong 3325$ ($\triangledown$), and $t = 1.5^{23} \cong 11222$ ($\circ$). Plotted is the scaled local concentration, $c(x,t) \, t^{1/3}$ (in arbitrary units) *vs.* $x/L(t)$.

For a random, equal-density, initial condition, this leads to a domain length growing as $dL/dt \propto |v_{AB} - v_{BA}|$, which from (1.12) is proportional to the density difference across the interfaces. These differences are typically of the order of the domain densities themselves, leading to $dL/dt \propto c_{AB} - c_{BA} \propto c(t)$. Since the typical concentration in a domain of length L is of order $1/\sqrt{L}$, the domain size should therefore grow as $L \sim t^{2/3}$, while the concentration should decay as $c(t) \sim t^{-1/3}$.

A more microscopic description emerges from the scaled microcanonical domain profile in which the length is rescaled by $t^{-2/3}$ and the density by $t^{1/3}$ (Fig. 1.3). This profile exhibits good data collapse over most of the domain, except near the trailing edge. The departure from 'bulk' scaling appears to stem from diffusive boundary layers of size $\sqrt{t}$. The relative extent of this boundary layer with respect to the domain length decreases as $t^{-1/6}$, in qualitative agreement with numerical simulations. Janowsky

[19] has recently obtained similar results for the domain profile, but gives a somewhat different interpretation.

1.3 Single-species reactions

The kinetics of *homogeneous* diffusion-controlled single-species annihilation, $A + A \to 0$, and coalescence, $A + A \to A$, is now relatively well understood. For spatial dimension $d > 2$, the kinetics is accounted for by the rate equation, which predicts that $c(t) \propto t^{-1}$. For $d \leq 2$, suitably modified rate equations and the Smoluchowski approach both predict that $c(t) \propto t^{-d/2}$, but with logarithmic corrections appearing for $d = 2$. In 1D, exact solutions, based on the image method [20], an occupation number formalism [21], or a mapping onto the kinetic Ising-Glauber model at zero temperature [22] provide definitive results for single-species annihilation. Accompanying the anomalous kinetics for $d \leq 2$ is a spatial 'ordering' in which the probability of finding particles at the typical separation is enhanced compared to a random distribution [14,23]. Similarly, for diffusion-controlled coalescence, exact solutions have been constructed based on analyses of interparticle distribution functions [14] or on analogies with the voter model [24] and related probabilistic formulations [25-27].

1.3.1 Heterogeneous diffusive motion

A generalization of single-species annihilation that exhibits a wide range of phenomenology is *heterogeneous* annihilation, $A_i + A_j \xrightarrow{K_{i,j}} 0$ [28]. Here A_i denotes the ith 'species', with diffusivity D_i, and the reaction rate matrix $K_{i,j}$ is a function of the diffusivities of the two reacting 'species'; we use the terminology of different species to describe a reaction that is actually single-species annihilation but with distinct rates for different reaction channels. This simple generalization has a surprisingly rich array of kinetics.

The rate equations can be adapted to account for the kinetics in the mean-field approximation. When the number of species is finite, the rate equations predict that the least mobile species decays as t^{-1}, while the other species decay more quickly, each with an associated exponent that depends on the diffusivity ratio between it and the slowest species.

When the diffusivities are drawn from a continuous distribution, the evolution of the diffusivity distribution, $P(D, t)$, is described by the integro-differential equation

$$\frac{\partial P(D,t)}{\partial t} = -P(D,t) \int_0^\infty dD'\,(D+D')\,P(D',t)$$

$$= -P(D,t)[D\mathcal{P}_0(t) + \mathcal{P}_1(t)], \tag{1.13}$$

where the kth moment $\mathcal{P}_k(t) \equiv \int_0^\infty dD\,D^k\,P(D,t)$. Thus the zeroth moment gives the particle concentration, $c(t) = \mathcal{P}_0(t)$, while the average diffusion coefficient $\langle D \rangle = \mathcal{P}_1(t)/\mathcal{P}_0(t)$. The solution to (1.13) is

$$P(D,t) = P(D,0)\exp\left[-D\int_0^t \mathcal{P}_0(t')dt' - \int_0^t \mathcal{P}_1(t')dt'\right]$$

$$= P(D,0)\sqrt{\mathcal{P}_0(t)}\,\exp\left[-D\int_0^t c(t')dt'\right], \tag{1.14}$$

where the second line is obtained by first integrating (1.13) over D to relate the moments $\mathcal{P}_1$ and $\mathcal{P}_0$.

When $P(D, t=0) \sim D^\mu$ as $D \to 0$, (1.14) can be analyzed by scaling. Under the assumption of power-law decays for the average concentration and diffusivity, $c \sim t^{-\alpha}$ and $\langle D \rangle \sim t^{-\beta}$ for $t \to \infty$, a natural scaling ansatz for the time-dependent diffusivity distribution is

$$P(D,t) \simeq t^{\beta-\alpha}\Phi(Dt^\beta). \tag{1.15}$$

Substituting this into (1.14) and applying consistency conditions, one finds [28] the basic exponents to be $\alpha = (2+2\mu)/(3+2\mu)$, $\beta = 1/(3+2\mu)$.

For $d \leq 2$, the Smoluchowski theory [12] is ideally suited for adaptation to heterogeneous $A_i + A_j \to 0$. As outlined in Sec. 1.2, in the Smoluchowski approach one computes the particle flux towards a reference absorbing particle of the rest of the background particles. In 1D, this background concentration is $c(x,t) = c_\infty \,\mathrm{erf}(x/\sqrt{4Dt})$, from which the particle flux to the absorber is $\phi = c_\infty \sqrt{D/\pi t}$. This is identified as an *effective reaction rate*, $\tilde{k}$.

As an illustrative and intriguing example, consider the 'impurity' problem, namely a background of identical particles with diffusivity D and concentration c, and relatively rare impurities of diffusivity D_I and concentration c_I in 1D. In 1D, the survival of an impurity is equivalent to the probability that a given Ising spin with zero-temperature Glauber dynamics does not flip. This specific problem has been raised in studies of domain-coarsening phenomena [29]. In the limit of $c_I \ll c$, the influence of background-impurity and impurity-impurity reactions can be neglected and the effective (Smoluchowski) rate equations are

$$\dot{c} \cong -2\tilde{k}_{BB}c^2 \sim -2\sqrt{\frac{2D}{\pi t}}c^2, \quad \dot{c}_I \cong -2\tilde{k}_{BI}c\,c_I \sim -2\sqrt{\frac{D+D_I}{\pi t}}c\,c_I. \quad (1.16)$$

Here $\tilde{k}_{BB}$ and $\tilde{k}_{BI}$ are the effective rates for background-background and background-impurity reactions, obtained from straightforwardly generalizing the Smoluchowski theory to particles with different diffusivities. From the first equation, the background concentration vanishes as $c = \sqrt{\pi/(32Dt)}$. (In comparison, $c(t)/c_{\text{exact}}(t) = \pi/2$.) A crucial element of the second equation is that the decay exponent of the impurity species is determined by the amplitude of $c(t)$. One finds $c_I(t) \sim t^{-\sqrt{(1+\epsilon)/8}}$, with $\epsilon = D_I/D$.

The special case of the stationary impurity ($\epsilon = 0$) merits emphasis. The above Smoluchowski theory gives the exponent of $c_I(t)$ as $1/\sqrt{8} \cong 0.35355\ldots$, while numerical simulations give 0.375 [28,29]. By a mapping to steady-state aggregation with a point monomer source, Derrida *et al.* [30] have recently shown that this exponent is, in fact, equal to 3/8. An intuitive understanding of this result is still lacking.

The nonuniversal behavior for the impurity decay has a counterpart in single-species coalescence. The impurity decay in coalescence can be simply analyzed, since the 'cage' enclosing a given impurity, defined as its nearest neighbors, evolves *only* by diffusion [31]. This three-body system of the impurity and its two nearest-neighbors can be transformed, in turn, to a single random walker that diffuses within an absorbing two-dimensional wedge whose opening angle depends on D/D_I. This gives a survival probability that decays as $t^{-\alpha}$, with $\alpha = \pi/\{2\cos^{-1}[\epsilon/(1+\epsilon)]\}$.

Such a rigorous mapping does not exist for impurity decay in annihilation, since the cage's evolution is an intrinsically many-body process. Nevertheless, the Smoluchowski theory is qualitatively identical for both annihilation and coalescence. In both cases, the mechanism underlying the nonuniversal impurity decay is the equivalence to the survival probability $S(t)$ of a diffusing particle inside an absorbing interval of length $L = (At)^{1/2}$. For this probability, we have $S(t) \propto t^{-\alpha(A/D)}$, i.e., the decay exponent is dependent on the dimensionless parameter A/D [32]. The Smoluchowski theory predictions turn out (fortuitously perhaps) to be quantitatively accurate for impurity decay in annihilation, but somewhat less accurate in the corresponding coalescence process. For both cases, however, the Smoluchowski approach provides a useful paradigm for treating this nonuniversal aspect of the decay kinetics.

One can also adapt the Smoluchowski approach to heterogeneous annihilation in 1D by incorporating a time-dependent reaction rate in the rate
equation (1.13). This leads to

$$\frac{\partial P(D,t)}{\partial T} = -P(D,t) \int_0^\infty dD' \sqrt{D+D'}\, P(D',T). \qquad (1.17)$$

The correspondence with the mean-field approach has been sharpened by
introducing the modified variable $T = 4\sqrt{t/\pi}$, to eliminate the time dependence on the r.h.s. of (1.17). While this equation does not appear to be
solvable, presumably exact exponent values can be obtained by replacing the
kernel by one with the same homogeneity degree, $\sqrt{D+D'} \to \sqrt{D}+\sqrt{D'}$,
for which a scaling analysis yields the exponents $\alpha = (2+2\mu)/(5+4\mu)$,
$\beta = 1/(5+4\mu)$.

In summary, the Smoluchowski theory provides a simple and surprisingly
comprehensive account for the kinetics of heterogeneous single-species annihilation, both in the mean-field limit and in 1D.

1.3.2 Ballistic annihilation

As mentioned in Sec. 1.1, ballistically driven reactions are relatively unex-
plored in spite of their relative simplicity and apparent richness. In such
systems, particles move with their initial velocities until a reaction occurs,
so that the initial condition determines the time dependence. Past work has
primarily been on the '$\pm$' model in 1D [33], where each particle velocity is
either $+v_0$ or $-v_0$ with equal probability. Part of the interest in this reaction
is its equivalence to the polynuclear growth model (Fig. 1.4) [34]. This is a
model of evolving 'positive' and 'negative' terraces which move at constant
speed and annihilate when oppositely oriented terraces meet.

For the $\pm$ ballistic reaction, the concentration decays as $c(t) \propto \sqrt{c(0)/vt}$.
In analogy with the diffusion-controlled case $A + B \to 0$, this result can
be understood by considering density fluctuations in a domain of length ℓ.
In such a region, there will typically be an imbalance $\delta n \simeq \sqrt{c(0)\ell}$ in the
number of right-moving and left-moving particles. After a time $t = \ell/v$,
only this residual fluctuation will remain and the concentration will be of
order $c(t) \simeq \delta n/\ell$. Expressing ℓ in terms of t then gives $c(t) \propto t^{-1/2}$. In the
following, extensions to more general velocity distributions will be presented.

1. Continuous velocity distribution

For a continuous zero mean initial velocity distribution, $P(v, t = 0)$, basic
quantities that characterize the reaction include the concentration, $c(t) =$

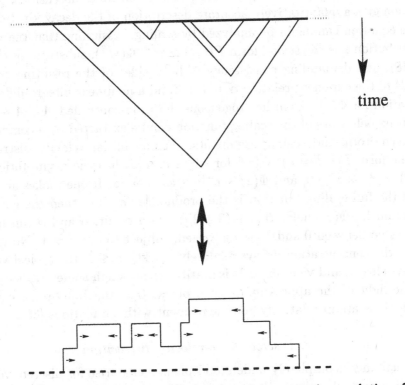

Fig. 1.4. Schematic illustration of the equivalence of the space-time evolution of the deterministic $\pm$ model and the polynuclear growth model of surface growth. The initial conditions of the two systems are equivalent. The velocities of the terraces in the polynuclear growth model representation are indicated.

$\int dv\, P(v,t) \sim t^{-\alpha}$ and the rms velocity, $v_{\mathrm{rms}} = [\int dv\, v^2 P(v,t)/c(t)]^{1/2} \sim t^{-\beta}$. From a mean-free path argument, the time between collisions for particles of (fixed) radius r and speed v_{rms} with concentration c is $t \sim 1/(cv_{\mathrm{rms}}r^{d-1})$, or $cv_{\mathrm{rms}} \propto t^{-1}$, leading to the scaling relation $\alpha + \beta = 1$. Since the particle lifetime is proportional to $1/v$, faster particles tend to annihilate more quickly, and the typical velocity decays in time. This parallels the behavior found in heterogeneous diffusion-controlled single-species annihilation.

The evolution of the velocity distribution in 1D is described by the Boltzmann equation [35]

$$\frac{\partial P(v,t)}{\partial t} = -kP(v,t) \int\limits_{-\infty}^{\infty} dv'\, |v - v'|\, P(v',t). \qquad (1.18)$$

In spite of its mean-field character, (1.18) and its d-dimensional generalization give a quantitatively accurate description of the decay kinetics. This rate equation can again be analyzed by scaling. Assuming that the velocity distribution has the scaled form $P(v,t) \simeq t^{\beta-\alpha}\Phi(vt^\beta)$, substituting this into (1.18), and demanding consistency of both sides of the resulting equation leads to the exponent relation $\alpha + \beta = 1$ and a nonlinear integro-differential equation for $\Phi(z)$. As in heterogeneous diffusion-controlled $A + A \to 0$, the limiting behaviors of the scaling function can be extracted by a combination of asymptotic and scaling arguments. For an initial velocity distribution of the form $P(v, t = 0) \propto |v|^\mu$ for small v, $\Phi(z)$ has the asymptotic forms $\Phi(z) \sim z^\mu$ as $z \to 0$, and $\Phi(z) \sim e^{-|z|/\beta}$ as $z \to \infty$. If one makes an ansatz that the full scaling function is the product $|z|^\mu e^{-|z|/\beta}$, then the governing equation for $\Phi(z)$ yields $\beta = 1/(3 + 2\mu)$. As a result, α and β can take on any value between 0 and 1 as μ is varied, subject to $\alpha + \beta = 1$. Notice that when the concentration decays relatively quickly, $\alpha \lesssim 1$, the typical velocity decays slowly, and vice versa. It is gratifying, although somewhat surprising in the light of the approximations involved, that the Boltzmann equation predictions are in relatively good agreement with simulations [35].

2. Discrete three-velocity distribution

Unusual and incompletely understood phenomena arise for discrete velocity distributions. A generic illustration is the trimodal distribution $P(v, t = 0) = p_+\delta(v - 1) + p_0\delta(v) + p_-\delta(v + 1)$, with $p_+ + p_0 + p_- = 1$ [36]. In the mean-field limit, the kinetics of the symmetric system with $p_+ = p_- \equiv p_\pm$ is described by the rate equations, $\dot{c}_\pm = -c_0 c_\pm - 2c_\pm^2$, $\dot{c}_0 = -2c_0 c_\pm$, with corresponding asymptotic behaviors

$$c_\pm(t) \sim \frac{1}{2}c_0(\infty)\, e^{-c_0(\infty)t}, \qquad c_0(t) \sim c_0(\infty)\exp\left[e^{-c_0(\infty)t}\right]. \qquad (1.19)$$

Here $c_0(\infty) = c_0(t = 0)\, e^{-2c_\pm(0)/c_0(0)}$. Thus the mobile particles decay exponentially in time, while a residue of stationary particles always remains whose concentration is vanishingly small if the initial concentration is relatively small.

However, from both nonrigorous approaches and an exact solution for the special case $p_+ = p_-$ [37,38], rather different behavior occurs in 1D. As a function of p_0, a transition occurs from a regime where the stationary particles persist, for $p_0 > 1/4$, to a regime where $c_0(t) \sim t^{-1}$ and $c_\pm(t)^{-1/2}$, for $p_0 < 1/4$. At a 'tricritical' point located at $p_0 = 1/4$, the concentrations of the mobile and stationary species decay as $t^{-2/3}$. While all asymptotic information is contained in the exact solution, the qualitative approaches are still instructive. The location of the tricritical point may be found by a stoichiometric argument [36]. Since half the stationary particles react with + particles, the fraction of + particles available to react with − particles is $p_+ - \frac{1}{2}p_0$. This is proportional to the number of +− annihilation events per unit length, $\mathcal{N}_{+-}$. Similarly, the relative number $\mathcal{N}_{0-}$ of 0− annihilation events per unit length equals $\frac{1}{2}p_0$. It is reasonable to assume that the relative number of annihilation events is proportional to the relative velocities of the collision partners, so that $\mathcal{N}_{+-}/\mathcal{N}_{0-} = 2$. Combining the resulting relation, $p_+ - \frac{1}{2}p_0 = p_0$, with the normalization condition, $2p_+ + p_0 = 1$, gives the exact location of the tricritical point: $p_0 = 1/4$, $p_\pm = 3/8$.

The two different exponent values for the decay of $c_0(t)$ and $c_\pm(t)$ in the case $c_0(0) < 1/4$ can also be understood by a probabilistic argument that applies in the limit $c_0(0) \to 0$. For an infinitesimal concentration of stationary 'impurities', the moving particles react among themselves with overwhelming probability and the background system reduces to the ± model, for which $c_\pm(t) \sim t^{-1/2}$. On the other hand, a stationary particle survives only if it is not annihilated by moving particles incident from either direction. Since the probabilities of each of these two events are independent, it follows that $c_0(t) \sim [c_\pm(t)]^2 \sim t^{-1}$ in the limit $c_0(0) \ll c_\pm(0)$.

When the initial concentrations of the three species are arbitrary, there are three 'phases'—regions of the phase diagram where a single species persists in the long-time limit (Fig. 1.5). As just discussed, the stationary species decays relatively quickly along the boundary between the + phase (where right-moving particles persist) and the − phase. The complementary situation of the decay of −'s along the +0 phase boundary also exhibits peculiar characteristics. For simplicity, consider an infinitesimal concentration of −'s in a background of equal concentrations of 0's and +'s. By a Galilean transformation, this is equivalent to 'fast impurities' in a symmetric ± background. A Lifshitz-type argument suggests that the survival probability of this fast impurity decays slower than exponentially but faster than a power law in time.

The basis of this argument is to consider a subset of configurations which give the dominant contribution to the impurity survival probability $S(t)$, but which are sufficiently simple to evaluate [36]. For the impurity to survive to time t, the background ± particles must annihilate *only* among themselves up to this time. On a space-time diagram, the dominant contribution to $S(t)$ stems from a sequence of ever larger self-annihilation triangles that just 'miss' the impurity world line (Fig. 1.6). The base of the nth triangle $x_n \propto [(v_0 + 1)/(v_0 - 1)]^n \equiv \beta^n$ and the number of triangles in this self-annihilation sequence up to time t is $N \simeq \ln t / \ln \beta$. Because the probability

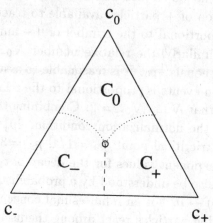

Fig. 1.5. Phase diagram of the 1D three-velocity model in the triangle defined by the relative concentrations of the three species. Along the broken line, $c_\pm(t) \sim t^{-1/2}$, while $c_0(t) \sim t^{-1}$. At the point marked by the small circle, $c_\pm(0) = 3/8$, $c_0(0) = 1/4$, all species decay as $t^{-2/3}$. Along the dotted lines, the nature of the decay is unknown, except very close to the extrema that correspond to the 'fast impurity' problem (see below). The symbols inside the triangle indicate the concentrations of the species dominant in the long-time limit.

of a particle annihilating with its nth neighbor asymptotically decreases as $n^{-3/2}$ [33], a self-annihilation triangle of base x_n occurs with probability $x_n^{-3/2}$. Finally $S(t)$ is the product of the occurrence probabilities of this self-annihilation triangle sequence,

$$S(t) \sim \prod_{n=1}^{\ln t/\ln \beta} (2x_0\beta^n)^{-3/2} \sim \exp[-3\ln^2 t/(4\ln \beta)]. \qquad (1.20)$$

This result has been confirmed numerically.

A natural continuation of the above line of modeling is to the four-velocity model with particle velocities $\pm v_1$ and $\pm v_2$, $v_2 > v_1$, and with relative concentrations c_1 and c_2. While the rate equations predict that the faster species decays as t^{-v_2/v_1} and the slower species decays as t^{-1}, the 1D system exhibits tricritical behavior reminiscent of the three-velocity model. Namely, there exists an initial condition, which depends on v_2/v_1, for which all four species appear to decay at the same power-law rate of approximately $t^{-0.72}$. A comprehensive understanding of these and more general discrete velocity systems is still lacking.

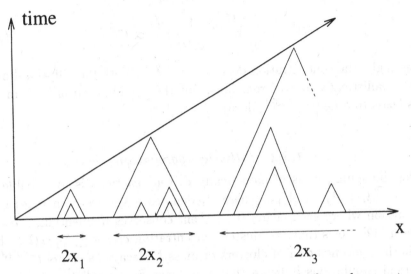

Fig. 1.6. World line of a fast impurity in a background of equal concentrations of $\pm$ particles. Successive triangles of self-annihilating background particles are indicated.

1.3.3 The stochastic $\pm$ model

As a final example of a single-species annihilation process that exhibits unusual kinetics, consider the 'stochastic' $\pm$ model, where particles execute biased diffusion with fixed bias (either right or left) for each particle [36]. For the deterministic $\pm$ model, particles with the same velocity never meet, by definition. However, in the stochastic $\pm$ model, the superimposed diffusion permits same-velocity particles to annihilate, a mechanism that leads to surprising behavior. This can be determined by dimensional analysis. If the particle diffusion coefficient is D, then the stochastic $\pm$ model is fully characterized by the initial concentration c_0, the velocity v_0, and D. From these parameters, the only variable combinations with the dimensions of concentration are, c_0, $1/(v_0 t)$, and $1/\sqrt{Dt}$. From basic considerations about the nature of the decay, the time-dependent concentration is anticipated to have the form

$$c(t) \propto (c_0)^\mu \left(\frac{1}{v_0 t}\right)^\nu \left(\frac{1}{\sqrt{Dt}}\right)^{1-\mu-\nu}. \tag{1.21}$$

The exponents μ and ν can be determined by requiring that $c(t)$ matches with (a) the diffusion-limited result $c(t) \to (Dt)^{-1/2}$ when $t < \tau_v \simeq D/v_0^2$, the crossover time below which drift can be ignored, and (b) the ballistic result $c(t) \rightarrow (c_0/v_0 t)^{1/2}$ for $t < \tau_D \simeq 1/(Dc_0^2)$, which is the time for adjacent particles to meet by diffusion. This matching gives

$$c(t) \sim \left(\frac{1}{v_0 t}\right)^{1/2} \left(\frac{1}{\sqrt{Dt}}\right)^{1/2} \propto t^{-3/4}. \tag{1.22}$$

Interestingly, the concentration decays as $t^{-1/2}$ for both the diffusion-limited and the ballistically limited reactions, but the combination of both mechanisms leads to a faster, $t^{-3/4}$, decay.

1.3.4 Ballistic aggregation

Another intriguing class of ballistically driven reactions is irreversible aggregation, $A_i + A_j \rightarrow A_{i+j}$, where A_i denotes a species with mass i. In the situation where each aggregation event conserves momentum, a scaling approach [39] shows that the cluster concentration $c(t) \equiv \sum_{k=1}^{\infty} c_k(t)$, where $c_k(t)$ is the concentration of clusters of mass k, decays as $c(t) \sim t^{-2d/(d+2)}$. Numerical simulations indicate that this mean-field prediction holds even in 1D. However, there are significant discrepancies in microscopic aspects of the reaction in 1D [40]. For example, the scaling approach gives $c_k(t) \sim \exp(-\text{constant} \times t^{1/3})$ for cluster mass k much less than the typical mass, while a Lifshitz tail argument gives $c_k(t) \sim t^{-1}$, a result that has been verified numerically [40].

The result for the cluster concentration can be reproduced by a mean-free-path argument that parallels the earlier 'microscopic' formulation of the rate equation. Consider a monomer-only initial condition in which each particle has the same speed but random direction. The momentum of an aggregate of mass m is proportional to $m^{1/2}$, since it is the sum of m random momenta. Consequently, the time between collisions at any stage of the reaction is $\Delta t \simeq 1/(cv\sigma)$, where c is the concentration, $v = p/m \propto m^{-1/2}$ is the typical velocity, and σ is the cross-section. In this time interval, the concentration decrement is of order $\Delta c \sim -c$. Thus $\Delta c/\Delta t \simeq -c/[1/(cv\sigma)]$. For a typical aggregate with mass $m \propto 1/c$, the r.h.s. can be rewritten in terms of the concentration only, using $v \sim c^{1/2}$ and $\sigma \sim c^{-(d-1)/d}$, leading to $c(t) \sim t^{-2d/(d+2)}$.

In the following, a complementary problem of ballistic aggregation, which models traffic flow on a 1D road with no passing, will be discussed [41]. Consider zero-size cars that move ballistically in one direction. We suppose that whenever a faster car or cluster overtakes a slower object, the larger final cluster assumes the velocity of the overtaken object. This reaction can be schematically represented as $A_{m_1,v_1} + A_{m_2,v_2} \rightarrow A_{m_1+m_2,\min\{v_1,v_2\}}$, where

A_{m_i,v_i} denotes a cluster with velocity v_i and which contains m_i cars. Scaling approaches, together with the statistical properties of the minimal random variable of a sample, determine basic system observables.

Let m and v be the typical cluster mass (or number of cars) and cluster velocity at time t. Without loss of generality, the minimal car velocity may be taken to be zero. The typical distance, ℓ, between clusters therefore grows as $\ell \sim vt$. Since the typical cluster mass is proportional to the typical intercluster distance, one also has $m \sim \ell \sim vt$. To find the typical velocity, one has to relate the cluster size to its velocity. Such a relation may be found exactly for a 'one-sided' problem in which the 'leading' car is placed at $x = 0$ and other cars are *only* in the domain $x < 0$. This leading car ultimately forms a cluster that includes all *consecutive* cars to its left whose initial velocities are larger than v. The probability that there are exactly k such cars equals $\Pi_-\Pi_+^k$, where $\Pi_+ = 1 - \Pi_- = \int_v^\infty P(v', t = 0)dv'$ is the probability that a car has velocity larger than v. The average number of cars in the cluster that ultimately forms is therefore given by $\langle m(v) \rangle = \sum_{k=1}^\infty k\Pi_-\Pi_+^k = \Pi_+/\Pi_-$.

For a power-law small-velocity tail of the initial velocity distribution, $P(v, t = 0) \propto v^\mu$ for $v \ll 1$ (with $\mu > -1$ for normalizability), $\langle m(v) \rangle \cong v^{-1-\mu}$. Under the assumption that this 'one-sided' result also applies to the original 'two-sided' problem and also using $m \sim vt$ one obtains

$$m \sim c^{-1} \sim t^\alpha, \quad \text{with } \alpha = \frac{\mu+1}{\mu+2}; \quad v \sim t^{-\beta}, \quad \text{with } \beta = \frac{1}{\mu+2}. \quad (1.23)$$

In analogy with ballistic annihilation with continuous velocities, the exponent relation $\alpha + \beta = 1$ is a consequence of the relation $c \sim 1/vt$.

An extension of the above reasoning allows one to solve for the mass and velocity distributions in the traffic model. As an illustration, consider the survival probability of a given car with velocity v, $S(v, t)$. Here 'survival' means that the car does not overtake traffic, but if overtaken still 'survives', i.e., a survivor leads a cluster of size $m \geq 1$. The survival probability can be found by considering the possible collisions of a car with initial velocity v and position x with slower cars. A collision with a slower v'-car does not occur up to time t if the interval $[x, x + (v - v')t]$ does not include the v'-car. For a continuous initial velocity distribution and a Poissonian initial spatial distribution, the probability that the v'-car is not in the interval $[x, x + (v - v')t]$ is $\exp\{-dv' P(v', t = 0)(v - v')t\}$. The probability that the initial car survives up to time t equals the product of these pair survival factors for every $v' < v$. Hence,

$$S(v,t) = \exp\left\{-t\int_0^v dv'(v-v')\,P(v',t=0)\right\},\qquad(1.24)$$

and the cluster velocity distribution is $P(v,t) = P(v,t=0)\,S(v,t)$.

For $P(v,t=0) \sim v^\mu$ as $v \to 0$, (1.24) gives the asymptotic velocity distribution as

$$P(v,t) \propto v^\mu \exp\left\{-\text{constant} \times tv^{\mu+2}\right\}.\qquad(1.25)$$

This universal form validates the scaling assumption that the asymptotic decay and the shape of the limiting distribution are determined solely by the exponent μ that characterizes the low-velocity tail of $P_0(v)$. From (1.25), the total concentration, $c(t) = \int_0^\infty dv\,P(v,t)$, and the average cluster velocity $\langle v(t)\rangle = \int dv\,v P(v,t)/\int dv P(v,t)$ are found to agree with (1.23). Extensions of this reasoning determine the complete distribution of clusters of mass m and velocity v.

1.4 Summary

The scaling approach is a simple yet powerful tool for analyzing the kinetics of simple reaction processes. In diffusion-limited two-species annihilation, a scaling analysis leads to an understanding of the coarsening mosaic of A and B domains from the initial density fluctuations. A relatively simple-minded adaptation of the rate equation approach reveals that the spatial distribution of reactants involves a multiplicity of scales that originates from the existence of a new length, ℓ_{AB}, the separation between AB nearest-neighbor pairs. Consideration of the density profile of a single domain provides a revealing picture of the 'internal' spatial organization of reactants. Scaling approaches are also useful in elucidating some of the unexpected features of two-species annihilation when the reactants undergo driven diffusion.

Although many aspects of 1D single-species reaction kinetics are exactly solvable, scaling approaches provide a relatively simple route to a comprehensive understanding. An attractive feature of the scaling formulation is that the apparently disparate processes of diffusion-controlled and ballistically controlled reactions may be analyzed in essentially identical manners. Although quantitative details depend on the specifics of a particular reaction, qualitative features are universal and are captured by a scaling description.

One intriguing aspect of single-species reactions, for which scaling appears to have limited utility, is the case of discrete diffusivity or velocity distributions. For example, in diffusive $A + A \rightarrow 0$ in 1D with an infinitesimal fraction of immobile particles in a background of equally mobile particles the density of the immobile particles decays as $t^{-3/8}$, compared to a decay of $t^{-1/2}$ for the background particles. Variations of this type of problem may be fruitful areas for additional investigation. For ballistically driven annihilation with three different velocity species in 1D, a wide range of kinetics arises, either 'critical', with two species decaying as $t^{-1/2}$ and the minority species decaying more quickly, or 'tricritical', where all three species decay as $t^{-2/3}$. While this latter behavior has been found by an exact solution, an intuitive understanding is still lacking and the full range of phenomenology appears ripe for further exploration. Finally, for combined ballistic and diffusive reactant motion, the concentration decays more rapidly than in the limiting situations where only one transport mechanism is operative. This interesting behavior can be accounted for by dimensional analysis. However, a microscopic theory has yet to be developed.

The author thanks Eli Ben-Naim, Slava Ispolatov, Paul Krapivsky, and François Leyvraz for pleasant collaborations that formed the basis of much of the work reviewed here. His initial work in this field was performed in collaboration with the late Kiho Kang who he still fondly remembers. The author is also grateful to Dani ben-Avraham, Charlie Doering, and Vladimir Privman for many instructive discussions, and to Paul Krapivsky for a critical reading of this chapter. Finally, he is grateful to the granting agencies that provided financial support for his work, including the ARO (through grant DAAH04-93-G-0021), the NSF (through grants INT-8815438 and DMR-9219845), and the Donors of The Petroleum Research Fund, administered by the American Chemical Society.

References

[1] A. A. Ovchinnikov and Ya. B. Zel'dovich, *Chem. Phys.* **28**, 215 (1978); S. F. Burlatskii and A. A. Ovchinnikov, *Russ. J. Phys. Chem.* **52**, 1635 (1978).

[2] D. Toussaint and F. Wilczek, *J. Chem. Phys.* **78**, 2642 (1983).

[3] K. Kang and S. Redner, *Phys. Rev. Lett.* **52**, 955 (1984); *Phys. Rev.* **A32**, 435 (1985).

[4] K. Lee and E. J. Weinberg, *Nucl. Phys.* **B246**, 354 (1984).

[5] P. Meakin and H. E. Stanley, *J. Phys.* **A17**, L173 (1984).

[6] G. Zumofen, A. Blumen and J. Klafter, *J. Chem. Phys.* **82**, 3198 (1985).

[7] L. W. Anacker and R. Kopelman, *Phys. Rev. Lett.* **58**, 289 (1987); D. ben-Avraham and C. R. Doering, *Phys. Rev.* **A37**, 5007 (1988); K. Lindenberg, B.

J. West and R. Kopelman, *Phys. Rev. Lett.* **60**, 1777 (1988); E. Clément, L. M. Sander and R. Kopelman, *Phys. Rev.* **A39**, 6455 (1989).

[8] M. Bramson and J. L. Lebowitz, *Phys. Rev. Lett.* **61**, 2397 (1988); M. Bramson and J. L. Lebowitz, *J. Stat. Phys.* **62**, 297 (1991); M. Bramson and J. L. Lebowitz, *J. Stat. Phys.* **65**, 941 (1991).

[9] B. P. Lee and J. L. Cardy, *Phys. Rev.* **E50**, 3287 (1994); B. P. Lee and J. L. Cardy, *J. Stat. Phys.* **80**, 971 (1995).

[10] L. Gálfi and Z. Racz, *Phys. Rev.* **A38**, 3151 (1988); Y.-E. L. Koo and R. Kopelman, *J. Stat. Phys.* **65**, 893 (1991); S. Cornell, M. Droz and B. Chopard, *Phys. Rev.* **A44**, 4826 (1991); M. Araujo, S. Havlin, H. Larralde and H. E. Stanley, *Phys. Rev. Lett.* **68**, 1791 (1992); H. Larralde, M. Araujo, S. Havlin and H. E. Stanley, *Phys. Rev.* **A46**, 855 (1992); E. Ben-Naim and S. Redner, *J. Phys.* **A25**, L575 (1992); S. Cornell and M. Droz, *Phys. Rev. Lett.* **70**, 3824 (1993); P. L. Krapivsky, *Phys. Rev.* **E51**, 4774 (1995).

[11] For recent reviews on two-species annihilation see, e.g., Ya. B. Zel'dovich and A. S. Mikhailov, *Sov. Phys. Usp.* **30**, 23 (1988); V. Kuzovkov and E. Kotomin, *Rep. Prog. Phys.* **51**, 1479 (1988); R. Kopelman, *Science* **241**, 1620 (1988); A. S. Mikhailov, *Phys. Reports* **184**, 307 (1989); A. A. Ovchinnikov, S. F. Timashev and A. A. Belyy, in *Kinetics of Diffusion Controlled Chemical Processes* (Nova Science Publishers, 1990); S. Redner and F. Leyvraz, *Fractals and Disordered Systems*, Vol. **2**, S. Havlin and A. Bunde, eds. (Springer, Heidelberg, 1994).

[12] M. V. Smoluchowski, *Z. Phys. Chem.* **92**, 215 (1917); S. Chandrasekhar, *Rev. Mod. Phys.* **15**, 1 (1943).

[13] See, e.g., G. H. Weiss and R. J. Rubin, *Adv. Chem. Phys.* **52**, 363 (1983), and references therein.

[14] C. R. Doering and D. ben-Avraham, *Phys. Rev.* **A38**, 3035 (1988); D. ben-Avraham, M. A. Burschka and C. R. Doering, *J. Stat. Phys.* **60**, 695 (1990).

[15] F. Leyvraz and S. Redner, *Phys. Rev. Lett.* **66**, 2168 (1991); *Phys. Rev.* **A46**, 3132 (1992). The importance of interparticle distributions in two-species annihilation was apparently first raised in P. Argyrakis and R. Kopelman, *Phys. Rev.* **A41**, 2121 (1990).

[16] See, e.g., L. D. Landau and E. M. Lifshitz, *Quantum Mechanics* (Pergamon Press, New York, 1977).

[17] S. A. Janowsky, *Phys. Rev.* **E51**, 1858 (1995).

[18] I. Ispolatov, P. L. Krapivsky and S. Redner, *Phys. Rev.* **E52**, 2540 (1995).

[19] S. A. Janowsky, *Phys. Rev.* **E52**, 2535 (1995).

[20] D. C. Torney and H. M. McConnell, *Proc. Roy. Soc. London* **A387**, 147 (1983); D. C. Torney and H. M. McConnell, *J. Phys.* **C87**, 1941 (1983).

[21] A. A. Lushnikov, *Sov. Phys. JETP* **64**, 811 (1986).

[22] Z. Rácz, *Phys. Rev. Lett.* **55**, 1707 (1985); J. G. Amar and F. Family, *Phys. Rev.* **A41**, 3258 (1990); V. Privman, *J. Stat. Phys.* **69**, 629 (1992).

[23] P. Argyrakis and R. Kopelman, *Phys. Rev.* **A41**, 2113 (1990).

[24] T. J. Cox and D. Griffeath, *Ann. Prob.* **14**, 347 (1986). For general results about the voter model, see, e.g., R. Durrett, *Lecture Notes on Particle Systems and Percolation* (Wadsworth and Brooks/Cole, Pacific Grove, CA, 1988).

[25] M. Bramson and D. Griffeath, *Z. Wahrsch. verw. Gebiete* **53**, 183 (1980).

[26] J. L. Spouge, *Phys. Rev. Lett.* **60**, 873 (1988).

[27] D. J. Balding, *J. Appl. Phys.* **25**, 733 (1988); D. J. Balding and N. J. B. Green, *Phys. Rev.* **A40**, 4585 (1989).

[28] P. L. Krapivsky, E. Ben-Naim and S. Redner, *Phys. Rev.* **E50**, 2474 (1994).

[29] B. Derrida, A. J. Bray and C. Godréche, *J. Phys.* **A27**, L357 (1994); D. Stauffer, *J. Phys.* **A27**, 5029 (1994); J. L. Cardy, *J. Phys.* **A28**, L19 (1995); B. Derrida, *J. Phys.* **A28**, 1481 (1995).

[30] B. Derrida, V. Hakim and V. Pasquier, *Phys. Rev. Lett.* **75**, 751 (1995).

[31] F. Leyvraz, unpublished; D. ben-Avraham, *J. Chem. Phys.* **88**, 941 (1988); M. E. Fisher and M. P. Gelfand, *J. Stat. Phys.* **53**, 175 (1988); M. Bramson and D. Griffeath, in *Random Walks, Brownian Motion and Interacting Particle Systems: A Festschrift in Honor of Frank Spitzer*, R. Durrett and H. Kesten, eds. (Birkhauser, 1991), and references therein.

[32] For a pedagogical account see, e.g., P. Krapivsky and S. Redner, *Am. J. Phys.* **64**, 546 (1996). References to original literature are contained therein.

[33] Y. Elskens and H. L. Frisch, *Phys. Rev.* **A31**, 3812 (1985); J. Krug and H. Spohn, *Phys. Rev.* **A38**, 4271 (1988).

[34] D. Kashchiev, *J. Crystal Growth* **40**, 29 (1977); C. H. Bennett, M. Büttiker, R. Landauer and H. Thomas, *J. Stat. Phys.* **24**, 419 (1981); M. C. Bartelt and J. W. Evans, *J. Phys.* **A26**, 2743 (1993).

[35] E. Ben-Naim, S. Redner and F. Leyvraz, *Phys. Rev. Lett.* **70**, 1890 (1993).

[36] P. Krapivsky, S. Redner and F. Leyvraz, *Phys. Rev.* **E51**, 3977 (1995). Note that a version of ballistic annihilation with a trimodal velocity distribution was introduced by W. S. Sheu, C. van den Broeck and K. Lindenberg, *Phys. Rev.* **A43**, 4401 (1991). However, the collision rules of this model were formulated to give behavior similar to that of the two-velocity model.

[37] J. Piasecki, *Phys. Rev.* **E51**, 5535 (1995).

[38] M. Droz, L. Frachebourg, J. Piasecki and P.-L. Rey, *Phys. Rev. Lett.* **75**, 160 (1995); *Phys. Rev.* **E51**, 5541 (1995).

[39] G. F. Carnevale, Y. Pomeau and W. R. Young, *Phys. Rev. Lett.* **64**, 2913 (1990).

[40] Y. Jiang and F. Leyvraz, *J. Phys.* **A26**, L179 (1993).

[41] E. Ben-Naim, P. L. Krapivsky and S. Redner, *Phys. Rev.* **E50**, 822 (1994).

·2

The coalescence process, $A+A \rightarrow A$, and the method of interparticle distribution functions

Daniel ben-Avraham

The kinetics of the diffusion-limited coalescence process, $A + A \rightarrow A$, can be solved exactly in several ways. In this chapter we focus on the particular technique of interparticle distribution functions (IPDFs), which enables the exact solution of some nontrivial generalizations of the basic coalescence process. These models display unexpectedly rich kinetic behavior, including instances of anomalous kinetics, self-ordering, and a dynamic phase transition. They also reveal interesting finite-size effects and shed light on the combined effects of internal and external fluctuations. An approximation based on the IPDF method is employed for analysis of the crossover between the reaction-controlled and diffusion-controlled regimes in coalescence when the reaction rate is finite.

2.1 Introduction

Reaction-diffusion systems are those in which the reactants are transported by diffusion [1,2]. Two fundamental time scales characterize these systems: (a) the diffusion time—the typical time between collisions of reacting particles, and (b) the reaction time—the time that particles take to react when in proximity. When the reaction time is much larger than the diffusion time, the process is reaction-limited. In this case the law of mass action holds and the kinetics is well described by classical rate equations. In recent years there has been a surge of interest in the less tractable case of diffusion-limited processes, where the reaction time may be neglected [3-8]. In this limit, local fluctuations in concentrations and fluctuations in number space (the local number of particles is discrete and cannot always be replaced by a continuous variable, such as concentration) dominate the kinetics [7-12]. The effect of fluctuations is most pronounced in low dimensions or in confined geometries. For this reason, much research has focused on diffusion-limited reactions in one dimension (1D).

An added bonus of working in 1D is that the models are simpler, allowing for more accurate computer simulation data, and series analyses and systematic approximations can be carried out to a farther degree. In some cases the 1D models are simple enough that they can be solved exactly. Examples are the one-species annihilation process, $A + A \to 0$, and coalescence, $A + A \to A$, [13-28], and even a two-species annihilation model, $A + B \to 0$, where same-species particles stick together [29]. Exactly solvable models not only yield invaluable physical insights but also serve as benchmark tests for approximation techniques and for numerical simulations [30].

Here, we are considering the one-dimensional process of diffusion-limited coalescence of one species, $A + A \to A$, along with the method of inter-particle distribution functions (IPDFs) that was introduced for its analysis [22]. The IPDF method is a powerful technique that yields exact results for the basic one-species coalescence process and for numerous variations and generalizations, including adding on the reverse reaction, $A \to A + A$, and input of A particles. In spite of the stark simplicity of the coalescence model it is surprisingly rich. The exact solutions reveal anomalous kinetics, critical ordering, and a nonequilibrium dynamic phase transition [22-25]. With the IPDF method one can solve exactly the coalescence model in finite lattices and study finite-size effects [26]. Exact analysis is also possible of a coalescence model with inhomogeneous initial conditions, which gives rise to Fisher waves [27]. Finally, the coalescence model with fluctuating external fields can be solved exactly by the IPDF method, providing one with a rare opportunity to study the combined effects of internal and external noise in nonequilibrium kinetics [28].

The chapter is organized as follows. In Sec. 2.2, we introduce the IPDF method. For clarity, and to connect with cellular automata [31,32] and numerical simulations, the reaction models are described on a spatial lattice (but in continuous time). The spatial continuous limit is described as well. In Sec. 2.3, we review specific instances of the coalescence model that have been solved exactly through the IPDF method, and we discuss the physical insights gained by these solutions. Section 2.4 is devoted to the coalescence process with finite reaction rate, where a crossover between reaction-limited and diffusion-limited behavior is possible. Unfortunately, there exists no exact treatment of the kinetics for this case. Instead, we present an approximation approach based on the IPDF method [33]. This approximation is sensitive enough to deal with the most salient features of crossover behavior. We conclude with a summary and discussion, in Sec. 2.5.

2.2 The IPDF method

The interparticle distribution function (IPDF) method was originally introduced for the solution of the diffusion-limited coalescence process, $A + A \rightarrow A$, in 1D [22]. This model and its generalizations will be described in detail in subsequent sections. Presently, we merely wish to introduce the main ingredients of the IPDF technique. For this purpose, it will suffice to consider a 1D lattice, with lattice spacing Δx, in which each lattice site can be either empty (o) or occupied by a single particle (•). The states of the sites evolve according to some dynamic rules that are prescribed by the model in question. We shall further assume that an ensemble average of the state of our system is translationally invariant. (This restriction is not absolutely necessary, as we will see in Sec. 2.3.5, but it makes our initial presentation simpler.)

We define the quantity $E_n(t)$ as the probability that a randomly chosen segment of n consecutive sites is empty, i.e., contains no particles. The E_n can be used to describe important characteristics of the distribution of particles (including the concentration and the distribution of interparticle distances) and to construct a closed kinetic equation describing the dynamic rules. For example, the probability that a site is occupied is $1 - E_1$. Thus, the density, or concentration of particles, is expressed as

$$c(t) = (1 - E_1)/\Delta x. \tag{2.1}$$

E_n gives the probability that sites 1 through n are empty, while E_{n+1} gives the probability that sites 1 through $n + 1$ are empty. The event that 1 through n are empty contains the event that 1 through $n + 1$ are empty. Thus, the probability that a segment of n sites is empty but that there is a particle at the adjacent site $n + 1$, is

$$\mathrm{Prob}(\overbrace{\mathrm{o} \cdots \mathrm{o}}^{n} \bullet) = E_n - E_{n+1}. \tag{2.2}$$

From the $E_n(t)$ we may also derive $p_n(t)$, the probability that the nearest neighbor to, say, the right of a given particle is n lattice spacings away, at time t. That is, p_1 is the probability that the nearest neighbor lies in the site next to the particle, p_2 is the probability that the nearest neighbor is two sites away, etc. The p_n are normalized, $\sum p_n = 1$, and the average distance between particles is the reciprocal of the concentration

$$\langle n\Delta x \rangle = \sum_{n=1}^{\infty} n p_n \Delta x = \frac{1}{c}. \tag{2.3}$$

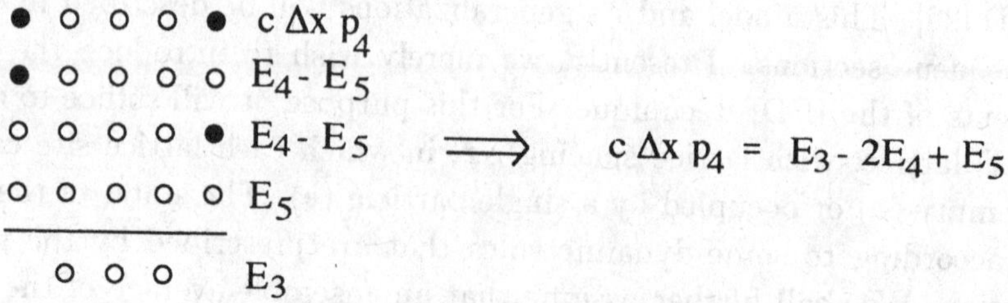

● ○ ○ ○ ● $c\,\Delta x\,p_4$

● ○ ○ ○ ○ $E_4 - E_5$

○ ○ ○ ○ ● $E_4 - E_5$ $\Longrightarrow$ $c\,\Delta x\,p_4 = E_3 - 2E_4 + E_5$

○ ○ ○ ○ ○ E_5

────────────

○ ○ ○ E_3

Fig. 2.1. Alternative derivation of (2.4): the probability of an empty three-site interval may be expressed as a sum of probabilities of five-site configurations.

Choose a lattice site at random. The probability that the next n sites are empty, E_n, may be written in terms of the p_n. The probability that the chosen point lies within a gap of length m is proportional to mp_m, which can be normalized with the help of (2.3), yielding the probability distribution $c\,\Delta x\,mp_m$. The probability that there are k lattice spacings until the next particle, given that the point is in the gap of length m, is $1/m$ if $1 \leq k \leq m$, and 0 otherwise. Thus, the (unconditional) probability that there are exactly k lattice spacings to the next particle is $\sum_{m=k}^{\infty} m^{-1} c\,\Delta x\,mp_m = c\,\Delta x \sum_{m=k}^{\infty} p_m$. Finally, the probability that the next n sites are empty, E_n, is the probability that $k > n$: $E_n = c\,\Delta x \sum_{k=n+1}^{\infty} \sum_{m=k}^{\infty} p_m$. This can be inverted to yield

$$c\,\Delta x\,p_n = E_{n-1} - 2E_n + E_{n+1}. \tag{2.4}$$

This equation does not apply for p_1, because E_n is not defined for $n = 0$. To obtain the probability that two adjacent sites are occupied, we simply enumerate all their possible states (the sum of the probabilities is 1) and use (2.2) to get

$$\text{Prob}(\bullet\bullet) = 1 - [\text{Prob}(\circ\circ) + \text{Prob}(\circ\bullet) + \text{Prob}(\bullet\circ)] = 1 - 2E_1 + E_2. \tag{2.5}$$

Then, p_1 is equal to the conditional probability $\text{Prob}(\bullet\bullet)$, given that the first site is occupied, so that

$$c\,\Delta x\,p_1 = 1 - 2E_1 + E_2. \tag{2.6}$$

In fact, a similar argument can be used to rederive (2.4); see Fig. 2.1.

For some models with simple dynamics, the various probabilities discussed above are sufficient to write down a closed kinetic equation that involves only the E_n.

2.2.1 The continuum limit

To achieve a spatial continuum limit we define the spatial coordinate $x = n\Delta x$. The probabilities $E_n(t)$ are replaced by the two-variable function $E(x, t)$ and, in the limit of $\Delta x \to 0$, (2.1) becomes

$$c(t) = -[\partial E(x, t)/\partial x]_{x=0}. \tag{2.7}$$

Likewise, the probabilities $p_n(t)$ are replaced by probability densities $p(x, t)$, which are related to the density of an empty interval, $E(x, t)$, through (see (2.4)):

$$c(t)p(x, t) = \partial^2 E(x, t)/\partial x^2. \tag{2.8}$$

We refer to the p_n, or to $p(x, t)$, as the 'interparticle distribution function', which is the origin of the name of our technique. The IPDF method enables one to compute $c(t)$, the time dependence of the concentration. The IPDFs themselves are studied because they convey more detailed information than the average density, and because of their relation to the Smoluchowski problem [34-43]. IPDFs were first introduced by Kopelman *et al.* [12,44,45] for the analysis of simulations and experimental results for diffusion-limited reactions and have since continued to attract considerable interest. Our approach in this chapter is often referred to as 'the method of empty intervals', after the E_n probabilities, but we prefer our nomenclature because of the important role played by the IPDFs.

2.3 The one-species coalescence model

Our basic model is a 1D coalescence process of point particles, $A + A \to A$. The IPDF method can also handle the reverse process, $A \to A + A$, as well as that involving a steady input of A particles. These process types are realized through four different actual processes, as described below [25].

Diffusion. Particles move randomly to the nearest lattice site with a hopping rate $2D/(\Delta x)^2$. The diffusion is symmetric and at a rate $D/(\Delta x)^2$ to the right and at the same rate to the left. On long length and time scales this yields normal diffusion with diffusion coefficient D.

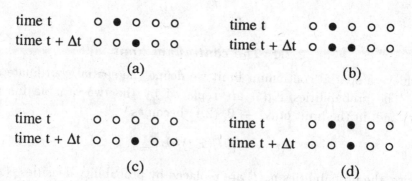

Fig. 2.2. Dynamic rules: (a) diffusion, (b) birth, (c) input, and (d) coalescence.

Birth. A particle gives birth to another at an adjacent site, at a rate $v/\Delta x$. This means a rate $v/2\Delta x$ for birth on each side of the original particle. Notice that while v is a constant (with units of velocity), the rate $v/\Delta x$ diverges in the continuum limit $\Delta x \to 0$. This is necessary because of the possible recombination of the newborn and the original particle, which also takes place at infinite rate when $\Delta x \to 0$. The birth process models the reverse reaction, $A \to A + A$.

Input. Any empty site spontaneously becomes occupied at a probability rate $R\Delta x$. Here R is the average number of particles input per unit length per unit time.

Coalescence. When a particle coincides with another through diffusion or birth, then it disappears. Symbolically, the coalescence process is $A + A \to A$. Because this coalescence reaction is infinitely fast, the overall process is diffusion-limited. In Sec. 2.4 we will consider the consequences of relaxing this condition.

Each of these processes—except coalescence—takes place independently of the others. The various processes are illustrated in Fig. 2.2.

To implement the IPDF method, we construct a closed kinetic equation for the evolution of the E_n. We consider the changes in E_n due to the different processes during a small time interval Δt.

Diffusion: We may have an empty segment of n sites and site $n+1$ occupied (with probability $E_n - E_{n+1}$; (2.2)). If the particle in the $(n+1)$th site hops into the empty segment the probability E_n decreases by $[D/(\Delta x)^2](E_n -$

E_{n+1}). Likewise, we may have an empty segment of $n-1$ sites and site n occupied (with probability $E_{n-1} - E_n$). If the particle at the edge hops to site $n+1$ then E_n increases by $[D/(\Delta x)^2](E_{n-1} - E_n)$. Therefore, the total change in E_n due to diffusion is

$$(\partial_t E_n)_{\text{diffusion}} = 2\frac{D}{(\Delta x)^2}(E_{n-1} - 2E_n + E_{n+1}), \qquad (2.9)$$

where the additional factor of 2 accounts for the possibility that both processes may take place at either side of the segment independently.

Birth: The birth process brings about a decrease in E_n in a way similar to the diffusion process. In the case of an empty segment of n sites and the $(n+1)$th site occupied, the particle at the edge may give birth to a particle inside the empty segment. Thus, the change in E_n due to birth (accounting for both sides of the segment) is

$$(\partial_t E_n)_{\text{birth}} = -\frac{v}{\Delta x}(E_n - E_{n+1}). \qquad (2.10)$$

Input: Whenever a particle is input into an empty n-site segment, E_n decreases. Since the rate of input into any of the n sites is $R\Delta x$, we have

$$(\partial_t E_n)_{\text{input}} = -R\,n\,\Delta x\,E_n. \qquad (2.11)$$

Coalescence: In a sense, the coalescence process is part of each of the three processes already discussed above. Coalescence may follow each of these processes immediately, without any obvious additional effect on the E_n. It affects our analysis by imposing a boundary condition, as follows. Because E_0 is not defined, (2.9) is valid only for $n > 1$. We need to consider $(\partial_t E_1)_{\text{diffusion}}$ separately. But according to (2.1), $\partial_t E_1 = -\partial_t c\Delta x$ is the rate of change in the *number* of particles. Coalescence may take place when either of two adjacent particles hops into its neighbor, decreasing the number of particles by one. Then, using the hopping rate and (2.5),

$$(\partial_t E_1)_{\text{diffusion}} = 2\frac{D}{(\Delta x)^2}(1 - 2E_1 + E_2). \qquad (2.12)$$

To make this consistent with (2.9) for the case of $n = 1$, we require the boundary condition $E_0 = 1$. Notice that (2.10) and (2.11) for the birth and input processes pose no additional constraints.

Combining all the different contributions to changes in E_n, we get

$$\partial_t E_n = 2\frac{D}{(\Delta x)^2}(E_{n-1} - 2E_n + E_{n+1}) - \frac{v}{\Delta x}(E_n - E_{n+1}) - R\,n\,\Delta x\,E_n. \quad (2.13)$$

with the boundary condition $E_0 = 1$. Another boundary condition, $E_\infty(t) = 0$, is always true, unless the density of particles is strictly zero. Finally, an initial condition may be derived from the configuration of the lattice at time $t = 0$. For example, an initially empty lattice corresponds to $E_n(0) = 1$, while a homogeneously random distribution of particles with density c_0 would lead to $E_n(0) = (c_0 \Delta x)^n$, etc.

Although (2.13) can be tackled through standard approaches for difference equations, it is convenient to pass to the continuum limit, as discussed in Sec. 2.2. We then have

$$\frac{\partial E(x,t)}{\partial t} = 2D\frac{\partial^2 E}{\partial x^2} + v\frac{\partial E}{\partial x} - RxE, \tag{2.14}$$

with boundary conditions $E(0,t) = 1$ and $E(\infty, t) = 0$. This yields itself to an exact solution for $E(x,t)$, which is then used to find $c(t)$ and $p(x,t)$.

2.3.1 Irreversible coalescence, $A + A \to A$

This is the case when $v = 0$ and $R = 0$ [22]. Equation (2.14) can be solved exactly for a variety of initial conditions. It is found that for generic initial distributions of particles (excluding the exotic cases of fractal and power-law distributions, where the initial concentration is zero) the system reaches a universal long-time asymptotic behavior. Thus,

$$c(t) \to 1/\sqrt{2\pi Dt}, \qquad \text{as } t \to \infty, \tag{2.15}$$

and the IPDF behaves according to $p(x,t) \to \frac{x}{4Dt} \exp(-\frac{x^2}{8Dt})$, as $t \to \infty$. This is better expressed in terms of the dimensionless interparticle distance $\xi = c(t)x$:

$$p(\xi, t) = c(t)p(x,t) \to \tfrac{1}{2}\pi\xi \exp\left(-\tfrac{1}{4}\pi\xi^2\right), \qquad \text{as } t \to \infty. \tag{2.16}$$

The IPDF is stationary in the long-time asymptotic limit. The concentration decay is anomalous (compare (2.15) with $c \sim 1/t$, predicted by classical rate equations) and the reaction induces self-ordering: for large ξ, $p(\xi)$ falls off more sharply than for a random homogeneous distribution of particles for which $p(\xi) = \exp(-\xi)$. In addition, there is an effective repulsion between particles, arising from the coalescence reaction, which results in a dramatic depletion of $p(\xi)$ near $\xi = 0$; see Fig. 2.3. The simple coalescence model serves as an example of *dynamic self-ordering* in a far-from-equilibrium system.

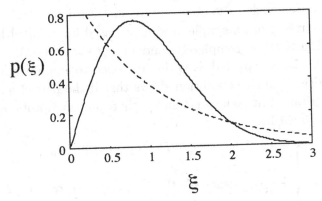

Fig. 2.3. IPDF for the simple coalescence process (solid line) compared to that of a completely random distribution (broken line).

2.3.2 Coalescence with input

This case is obtained with $v = 0$ but $R > 0$ [23]. It leads to a stationary distribution of particles that is, again, independent of initial conditions. The steady-state density is

$$c_s(R, D) = \frac{|\mathrm{Ai}'(0)|}{\mathrm{Ai}(0)} \left(\frac{R}{2D}\right)^{1/3} \approx 0.72901 \left(\frac{R}{2D}\right)^{1/3}, \qquad (2.17)$$

where $\mathrm{Ai}(z)$ is the Airy function and $\mathrm{Ai}'(z) = d\mathrm{Ai}(z)/dz$. The fact that c_s is proportional to $R^{1/3}$ instead of the mean-field $c_s \sim R^{1/2}$ shows that diffusion-limited coalescence is anomalous and behaves *effectively* like a three-particle reaction rather than the two-particle reaction that it really is. The stationary distribution of particles is

$$p_s(x) = \left(\frac{R}{2D}\right)^{1/3} \frac{\mathrm{Ai}''[(R/2D)^{1/3}x]}{|\mathrm{Ai}'(0)|}. \qquad (2.18)$$

Thus, effective repulsion between particles still takes place and $p_s(x) \sim x$ for small x, even for arbitrarily large input rates. On the other hand, the tail of the distribution falls off less sharply, as $\exp(-x^{3/2})$, rather than $\exp(-x^2)$ as in coalescence without input.

The steady state of the coalescence model with input serves as an example of *static self-ordering* in a far-from-equilibrium system.

2.3.3 Reversible coalescence, $A + A \leftrightarrow A$

This case includes the reverse reaction, $A \to A + A$ ($v > 0$) [24]. Since this is enough to sustain a nonempty steady state, we may neglect input ($R = 0$). With the back reaction present the system approaches true equilibrium. The exact solution yields the steady-state distribution $c_s = v/2D$, which can also be derived from a simple argument based on detailed balance. At equilibrium, the IPDF is completely random, $p(\xi) = \exp(-\xi)$.

While all this is as expected, it is the *approach* to equilibrium that holds the real surprise. The exact solution shows that if the particles are initially randomly distributed at concentration c_0, then the asymptotic approach to equilibrium is given by

$$c(t) - c_s \sim \begin{cases} -(c_s - 2c_0)\exp[-2Dc_0(c_s - c_0)t]\,, & c_0 < c_s/2, \\[2mm] -\dfrac{1}{\sqrt{2\pi Dt}}\exp(-\tfrac{Dc_s^2}{2}t)\,, & c_0 = c_s/2, \\[2mm] \dfrac{c_s^{-2} - (c_s - 2c_0)^{-2}}{\sqrt{\pi}(Dt)^{3/2}}\exp(-\tfrac{Dc_s^2}{2}t)\,, & c_0 > c_s/2. \end{cases} \qquad (2.19)$$

Thus, there is a dynamic phase transition in the relaxation time of the system: if the initial concentration is larger than $c_s/2$, the relaxation time is constant, $\tau = 2/(Dc_s^2)$, but when the initial concentration is smaller than $c_s/2$, the relaxation time becomes dependent upon c_0, $\tau = 1/[2Dc_0(c_s - c_0)]$; see Fig. 2.4.

2.3.4 Finite-size lattices

The coalescence process on lattices of finite size can also be exactly analyzed through the IPDF method [26]. Suppose that one has a lattice of length L. The boundary condition $E(\infty, t) = 0$ is simply replaced by $E(L, t) = 0$. The rest of the analysis is unaffected by this change, other that some care must be taken in imposing periodic boundary conditions, so as to maintain a translation-invariant distribution of particles at all times.

Doering and Burschka [26] have explored in this fashion what becomes of the phase transition in the relaxation dynamics discussed above, when the system is finite. They have solved (2.14) with the boundary condition $E(L, t) = 0$ through a straightforward spectral decomposition. The equilibrium concentration shifts to $c_s = (v/2D)/[1 - e^{-vL/2D}]$ and the IPDF deviates from a clean Poisson distribution in a similar way. The full solution reveals that the approach to equilibrium is eventually dominated by the finite size of the lattice. However, the initial approach to equilibrium matches that of an infinite system. Defining a crossover time between these

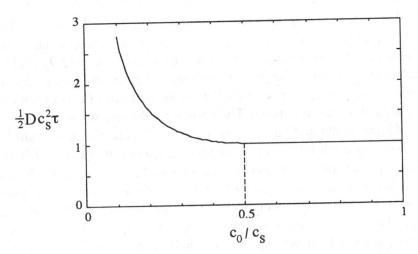

Fig. 2.4. Dynamic phase transition in the relaxation time of reversible coalescence.

two regimes, T_c, Doering and Burschka show that $T_c \sim (L^2/2D)\ln(1/\epsilon)$, where $\epsilon = (c_0 - \frac{1}{2}c_s)/\frac{1}{2}c_s$ is a measure of how far the system is from criticality. Thus, their analysis leads to the interesting conclusion that even in a finite system there exist arbitrarily long time scales as $\epsilon \to 0$.

2.3.5 Inhomogeneous systems and external noise

The coalescence process can also be exactly analyzed under inhomogeneous conditions, for which there is no translation invariance [27]. The idea is to extend the concept of the empty interval probability to $E(x, y, t)$, the probability that there are no particles in the interval $x < z < y$ at time t. With this new definition, the density of particles at x is

$$c(x, t) = -\left[\partial E(x, y, t)/\partial y\right]_{y=x}, \tag{2.20}$$

and the conditional joint probability that there are particles at x and y but none in between is

$$p(x, y, t) = -\partial^2 E(x, y, t)/(\partial x \partial y). \tag{2.21}$$

A specific example is the reversible coalescence process with inhomogeneous initial conditions for the distribution of particles. The partial differential equation for the empty interval probability becomes

$$\frac{\partial E(x,y,t)}{\partial t} = D\frac{\partial^2 E}{\partial x^2} + D\frac{\partial^2 E}{\partial y^2} - \frac{v}{2}\frac{\partial E}{\partial x} + \frac{v}{2}\frac{\partial E}{\partial y}, \qquad (2.22)$$

on the half-space $x < y$, and the boundary condition, $E_0 = 1$, becomes $\lim_{y\downarrow x \text{ or } x\uparrow y} E(x,y,t) = 1$. A situation of interest occurs when initially the half-space $x < 0$ has the equilibrium concentration, $v/2D$, while the other half-space $x > 0$ is empty. In the classical mean-field approximation such an initial state gives rise to Fisher waves—a front of stationary shape propagating at a constant speed. The present model enables one to study the effects of internal fluctuations on Fisher waves. This has been the subject of much recent research [46-51]. For the model at hand, it is found that there is propagation at speed $v/2$, but in addition there is a widening of the front: the front's width increases as $\sqrt{Dt}$ [27]. The reversible coalescence process has been recently studied numerically in higher dimensions, by Riordan *et al.* [51]. They find that when the dimension is above $d = 3$ the process is well described by mean-field rate equations, exactly as for classical Fisher waves. For $d = 2$, the wavefronts widen with time as $t^{0.27}$, similarly to the exact solution in 1D. In $d = 3$ the slow widening of the front may be interpreted to be logarithmic in time, reinforcing the feeling that $d = 3$ is the critical dimension.

The formalism discussed above allows also the study of external spatial and temporal fluctuations. For example, diffusion of the particles could be made space dependent, $D = D(x)$, or the rate of the reverse process could be made time dependent, $v = v(t)$, etc., providing one with a unique opportunity to study the combined effects of internal and external noise. A specific example of a dichotomous external noise, where v assumes one of two possible values according to a Markovian probability jump, has been studied by Doering [28]. He shows that the nonequilibrium steady state possesses new microscopic length scales, demonstrating that spatially homogeneous external noise can introduce spatial correlations.

2.3.6 Correlation functions

As a final example of the wealth of exact results that can be obtained for coalescence, we consider the two-point correlation function of the simple coalescence process. The n-point correlation functions may be derived by considering the probability of finding n disjoint intervals empty [28]. Let $E^{(n)}(x_1,y_1,x_2,y_2,\ldots,x_n,y_n,t)$ be the probability of finding all intervals $[x_i,y_i]$ $(i = 1,2,\ldots,n)$ empty at time t. The n-point correlation function, that is, the joint probability density of finding particles at positions $x_1, x_2, \ldots, x_n$ at time t is

$$
c^{(n)}(x_1, \ldots, x_n, t) =
$$
$$
(-1)^n \frac{\partial^n}{\partial y_1 \cdots \partial y_n} \left[E^{(n)}(x_1, y_1, \ldots, x_n, y_n, t) \right]_{y_1 = x_1, \ldots, y_n = x_n}. \quad (2.23)
$$

To obtain the two-point correlation function it is sufficient to consider $E^{(2)}$, which satisfies the equation (for simple coalescence) [28]

$$
\frac{\partial E^{(2)}}{\partial t} = D \left(\frac{\partial^2}{\partial x_1^2} + \frac{\partial^2}{\partial y_1^2} + \frac{\partial^2}{\partial x_2^2} + \frac{\partial^2}{\partial y_2^2} \right) E^{(2)}, \quad (2.24)
$$

with the boundary conditions $\lim_{y_1 \downarrow x_1} E^{(2)}(x_1, y_1, x_2, y_2, t) = E^{(1)}(x_2, y_2, t)$, $\lim_{y_2 \downarrow x_2} E^{(2)}(x_1, y_1, x_2, y_2, t) = E^{(1)}(x_1, y_1, t)$, $\lim_{x_2 \downarrow y_1} E^{(2)}(x_1, y_1, x_2, y_2, t) = E^{(1)}(x_1, y_2, t)$, and $\lim_{x_1 \downarrow -\infty} E^{(2)} = \lim_{y_2 \uparrow +\infty} E^{(2)} = 0$; $E^{(1)} = E$ is the usual density of single empty intervals discussed above. Burschka *et al.* [52] obtain the elegant solution

$$
E^{(2)}(x_1, y_1, x_2, y_2) = E^{(1)}(x_1, y_1) E^{(1)}(x_2, y_2) - E^{(1)}(x_1, x_2) E^{(1)}(y_1, y_2)
$$
$$
+ E^{(1)}(x_1, y_2) E^{(1)}(y_1, x_2), \quad (2.25)
$$

where, for brevity, we have omitted the time dependence. For the long-time asymptotic limit of simple coalescence we have $E^{(1)}(x, y, t) = \mathrm{erfc}(\frac{1}{2}\sqrt{\pi}\xi)$, $\xi = c(t)(y - x)$, and so from (2.23) and (2.25) one gets the two-point correlation function [52]:

$$
c^{(2)}(\xi) = c^2 [1 - e^{-\frac{1}{2}\pi\xi^2} + \tfrac{1}{2}\pi\xi e^{-\frac{1}{4}\pi\xi^2} \mathrm{erfc}(\tfrac{1}{2}\sqrt{\pi}\xi)]. \quad (2.26)
$$

Notice that for large distances the correlation is lost, $c^{(2)}(\xi \to \infty) \to c^2$, and that $c^{(2)}$ is a monotonically increasing function of ξ; see Fig. 2.5.

It is interesting to compare the exact result of (2.26) with the one obtained assuming that the positions of the particles form a renewal process determined by the (exact) IPDF. Assuming a renewal process, the two-point correlation function is related to the IPDF through $c^{(2)}(x) = cp(x) + \int_0^x p(x')c^{(2)}(x - x')dx'$. Thus, $c^{(2)}$ may be obtained by means of a Laplace transform, $\tilde{c}^{(2)} = c\tilde{p}/(1 - \tilde{p})$, where the tilde denotes a Laplace-transformed function and, using the long-time asymptotic result for simple coalescence, $p(x) = (x/4Dt)\exp(-x^2/8Dt)$. For $x \ll 1/c$ (or $\xi \ll 1$), we obtain $c^{(2)}(\xi) = c^2(\frac{1}{2}\pi\xi - \frac{1}{12}\pi^2\xi^3 + \ldots)$. The first term agrees with the exact result (2.26), but the neglect of correlations is apparent in the higher-order terms. More interestingly, for $\xi \gg 1$

$$
c^{(2)}(\xi) \sim c^2 - c^2\sqrt{\pi}e^{-a\sqrt{\pi}\xi} \cos(b\sqrt{\pi}\xi + \phi), \quad (2.27)
$$

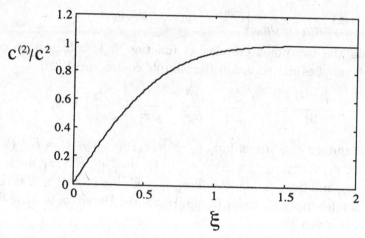

Fig. 2.5. Two-point correlation function for the simple coalescence process in the long-time limit (2.26).

where $a \approx 1.35481$ and $b \approx 1.99147$ are the real and imaginary parts of the zeros of the complementary error function, $\text{erfc}(-a \pm bi) = 0$, that are closest to the imaginary axis; $\tan \phi = -b/a$. Thus, the approach to the long-distance limit of $c^{(2)} \sim c^2$ is exponential, accompanied by sinusoidal oscillations. Such oscillations are typical of two-point correlation functions of hard-sphere gases. Here the effective repulsion between particles gives rise to the oscillations. What is interesting is that the asymptotic approach to the exact result (2.26), $c^{(2)} \sim c^2 - 2c^2\xi^{-2} \exp(-\frac{1}{2}\pi\xi^2)$, is much faster and *has no oscillations*. The correlations between adjacent IPDFs (taken into account in the exact derivation) conspire to smooth out the tail of the two-point correlation function.

2.4 The coalescence process with finite reaction rate

In spite of the richness of exact results obtainable for the coalescence process, several important generalizations remain largely unsolved. One such generalization is the case of finite reaction rates [33,53-55]. Here we study this by means of an IPDF-based approximation [56] in the spirit of Kirkwood.

Suppose that when a particle hops onto an occupied site, coalescence takes place with probability k, while with probability $1 - k$ the particle is reflected back to its original position and no reaction takes place. The probability k controls the rate of the coalescence reaction $A + A \to A$. When $k = 0$ the reaction rate is zero, and the particles merely diffuse, bouncing off one another. In the other extreme, when $k = 1$, reactions are immediate, i.e., the reaction rate is infinite. This is the purely diffusion-limited case. When k is very small but finite, one expects to find a regime dominated by the slow reaction rate. This is the classical, reaction-limited case. The model is of interest because of the crossover between the two latter regimes.

When the reaction probability $k < 1$, hopping out of the edge of an interval is not always possible (see Sec. 2.3). If the target site is occupied then with probability $1 - k$ hopping (and coalescence) are disallowed. To account for this effect we require the probability of finding intervals containing $n - 1$ consecutive empty sites followed by two occupied sites. We use the approximation [33]

$$\mathrm{Prob}(\overbrace{\circ \circ \cdots \circ}^{n-1} \bullet\bullet) \approx \frac{\mathrm{Prob}(\overbrace{\circ \circ \cdots \circ}^{n-1} \bullet)\mathrm{Prob}(\bullet\bullet)}{\mathrm{Prob}(\bullet)}$$

$$= \frac{(1 - 2E_1 + E_2)(E_{n-1} - E_n)}{1 - E_1}. \qquad (2.28)$$

With this approximation the evolution equation for arbitrary k is

$$\partial_t E_n = \frac{2D}{(\Delta x)^2}(E_{n-1} - 2E_n + E_{n+1}) - \frac{2D}{(\Delta x)^2}(1-k)\frac{1 - 2E_1 + E_2}{1 - E_1}(E_{n-1} - E_n), \qquad (2.29)$$

where the correction due to failed coalescence attempts is represented by the last term. Notice that when $k = 1$ this properly reduces to (2.9). For the special case of $n = 1$, (2.12) is still valid, so that one derives the usual boundary condition, $E_0(t) = 1$ (and, of course, $E_\infty(t) = 0$).

We argue that, in spite of the approximation involved, (2.29) is asymptotically correct both in the early- and long-time regimes, and hence it may provide a reasonable interpolation for the intermediate-time regime. If the starting configuration of the system is random, then $E_n = E_1^n$ and the state of consecutive intervals is uncorrelated, regardless of how large $c(0)$ may be. This situation will persist, and (2.29) will hold, until the concentration drop is noticeable (i.e., the end of the early-time regime) since only reactions induce correlations. (In fact, diffusion randomizes the system.) After very long times, on the other hand, the concentration of particles becomes very small. As a result, adjacent occupied sites become extremely rare and the correction term in (2.29) eventually becomes negligible.

2.4.1 Approximate integration of the evolution equation

Equation (2.29) can be integrated directly by iteration, or one may first pass to the continuum limit and then proceed through a variety of methods suitable for differential equations. Here we propose an alternative approach based on an approximation of (2.29). This approximation is a mere *mathematical* convenience, designed to enable us to obtain a solution in closed form.

We first sum (2.29) over the index n, from 1 to ∞, to yield

$$\partial_t \sum_{n=1}^{\infty} E_n = \frac{2D}{(\Delta x)^2}(1 - E_1) - \frac{2D}{(\Delta x)^2}(1 - k)\frac{1 - 2E_1 + E_2}{1 - E_1}, \qquad (2.30)$$

where we have used the usual boundary conditions. The r.h.s. can be expressed as a function of E_1 only, with the help of (2.12). For the l.h.s., we make the approximation $\sum E_n \approx A/(1 - E_1)$, where A is a constant. The justification for this is that, in the long-time asymptotic limit, $E_n \approx 1$ for all n up to a characteristic $\langle n \rangle = 1/(1 - E_1)$; it then falls sharply to zero for $n > \langle n \rangle$. More precisely, in the long-time asymptotic limit the reaction proceeds as if k were effectively 1, in which case we know that $A = 2/\pi$ exactly. The approximation lies in the fact that we assume A to be constant *at all times*. Indeed, the variation in A is quite small; at the beginning of the process, when the distribution is random, $A = 1$. Let us then assume that $A = 2/\pi$ (to match the exact long-time asymptotic solution) holds true throughout the process. Equation (2.30) then becomes

$$\frac{d}{d\tau}\left(\frac{2}{\pi C}\right) = C + \left(\frac{1 - k}{kC}\right)\frac{dC}{d\tau}, \qquad (2.31)$$

where $\tau \equiv 2Dt/(\Delta x)^2$ is a dimensionless time variable and $C \equiv 1 - E_1$. The solution is

$$C = \left[1 - k + \sqrt{\left(\frac{2k}{\pi C_0} + 1 - k\right)^2 + \frac{4k^2}{\pi}\tau}\right] \bigg/ \left[2\left(\frac{k}{\pi C_0^2} + \frac{1 - k}{C_0} + k\tau\right)\right], \qquad (2.32)$$

where $C_0 \equiv C(t = 0)$. This shows that the intermediate-time regime is merely an interpolation between a classical decay, $C \sim 1/t$, and a diffusion-limited decay, $C \sim 1/\sqrt{t}$. Analysis of the solution yields two crossover times: $\tau_1 = 2/(\pi C_0^2) + (1 - k)/(kC_0)$, the crossover time between the early regime—when reactions are rare—and the intermediate regime, and $\tau_2 = \pi(1-k)^2/(4k^2)$, the crossover time between the intermediate regime and the long-time asymptotic, diffusion-controlled, behavior.

The agreement between the approximate result for the concentration (2.32) obtained by using IPDFs (which can be computed from the E_n and (2.6)) and that from computer simulations is quite good [33].

2.4.2 Back reactions and input

Coalescence at finite reaction rates can be also studied in the case of back reactions and input. It is important to examine such situations because they may impose a stricter test on the validity of our approximation.

In the case of back reactions, the evolution equation (with the approximation (2.28)) is

$$\partial_t E_n = \frac{2D}{(\Delta x)^2}(E_{n-1} - 2E_n - E_{n+1})$$
$$-\frac{2D}{(\Delta x)^2}(1-k)\frac{1 - 2E_1 + E_2}{1 - E_1}(E_{n-1} - E_n)$$
$$-\frac{v}{\Delta x}(E_n - E_{n+1}). \tag{2.33}$$

In the steady state, the l.h.s. is equal to zero and (2.33) becomes a recursion relation for the E_n. The solution is a state of maximum entropy, $E_n = E_1^n$, since with the back reaction the system arrives at a true equilibrium state. Taking into account the usual boundary conditions we find

$$E_n = \left(\frac{2Dk}{2Dk + v\Delta x}\right)^n, \quad \text{and} \quad c_s = \frac{1 - E_1}{\Delta x} = \frac{v}{2Dk + v\Delta x}. \tag{2.34}$$

Notice that although (2.33) contains an approximation, the corresponding steady-state equation is exact: when the IPDF is completely random the approximation of (2.28) becomes exact. This is well confirmed by simulations.

In the case of particle input the evolution equation is

$$\partial_t E_n = \frac{2D}{(\Delta x)^2}(E_{n-1} - 2E_n - E_{n+1}) \tag{2.35}$$
$$-\frac{2D}{(\Delta x)^2}(1-k)\frac{1 - 2E_1 + E_2}{1 - E_1}(E_{n-1} - E_n) - Rn\Delta x E_n,$$

where we have used the same Kirkwood approximation as before. Here the steady-state limit is less simple than for back reactions. It is most easily derived by considering the continuum limit of the steady-state equation:

$$0 = 2D\partial_x^2 E - 2D(1-k)\omega\partial_x E - xRE, \tag{2.36}$$

where $\omega = [\partial_x^2 E]_{x=0}/[\partial_x E]_{x=0} = -[\partial_x^2 E]_{x=0}/c_s$ is a constant. We determine ω from the discrete steady state equation for $n=1$,

$$0 = \frac{2D}{(\Delta x)^2}k(1 - 2E_1 + E_2) - \Delta x R E_1. \tag{2.37}$$

This simply equates the rates of input events and coalescence events in the steady state. The continuum limit of (2.37) is too drastic in that it yields zero for each of these rates (and also $\omega = 0$).

A somewhat inelegant, but effective, way around this is to retain $\Delta x R$ as finite, so that $\omega = -R\Delta x/(2Dkc_s) \neq 0$. Then, the solution to (2.36) with the usual boundary conditions, $E(0,t)=1$ and $E(\infty,t)=0$, is

$$E(x) = \exp\left(-\frac{x}{\kappa c_s}\right)\frac{\mathrm{Ai}[r^{1/3}x + r^{-2/3}/(\kappa c_s)^2]}{\mathrm{Ai}[r^{-2/3}/(\kappa c_s)^2]}, \tag{2.38}$$

where $\kappa = 4Dk/[(1-k)R\Delta x]$ and $r = R/(2D)$. From the relation $c_s = -[\partial_x E]_{x=0}$, we then obtain a transcendental equation for the steady-state concentration, c_s:

$$c_s = \frac{1}{\kappa c_s} - r^{1/3}\frac{\mathrm{Ai}'[r^{-2/3}/(\kappa c_s)^2]}{\mathrm{Ai}[r^{-2/3}/(\kappa c_s)^2]}. \tag{2.39}$$

In Fig. 2.6, we plot c_s as a function of the reaction probability k, for fixed r ($r^{1/3} = 0.04$), as obtained from computer simulations and from numerical integration of the discrete steady-state equation. The agreement between simulations and theory is quite good. Also, for the range shown, the agreement between (2.39) and the numerical integration is better than 4%.

When $c \ll r^{1/3}/\kappa$ one derives from (2.39)

$$c_s = \frac{1}{2}\left(\tilde{c}_s + \sqrt{\tilde{c}_s^2 + \frac{2(1-k)\Delta x}{k}r}\right), \qquad \tilde{c}_s = -\frac{\mathrm{Ai}'(0)}{\mathrm{Ai}(0)}r^{1/3}, \tag{2.40}$$

where $\tilde{c}_s$ is the steady-state concentration when $k=1$ (and is exact). Thus, we observe a crossover from the diffusion-limited behavior, $c_s \sim \tilde{c}_s \sim (R/2D)^{1/3}$, to the classical result, $c_s \sim \frac{1}{2}\sqrt{(1-k)R\Delta x/kD}$, as $k \to 0$.

While the approximation approach reviewed here seems to describe reasonably well the crossover behavior between the diffusion-limited and reaction-limited regimes for coalescence with finite reaction rates, it is important

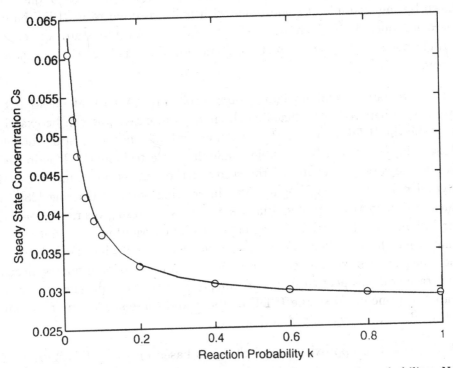

Fig. 2.6. Stationary concentration as a function of the reaction probability. Numerical integration results (solid line) are compared to simulations (circles).

to remember that it is merely an approximation. A phenomenological approximation advanced by Hoyuelos and Mártin [54] yields similar results. Privman *et al.* [55] have suggested a different approximation scheme, which, according to recent findings [57] yields more accurate long-time asymptotic results. Finally, the approximation method presented here may be systematically improved by introducing the Kirkwood truncation ansatz at a higher level of the evolution rate equations.

2.5 Summary and discussion

In summary, the IPDF technique enables one to obtain exact solutions for a large class of diffusion-limited coalescence models in 1D. These exact solutions reveal a breadth of phenomena that provide invaluable physical insight into nonequilibrium dynamics. They also serve for benchmark testing of numerical approximations and simulation algorithms. Moreover, the IPDF method may form the basis of new promising approximation techniques, as shown by the example of coalescence with finite reaction rate. IPDF-based approximations have been successfully used also for the study of many-particle coalescence $nA \rightarrow mA$ [58], and for the analysis of the contact process in 1D [59].

There remain several intriguing open problems. A natural question to pose is whether we have exhausted the nontrivial cases that can be exactly solved by the IPDF method. Perhaps one can find other interesting models whose solutions would be possible using the same technique? This indeed may be the case. For example, the study of the joint effect of internal and external noise has barely begun. The interesting work of Doering [28], in which he demonstrates that spatially homogeneous noise gives rise to spatial correlations, could certainly be extended to inhomogeneous external noise. What would be the effect of a spatially dependent diffusion coefficient, as when the particles are subject to a random potential, or of an inhomogeneous input rate (with respect to space, or time, or both!)? Such problems may be treated in principle with the IPDF method, and some specific models may be tractable and physically illuminating.

With regard to approximation techniques based on the IPDF method, a very important question is whether any of these approximate approaches can be systematically improved; this would certainly seem to be the case. One possibility is to use the joint density probabilities of Sec. 2.3.6, pushing the truncation approximation to a higher level. It would be interesting to see if such techniques were sensitive enough to predict the logarithmic corrections in the three-body coalescence processes $3A \rightarrow mA$ ($m = 1$ or 2), for example, [58,60-62].

But, more importantly, IPDF-based approximation methods can be applied to a much wider class of models, almost without restriction. An interesting application would be to the annihilation process, $A+A \rightarrow 0$. Since the density of particles as a function of time is known exactly, this could serve as a test for the IPDF-based approximation. Parent and L'Heureux [63] have applied an IPDF approximation method to another prototype problem: diffusion-limited coalescence of A particles with spontaneous conversion between two types of particles, $A \leftrightarrow B$. They have also studied an IPDF-based approximation for coalescence in two dimensions (the upper critical dimension for this process) that is sensitive enough to capture the expected logarithmic corrections [64].

This chapter touches upon work carried out over the years with mentors, colleagues and students: M. Burschka, C. Doering, S. Havlin, F. Leyvraz, J. Lin, R. Kopelman, S. Redner, G. Weiss, and D. Zhong. The author is very grateful to them for the deeply instructive and wonderfully enjoyable collaborations. He also thanks M. Burschka, C. Doering, and W. Horsthemke, for making their results on the two-point correlation function available to him prior to publication.

References

[1] K. J. Laidler, *Chemical Kinetics* (McGraw-Hill, New York, 1965).
[2] S. W. Benson, *The Foundations of Chemical Kinetics* (McGraw-Hill, New York, 1960).
[3] N. G. van Kampen, *Stochastic Processes in Physics and Chemistry* (North-Holland, Amsterdam, 1981).
[4] H. Haken, *Synergetics* (Springer, Berlin, 1978).
[5] G. Nicolis and I. Prigogine, *Self-Organization in Non-Equilibrium Systems* (Wiley, New York, 1980).
[6] T. M. Liggett, *Interacting Particle Systems* (Springer, New York, 1985).
[7] K. Kang and S. Redner, *Phys. Rev.* A32, 435 (1985); V. Kuzovkov and E. Kotomin, *Rep. Prog. Phys.* 51, 1479 (1988).
[8] *J. Stat. Phys.* 65, nos. 5/6 (1991): this issue contains the proceedings, *Models of Non-Classical Reaction Rates*, of a conference held at NIH (March 25-27, 1991) in honor of the 60th birthday of G. H. Weiss.
[9] D. ben-Avraham, *J. Stat. Phys.* 48, 315 (1987); *Phil. Mag.* B56, 1015 (1987).
[10] D. Toussaint and F. Wilczek, *J. Chem. Phys.* 78, 2642 (1983).
[11] G. Zumofen, A. Blumen and J. Klafter, *J. Chem. Phys.* 82, 3198 (1985).
[12] R. Kopelman, *J. Stat. Phys.* 4, 185 (1986); *Science* 241, 1620 (1988).
[13] M. Bramson and D. Griffeath, *Ann. Prob.* 8, 183 (1980); *Z. Wahrsch. Geb.* 53, 183 (1980).
[14] D. C. Torney and H. M. McConnell, *Proc. Roy. Soc. Lond.* A387, 147 (1983).
[15] L. Peliti, *J. Phys.* A19, L365 (1985).
[16] Z. Rácz, *Phys. Rev. Lett.* 55 1707 (1985).
[17] A. A. Lushnikov, *Phys. Lett.* A120, 135 (1987).
[18] J. L. Spouge, *Phys. Rev. Lett.* 60, 871 (1988).
[19] H. Takayasu, I. Nishikawa and H. Tasaki, *Phys. Rev.* A37, 3110 (1988).
[20] E. Clément, R. Kopelman and L. Sander, *Chem. Phys.* 180, 337 (1994).
[21] V. Privman, *J. Stat. Phys.* 69, 629 (1992); *J. Stat. Phys.* 72, 845 (1993).
[22] C. R. Doering and D. ben-Avraham, *Phys. Rev.* A38, 3035 (1988).
[23] C. R. Doering and D. ben-Avraham, *Phys. Rev. Lett.* 62, 2563 (1989).
[24] M. A. Burschka, C. R. Doering and D. ben-Avraham, *Phys. Rev. Lett.* 63, 700 (1989).
[25] D. ben-Avraham, M. A. Burschka, and C. R. Doering, *J. Stat. Phys.* 60, 695 (1990).
[26] C. R. Doering and M. A. Burschka, *Phys. Rev. Lett.* 64, 245 (1990).
[27] C. R. Doering, M. A. Burschka and W. Horsthemke, *J. Stat. Phys.* 65, 953 (1991).
[28] C. R. Doering, *Physica* A188, 386 (1992).
[29] V. Privman, A. M. R. Cadilhe and M. L. Glasser, *J. Stat. Phys.* 81, 881 (1995).

[30] J. C. Lin, C. R. Doering and D. ben-Avraham, *Chem. Phys.* **146**, 355 (1990).

[31] H. Berryman and D. Franceschetti, *Phys. Lett.* **A136**, 348 (1989).

[32] D. Dab, A. Lawniczak, J.-P. Boon and R. Kapral, *Phys. Rev. Lett.* **64**, 2462 (1990).

[33] D. Zhong and D. ben-Avraham, *J. Phys.* **A28**, 33 (1995).

[34] G. H. Weiss, R. Kopelman and S. Havlin, *Phys. Rev.* **A39**, 1620 (1989).

[35] D. ben-Avraham and G. H. Weiss, *Phys. Rev.* **A39**, 466 (1989).

[36] J. G. Amar and F. Family, *Phys. Rev.* **A41**, 3258 (1990).

[37] H. Taitelbaum, R. Kopelman, G. H. Weiss and S. Havlin, *Phys. Rev.* **A41**, 3116 (1990).

[38] S. Havlin, H. Larralde, R. Kopelman and G. H. Weiss, *Physica* **A169**, 337 (1990).

[39] S. Redner and D. ben-Avraham, *J. Phys.* **A23**, L1169 (1990).

[40] R. Schoonover, D. ben-Avraham, S. Havlin, R. Kopelman and G. H. Weiss, *Physica* **A171**, 232 (1991).

[41] S. Havlin, R. Kopelman, R. Schoonover and G. H. Weiss, *Phys. Rev.* **A43**, 5228 (1991).

[42] G. H. Weiss and J. Masoliver, *Physica* **A174**, 209 (1991).

[43] G. H. Weiss, *Physica* **A192**, 617 (1993).

[44] R. Kopelman, S. J. Parus and J. Prasad, *J. Chem. Phys.* **128**, 209 (1988).

[45] P. Argyrakis and R. Kopelman, *Phys. Rev.* **A41**, 2114 (1990).

[46] A. Lemarchand, H. Lemarchand, E. Sulpice and E. Mareschal, *Physica* **A188**, 277 (1992).

[47] D. Gruner, R. Kapral and A. Lawniczak, *J. Chem. Phys.* **99**, 3938 (1993).

[48] A. Lemarchand, A. Lesne, A. Perera, M. Moreau and M. Mareschal, *Phys. Rev.* **E48**, 1568 (1993).

[49] H. Breuer, W. Huber and F. Petruccione, *Physica* **D73**, 259 (1994).

[50] D. ben-Avraham, F. Leyvraz and S. Redner, *Phys. Rev.* **E50**, 1843 (1994).

[51] J. Riordan, C. R. Doering and D. ben-Avraham, *Phys. Rev. Lett.* **75**, 565 (1995).

[52] M. A. Burschka, C. R. Doering and W. Horsthemke, unpublished.

[53] L. Braunstein, H. O. Mártin, M. D. Grynberg and H. E. Roman, *J. Phys.* **A25**, L255 (1992).

[54] M. Hoyuelos and H. O. Mártin, *Phys. Rev.* **E48**, 3309 (1993).

[55] V. Privman, C. R. Doering and H. L. Frisch, *Phys. Rev.* **E48**, 846 (1993).

[56] See, for example, A. D. McQuarrie, *Statistical Mechanics*, p. 267 (Harper and Row, 1976).

[57] H. Simon, *J. Phys.* **A28**, 6585 (1995).

[58] D. ben-Avraham, *Phys. Rev. Lett.* **71**, 3733 (1993); D. ben-Avraham and D. Zhong, *Chem. Phys.* **180**, 329 (1994).

[59] E. Ben-Naim and P. Krapivsky, *J. Phys.* **A27**, L481 (1994).

[60] V. Privman and M. D. Grynberg, *J. Phys.* **A25**, 6567 (1992).

[61] P. Krapivsky, *Phys. Rev.* **E49**, 3233 (1994).

[62] B. P. Lee, *J. Phys.* **A27**, 2633 (1994).

[63] R. Parent and I. L'Heureux, *Phys. Lett.* **A189**, 154 (1994).

[64] R. Parent and I. L'Heureux, private communication.

3

Critical phenomena at absorbing states

Ronald Dickman

Continuous phase transitions from an absorbing to an active state arise in diverse areas of physics, chemistry and biology. This chapter reviews the current understanding of phase diagrams and scaling behavior at such transitions, and recent developments bearing on universality.

3.1 Introduction

Stochastic processes often possess one or more *absorbing states*—configurations with arrested dynamics, admitting no escape. Phase transitions between an absorbing state and an active regime have been of interest in physics since the late 1950s, when Broadbent and Hammersley introduced directed percolation (DP) [1-8]. Subsequent incarnations include Reggeon field theory [9-13], a high-energy model of peripheral interest to most condensed matter physicists, and a host of more familiar problems such as autocatalytic chemical reactions, epidemics, and transport in disordered media [14-18]. For the simpler examples—Schlögl's models, the contact process, and directed percolation itself—many aspects of critical behavior are well in hand [3-8,13-15,18]. In the mid-1980s absorbing-state transitions found renewed interest due to the catalysis models devised by Ziff and others [19-23], and to a proposed connection with the transition to turbulence [91]. A further impetus has been the ongoing quest to characterize universality classes for these transitions. Parallel to these developments, probabilists studying *interacting particle systems* have established a number of fundamental theorems for models with absorbing states [24-27].

Interest in the influence of kinetic rules on the phase diagram has spawned many models over the last decade; the majority must go unmentioned here [28]. We focus instead on examples illuminating the most salient aspects

51

of generic or universal properties. Limitations of space also preclude any detailed discussion of methodologies (see Sec. 3.7 for a list of the principal ones).

The remainder of this chapter is organized as follows. In Sec. 3.2, we describe the contact process, the simplest model with an absorbing-state transition. Section 3.3 contains a brief discussion of catalysis models. Arguments underlying the universality principle for absorbing-state transitions (sometimes called the 'DP conjecture') are sketched in Sec. 3.4. Nonuniversal critical behavior in models with multiple absorbing states is taken up in Sec. 3.5, followed in Sec. 3.6 by a review of models with a new kind of critical behavior, epitomized by branching annihilating random walks.

3.2 The contact process

The *contact process* (CP) and directed percolation (essentially a simultaneously updated version of the CP), are to absorbing-state transitions what the Ising model is to equilibrium critical phenomena: they are the minimal examples of their respective classes. The CP was proposed by Harris as a model of an epidemic or, more generally, of competition between two elementary processes—local self-replication, and spontaneous annihilation of discrete entities ('particles')—in a spatially distributed population [17]. It is a continuous-time Markov process in which each site i of a lattice (typically the d-dimensional hypercubic lattice, $\mathbf{Z}^d$), is either vacant or occupied (denoted as $\sigma_i = 0$, or 1, respectively, multiple occupancy being forbidden). An empty site with n occupied neighbors becomes occupied at a rate $\lambda n/q$, where q is the coordination number of the lattice, while occupied sites are vacated at unit rate. The rates for the one-dimensional (1D) CP are summarized in Fig. 3.1. Since particles can only be created by other particles, the vacuum is absorbing. Long-time survival of the population depends on the creation rate: extinction is certain if λ is too small. The boundary between persistence and extinction is marked by a critical point.

Under what conditions does the population survive? A precise answer is available, but it is instructive to begin with a simple mean-field theory (essentially Schlögl's model [16] of a well-stirred autocatalytic chemical system). Treating each site as statistically independent, and assuming spatial homogeneity, the particle density $\rho(t)$ is governed by

$$\frac{d\rho}{dt} = (\lambda - 1)\rho - \lambda\rho^2. \tag{3.1}$$

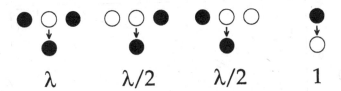

Fig. 3.1. Transition rates for the CP in 1D.

For $\lambda \leq 1$ the only stationary solution is the vacuum, $\bar{\rho} = 0$, but for $\lambda > 1$ there is also an active state with $\bar{\rho} = 1 - \lambda^{-1}$. It is natural to regard the stationary particle density, $\bar{\rho}$, as the order parameter for the transition at $\lambda_c = 1$. If we let $\Delta \equiv \lambda - \lambda_c$, then we may introduce the order-parameter exponent β in the usual manner: $\bar{\rho} \propto \Delta^{\beta}$ ($\Delta > 0$). In mean-field theory, $\beta = 1$.

Two fundamental rigorous results on the CP are the existence of a phase transition (for all dimensions $d \geq 1$) and the fact that the transition is continuous [24-27]. One also knows that, starting from a finite population, the survival probability and density die out exponentially for $\lambda < \lambda_c$, while for $\lambda > \lambda_c$ the radius of the populated region grows $\propto t$. (An active steady state exists only in the infinite-size limit, since any finite system will eventually be trapped in the vacuum! Even in systems of modest size, however, the quasi-stationary state is sufficiently long lived to permit detailed study in simulations.) There are no exact results for λ_c (there are bounds [24-25]), but simulations and series analyses give $\lambda_c \simeq 3.2978$ for $d = 1$, while β takes values near 0.277 and 0.58 for $d = 1$ and 2, respectively. Exponents assume their mean-field values for $d \geq d_c = 4$, the upper critical dimension. Figure 3.2 depicts the stationary density as a function of λ in the 1D CP. (The data are from a series calculation [29].)

It has been known for some time that the critical exponents for the contact process in d dimensions are the same as for directed percolation (DP) in $d+1$ dimensions, the oriented axis in DP corresponding to time in the CP [14-15]. Thus we may take over a host of results on critical behavior; only the highlights are summarized here.

Two essential aspects of the stationary state are the spatial correlation function

$$C(r) \equiv \langle \sigma_r(t)\sigma_0(t) \rangle - \bar{\rho}^2, \tag{3.2}$$

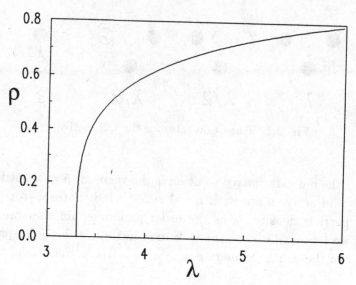

Fig. 3.2. Stationary density in the 1D contact process.

Table 3.1.

Critical exponents for the contact process in 1D. The figures in parentheses indicate uncertainties

β	0.2769(2)
$\nu_\parallel$	1.736(1)
$\nu_\perp$	1.0972(6)
δ	0.1597(3)
η	0.317(2)
z	1.272(7)

and the temporal correlation function

$$C_s(t) \equiv \langle \sigma_r(t_0 + t)\sigma_r(t_0) \rangle - \bar{\rho}^2. \tag{3.3}$$

For $\lambda \neq \lambda_c$, correlations are expected to decay exponentially, with $C(r) \propto e^{-r/\xi}$, and $C_s(t) \propto e^{-t/\tau}$ defining the correlation length ξ and relaxation time τ. The exponents characterizing the divergence of correlations at λ_c are defined thus: $\xi \propto \Delta^{-\nu_\perp}$ and $\tau \propto \Delta^{-\nu_\parallel}$, where $\nu_\parallel = 1$ in the mean-field analysis of (3.1). Critical exponents for the 1D contact process are collected in Table 3.1.

3.2.1 Time-dependent behavior

Consider the evolution of the CP starting with only a single particle, at the origin. (We are interested in the set of all such trials.) Two important features of the evolution from this near-vacuum state are the survival probability $P(t)$, and the mean number of particles $n(t)$ at time t: $n(t) = \sum_r \langle \sigma_r \rangle_t$. Here $\langle \cdots \rangle_t$ represents the mean over *all* trials, including those reaching the vacuum before time t. The spreading of the population is measured by

$$R^2(t) \equiv \frac{1}{n(t)} \sum_r r^2 \langle \sigma_r \rangle_t . \tag{3.4}$$

In the subcritical regime ($\lambda < \lambda_c$), both $P(t)$ and $n(t)$ decay exponentially, and $R^2 \propto t$. In the supercritical regime $P(t) \to P_\infty > 0$ as $t \to \infty$. In fact, the ultimate survival probability, P_∞, equals the stationary particle density $\bar{\rho}$ [14,24,25,30]. The active region has a finite density, and spreads into the vacuum at a constant rate, so that $n(t) \propto t^d$ and $R^2(t) \propto t^2$ in the supercritical regime. Just at the critical point the process dies with probability one, but the mean lifetime diverges. In the absence of a characteristic time scale, the asymptotic evolution follows power laws, conventionally written as $P(t) \propto t^{-\delta}$, $n(t) \propto t^\eta$, and $R^2(t) \propto t^z$. We shall refer to δ, η, and z as 'spreading exponents'. A more complete description includes finite-time corrections to scaling, for example,

$$P(t) \propto t^{-\delta}(1 + at^{-\theta} + bt^{-\delta'} + \cdots). \tag{3.5}$$

(It appears that $\theta = 1$ for the 1D CP [4-6,31].)

In a system of size L, power-law spreading obtains only for $\sqrt{R^2(t)} \ll L$. Simulations must be conducted accordingly. Nevertheless, such 'time-dependent' studies tend to be more effective in determining λ_c and the spreading exponents than are steady-state simulations, which are employed to find static properties such as $\bar{\rho}$ and β. (In steady-state simulations small errors in λ_c can lead to sizable errors in β. Large systems and finite-size scaling analysis are usually needed.)

Grassberger and de la Torre propounded a scaling hypothesis for absorbing-state transitions, embodied in the expressions [14]:

$$P(t) \simeq t^{-\delta} \phi(\Delta t^{1/\nu_\parallel}), \tag{3.6}$$

and

$$\rho(x, t) \simeq t^{\eta - zd/2} F(x^2/t^z, \Delta t^{1/\nu_\parallel}), \tag{3.7}$$

where $\rho(x,t)$ is the mean particle density. The scaling hypothesis leads to several exponent relations, which are supported by numerical results, e.g., $z = 2\nu_\perp/\nu_\parallel$, and $\delta = \beta/\nu_\parallel$, as well as hyperscaling relations, such as

$$4\delta + 2\eta = zd, \tag{3.8}$$

expected to hold for $d \leq 4$. Extensions of (3.8) have been derived recently [32,33].

While much of the above will be familiar to readers acquainted with directed percolation, we review these results because they comprise the generic critical behavior of models with absorbing states (see Sec. 3.4). Thus it is worth noting that a number of variants of the CP have been explored in detail, confirming the DP critical behavior. These include the θ-contact process, in which (in 1D) the rate for filling a vacant site with two occupied neighbors is $\theta\lambda$ while the creation rate at a site with only one occupied neighbor is λ [34-36], and the diffusive CP, in which particles execute nearest-neighbor hopping at a rate $\mathcal{D}$ [37-39]. (In the limit $\mathcal{D} \to \infty$ one expects to recover mean-field behavior.) Generalized CPs with multiparticle annihilation rules [87] and multiparticle creation rules [60,88] provide further examples of DP-like criticality. Devising a single-component model with a *first-order* transition requires some feature that destabilizes low-density states [89]. A relatively simple instance is the *triplet creation model* [88], which has a discontinuous transition at high (but finite) diffusion rates, and a continuous (DP) transition for smaller $\mathcal{D}$. Critical scaling in *three-state* systems such as voter and lattice predator-prey models has yet to be studied in detail [85,86].

3.3 Catalysis models

The simple model introduced by Ziff, Gulari, and Barshad (ZGB) for the oxidation of carbon monoxide (CO) on a catalytic surface has become one of the most widely studied models in nonequilibrium statistical physics [19,40,41]. The sites of the lattice ($\mathbf{Z}^2$ or some other two-dimensional structure), may be vacant, or occupied by oxygen (O) or carbon monoxide (CO). Each step of the process involves adsorption and, if possible, reaction. An adsorbing species is chosen, CO with probability Y, O_2 with probability $1 - Y$. (Y reflects the relative abundances of CO and O_2 in the gas surrounding the catalyst.) In a CO event, a site $\mathbf{x}$ is chosen, and if it is vacant, CO is provisionally adsorbed there (if $\mathbf{x}$ is occupied, the step terminates with the configuration unchanged). If none of the neighbors harbor O, the CO

Table 3.2.

Processes in the ZGB model. 'A' represents CO, 'B' oxygen

Event	ΔX	Rate
$B_2 + 2v \rightarrow 2B_{ads}$	2	$(1-Y)\theta_{vv}Q(\tilde{A}, \tilde{A}\|v,v)$
$B_2 + 2v \rightarrow B_{ads} + 2v$	2	$2(1-Y)\theta_{vv}Q(\tilde{A}, A\|v,v)$
$B_2 + 2v \rightarrow 4v$	2	$(1-Y)\theta_{vv}Q(A, A\|v,v)$
$A + v \rightarrow A_{ads}$	-1	$Y\theta_v[1 - Q(B\|v)]$
$A + v \rightarrow 2v$	-1	$Y\theta_v Q(B\|v)$

molecule remains at $\mathbf{x}$; otherwise the CO and a neighboring O (selected at random, if need be) react, liberating CO_2, so that both $\mathbf{x}$ and the site that held O become vacant. O_2 events follow similar rules, except that adsorption only occurs when both sites of a nearest-neighbor pair $\{\mathbf{x}, \mathbf{y}\}$ are vacant. Reactions between newly adsorbed O atoms and neighboring COs proceed independently, each O remaining (reacting) if none (one or more) of its neighbors has a CO molecule. The reaction rate is infinite, in that all CO-O pairs must react before anything else can happen.

If every site becomes occupied by the same kind of molecule, the reaction ceases, and the system is said to be 'poisoned'. Simulations reveal that for $Y < y_1 = 0.39065(10)$ the lattice is eventually poisoned with O, while for $Y > y_2 = 0.52560(1)$ there is CO poisoning [19,21,42]. For $y_1 < Y < y_2$, there is an active steady state, characterized by a nonzero vacancy density. The transition from the O-poisoned to the reactive state is continuous, while the CO transition is strongly discontinuous: as Y is increased beyond y_2 the CO coverage jumps from a few per cent to unity, and the CO_2 production rate, which grows steadily with Y in the active phase, falls to zero.

That the ZGB model has an active state at all derives from the geometrical difference between CO and O_2 adsorption, as the simpler AB monomer-monomer reaction (similar rules, but with both species requiring only a single site), has no active state [43]. Ignoring trivial cases in which the chosen sites are occupied, each event leads to one of five outcomes. The associated changes, ΔX, in X, which measures the excess of adsorbed O over CO, are listed in Table 3.2, where the quantity $Q(\tilde{A}, A|v,v)$ represents the conditional probability that none of the neighbors of $\mathbf{x}$ is occupied by an A and that at least one neighbor of $\mathbf{y}$ does harbor an A, given that neighboring sites $\mathbf{x}$ and $\mathbf{y}$ are both vacant, and so on. (Generally, in $Q(\cdots|v,v)$, the argument $\tilde{A}$ means that none of the neighbors of a vacant site is occupied by an A, and the argument A means that at least one of the neighbors is occupied by an A.) $Q(A|v)$ denotes the conditional probability that at least one neighbor of a vacant site is occupied by an A; θ_{vv} denotes the probability that both sites of a nearest-neighbor pair are vacant.

We may regard $X(t)$ as a random walk with transition probabilities:

$$P(X \rightarrow X + 2) = (1 - Y)\theta_{vv}, \qquad P(X \rightarrow X - 1) = Y\theta_v. \qquad (3.9)$$

A steady state can only exist in the absence of drift, i.e., $P(X \rightarrow X - 1) = 2P(X \rightarrow X + 2)$, or

$$Y/(1 - Y) = 2P(v|v), \qquad (3.10)$$

where $P(v|v) = \theta_{vv}/\theta_v$ is the conditional probability that a particular site is vacant, given that one of its neighbors is. Since $P(v|v) \leq 1$, there can be no steady state for $Y > 2/3$. The lattice must become poisoned with CO when $Y \geq y_2$, for some $y_2 \leq 2/3$. The coverage will drift in the other direction, towards O poisoning, if $Y/(1 - Y) < 2P(v|v)$. Since new vacancies can only appear at sites adjacent to existing ones, we should expect $P(v|v)$ to remain nonzero even as $\theta_v \rightarrow 0$. It follows that the lattice will become poisoned with O when $Y \leq y_1$, for some $y_1 > 0$.

To see why CO poisoning is discontinuous, suppose most (but not all), sites are occupied by CO. O_2 can adsorb only at rare vacancy pairs, and O atoms that are actually adsorbed will not remain long before reacting. Every vacancy, by contrast, is vulnerable to occupation by CO. Hence configurations with most sites occupied by CO are unstable to complete CO saturation, and there is a gap in the range of stationary CO coverages. On the other hand, a state with high O coverage is not necessarily unstable to O poisoning, because isolated vacancies are not susceptible to filling by O—only pairs are. Again, each vacancy can adsorb CO, which will typically react with an adjacent O (if the O-coverage is high), permitting vacancy clusters to grow, much as clusters of particles grow in the CP. So states with $\theta_O \simeq 1$ can be stable, and the transition to the O-poisoned lattice is continuous.

The simplest mean-field description of the ZGB predicts the CO transition, but not O poisoning. The main features of the phase diagram are reproduced in a pair-level mean-field analysis [44]. Theoretical arguments and numerical studies indicate that the critical behavior is that of $((2 + 1)$-dimensional) directed percolation [20,21,45,46]. Unfortunately O poisoning has not been observed experimentally, since in all the systems studied thus far an O-saturated surface still includes CO adsorption sites.

CO poisoning, by contrast, has been seen in numerous experiments [47]. At low temperatures the reaction rate R falls discontinuously as the CO fraction of the gas phase increases beyond a threshold value. With increasing temperature the gap ΔR narrows, eventually disappearing in a critical point. The temperature dependence of ΔR reflects thermally activated CO desorption. This process can be incorporated in the model by adding the event $CO_{ads} \to CO_g + v$ at rate k. With CO desorption there is no longer a CO-poisoned state, but for small k there is a discontinuous transition between low- and high-CO coverages [42,48]. In this case the critical behavior is not connected with DP (neither phase is absorbing) but, rather, is expected to belong to the Ising class [48].

Under perturbations such as finite reaction rate, CO desorption, anisotropic adsorption probabilities, and diffusion of one or both species, O poisoning remains a DP-like transition [49-51]. While the phase diagram and critical behavior in the ZGB model are reasonably well understood, several puzzles remain. Motivated by the observation that catalytic surfaces are often fractal-like, Albano studied the model on percolation clusters, and found that *both* the CO and O poisoning transitions are continuous [52]. Similar results are found on DLA clusters [53] and on randomly diluted lattices [54]. On the Sierpiński gasket, however, CO poisoning remains [55] first-order! An explanation for this difference was proposed by ben-Avraham [56], based on the role of so-called 'red bonds' that provide the sole connection between 'blobs' of multiply connected sites. Several recent studies have focused on the interface between reactive and CO-poisoned regions, placing its dynamics in the KPZ class [57,58].

3.4 Universality

That the same critical-exponent values are found at absorbing-state transitions in diverse models is strong evidence of universality. According to the *DP conjecture*, almost all such transitions belong to the DP universality class [20,59,60,92]. For the benefit of skeptics, a more precise statement follows. In the absence of any special conservation laws, continuous transitions between an active and an absorbing state of a many-particle system with local interactions and a scalar order parameter are generically of the DP type. Here 'local' means that the probability of an event at site $\mathbf{x}$ depends only on the configuration within some finite radius of $\mathbf{x}$, or that the influence of the configuration at $\mathbf{y}$ on events at $\mathbf{x}$ falls off sufficiently rapidly with $|\mathbf{x}-\mathbf{y}|$. The required decay rate (presumably a power law), has not been determined, but it is clear that if one takes the creation probability to be proportional to the *global* density (as in a long-range CP, for example), one will observe mean-field behavior. 'Generically' means 'for almost all parameter values'. The term 'special conservation law' should seem less mysterious once an example has been considered; see Sec. 3.6. In any event, models like the CP and ZGB satisfy all the criteria for DP universality.

The idea underlying universality is that many different models have the same effective evolution when viewed on large length scales and long time scales. A minimal continuum description of the CP and allied models takes the form [59]

$$\frac{\partial \rho(x,t)}{\partial t} = -a\rho(x,t) - b\rho^2 - c\rho^3 + \cdots + D\nabla^2\rho + \eta(x,t), \qquad (3.11)$$

where η is a Gaussian noise with covariance

$$\langle \eta(x,t)\eta(x',t') \rangle \propto \rho(x,t)\delta(x-x')\delta(t-t'), \qquad (3.12)$$

so that it respects the absorbing state. Equation (3.11) describes a field of diffusively coupled Malthus-Verhulst processes; without noise, it is a local mean-field theory or *reaction-diffusion equation*. The detailed dependence of the coefficients on microscopic parameters like λ is not particularly significant, but one expects on general grounds that a depends linearly on Δ, the distance from the critical point, and that b is positive. (In the mean-field approximation, the critical point is at $a = 0$.) The nature of solutions to (3.11) and its connection to the CP have been put on a firm basis [61]; numerical integration yields DP critical exponents [62]. Long before the last-mentioned studies, Janssen used field-theoretic renormalization group methods to show that the cubic and higher-order terms are irrelevant to critical behavior provided $b > 0$ [59]. Equation (3.11) also serves as the starting point for an ϵ-expansion of the critical exponents about $d = 4$. Finally, since (3.11) is the field theory of directed percolation and the CP, the DP conjecture amounts to asserting that (subject to the provisos noted above), all absorbing-state transitions have the same set of relevant terms in their continuum limit. This seems highly plausible, but it must be remarked that an appropriate continuum description for certain models has yet to be formulated [60,63]. The DP conjecture does not rule out higher-order transitions, but these require fine-tuning of the parameters, so that (for example) a and b vanish simultaneously [64].

3.5 Multiple absorbing configurations

A multiplicity of absorbing configurations arises naturally in catalysis modeling. The dimer-trimer model, for example, features adsorption and reaction as in the ZGB model, but with CO (O_2) replaced by molecules requiring two (three) sites for adsorption [23]. In this model, and similarly in the dimer-dimer model [22], any configuration having only isolated vacancies is absorbing. Several 1D models with multiple absorbing configurations have also been studied [32,65,66]; we shall focus on the *pair contact process* (PCP), which evolves via the following rules. Choose a site i at random; do nothing unless both i and $i+1$ are occupied. If they are, then with probability p remove both particles; with probability $1-p$, choose site $i-1$ or site $i+2$ (with equal likelihood), and if this site is vacant, place a new particle there. The PCP, like the basic contact process, exhibits a continuous transition between an active and an absorbing state. But now any configuration devoid of nearest-neighbor pairs is absorbing; their number, on a lattice of N sites, exceeds $2^{N/2}$. In the PCP, and in all models with multiple absorbing configurations studied so far, the static exponents β, $\nu_{\parallel}$, and $\nu_{\perp}$ assume DP values [67], just as one expects on the basis of the DP conjecture.

In applying time-dependent simulations (of great utility for the CP, etc.), we are now presented with a vast choice of initial states. In the PCP, for example, we begin with a single nearest-neighbor pair; the rest of the lattice may be assigned any sequence (random or periodic) of vacancies and isolated particles. Remarkably, the spreading exponents vary continuously with the particle density ϕ! The class of 'natural' absorbing configurations (with particle density ϕ_{nat}), comprises those generated by the critical process, starting far from the absorbing state (all sites occupied, in the PCP). In simulations using natural initial configurations, the spreading exponents δ, η, z assume DP values. In fact, the exponents are insensitive to changes in short-range correlations provided the particle density remains ϕ_{nat}. Non-DP exponents are found for other densities. For example, the PCP with $\phi = 0$ has $\delta = 0.250(5)$ and $\eta = 0.215(5)$, while for $\phi = 0.432$ the exponents are $\delta = 0.0955(5)$ and $\eta = 0.38(1)$. A high particle density favors spreading, hence a slower decay of the survival probability (smaller δ), and a larger η, reflecting more rapid population growth.

Studies of the PCP and similar 1D models show δ and η varying linearly with ϕ over a wide range. The change in z is much smaller. The hyperscaling relation, (3.8), is violated by the tunable exponents, but the scaling argument can be extended to the case of exponents that depend on the initial density [32]. The scaling hypothesis of (3.6) and (3.7) remains valid, but for nonnatural initial states it is no longer true that $P_\infty \propto \bar{\rho}$, and one is led to introduce a new exponent, β', governing the ultimate survival probability: $P_\infty \propto \Delta^{\beta'}$. The scaling relation $\beta' = \delta\nu_\parallel$ implies that β' also depends on the initial density. Retracing the argument of Grassberger and de la Torre, one arrives at a generalized hyperscaling relation,

$$2\left(1 + \frac{\beta}{\beta'}\right)\delta + 2\eta = zd, \tag{3.13}$$

which has been verified numerically.

Simulations also indicate that $\delta + \eta$, the exponent governing the population growth in *surviving* critical trials, is independent of ϕ. (The data are also consistent with z's being independent of ϕ.) This suggests that asymptotic properties of surviving trials are independent of the initial density. One should also note that as $t \to \infty$ only a negligible fraction of a surviving cluster is actually in contact with the external density ϕ. In equilibrium critical phenomena, the appearance of continuously variable exponents is usually associated with a *marginal parameter*, one that does not change under a renormalization group transformation. One may view the initial density ϕ as playing an analogous role; it retains its value as the evolution is viewed on larger length and time scales, and affects the likelihood of survival no matter how large t is, since the process must repeatedly invade new territory to survive. Efforts to place these notions on a firmer basis are under way [68].

3.6 Branching annihilating random walks

Conservation laws have profound consequences in dynamics. This is familiar from equilibrium: a system with a conserved order-parameter density belongs to a dynamic universality class distinct from that of its nonconserving counterpart [69]. In this section we discuss absorbing-state transitions that are affected in a fundamental way by a conservation law.

In the *branching annihilating random walk* (BAW) model [70], particles hop on a lattice, annihilate upon contact, and produce 'offspring' (further particles) at neighboring sites. At each step of the process a particle (located, say, at site i), is selected at random; it either hops (with probability p), to a randomly chosen neighboring site, or (with probability $1-p$), creates n new particles, which are placed (one each) at the site(s) nearest i. Placement of offspring is random, if there is a choice. For example, if $n = 1$ the new particle appears at a randomly chosen nearest neighbor; in 1D, when

$n = 2(4)$, the new particles appear at the two (four) sites nearest the parent. Whenever, due either to hopping or to branching, a particle arrives at an occupied site, pairwise annihilation ensues. The vacuum is absorbing, just as in the contact process. In general, there is an active stationary state for small p, the density vanishing continuously as $p \nearrow p_c$. The exception is $n = 2$ in 1D, which has no active state for $p > 0$ [71,72]. This case is closely related to the diffusive annihilation process $A + A \to 0$ considered in Chs. 1, 2, 8.

Notice a basic difference between even- and odd-n BAWs: in the former, particle number is conserved modulo 2 ('parity conservation'). For even-n BAWs, the vacuum is accessible only if the particle number is even. Odd-n BAWs are not restricted in this manner, and it comes as no surprise that they belong to the DP universality class [73]. Even-n BAWs, however, exhibit *non-DP* critical behavior. A revealing observation is that minute violations of parity conservation (in the form of a very small rate for single-particle annihilation) throw the model back into the DP class [74]. Equally crucial is that conservation be *local*: a model with global (but not local) parity conservation shows DP exponents [75].

Jensen [76] derived precise exponent estimates for an even-n BAW ($n = 4$, in 1D): $\beta = 0.93(5)$; $\delta = 0.286(1)$; $\eta = 0.000(1)$; $z = 1.143(3)$; $\nu_\| = 3.25(10)$. (The critical value of p is 0.7215(5).) They are quite far from their DP counterparts, and remarkably, are consistent with simple rational values: $\delta \simeq 2/7$; $z \simeq 8/7$; $\nu_\| \simeq 13/4$; and $\beta \simeq 13/14$. (They also satisfy the hyperscaling relation $4\delta + 2\eta = zd$.) These exponents hold when there are an even number of particles. If the particle number is odd, all trials survive indefinitely ($P(t) = P_\infty = 1$), and so the only sensible choice for δ and for β' is zero. In this case Jensen's simulations yield $\eta = 0.282(5)$ (consistent with 2/7), and z the same as quoted above. The generalized hyperscaling relation (3.13) holds in the form

$$\delta + \beta/\nu_\| + \eta = zd/2. \tag{3.14}$$

The fact that the correlation-time exponent, $\nu_\|$, is nearly twice the DP value parallels the situation for the Ising model: the dynamic critical exponent $z \approx 2$ for nonconserving dynamics in two dimensions, while $z \approx 4$ if magnetization is conserved [69].

Several other models appear to show the same critical behavior as parity-conserving BAWs. In fact, the first examples of this class were devised some time ago by Grassberger, Krause and von der Twer [63]. Their models are 1D stochastic cellular automata in which the state (occupied or not) of site i at time t depends on the states of $i-1$, i, and $i+1$ at time $t-1$. The rules for

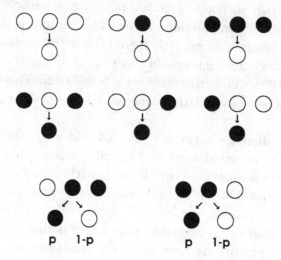

Fig. 3.3. Rules for the stochastic cellular automaton of Grassberger *et al.* [63]. Transitions in the upper two rows are deterministic. Branching probabilities are indicated for the bottom row.

their 'model A' are illustrated in Fig. 3.3. Inspection reveals two absorbing 2-cycles, checkerboard space-time patterns that differ by a unit shift. If we regard a string of sites with alternating occupancy ($\cdots\bullet\circ\bullet\circ\bullet\cdots$) as a domain, then any pair of like neighbors is a domain wall or kink. These play the role of particles in the BAW: when all kinks are annihilated and the system consists of a single domain, it is trapped in an absorbing 2-cycle. It therefore seems appropriate to refer to this model as the *kink cellular automaton* (KCA). Viewed in terms of kinks, the evolution comprises the elementary events of hopping, pairwise annihilation, and $n = 2$ branching. As the parameter p is increased, not only does branching occur more frequently, but an effective repulsion, $\propto p(1 - p)$, between kinks at neighboring sites comes into play. (In this way the KCA can evade the vacuum, unlike the $n = 2$ BAW.) The dynamics of the KCA are essentially the same as for parity-conserving BAWs (domain walls can only be created or destroyed in pairs!) so their critical behavior should be the same.

Simulations of the KCA model place the critical point at $p_c \simeq 0.540$, and yield $\beta = 0.94(6)$ and $\nu_{\parallel} = 3.3(2)$ [63]. Time-dependent studies starting with a single kink (the initial configuration $\cdots\bullet\circ\bullet\circ\circ\bullet\circ\bullet\circ\cdots$), yield $\eta \simeq 0.27$. These exponents agree quite nicely with those found for the $n = 4$ BAW, supporting the expectation of a common critical behavior in the two models. It seems reasonable to conclude that the KCA model and parity-conserving BAWs belong to the same universality class, distinct from DP, which might be called 'critical domain dynamics'.

Two further examples of this new class have been studied. The *interacting monomer-dimer model* (IMD) is a 1D version of the ZGB model, except that a pair of monomers cannot reside at adjacent sites, and a dimer (BB) cannot adsorb adjacent to a B on the lattice [77]. (In other words, the sequences AA and BBB are disallowed. However, A is allowed to adsorb between an A and a B: the immediate reaction eliminates the possibility of an AA pair.) There are two absorbing configurations—perfect domains of alternating A-filled and vacant sites—that differ by a unit shift. The transition to the absorbing state is continuous, occurring via pairwise annihilation of domain walls. The second example [78] is a (ferromagnetic) kinetic Ising model (in 1D) with spin flips (Glauber dynamics) at zero temperature in competition with nearest-neighbor exchange (Kawasaki dynamics) operating at infinite temperature. The former process corresponds to diffusion and pairwise annihilation of domain walls, the latter to branching with $n = 2$. As the rates for spin flips and exchange are varied, there is a continuous transition between a single-domain and multi-domain state. Simulations of these models yield exponents consistent with those of $n = 4$ BAWs and of the KCA.

The key feature common to these four examples of critical domain dynamics is local conservation of the particle number, modulo 2. ('Particle' means the entity representing the order parameter: a random walker in BAWs, a kink or domain wall in the KCA and the IMD.) The approach to the absorbing state occurs through the expansion of a single domain to fill the system. Close to the critical point, the particle density is small, and domains expand via diffusive encounters between particles that were initially well separated. This highlights the significance of local, as opposed to merely global, parity conservation. Critical domain dynamics is presumably the generic critical behavior for models obeying local parity conservation, just as DP is generic for models free of this constraint. Of course, a wider range of models needs to be explored before one can have confidence in this conjecture. There are also some fundamental open questions regarding models with parity conservation. First, all the models studied so far are 1D. What happens in higher dimensions? It appears likely that $d_c = 2$ for even-n BAWs, i.e., that they exhibit mean-field critical behavior in two or more dimensions [79]. The IMD and kinetic Ising models are readily extended to higher dimensions, where the approach to the absorbing state again depends on the expansion of one domain. But domain dynamics in two or more dimensions looks nothing like a BAW: the evolution takes place along extended structures that cannot be put in correspondence with particles. So there is no reason to expect domain models like the IMD to be in the same class as BAWs in two or more dimensions.

Another piece missing from the picture of critical domain dynamics is a continuum representation. As pointed out by Grassberger and coworkers [63], simple field theories incorporating parity conservation have DP-like critical behavior, which is obviously wrong! (The terms responsible for conserving parity are simply *irrelevant*.) A useful continuum model, analogous to (3.11), that provides the foundation for understanding universality has yet to be devised for parity-conserving models.

3.7 Summary

We have surveyed, without delving into great detail, the phase diagrams and critical properties of models with absorbing states. Readers interested in knowing more about the methods used in studying absorbing-state transitions may find the following (hopelessly incomplete) list of some use: Monte Carlo simulation and finite-size scaling analysis [31,66,80]; cluster mean-field theories [44], and associated coherent-anomaly-method analysis [81,82]; series expansions [3-6,13,29,83]; numerical analysis of the master equation [84]; renormalization group analysis [11,18,59].

While some of the simpler absorbing-state transitions are reasonably well understood, critical spreading in models with multiple absorbing configurations and parity-conserving systems present many avenues for further investigation, at the microscopic level and in the continuum limit.

The author is grateful to Iwan Jensen and Adriana Gomes Moreira for helpful comments.

References

[1] S. R. Broadbent and J. M. Hammersley, *Proc. Camb. Phil. Soc.* **53**, 629 (1957).

[2] J. Blease, *J. Phys.* C10, 917; 923; 3461 (1977).

[3] J. W. Essam and K. De'Bell, *J. Phys.* A14, L459 (1981).

[4] R. J. Baxter and A. J. Guttmann, *J. Phys.* A21, 3193 (1988).

[5] J. W. Essam, K. De'Bell, J. Adler and F. M. Bhatti, *Phys. Rev.* B33, 1982 (1986).

[6] J. W. Essam, A. J. Guttmann and K. De'Bell, *J. Phys.* A21, 3815 (1988).

[7] W. Kinzel, *Z. Physik* B58, 229 (1985).

[8] G. Deutscher, R. Zallen and J. Adler, eds., *Percolation Structures and Processes.* Ann. *Israel Phys. Soc.*, Vol. **5** (Hilger, Bristol, 1983).

[9] V. N. Gribov, *Sov. Phys. JETP* **26**, 414 (1968); V. N. Gribov and A. A. Migdal, *Sov. Phys. JETP* **28**, 784 (1969).

[10] H. D. I. Abarbanel, J. B. Bronzan, R. L. Sugar and A. R. White, *Phys. Rep.* **21**, 119 (1975).

[11] P. Grassberger and K. Sundermeyer, *Phys. Lett.* **B77**, 220 (1978).

[12] M. Moshe, *Phys. Rep.* **C37**, 255 (1978).

[13] R. C. Brower, M. A. Furman and M. Moshe, *Phys. Lett.* **B76**, 213 (1978).

[14] P. Grassberger and A. de la Torre, *Ann. Phys. (NY)* **122**, 373 (1979).

[15] J. L. Cardy and R. L. Sugar, *J. Phys.* **A13**, L423 (1980).

[16] F. Schlögl, *Z. Phys.* **B58**, 229 (1985).

[17] T. E. Harris, *Ann. Prob.* **76**, 1122.

[18] D. Elderfield and D. D. Vvedensky, *J. Phys.* **A18**, 2591 (1985); D. Elderfield and M. Wilby, *J. Phys.* **A20**, L77 (1987).

[19] R. M. Ziff, E. Gulari and Y. Barshad, *Phys. Rev. Lett.* **56**, 2553 (1986).

[20] G. Grinstein, Z.-W. Lai and D. A. Browne, *Phys. Rev.* **A40**, 4820 (1989).

[21] I. Jensen, H. C. Fogedby and R. Dickman, *Phys. Rev.* **A41**, 3411 (1990).

[22] E. V. Albano, *J. Phys.* **A25**, 2557 (1992).

[23] J. Köhler and D. ben-Avraham, *J. Phys.* **A24**, L621 (1991); D. ben-Avraham and J. Köhler, *J. Stat. Phys* **65**, 839 (1992).

[24] T. M. Liggett, *Interacting Particle Systems* (Springer, New York, 1985).

[25] R. Durrett, *Lecture Notes on Particle Systems and Percolation* (Wadsworth, Pacific Grove, 1988).

[26] C. Bezuidenhout and G. Grimmett, *Ann. Prob.* **18**, 1462 (1990).

[27] N. Konno, *Phase Transitions of Interacting Particle Systems* (World Scientific, Singapore, 1994).

[28] For further details see: J. Marro and R. Dickman, *Nonequilibrium Phase Transitions and Critical Phenomena* (Cambridge University Press, Cambridge), in preparation.

[29] R. Dickman and I. Jensen, *Phys. Rev. Lett.* **67**, 2391 (1991); I. Jensen and R. Dickman, *J. Stat. Phys.* **71**, 89 (1993).

[30] Equality of P_∞ and $\bar\rho$ is a special feature of the contact process, resulting from its so-called 'self-duality'. For models with unique absorbing configurations, the two quantities are not necessarily equal, but they have the same critical exponent, β.

[31] P. Grassberger, *J. Phys.* **A22**, 3673 (1989).

[32] J. F. F. Mendes, R. Dickman, M. Henkel and M. C. Marques, *J. Phys.* **A27**, 3019 (1994).

[33] R. Dickman and A. Yu. Tretyakov, *Phys. Rev.* **E52**, 3218 (1995).

[34] R. Durrett and D. Griffeath, *Ann. Prob.* **11**, 1 (1983).

[35] M. Katori and N. Konno, *J. Phys.* **A26**, 6597 (1993); M. Katori and N. Konno, in K. Kawasaki and M. Suzuki, eds., *Formation, Dynamics and Statistics of Patterns* (World Scientific, Singapore, 1993).

[36] I. Jensen and R. Dickman, *Physica* **A203**, 175 (1994).

[37] I. Jensen and R. Dickman, *J. Phys.* **A26**, L151 (1993).

[38] M. Katori and N. Konno, *Physica* **A186**, 578 (1992).

[39] M. Katori, *J. Phys.* **A27**, 7327 (1994).

[40] J. W. Evans, *Langmuir* **7**, 2514 (1991).

[41] V. P. Zhdanov and B. Kasemo, *Surf. Sci. Rep.* **20**, 111 (1994).

[42] R. M. Ziff and B. J. Brosilow, *Phys. Rev.* **A46**, 4630 (1992); B. J. Brosilow and R. M. Ziff, *Phys. Rev.* **A46**, 4534 (1992).

[43] K. A. Fichthorn and R. M. Ziff, *Phys. Rev.* **B34**, 2038 (1986); E. Clément,

P. Leroux-Hugon and L. M. Sander, *Phys. Rev. Lett.* **67**, 1661 (1991); P. L. Krapivsky, *J. Phys.* **A25**, 5831 (1992).

[44] R. Dickman, *Phys. Rev.* **A34**, 4246 (1986); P. Fischer and U. M. Titulaer, *Surf. Sci.* **221**, 409 (1989); M. Dumont, P. Dufour, B. Sente and R. Dagonnier, *J. Catal.* **122**, 95 (1990).

[45] B. Chopard and M. Droz, *J. Phys.* **A21**, 205 (1988).

[46] J. Mai and W. von Niessen, *Phys. Rev.* **A44**, R6165 (1991).

[47] M. Ehsasi, M. Matloch, O. Frank, J. H. Block, K. Christmann, F. S. Rys and W. Hirschwald, *J. Chem. Phys.* **91**, 4949 (1989).

[48] T. Tomé and R. Dickman, *Phys. Rev.* **E47**, 948 (1993).

[49] D. Considine, H. Takayasu and S. Redner, *J. Phys.* **A23**, L1181 (1990); J. Köhler and D. ben-Avraham, *ibid.* L141.

[50] B. Yu, D. A. Browne and P. Kleban, *Phys. Rev.* **A43**, 1770 (1991).

[51] I. Jensen and H. C. Fogedby, *Phys. Rev.* **A42**, 1969 (1991).

[52] E. V. Albano, *J. Phys.* **A23**, L545 (1990).

[53] J. Mai, A. Casties and W. von Niessen, *Chem. Phys. Lett.* **211**, 197 (1993).

[54] J.-P. Hovi, J. Vaari, H. P. Kaukonen and R. M. Nieminen, *Comp. Mat. Sci.* **1**, 33 (1993).

[55] A. Yu. Tretyakov and H. Takayasu, *Phys. Rev.* **A44**, 8388 (1991).

[56] D. ben-Avraham, *J. Phys.* **A26**, 3725 (1993).

[57] R. Ziff, *J. Chem. Phys.* **98**, 674 (1993).

[58] J. W. Evans and T. R. Ray, *Phys. Rev.* **50**, 4302 (1994).

[59] H. K. Janssen, *Z. Physik* **B42**, 151 (1981).

[60] P. Grassberger, *Z. Physik* **B47**, 365 (1982).

[61] C. Mueller and R. Tribe, *Prob. Th. Rel. Fields* **100**, 131 (1994).

[62] R. Dickman, *Phys. Rev.* **E50**, 4404 (1994).

[63] P. Grassberger, F. Krause and J. von der Twer, *J. Phys.* **A17**, L105 (1984); P. Grassberger, *J. Phys.* **A22**, L1103 (1989).

[64] T. Ohtsuki and T. Keyes, *Phys. Rev.* **A35**, 2697 (1987).

[65] I. Jensen, *Phys. Rev. Lett.* **70**, 1465 (1993).

[66] I. Jensen and R. Dickman, *Phys. Rev.* **E48**, 1710 (1993).

[67] I. Jensen, *Int. J. Mod. Phys.* **B8**, 3299 (1994).

[68] M. A. Muñoz, G. Grinstein, R. Dickman and R. Livi, *Phys. Rev. Lett.* **76**, 451 (1996).

[69] P. C. Hohenberg and B. I. Halperin, *Rev. Mod. Phys.* **49**, 435 (1977).

[70] M. Bramson and L. Gray, *Z. Warsch. verw. Gebiete* **68**, 447 (1985).

[71] A. Sudbury, *Ann. Prob.* **18**, 581 (1990).

[72] H. Takayasu and N. Inui, *J. Phys.* **A25**, L585 (1992).

[73] I. Jensen, *Phys. Rev.* **E47**, 1 (1993).

[74] I. Jensen, *J. Phys.* **A26**, 3921 (1993).

[75] N. Inui, A. Yu. Tretyakov, and H. Takayasu, *J. Phys.* **A28**, 1145 (1995).

[76] I. Jensen, *Phys. Rev.* **E50**, 3623 (1994).

[77] M. H. Kim and H. Park, *Phys. Rev. Lett.* **73**, 2579 (1994).

[78] N. Menyhárd, *J. Phys.* **A27**, 6139 (1994).

[79] H. Takayasu, and A. Yu. Tretyakov, *Phys. Rev. Lett.* **68**, 3060 (1992).

[80] T. Aukrust, D. A Browne and I. Webman, *Europhys. Lett.* **10**, 249 (1989); *Phys. Rev.* **A41**, 5294 (1990).

[81] N. Konno and M. Katori, *J. Phys. Soc. Jpn* **59**, 1581 (1990).

[82] A. L. C. Ferreira and S. K. Mendiratta, *J. Phys.* **A26**, L145 (1993); N. Inui, *Phys. Lett.* **A184**, 79 (1994).

[83] R. Dickman and M. A. Burschka, *Phys. Lett.* **A127**, 132 (1988); R. Dickman, *J. Stat. Phys.* **55**, 997 (1989).

[84] B. Yu and D. A. Browne, *Phys. Rev.* **E49**, 3496 (1994).

[85] K. Tainaka, *Phys. Lett.* **A176**, 303 (1993).

[86] K. Tainaka and S. Fukazawa, *J. Phys. Soc. Jpn* **61**, 1891 (1992).

[87] R. Dickman, *Phys. Rev.* **B40**, 7005 (1989); *Phys. Rev.* **A42**, 6985 (1990).

[88] R. Dickman and T. Tomé, *Phys. Rev.* **A44**, 4833 (1991).

[89] R. Bidaux, N. Boccara and H. Chaté, *Phys. Rev.* **A39**, 3094 (1989).

[90] G. Ódor and G. Szabó, *Phys. Rev.* **E49**, R3555 (1994).

[91] Y. Pomeau, *Physica* **D23**, 3 (1986).

[92] P. Grassberger, *J. Stat. Phys.* **79**, 13 (1995).

Part II: Kinetic Ising Models

Editor's note

Kinetic Ising models in 1D provide a gallery of exactly solvable systems with nontrivial dynamics. The emphasis has traditionally been on their exact solvability, although much attention has also been devoted to models with conservation laws that have to be treated by numerical and approximation methods.

Chapter 4 reviews these models with emphasis on steady states and the approach to steady-state behavior. Chapter 5 puts the simplest 1D kinetic Ising models into a wider framework of the evaluation of dynamical critical behavior, analytically, in 1D, and numerically, for general dimension. Finally, Ch. 6 describes low-temperature nonequilibrium properties such as domain growth and freezing.

For a general description of dynamical critical behavior, not limited to 1D, as well as an excellent review and classification of various types of dynamics, the reader is directed to the classical work [1]. Certain probabilistic cellular automata are equivalent to kinetic Ising models. One such example can be found in Ch. 8. Finally, we note that a kinetic Ising model with nonconserved (Glauber) dynamics allows exact solution in a quenched *random field* [2].

[1] P. C. Hohenberg and B. I. Halperin, *Rev. Mod. Phys.* **49**, 435 (1977).
[2] G. Forgacs, D. Mukamel and R. A. Pelcovits, *Phys. Rev.* **B30**, 205 (1984). Certain generalizations for the *XY*-chain are also presented there.

4

Kinetic Ising models with competing dynamics: mappings, correlations, steady states, and phase transitions

Zoltán Rácz

In this chapter we give a brief review of one-dimensional (1D) kinetic Ising models that display nonequilibrium steady states. We describe how to construct such models, how to map them onto models of particle and surface dynamics, and how to derive and solve (in some cases) the equations of motion for the correlation functions. In the discussion of particular models, we focus on various problems characteristically occurring in studies of nonequilibrium systems such as the existence of phase transitions in 1D, the presence or absence of the fluctuation-dissipation theorem, and the derivation of the Langevin equations for mesoscopic degrees of freedom.

4.1 Introduction

The Ising model is a static, equilibrium, model. Its dynamical generalization was first considered by Glauber [1] who introduced the single-spin-flip kinetic Ising model for describing relaxation towards *equilibrium*. Kawasaki [2] then constructed a spin-exchange version of spin dynamics with the aim of studying such relaxational processes in the presence of conservation of magnetization. Other conservation laws were introduced soon afterwards by Kadanoff and Swift [3] and thus the industry of kinetic Ising models was born.

The value of these models became apparent towards the end of the 1960s and the beginning of the 1970s when ideas of universality in static and dynamic critical phenomena emerged. Kinetic Ising models were simple enough to allow extensive analytical (series-expansion) and numerical (Monte Carlo) work, which was instrumental in determining critical exponents and checking universality.

Special attention has always been paid to 1D versions of kinetic Ising models, which could often be solved exactly and thus have provided a testing ground for theories applied to higher-dimensional systems. The main results pertaining to the 1D case are reviewed in this Part. In disguise, however, various versions of these models can be found in other Parts as well.

The 1980s saw the emergence of new types of kinetic Ising model. Once the problem of the equilibrium critical phase transitions had been sorted out, attention turned to nonequilibrium steady states and, in particular, to phase transitions (pattern formation) occurring in them. Since a fundamental theory of nonequilibrium steady states does not exist, these problems have been approached by following a traditional recipe: construct deterministic phenomenological equations, study the stability of their solutions, and add noise to find the effect of fluctuations [4]. An alternative and somewhat more microscopic approach was the introduction of kinetic Ising models in which the nonequilibrium steady states were produced by the spins' being driven by external fields and by contacts with several heat baths at different temperatures.

The first nonequilibrium steady-state kinetic Ising model in 1D was studied by Hill [5] in connection with interacting enzyme molecules. In present-day terminology, his model is a two-temperature, single spin-flip kinetic Ising model with two heat baths attached to every spin, each bath generating spin flips at a different temperature [6]. A kinetic Ising model corresponding to a driven lattice gas was introduced next [7]. Then models with competition either between spin flips and spin exchanges [8-12] or between spin flips and bond flips [13] were constructed, and spatially varying heat-bath temperatures [6,14] were also studied. The possibilities of creating models with competing dynamics are, of course, limitless and, looking at the number of existing models [15], one may feel somewhat overwhelmed. The reason for the abundance appears to be partly the experimental motivation (see Part VII) and partly the hope that by solving a sufficient number of these models an understanding of the distinguishing features of nonequilibrium steady states will emerge. Here the role of 1D models should be emphasized: exact solutions are highly desirable since they are not biased by equilibrium notions. Thus, important nonequilibrium features do not get lost, as often happens when approximation methods coming from equilibrium theories are applied to complicated models in higher dimensions.

To review an abundant field without a theory that would give a framework to the discussion is rather difficult. We would like to avoid an 'enumerate and comment' type of treatment and would like, instead, to provide an introductory guide to the models and methods used and to problems with which one is faced in this field. Thus, we shall start with the description of a few simple competing spin dynamics processes that produce nonequilibrium steady states and will demonstrate that these models are quite versatile in the sense that they can be mapped onto various lattice-gas and surface-evolution models. Then we show how to derive the equations for the correlation functions and find examples when the hierarchy of equations decouples and the model can be solved exactly. Along the way, we shall come across a number of problems recurring in the studies of far-from-equilibrium systems. The presence of phase transitions in 1D will be the dominating topic but the sensitivity of the results to the details of the dynamics, the 'violation' of the fluctuation-dissipation theorem, and the problem of derivation of Langevin equations for mesoscopic degrees of freedom will also be discussed. No complete solution exists to any of these problems. The simplicity and analytical basis of the treatment, however, may be illuminating to both newcomers and seasoned workers in this field.

Owing to limited space, the review in this chapter is necessarily incomplete and, furthermore, it is somewhat biased towards the author's own work. It is believed, however, that taken together with the other chapters in this book, one can get a fairly detailed view of what has been accomplished in connection with the 1D kinetic Ising models.

4.2 Master equation and mappings

The state $\{\sigma\} \equiv \{\ldots, \sigma_i, \sigma_{i+1}, \ldots\}$ of a 1D kinetic Ising model at time t is specified by stochastic Ising variables $\sigma_i(t) = \pm 1$ assigned to lattice sites $i = 1, 2, \ldots, N$. The interaction between nearest-neighbor spins is $-J\sigma_i\sigma_{i+1}$ and periodic boundary conditions ($\sigma_{N+1} = \sigma_1$) are usually assumed. The dynamics of the system is generated by contacts with several heat baths (labeled by α) at temperatures T_α; each of these heat baths tries to bring the system to equilibrium at temperature T_α by flips and rearrangements of the spins. Let the action of the αth heat bath on the ith spin be represented as $\{\sigma\} \rightarrow \Gamma_i^\alpha\{\sigma\}$, and assume that the rate of this process is given by $w_i^\alpha(\{\sigma\})$. The rate of the reverse process, $\Gamma_i^\alpha\{\sigma\} \rightarrow \{\sigma\}$, can be denoted as $w_i^\alpha(\Gamma_i^\alpha\{\sigma\})$ and the dynamics is then described by the following master equation for the probability distribution $P(\{\sigma\}, t)$:

$$\frac{\partial P(\{\sigma\}, t)}{\partial t} = \sum_\alpha \sum_{i=1}^N \left[w_i^\alpha(\Gamma_i^\alpha\{\sigma\}) \, P(\Gamma_i^\alpha\{\sigma\}, t) - w_i^\alpha(\{\sigma\}) \, P(\{\sigma\}, t) \right]. \quad (4.1)$$

The assumption that the dynamics is generated by heat baths means that the rates of the dynamical processes are each constrained to satisfy detailed balance at the appropriate temperatures:

$$w_i^\alpha(\{\sigma\}) \, P_\alpha^{\mathrm{eq}}(\{\sigma\}) = w_i^\alpha(\Gamma_i^\alpha\{\sigma\}) \, P_\alpha^{\mathrm{eq}}(\Gamma_i^\alpha\{\sigma\}), \qquad (4.2)$$

where P_α^{eq} is the equilibrium distribution of the Ising model at temperature T_α.

We shall consider three different types of heat bath, generating single spin flips, spin exchanges, and bond flips. The first one ($\alpha = 1$) produces single spin flips, i.e., $\Gamma_i^1\{\sigma\}$ differs from $\{\sigma\}$ in that the ith spin is flipped. The most general form of the rate of this process that satisfies detailed balance and depends only on σ_i and its neighbors can be written as [1]

$$w_i^1(\sigma) = \frac{1}{2\tau_1} \left[1 - \frac{\gamma}{2} \sigma_i \left(\sigma_{i+1} + \sigma_{i-1} \right) \right] \left(1 + \delta \sigma_{i+1} \sigma_{i-1} \right). \qquad (4.3)$$

Without any other heat baths, (4.1) and (4.3) define the Glauber model [1], which relaxes to the equilibrium state of the Ising model at temperature T_1 provided $\gamma = \tanh(2J/k_B T_1)$. The time scale for flips is set by τ_1 and the parameter δ is arbitrary apart from the restriction that it must lie in the interval $[-1, 1]$. The two choices we consider below are $\delta = 0$, which defines the exactly solvable case [1,16], and $\delta = \tanh^2(J/k_B T_1)$ which was introduced by Kimball [17] and has been used widely in connection with phase transitions in nonequilibrium steady states.

Before introducing the competing dynamical processes, let us consider various interpretations of the spin-flip dynamics. The Ising model can be viewed as a lattice gas model. The particles can either be identified with the spins pointing in a given direction (e.g., with $\downarrow$) [7] or with the domain walls between the regions of up and down spins [13]; see Fig. 4.1. In the first case, the flips correspond to the creation or annihilation of particles, the rates being dependent on the number of particles at neighboring sites. Thus the Glauber dynamics describes a system of interacting particles in equilibrium with a particle reservoir. The particles, however, do not move along the line; they just evaporate from or are deposited onto it.

The particle dynamics is more interesting in the second case, when particles are identified with domain walls. Then a flip may move a particle by one site (expressed as $\uparrow\downarrow\downarrow \Leftrightarrow \uparrow\uparrow\downarrow$ or $\bullet\cdot\cdot \Leftrightarrow \cdot\cdot\bullet$), may create two particles at neighboring empty sites, or may result in the annihilation of two neighboring particles (expressed as $\uparrow\uparrow\uparrow \Leftrightarrow \uparrow\downarrow\uparrow$ or $\cdot\cdot\cdot \Leftrightarrow \cdot\bullet\bullet$). The rates for moving left or right are equal to $1 - \delta$ (in units of $(2\tau_1)^{-1}$), thus the particles execute random walks. Whenever they meet they annihilate at a rate

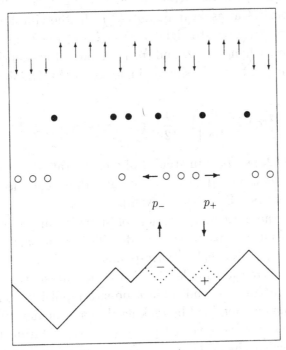

Fig. 4.1. Mappings among spin, particle, and surface configurations. Particles correspond either to down spins (○) or to domain walls (●), and the slope of the surface is given by the value of the spin variable at the given site. The surface dynamics of deposition and evaporation is equivalent to moving particles (down spins) to right or left at the corresponding rates of p_+ or p_-.

$(1+\gamma)(1+\delta)$, and particles are created in pairs at a rate $(1-\gamma)(1+\delta)$. One can see that pairs are not created in the limit $T \to 0$ $(\gamma \to 1)$; thus the $T = 0$ Glauber dynamics becomes equivalent to diffusion-limited annihilation [13].

Note that there is a simple experimental realization of the above dynamics in *trans*-polyacetylene, where photoexcitation can generate soliton-antisoliton pairs [18]. At elevated temperatures, the solitons and antisolitons each execute a random walk under the influence of thermal fluctuations and, since solitons and antisolitons alternate along the chain, every meeting of two excitations leads to annihilation. This system is discussed in more detail in [13].

We emphasize here once more that, without any competing process, the steady state is an equilibrium state independent of the interpretations of the elementary dynamical steps.

In order to introduce a competing dynamical process, we now consider a second heat bath at a temperature $T_2 \neq T_1$ that is assumed to induce exchanges of spins (Kawasaki dynamics [2]). In the simplest case, the exchanges are between nearest-neighbor sites i and $i + 1$, i.e., $\Gamma_i^2\{\sigma\}$ differs from $\{\sigma\}$ by the exchange of σ_i and σ_{i+1}. A rate of exchange that satisfies detailed balance (4.2) and is similar in form to the Glauber rate (4.3) is given by

$$ w_i^2(\sigma) = \frac{1}{2\tau_2} \left[1 - \frac{1}{2}\gamma_2 \left(\sigma_{i-1}\sigma_i + \sigma_{i+1}\sigma_{i+2} \right) \right], \tag{4.4} $$

where $\gamma_2 = \tanh(2J/k_B T_2)$. In studies of steady states, it is often assumed that the exchanges are random ($T_2 = \infty$) and thus that the actual form of (4.4) is immaterial (see Sec. 4.4 for examples).

The particle-dynamics interpretation of spin exchanges is particle hopping, provided that the particles are identified with spins pointing in a given direction ($\uparrow\downarrow\uparrow\uparrow \Leftrightarrow \uparrow\uparrow\downarrow\uparrow$ corresponds to $. \, o \, . \, . \Leftrightarrow . \, . \, o \, .$). In this picture, the Kawasaki model describes the diffusion of particles with nearest-neighbor attraction and infinite on-site repulsion. The driven lattice gas [7] can now be obtained by making the rate for the exchange $\uparrow\downarrow \Rightarrow \downarrow\uparrow$ different from that for the reversed process $\downarrow\uparrow \Rightarrow \uparrow\downarrow$. Within the formalism of heat baths, this can be done by introducing an inhomogeneous field that drives the up and down spins in different directions, e.g., an extra energy term $-\sum_i H_i\sigma_i$ with a field $H_i \sim i$ can be added. The constraint of detailed balance can be satisfied by various forms of exchange rates [7]. Often used examples are $\tau_2 w_i^2 = 1/(1 + e^x)$ or the Metropolis rate $\tau_2 w_i^2 = \min(1, e^x)$ where $x = \delta E/k_B T_2$, δE being the change of energy due to the exchange of spins.

As can be seen from Fig. 4.1, the spin-exchange dynamics can also describe a model of surface evolution provided the exchanges $\downarrow\uparrow \Rightarrow \uparrow\downarrow$ and $\uparrow\downarrow \Rightarrow \downarrow\uparrow$ are identified with deposition and evaporation events. This mapping is discussed in detail in Sec. 4.6.

If the domain walls are identified with particles then the spin exchanges correspond to more complicated particle dynamics. Pairs of particles may move together ($\uparrow\uparrow\downarrow\uparrow \Leftrightarrow \uparrow\downarrow\uparrow\uparrow$ or $. \, . \, \bullet \, . \, \bullet \, . \Leftrightarrow . \, \bullet \, . \, \bullet \, . \, .$), one particle may give birth to two others, or three particles may coagulate into one ($\uparrow\uparrow\downarrow\downarrow \Leftrightarrow \uparrow\downarrow\uparrow\downarrow$ or $. \, . \, \bullet \, . \, . \Leftrightarrow . \, \bullet \, . \, \bullet \, . \, \bullet \, .$). This mapping between spin and particle dynamics will be discussed in connection with branching annihilating random walks in Sec. 4.4.

The third heat bath that we shall consider generates bond flips, i.e., $\Gamma_i^3\{\sigma\}$ is obtained from $\{\sigma\}$ by cutting the spin chain between sites i and $i + 1$ and flipping all the spins on sites $j > i$. Although this is a nonlocal process in the spin language, it corresponds to a local process in the domain wall $\equiv$ particle picture, namely it describes the creation or annihilation of single particles ($\uparrow\uparrow\downarrow\downarrow \Leftrightarrow \uparrow\uparrow\uparrow\uparrow$ or $. . \bullet . . \Leftrightarrow . . . .$). Bond-flip rates satisfying detailed balance at T_3 can easily be constructed. For the purpose of describing diffusion-limited annihilation in the presence of particle sources (Sec. 4.7), it will be sufficient to consider the case $T_3 = \infty$, when the rate becomes $w_i^3(\sigma) = 1/\tau_3$.

Having introduced the elementary dynamical processes, one can start to investigate the consequences of competition between pairs of the above processes and this will be the topic of the following sections. A general remark applying to all cases we study is that the stationary states produced by competing dynamics are always nonequilibrium steady states: there is energy flow through the systems since the temperatures of the heat baths are different. Note that a driven lattice gas can also be considered to be in contact with two heat baths, one of them attached to the spin-spin interaction while the other connected to the driving field. In the steady state, there is an energy flow from the 'driving' heat bath to the 'internal' heat bath.

We conclude our discussion of the master equation and the mappings by noting that there is an alternative formulation of the master equation through the 'quantum spin-chain' formalism [19-20]. This formalism is useful for identifying exactly solvable cases and simplifying some of the calculations. In the simple cases we have considered, however, this formalism does not lead to a more economical treatment and we shall not discuss it.

4.3 Equations of motion for the correlation functions

A complete solution of a problem of competing dynamics consists of solving the master equation (4.1) for an arbitrary initial distribution, $P(\{\sigma\}, 0)$, and then calculating the time-dependent average of a physical quantity of interest, $A(\{\sigma\})$, e.g., $\sigma_i \sigma_j$, as

$$\langle A \rangle = \sum_{\{\sigma\}} A(\{\sigma\}) P(\{\sigma\}, t). \tag{4.5}$$

In reality, this program can be carried out only in exceptional cases [16] and the usual course of action is to derive an equation for $\langle A \rangle$ by taking a time derivative of equation (4.5) and using the master equation for evaluating $\partial P(\{\sigma\}, t)/\partial t$ on the right-hand side. This calculation results in averages of the type $\langle \sum_{i,\alpha} A w_i^\alpha \rangle$. In the trivial cases, these averages are proportional to $\langle A \rangle$ and one obtains the time evolution of $\langle A \rangle$ from a linear differential

equation. Such simplifications do occur, for example for the total magnetization in the Glauber model [1] and in some versions of the flip-and-exchange
models [21] (see Sec. 4.4).

The averages $\langle \sum_{i,\alpha} A w_i^\alpha \rangle$ usually contain terms $\langle B \rangle$ other than $\langle A \rangle$. Then
an equation for $\partial \langle B \rangle / \partial t$ must be derived, and, more often than not, an
infinite set of physical quantities are found to be coupled. If only a given type
of low-order correlation (e.g., the local magnetizations at different sites $\langle \sigma_j \rangle$
or the two-spin correlations $\langle \sigma_i \sigma_j \rangle$) is coupled, then the set of equations can
be solved easily provided that the system is either translationally invariant
(see the following sections) or can be divided into translationally invariant
sublattices [22].

Generally, problems become unsolvable when correlations of more than
two spins appear in the hierarchy of coupled equations. Within the field of kinetic Ising models, we know only two counter-examples. The first one is the
Glauber model with Kimball's flip rate [17], (4.3) with $\delta = \tanh^2(J/k_B T_1)$.
In this case, the one- and three-spin correlations couple to each other but
all the higher-order correlations decouple and the equations can be solved
[17]. The second example is the fast-exchange limit of the flip-and-exchange
model, where the higher-order correlations decouple [8]. This limit is discussed in the following section.

4.4 Flip-and-exchange models: phase transitions in 1D

An interesting feature of nonequilibrium steady states is that they may
display long-range order in 1D even though the interactions are of nearest-
neighbor type. In order to understand how a symmetry-breaking phase
transition may occur, let us start by solving the flip-and-exchange model
[8] in which spin flips at temperature T compete with spin exchanges at
temperature $T = \infty$. In this case, the rate for exchanges is $w_i^2(\sigma) = (2\tau_2)^{-1}(1 - \sigma_i \sigma_{i+1})$ while the rate for flips is given by (4.3). Then the
first equation in the hierarchy of correlation functions takes the form

$$\tau_1 \partial \langle \sigma_i \rangle / \partial t = -\langle \sigma_i \rangle + \tfrac{1}{2}\gamma(1 + \delta)\big(\langle \sigma_{i+1} \rangle + \langle \sigma_{i-1} \rangle\big)$$

$$-\delta \langle \sigma_{i+1} \sigma_i \sigma_{i-1} \rangle + (\tau_1/\tau_2)\big(\langle \sigma_{i-1} \rangle - 2\langle \sigma_i \rangle + \langle \sigma_{i+1} \rangle\big). \quad (4.6)$$

In the above equation, one can clearly see the contributions from the spin-
flip (relaxation) and spin-exchange (diffusion) dynamics. Without diffusion
$(\tau_1/\tau_2 \to 0)$, one has the equilibrium steady state and there is no ordering
in the system. One would expect that the $T = \infty$ diffusion cannot change
this since it merely disorders the system further. This intuition, however,

fails as we shall demonstrate by calculating the correlation length, ξ, for the $\delta = 0$ case and showing that ξ increases if the diffusion is switched on. The calculation can be done [21,23] since, for $\delta = 0$, the two-spin correlations, $\langle \sigma_i \sigma_j \rangle$, satisfy a closed set of equations. For $j > i + 1$, one finds

$$\frac{\partial \langle \sigma_i \sigma_j \rangle}{\partial t} = -2 \left(\frac{1}{\tau_1} + \frac{2}{\tau_2} \right) \langle \sigma_i \sigma_j \rangle + \left(\frac{\gamma}{2\tau_1} + \frac{1}{\tau_2} \right) \left(\langle \sigma_{i+1} \sigma_j \rangle + \langle \sigma_i \sigma_{j+1} \rangle \right.$$
$$\left. + \langle \sigma_{i-1} \sigma_j \rangle + \langle \sigma_i \sigma_{j-1} \rangle \right), \tag{4.7}$$

while the equation for $j = i + 1$ is slightly different:

$$\frac{\partial \langle \sigma_i \sigma_{i+1} \rangle}{\partial t} = -2 \left(\frac{1}{\tau_1} + \frac{1}{\tau_2} \right) \langle \sigma_i \sigma_{i+1} \rangle$$
$$+ \left(\frac{\gamma}{2\tau_1} + \frac{1}{\tau_2} \right) \left(\langle \sigma_i \sigma_{i+2} \rangle + \langle \sigma_{i-1} \sigma_{i+1} \rangle \right) + \frac{\gamma}{\tau_1}. \tag{4.8}$$

Assuming that the steady state is translationally invariant, the time-independent solution, $r_n = \langle \sigma_i \sigma_{i+n} \rangle_{t \to \infty}$, of the above equations is found [21,23] by substituting the exponentially decaying form $r_n = r_1 \exp[-(n-1)/\xi]$. Equation (4.7) then gives ξ while r_1 is obtained from (4.8). The correlation length is found to have the functional form of the equilibrium Ising model

$$\xi = \xi_{\text{Ising}}(\bar{\gamma}) = -1/\ln \left[\left(1 - \sqrt{1 - \bar{\gamma}^2} \right) \Big/ \bar{\gamma} \right], \tag{4.9}$$

$\gamma = \tanh(2J/k_B T_1)$ being replaced by $\bar{\gamma} = \tanh(2J/k_B \bar{T}_1)$, where $\bar{\gamma} = (\gamma + 2\tau_1/\tau_2)/(1 + 2\tau_1/\tau_2)$. The surprise here is that $\bar{\gamma} > \gamma$ for any nonzero τ_1/τ_2 and thus the effective temperature $\bar{T}_1$ is smaller than the equilibrium temperature T_1 for spin flips. Thus the correlation length has indeed increased as the result of random exchanges of the spins. One can also see that $\bar{\gamma} \to 1$, and, consequently, $\xi \to \infty$, as the rate of exchanges increases $(\tau_1/\tau_2 \to \infty)$.

This diverging correlation length can be understood as the result of nonlocality of the dynamics in the limit $\tau_1/\tau_2 \to \infty$. The spin exchanges rearrange the spins completely between two spin flips and thus the neighborhood of a spin which is being flipped consists of randomly selected spins from any part of the system. In other words, the information about the energetics gets carried to arbitrary distance by the exchanges. The condition that the flipping spin sees an average of the rest of the system is nothing other than the usual assumption of mean-field theories. Since mean-field theories yield phase transitions in low dimensions, it is not surprising that one may find [8] phase transitions in the 1D flip-and-exchange models in the limit $\tau_1/\tau_2 \to \infty$.

The existence of 1D ordering, however, is a bit more subtle; it depends on the details of the dynamics. The relevance of the form of the spin-flip rate can be understood by looking at (4.6). The equation becomes linear for the Glauber rate with $\delta = 0$ and thus the possibility of a phase transition for which the magnetization is the order parameter is excluded [24]. The flips and exchanges must couple to produce long-range order and this can happen only through higher-order correlations. Consequently, a nonlinear term must be present in (4.6) in order that the mean-field type of local equilibrium can be established. Indeed, if $\delta \neq 0$ then the mixing in the mean-field limit $\tau_1/\tau_2 \to \infty$ results [8] in the decoupling of the three-spin correlations, $\langle \sigma_{i+1}\sigma_i\sigma_{i-1} \rangle \to \langle \sigma_{i+1} \rangle \langle \sigma_i \rangle \langle \sigma_{i-1} \rangle$, and the steady-state value of the homogeneous magnetization, $m = \langle \sigma_i \rangle$, satisfies the familiar mean-field equation

$$[1 - \gamma(1 + \delta)]m + \delta m^3 = 0, \qquad (4.10)$$

yielding a second-order transition at $\gamma_c = 1/(1 + \delta)$.

It should be noted that (4.10) has been obtained using a spin-flip probability (4.3) containing nearest-neighbor spins only. Applying flip rates containing next-neighbor spins as well, one can easily construct [9] models that display first-order transitions in the mean-field limit considered above.

The mean-field nature of the transition can be viewed as resulting from the generation of effective interactions that are of infinite range. This picture has been confirmed by studies [11] of a version of the flip-and-exchange model in which the random exchanges were of infinite range, i.e., the spin exchanges took place between randomly chosen sites. A mean-field type of phase transition was found to take place even at finite τ_1/τ_2 and one could see that the finite-size scaling of the fluctuations of the magnetization coincided with those of the Ising model with each spin interacting with every other equally.

The flip-and-exchange model can also be modified [12] so that the competing processes generate effective interactions that decay with distance as a power law $V_{\text{eff}}(r) \sim r^{-1-\sigma}$. In order to have this, spin flips at temperature T must compete with random ($T = \infty$) spin exchanges that are Levy flights of dimension σ (the latter means that the probability that two spins at a distance r are exchanged decays with distance as $r^{-1-\sigma}$). Similar results appear to be valid in higher dimensions as well [25], with the effective interactions decaying as $V_{\text{eff}}(r) \sim r^{-d-\sigma}$ in dimension d.

It is clear from the examples considered above that nonlocal dynamics may generate long-range effective interactions and, consequently, the occurrence of phase transitions in these systems can be understood. There have been claims [26], however, that the flip-and-nearest-neighbor-exchange model shows a ferromagnetic phase transition at finite τ_1/τ_2 as well. Although analysis of finite-size effects and Monte Carlo study of the distribution function of the magnetization fluctuations indicate [27] that a ferromagnetic transition does not take place at the temperatures studied in [26], it nevertheless remains an intriguing question whether local interactions combined with local dynamics could produce a symmetry-breaking transition in 1D. It is known and understood for two-dimensional systems that *local interactions* with *local anisotropic dynamics* (two-temperature diffusive dynamics) may produce long-range effective interactions [28] and thus long-range order may be induced in systems where order would be excluded in equilibrium systems ($d = 2$ spherical or XY models [29,30]). Unfortunately, the anisotropy of local dynamics is essentially two dimensional, and thus long-range order in a 1D system with local interactions and local dynamics must be stabilized by a mechanism yet to be discovered.

Up to this point, we have discussed symmetry-breaking phase transitions. Kinetic Ising models, however, can also be models of another type of phase transition where the transition occurs between *absorbing* and *active* states without breaking any symmetry.

The best known example of *absorbing-state* $\leftrightarrow$ *active-state* transition is directed percolation [31], which is a particular example of so-called branching annihilating random walks [32]. In such models, particles execute random walks at a rate r_w, annihilate upon meeting, and produce k offspring at a rate r_o. Depending on the value of k, a transition between an *absorbing state* (zero density of particles) and an *active state* (finite density of particles) may occur as r_w/r_o is varied. It appears that the transitions that have been studied to date belong to two universality classes. For k odd, one finds the universality class of directed percolation [33], while a class distinct from this seems to emerge for k even [34-36].

The random-walk and annihilation parts of the above type of model are equivalent to the $T = 0$ Glauber dynamics and the production of offspring can be described as due to the flips of several spins. Thus branching annihilating random walks can be described as kinetic Ising models with competition between $T = 0$ spin flips and some more complicated multiple-spin processes. This was first noted in [37] for the case of $k = 2$, which is particularly simple since the offspring production can be described by $T = \infty$ nearest-neighbor spin-exchanges: $\uparrow\uparrow\downarrow\downarrow \Rightarrow \uparrow\downarrow\uparrow\downarrow$ or $..\bullet.. \Rightarrow .\bullet.\bullet.\bullet.$, as discussed in Sec. 4.2. The above mapping makes possible a detailed study [37,38] of the $k = 2$ case and allows one to draw a rather firm conclusion: its universality class is distinct from that of directed percolation.

An important point to observe in the above *absorbing-state* ↔ *active-state* transition is that no symmetry breaking occurs although a symmetry-breaking transition (*ferromagnet* ↔ *paramagnet*) does take place in the underlying spin system. The reason for this is that every state of given particle distribution corresponds to two spin configurations, one of which is obtained from the other by changing the signs of all the spins. Thus the breaking of the up-down symmetry in the spin model does not have any symmetry-breaking consequences for the particle system.

As one might imagine, there are many examples of *absorbing-state* ↔ *active-state* type transitions in systems where both interactions and dynamics are local. Although these phase transitions do occur in 1D, this should not surprise us since no symmetry breaking takes place and thus none of our notions about the existence of long-range order is violated.

4.5 Fluctuation-dissipation theorem

We close the discussion of flip-and-exchange models by adding a few remarks to the debate about the fluctuation-dissipation theorem in nonequilibrium steady states. The equilibrium version of this theorem states, in its simplest (static) form, that fluctuations of a quantity, say the fluctuations of the magnetization, $(\langle M^2 \rangle - \langle M \rangle^2)$, in the Ising model, are related to the response of this quantity to its conjugate field, $\chi = \partial \langle M/N \rangle / \partial H$, by

$$k_B T \chi = N^{-1}(\langle M^2 \rangle - \langle M \rangle^2). \tag{4.11}$$

This theorem is extremely useful. On the one hand it helps field-theoretic studies of fluctuations by simplifying diagrammatic expansions, while on the other hand it also gives a powerful checking procedure for Monte Carlo simulations. The fluctuations and susceptibilities can be measured independently, and violation of this theorem indicates that the system has not reached equilibrium.

The existence of similar theorems for nonequilibrium steady states would be equally useful. Although general considerations show [39] that fluctuation-dissipation type theorems do exist for nonequilibrium steady states as well, the construction of the physical quantities that satisfy those theorems needs the steady-state distribution function, i.e., it requires knowledge of the complete solution of the problem. This is not satisfactory from a practical point of view. The dissatisfaction with the fact that simple physical quantities usually do not satisfy the fluctuation-dissipation theorem has developed into a terminology of the 'violation of the fluctuation-dissipation theorem'.

Though it may be technically incorrect, the meaning of this terminology is clear: the equilibrium version of the fluctuation-dissipation theorem is not satisfied.

Of course, there are nonequilibrium steady states where the fluctuation-dissipation theorem holds for ordinary physical quantities. We shall demonstrate this by showing [21] that, in the limit $H \to 0$, (4.11) is satisfied in the steady state of the flip-and-exchange model in which Glauber flips ($\delta = 0$ in (4.3)) at temperature T compete with nearest-neighbor exchanges at $T = \infty$. This model has been discussed in Sec. 4.4, where the steady-state two-spin correlations $r_n = \langle \sigma_i \sigma_{i+n} \rangle$ were calculated. Since $\langle M^2 \rangle$ can be expressed through r_n, the calculation of the r.h.s. of equation (4.11) is straightforward:

$$\frac{1}{N}(\langle M^2 \rangle - \langle M \rangle^2) = 1 + 2 \sum_{n=1}^{\infty} r_n = 1 + \frac{2r_1}{1 - e^{-1/\xi}} . \qquad (4.12)$$

The calculation of the susceptibility on the l.h.s. of (4.11) is more complicated. We have to introduce a small magnetic field H, find the new transition rates, and obtain the steady-state value of the magnetization $\langle M \rangle = N\langle \sigma_i \rangle$. As a first step, we must examine how the flip rate and the exchange rate are affected by the magnetic field H. The rate of spin exchanges can be assumed to be unaffected since the change of energy caused by an exchange $\uparrow\downarrow \Rightarrow \downarrow\uparrow$ is independent of H. The spin-flip rate, however, should be modified since a flip involves an extra energy change due to H. A simple form of the flip rate that satisfies detailed balance has been suggested by Glauber [1]:

$$w_i^{1h}(\{\sigma\}) = \frac{1}{2\tau_1} \left[1 - \frac{1}{2}\gamma\sigma_i \left(\sigma_{i+1} + \sigma_{i-1} \right) \right] (1 - \sigma_i \tanh h), \qquad (4.13)$$

where $h = H/k_B T$. Next, one derives an equation similar to (4.6), assumes homogeneity of the steady state, and finds the following equation for $\langle \sigma_i \rangle = m$:

$$\tau_1 \dot{m} = -(1 - \gamma)m + [1 - \gamma r_1(t)] \tanh h \quad . \qquad (4.14)$$

In the limit $h \to 0$, one finds the steady-state value of m by setting $\dot{m} = 0$ and replacing $r_1(t)$ with the steady-state value of r_1 calculated for the system without the field. As a result, the susceptibility is obtained as

$$k_B T \chi = \lim_{H \to 0} (m/H) = (1 - \gamma r_1)/(1 - \gamma). \qquad (4.15)$$

It is a matter of simple algebra now to show that the right-hand sides of (4.12) and (4.15) are equal and thus that the fluctuation-dissipation theorem is satisfied for the magnetization [21].

This result may be a consequence of the fact that the magnetization is a special variable in the flip-and-exchange model since one of the competing processes conserves magnetization. The above calculation alone, of course, does not allow one to draw a general conclusion. Similar studies, however, may give enough insight to answer the simplest question about the nonequilibrium fluctuation-dissipation theorem: what are the features of the rates of competing processes that determine whether the equilibrium version of the fluctuation-dissipation theorem remains valid for a given quantity such as the magnetization?

4.6 Surface evolution: derivation of the KPZ equation

Our aim in this section is to demonstrate the usefulness of the mapping between spin and surface configurations. We do this by showing that the equations of motion for the spin correlation functions can be used to derive Langevin equations for the surface evolution in an approximate but rather unambiguous way.

In order to explain the idea and demonstrate how it works let us consider a single-step model of surface evolution [40,41]. The surface shown in Fig. 4.1 can be characterized by height differences $h_i - h_{i-1}$ that take on values ± 1; hence the description [41] in terms of Ising spins $h_i - h_{i-1} = \sigma_i = \pm 1$. The elementary steps of the surface dynamics are shown in Fig. 4.1 where one can also see that the surface model is equivalent to a driven lattice gas [7]. In terms of spin variables, the dynamics consists of exchanges of neighboring spins (Kawasaki dynamics), with unequal rates for $\uparrow\downarrow\Rightarrow\downarrow\uparrow$ and $\downarrow\uparrow\Rightarrow\uparrow\downarrow$ when the deposition and evaporation rates are different ($p_+ \neq p_-$). The rate of exchange of spins σ_i and σ_{i+1} is a variant of (4.4) and can be parametrized as

$$w_i^2(\sigma) = [1 - \sigma_i\sigma_{i+1} + \lambda(\sigma_{i+1} - \sigma_i)]/(4\tau_2), \qquad (4.16)$$

where $\lambda = p_+ - p_-$.

Since the average slope of the interface is given by the average of the spin variable, $\langle h_i - h_{i-1} \rangle = \langle \sigma_i \rangle$, we start by deriving an equation for $\langle \sigma_i \rangle$. From (4.16) we obtain

$$2\tau_2 \partial \langle \sigma_i \rangle/\partial t = \langle \sigma_{i-1} \rangle - 2\langle \sigma_i \rangle + \langle \sigma_{i+1} \rangle + \lambda\langle \sigma_i(\sigma_{i+1} - \sigma_{i-1}) \rangle. \qquad (4.17)$$

After coarse-graining and taking the continuum limit, one finds that the slope of the surface, $v = \partial h(x,t)/\partial x$, satisfies the equation

$$\bar{\tau}_2 \frac{\partial}{\partial t}\langle v \rangle = \frac{\partial^2}{\partial x^2}\langle v \rangle + \bar{\lambda}\langle v \frac{\partial}{\partial x} v \rangle, \tag{4.18}$$

where $\bar{\tau}_2$ is a coarse-grained time scale, $\bar{\lambda}$ is a coarse-grained value of $\lambda = p_+ - p_-$, and the higher derivatives that result from taking the continuum limit have been omitted.

Let us imagine now that there is a Langevin equation for the coarse-grained variable v. Then equation (4.18) is just the deterministic part of that equation after averaging over the noise. Since the spin-exchange dynamics implies that $\sum \sigma_i = \int(\partial h/\partial x)dx$ is conserved, the noise must conserve v. Consequently, the simplest Langevin equation that produces (4.18) is the Burgers' equation:

$$\bar{\tau}_2 \frac{\partial v}{\partial t} = \frac{\partial^2 v}{\partial x^2} + \bar{\lambda}v\frac{\partial v}{\partial x} + \frac{\partial \eta}{\partial x}, \tag{4.19}$$

which can be obtained by taking a spatial derivative of the Kardar-Parisi-Zhang (KPZ) equation [42]

$$\bar{\tau}_2 \frac{\partial h}{\partial t} = \frac{\partial^2 h}{\partial x^2} + \frac{\bar{\lambda}}{2}\left(\frac{\partial h}{\partial x}\right)^2 + \eta, \tag{4.20}$$

where η is the Gaussian white noise. Thus, we have derived the KPZ equation for the single-step model and arrive at the conclusion that this surface dynamics is in a universality class containing the KPZ model for $\bar{\lambda} \neq 0$, while it is in the Edwards-Wilkinson universality class [43] for $\lambda = 0$ ($\bar{\lambda} = 0$ follows from $\lambda = 0$ by symmetry). This is indeed observed in simulations of the single-step model [40,41].

Coarse graining is, of course, a nontrivial procedure and may lead to both the vanishing and the emergence of terms that are not present in the original equations. Nevertheless, the type of derivation described above is useful since it results in an equation in which all the deterministic terms have the right symmetry and are consistent with the conservation laws present in the system.

4.7 Diffusion-limited annihilation in the presence of particle sources

In this section we demonstrate that nonlocal spin processes such as bond-flip dynamics may produce closed equations for the two-spin correlations and, furthermore, that the equations for the correlation functions in general allow the calculation of both the steady-state correlations and the relaxational spectrum of the appropriate fluctuations. This will be seen by calculating the steady-state particle density $\langle n \rangle$ and the relaxation time τ of density fluctuations in a model where particles move diffusively and annihilate irreversibly and a steady source of particles is present.

This process can be viewed as a 1D model of aerosol formation where the aggregation centers are generated by photo-oxidation and sedimentation makes the larger clusters disappear from the system. A detailed discussion of the correspondence based on universality arguments is contained in [13].

The particles of this model are identified with domain walls of the spin chain. Accordingly, the diffusion-limited annihilation is described by the $T = 0$ Glauber model with $\delta = 0$ in the flip rate (4.3) while the source of particles is represented by bond flips as discussed in Sec. 4.2. We shall use the $T = \infty$ version of the bond-flip rate as given by $w_i^3(\sigma) = 1/\tau_3$.

Since we are interested in the particle density, which is expressed through nearest-neighbor correlations as $\langle n \rangle = (1 - \langle \sigma_i \sigma_{i+1} \rangle)/2$, we have to derive the equations for the two-spin correlations. Assuming that the initial state is translationally invariant, $\langle \sigma_i \sigma_{i+k} \rangle$ at all times depends only on k and one finds that $\langle \sigma_i \sigma_{i+k} \rangle \equiv r_k$ satisfies the following equation for $k > 0$,

$$\tau_1 \dot{r}_k = r_{k-1} - 2r_k + r_{k+1} - 2(1 + hk)r_k , \qquad (4.21)$$

where h is the ratio of the characteristic time scales of spin-flip and bond-flip processes, $h = \tau_1/\tau_3$. For $k = 0$, one has $r_0 = 1$, while $r_k = r_{-k}$ for $k < 0$. It is remarkable that (4.21) is a closed set of equations for two-spin correlations in spite of the nonlocality of bond flips in the spin language.

The steady-state value of r_k, $\bar{r}_k$, is obtained by setting $\dot{r}_k = 0$ and using known recurrence relations for the Bessel functions J_ν of the first kind of order ν [44]. The result is $\bar{r}_k = C J_{k+h^{-1}}(h^{-1})$, C being determined from the condition $r_0 = 1$. Thus the steady-state density is given by

$$\langle n \rangle = \tfrac{1}{2}[1 - J_{1+h^{-1}}(h^{-1})/J_{h^{-1}}(h^{-1})], \qquad (4.22)$$

and the small-h limit is found to scale as $\langle n \rangle \sim h^{1/3}$.

Now, in order to find the relaxational spectrum of homogeneous density fluctuations, we assume that perturbations decay exponentially and seek solutions to (4.21) in the form $r_k = \bar{r}_k + q_k e^{-2\zeta t/\tau_1}$. Since (4.21) is linear, the equation for q_k is found to be the same as the equation for $\bar{r}_k$ but with $1 + hk$ replaced by $1 - \zeta + hk$. Thus q_k is obtained as $q_k = \bar{C} J_{k+(1-\zeta)h^{-1}}(h^{-1})$, and the condition $q_0 = 0$, which follows from $r_0 = \bar{r}_0 = 1$, provides an equation that determines the possible values of ζ, $J_{(1-\zeta)h^{-1}}(h^{-1}) = 0$. The analysis

[13] of this equation leads to the conclusions that all ζ's are positive and that the smallest ζ scales as $\zeta_{min} \sim h^{2/3}$ for small h. Thus the relaxation time of the slowest mode diverges as $\tau \sim h^{-2/3}$ in the $h \to 0$ limit.

The results $\langle n \rangle \sim h^{1/3}$ and $\tau \sim h^{-2/3}$ are interesting because a scaling theory of aggregation [45] predicts that the exponents in $\langle n \rangle \sim h^{1/\delta}$ and in $\tau \sim h^{-\Delta}$ are related by the scaling law $\Delta + 1/\delta = 1$. As we can see, this relationship is indeed verified for a 1D model.

4.8 Concluding remarks

As we can see from the examples discussed above, 1D kinetic Ising models are useful for studying general questions of nonequilibrium statistical physics as well as for solving particular problems that may have experimental relevance. We hope that it has also become clear that the mappings to particle and surface dynamics are instrumental in opening avenues for unified treatments of various phenomena.

The author is grateful to K. E. Bassler, B. Bergersen, M. Droz, D. Liu, N. Menyhárd, M. Plischke, B. Schmittmann, P. Tartaglia, T. Tél, H.-J. Xu, and R. K. P. Zia for helpful discussions and collaborations which were important in forming his understanding of the problems described in this chapter. His work was supported by the Hungarian Academy of Sciences Grant OTKA 2090, and by EC Network Grant ERB CHRX-CT92-0063.

References

[1] R. J. Glauber, *J. Math. Phys.* **4**, 294 (1963).
[2] K. Kawasaki, *Phys. Rev.* **145**, 224 (1966).
[3] L. P. Kadanoff and J. Swift, *Phys. Rev.* **165**, 310 (1968).
[4] M. C. Cross and P. C. Hohenberg, *Rev. Mod. Phys.* **65**, 851 (1994).
[5] T. L. Hill, *J. Chem. Phys.* **76**, 1122 (1982).
[6] P. L. Garrido, A. Labarta and J. Marro, *J. Stat. Phys.* **49**, 551 (1987).
[7] S. Katz, J. L. Lebowitz and H. Spohn, *J. Stat. Phys.* **34**, 497 (1984); for a review see B. Schmittmann and R. K. P. Zia, *Statistical Mechanics of Driven Diffusive Systems. Phase Transitions and Critical Phenomena*, Vol. **17**, C. Domb and J. L. Lebowitz, eds. (Academic Press, New York, 1995).
[8] A. DeMasi, P. A. Ferrari and J. L. Lebowitz, *Phys. Rev. Lett.* **55**, 1947 (1985); *J. Stat. Phys.* **44**, 589 (1986).
[9] J. M. Gonzalez-Miranda, P. L. Garrido, J. Marro and J. L. Lebowitz, *Phys. Rev. Lett.* **59**, 1934 (1987).
[10] J.-S. Wang and J. L. Lebowitz, *J. Stat. Phys.* **51**, 893 (1988); J.-S. Wang, K. Binder and J. L. Lebowitz, *J. Stat. Phys.* **56**, 783 (1989).

[11] M. Droz, Z. Rácz and P. Tartaglia, *Phys. Rev.* A41, 6621 (1990); *Physica* A177, 401 (1991).

[12] B. Bergersen and Z. Rácz, *Phys. Rev. Lett.* 67, 3047 (1991).

[13] Z. Rácz, *Phys. Rev. Lett.* 55, 1707 (1985).

[14] H. W. Blöte, J. R. Heringa, A. Hoogland and R. K. P. Zia, *J. Phys.* A23, 3799 (1990).

[15] Some of the models with competing dynamics do not have 1D versions. Interesting examples are models in which spin exchanges are field driven in one direction and heat-bath driven in the perpendicular directions: S. Katz, J. L. Lebowitz and H. Spohn, *J. Stat. Phys.* 34, 497 (1984); B. Schmittmann, *Int. J. Mod. Phys.* B4, 2269 (1989), or models in which spin exchanges in different directions are governed by heat baths at different temperatures: P. L. Garrido, J. L. Lebowitz, C. Maes and H. Spohn, *Phys. Rev.* A42, 1954 (1990); B. Schmittmann and R. K. P. Zia, *Phys. Rev. Lett.* 66, 357 (1991).

[16] A general solution of the Glauber model is given by B. U. Felderhof, *Rep. Math. Phys.* 1, 215 (1970).

[17] J. C. Kimball, *J. Stat. Phys.* 21, 289 (1979).

[18] W. P. Su and J. R. Schriffer, *Proc. Natl. Acad. Sci. USA* 77, 5626 (1980); J. Orenstein and G. L. Baker, *Phys. Rev. Lett.* 49, 1043 (1982).

[19] E. Siggia, *Phys. Rev.* B16, 2319 (1979).

[20] F. C. Alcaraz, M. Droz, M. Henkel and V. Rittenberg, *Ann. Phys. (NY)* 230, 250 (1994); G. M. Schütz, *J. Stat. Phys.* 79, 243 (1995).

[21] M. Droz, Z. Rácz and J. Schmidt, *Phys. Rev.* A39, 2141 (1989).

[22] Z. Rácz and R. K. P. Zia, *Phys. Rev.* E49, 139 (1995).

[23] M. Q. Zhang, *J. Stat. Phys.* 53, 1217 (1988).

[24] There are a number of models with parallel and sequential dynamics where the possibility of symmetry breaking can be excluded because of the linearity of the evolution equations. See, e.g., I. Kanter and D. S. Fisher, *Phys. Rev.* A40, 5327 (1989).

[25] H.-J. Xu, B. Bergersen and Z. Rácz, *Phys. Rev.* E47, 1520 (1993).

[26] J. J. Alonso, J. Marro and J. M. González-Miranda, *Phys. Rev.* E47, 885 (1993).

[27] Z. Rácz, unpublished.

[28] B. Schmittmann, *Europhys. Lett.* 24, 109 (1993).

[29] K. E. Bassler and Z. Rácz, *Phys. Rev. Lett.* 73, 1320 (1994).

[30] K. E. Bassler and Z. Rácz, *Phys. Rev.* E52, R9 (1995).

[31] W. Kinzel and J. M. Yeomans, *J. Phys.* A14, L163 (1981).

[32] M. Bramson and L. Gray, *Z. Wahrsch. verw. Gebiete* 68, 447 (1985).

[33] P. Grassberger, *Z. Physik* B47, 365 (1982).

[34] P. Grassberger, F. Krause and T. von der Twer, *J. Phys.* A17, L105 (1984); P. Grassberger, *J. Phys.* A22, L1103 (1989).

[35] H. Takayasu and A. Yu. Tretyakov, *Phys. Rev. Lett.* 68, 3060 (1992).

[36] I. Jensen, *J. Phys.* A26, 3921 (1993); *Phys. Rev.* E50, 3623 (1994).

[37] N. Menyhárd, *J. Phys.* A27, 6139 (1994).

[38] N. Menyhárd and G. Ódor, *J. Phys.* A28, 4505 (1995).

[39] R. Graham, *Z. Physik* 26, 397 (1977).

[40] P. Meakin, P. Ramanlal, L. M. Sander and R. C. Ball, *Phys. Rev.* A34, 5081 (1986).

[41] M. Plischke, Z. Rácz and D. Liu, *Phys. Rev.* B35, 3485 (1987).

[42] M. Kardar, G. Parisi and Y. C. Zhang, *Phys. Rev. Lett.* 56, 889 (1986).

[43] S. F. Edwards and D. R. Wilkinson, *Proc. Roy. Soc. London* **A381**, 17 (1982).

[44] M. Abramowitz and I. A. Stegun, eds., *Handbook of Mathematical Functions* (Dover, New York, 1965).

[45] Z. Rácz, *Phys. Rev.* **A32**, 1129 (1985).

5

Glauber dynamics of the Ising model

Nobuyasu Ito

Recent results for the Glauber-type kinetic Ising models are reviewed in this chapter. Exact solutions for chains and simulational results for the dynamical exponents for square and cubic lattices are given.

5.1 Introduction

A study on the dynamical behavior of the Ising model [1] must begin with the introduction of a temporal evolution rule, because the Ising model itself does not have any a priori dynamics naturally introduced from the kinetic theory. Various kinds of dynamics are possible and some are useful to describe and predict physical phenomena or to make simulation studies of the equilibrium state. The Ising model with an appropriately defined temporal evolution rule is called the *kinetic Ising model*.

The statistical mechanical studies of the dynamical behavior in and around the equilibrium state started in the 1950s. During that decade, theoretical and computational developments provided a breakthrough and advanced such studies. The Kubo theory [2] and its successful application established the linearly perturbed regime around the equilibrium state generally treated by methods of statistical mechanics. It gave a means of investigating the dynamic behavior of macroscopic systems. Another great advance in that decade was the application of computing machines to statistical physics [3]. Dynamical Monte Carlo (MC) simulation on computers gave rise to the problem of computational efficiency, which is related to the dynamical behavior of the system, although this aspect became clear rather recently, in the 1980s. This chapter is a brief review of recent studies on kinetic Ising models mainly motivated by such simulational aspects.

93

In the 1960s, a remarkable development was achieved in the study of the dynamics of critical phenomena near a second-order transition point, i.e., dynamical critical phenomena [4]. Numerous experiments have been found to be described by the kinetic Ising model.

Several characteristics of kinetic Ising models have been identified. One is whether the order parameter (magnetization) is conserved. A dynamics that conserves the order parameter is called a *Kawasaki dynamics* and used to model the dynamic behavior of, e.g., solutions, liquids, alloys. A non-conserving dynamics is called a *Glauber dynamics*, and models magnetic materials, for example. These dynamics are named after the first pioneers [5,6]. They are in different dynamical universality classes. From the simulational viewpoint, the Glauber dynamics is more efficient if one's interest is only to study the equilibrium state. The conservation constraint makes relaxation slower in the Kawasaki case.

Another categorization is according to whether the temporal evolution is made locally. When a next spin state is determined by updating a finite number of spins, the dynamics is called 'local'. A nonlocal dynamics updates an unbounded number of spins each time. A single-spin flip dynamics is local, and a cluster-type dynamics is nonlocal at some parameter range near critical point (that is, in the scaling region). Different local dynamics are considered to be in the same dynamical universality class if their conservation laws are the same. A nonlocal dynamics can achieve a faster critical relaxation. The Swendsen-Wang [7] and Wolff dynamics [8] are simple and useful and can be more efficient than a local dynamics; here the 'efficiency' is in the computational simulational sense [9]. Such a dynamics is called an accelerated dynamics (or algorithm) [10].

Most of the effort has been devoted to the stochastic dynamics described in the next section; the above introduction applies to the stochastic case. We just mention here that deterministic dynamics are also challenging, and that some success has been achieved. An interesting example of such a dynamics is the 'Q2R' microcanonical simulation [11].

In this chapter, an introduction to the kinetic Ising model with Glauber-type local (e.g., single-spin-flip) dynamics is given, with a review of some recent results. There have been numerous accomplishments in this model, and more than one chapter of this book addresses it. The emphasis here will be on exact and simulational results. Some good reviews are already available [12,13]. The topics here are selected mainly from the computational physics viewpoint; some results, solutions, and arguments will be pedagogical in the real physical phenomenological sense. Our review is intended to serve as a guide to efficient computer simulations. Several kinds of dynamics, solved analytically for chains, provide ideas about flip-rule dependence, with implications for the critical dynamics of square- and cubic-lattice Ising models, relevant to the study of real physical systems.

5.1.1 Stochastic dynamics

A ferromagnetic Ising model on any finite lattice without external field is considered in the following. The energy of each configuration $\{\sigma = \pm 1\}$ is denoted by

$$E(\{\sigma\}) = -J \sum_{i,j-\text{nearest neighbors}} \sigma_i \sigma_j \quad (J > 0), \tag{5.1}$$

where i, j run over the lattice sites. This energy may be called the Hamiltonian function/operator of the system, but it does not imply any temporal evolution. In this sense, the Ising dynamics does not have a kinetic foundation. It is the model to study the thermal equilibrium distribution over the configuration space, $P_{\text{eq}}(\{\sigma\})$. So, a dynamics of distribution function which converges to the equilibrium distribution is often used to define the kinetic Ising model. The distribution function at time t is denoted by $P(\{\sigma\}; t)$. Then a stochastic dynamics is described using a fixed set of transition probabilities, $w(\{\sigma\} \to \{\sigma'\})$, normalized according to

$$\sum_{\{\sigma'\}} w(\{\sigma\} \to \{\sigma'\}) = 1. \tag{5.2}$$

For the continuous time case,

$$\frac{1}{\alpha} \frac{dP(\{\sigma\}; t)}{dt} = \sum_{\{\sigma'\}} w(\{\sigma'\} \to \{\sigma\}) P(\{\sigma'\}; t)$$
$$- \sum_{\{\sigma'\}} w(\{\sigma\} \to \{\sigma'\}) P(\{\sigma\}; t) = (\hat{W} - 1)P, \tag{5.3}$$

where α is a constant that sets the time scale. For the discrete time case,

$$P(\{\sigma\}; t+1) = P(\{\sigma\}; t) + \sum_{\{\sigma'\}} w(\{\sigma'\} \to \{\sigma\}) P(\{\sigma'\}; t)$$
$$- \sum_{\{\sigma'\}} w(\{\sigma\} \to \{\sigma'\}) P(\{\sigma\}; t) = \hat{W} P, \tag{5.4}$$

where the time t is assumed to be integer-valued. A linear operator $\hat{W}$ on the configuration space is introduced as a matrix of the transition probabilities. The difference between continuous and discrete time is simply in whether the spectra values are exponentiated or not, so it is not essential to make this distinction [14]; see below. In the following, only a discrete-time dynamics is treated, because it is simpler than a continuous-time one when later we consider the equivalent static model, and also because it is widely used in MC simulations.

A sufficient condition that a set of w's gives a dynamics converging to an equilibrium distribution is the detailed balance condition in combination with the ergodicity condition [15]. The detailed balance condition requires that the incoming flow and outgoing flow are the same for any two configuration $\{\sigma\}$ and $\{\sigma'\}$ in the equilibrium state, i.e., $w(\{\sigma'\} \to \{\sigma\})P_{eq}(\{\sigma'\}) = w(\{\sigma\} \to \{\sigma'\})P_{eq}(\{\sigma\})$. It is the condition that the transition probability is self-adjoint if regarded as a linear operator on the vector space spanned by the configurations, with an inner product whose weight is the equilibrium distribution. Ergodicity requires two conditions, the connectedness property and the aperiodicity property. Connectedness requires that any configuration can be reached from any other configuration with finite probability. Aperiodicity requires that the configuration space is not divided into subspaces that are visited cyclically; it is necessary for discrete-time dynamics. These conditions are mathematically stated by using the set I of positive integers defined for any two configurations $\{\sigma\}$ and $\{\sigma'\}$:

$$I(\{\sigma\} \to \{\sigma'\}) \equiv \{n, \text{such that } n \text{ is a positive integer and}$$
$$\sum_{\{\sigma\}_1} \cdots \sum_{\{\sigma\}_{n-1}} w(\{\sigma\} \to \{\sigma\}_1)$$
$$\times w(\{\sigma\}_1 \to \{\sigma\}_2) \cdots w(\{\sigma\}_{n-1} \to \{\sigma'\}) \neq 0\}. \quad (5.5)$$

Thus, connectedness is the condition that this set is not empty for any two configurations $\{\sigma\}$ and $\{\sigma'\}$. Aperiodicity is the condition that the largest common divisor of $I(\{\sigma\} \to \{\sigma\})$ is 1 for any configuration $\{\sigma\}$ [16]. The ergodicity condition can be satisfied even in the ordered phase, for a finite lattice. When the lattice size goes to infinity, some spectral eigenvalues become 1 at and below the critical temperature; see [17] for details.

5.1.2 Relaxation

When these conditions are satisfied, the linear operator $\hat{W}$ in the master equation (5.4) has discrete and real eigenvalues, and furthermore there is only a single occurrence of the eigenvalue 1, corresponding to the equilibrium distribution, P_{eq}. Each eigenvalue $\lambda_\ell \neq 1$, with eigenvector P_ℓ, corresponds to a relaxation time $\tau_l = 1/\log(1/\lambda)$ and a corresponding relaxation mode. The equilibrium distribution, to be labeled $\ell = 0$, does not decay. With these modes, the evolution of the distribution can be expressed as

$$P(\{\sigma\};t) = \sum_\ell p_\ell(t)P_\ell(\{\sigma\}) = \sum_\ell \lambda_\ell^t p_\ell(0)P_\ell(\{\sigma\})$$

$$= p_0(0)P_{eq}(\{\sigma\}) + \sum_{\ell \neq 0} \exp(-t/\tau_\ell)p_\ell(0)P_\ell(\{\sigma\}), \qquad (5.6)$$

where $p_\ell(t)$ denotes the amplitude of P_ℓ in $P(t)$ [18]. The relaxation of any physical quantity can be formally expressed using these P_ℓ and λ_ℓ [17,19-21]. The initial configuration determines the amplitude of each mode. If it does not include some decay modes, the corresponding decay rates do not contribute. Furthermore the observed quantity is usually an averaged physical quantity or its distribution, not the distribution over configuration space. So if the observed quantity becomes zero for some modes, the corresponding decay rates are not observed even when they are present in the initial distribution.

As noted in the following section, exact solutions or rigorous proofs can often treat each relaxation mode and each relaxation time separately. Perturbative, simulational, or experimental studies usually observe relaxations of physical quantities that show behavior averaged over many modes.

Two kinds of relaxation are to be distinguished. One is the relaxation from a given initial distribution different from the equilibrium distribution. Examples include a random or an all-aligned configuration, corresponding to infinitely high or zero temperature respectively. Such relaxation is called *nonequilibrium* or *nonlinear*. When the initial distribution is the equilibrium distribution, it is called *equilibrium* or *linear* relaxation. The terms 'linear' and 'nonlinear' are from linear-response theory [2]. In the equilibrium relaxation case, the relaxation of the two-time correlation function of quantities A and B, $C_{AB}(t) = \langle A(0)B(t) \rangle - \langle A(0) \rangle \langle B(0) \rangle$, is relevant. Here $\langle \cdot \rangle$ denotes an average in equilibrium, so a uniform translation in time does not change this average. The equilibrium and nonequilibrium relaxations show different scaling behaviors at the critical region.

The stochastic dynamics considered in this section was introduced when applications of the digital computer to condensed matter physics were sought in the 1950s. The equilibrium distribution is almost always of sharp Gaussian form, so the importance-sampling method of dynamical MC, simulating such a master equation, works efficiently. When we can make the simulation order of magnitude longer than the relaxation time, which is finite for a finite system, the convergence speed is $(\text{CPU time})^{-1/2}$. This complexity cannot be considered efficient, but at least its order is algebraic.

It is sometimes fruitful to regard the temporal direction as a spatial axis. The transition probabilities are nonnegative, so it is always possible to interpret the temporal evolution rule as an energetic interaction with some constraint between the states at time t and $t + 1$. The local dynamics reduces to a local interaction. This transcription of a kinetic model to an equilibrium static model maps d-dimensional dynamics to $(d + 1)$-dimensional statics. This dynamics-statics transcription is useful for solving some kinetic Ising chains and for studying the nature of two- or three-dimensional kinetic models.

5.1.3 Critical dynamics

Dynamical critical phenomena are characterized by the divergence near the critical point of the relaxation time τ as $\tau \sim \xi^z$, where ξ denotes the spatial correlation length; z is called the dynamical critical exponent. The value of this exponent may be conjectured to be 2 for local Glauber dynamics [22]. Indeed, it takes about ξ^2 time steps to propagate fluctuations through correlated regions of size ξ under diffusive dynamics. This conjectured value turns out to be true for chain and mean-field cases. More generally, however, correlated clusters are fractal rather than compact and dynamic correlations will propagate differently on such clusters [23]. So z deviates from 2 in two- and three-dimensional kinetic Ising models. This simple argument gives some indication why the system shows critical behavior also in its dynamics.

The dynamical critical phenomena are characterized by the scaling relation between t and $\epsilon^{-z\nu}$ where ϵ and ν denote $T - T_c$ and the correlation length exponent [24]. The relaxation time is not unique: it may depend on the observed quantity; its equilibrium value may be different from its nonequilibrium value. But z is expected to be the same. For the equilibrium case, it was observed that the relaxation time depends only on the parity of the physical quantity under an all-spin inversion [25] and that the exponent depends only on the locality of the dynamics [26]. In the next section, we will see for exact solutions for chains how the relaxation time changes for different, though similar, dynamics.

Now assume that A is a quantity with static exponent ω with respect to ϵ. The nonequilibrium relaxation can be analyzed from the scaling form of the relaxation function $A(\epsilon, L, t)$ as

$$A(\epsilon, L, t) - A(\epsilon, L, \infty) = L^{-\omega/\nu} f(\epsilon L^{1/\nu}, t L^{-z}), \qquad (5.7)$$

where L and f denotes the system size and scaling function, respectively. At the critical point, $\epsilon = 0$, the relaxation of A obeys the power-law decay $t^{-\omega/z\nu}$, for times $t \ll L^z$ [27].

5.2 Kinetic Ising chains

Since Glauber found the first exact solution for the simplest kinetic Ising chain [5], exact solutions have been found for many other kinetic Ising chains. The original Glauber's model involves randomly selected single-spin flips with heat-bath type transition probability, in continuous time. Other solved models include various kinds of flip sequence and some multi-spin flip dynamics. These are found by generalization of Glauber's method and by analyzing the mathematical structure behind his model. In this section, kinetic Ising chains are reviewed.

It is proved that the dynamical exponent z is 2 for the simplest Glauber dynamics of the Ising chain. The dynamical universality class in the one-dimensional (1D) case has a tricky feature. When the coupling constant alternates, the value of z changes continuously depending on the ratio of the two coupling constant [28]; however, the $z = 2$ universality is expected to be important for Glauber dynamics [29]. In the following, we present some solutions of the kinetic Ising chain, and the emphasis here will not be on the exponent $z = 2$ but on how the relaxation changes when we change the kinetics, especially within the simulationally realistic range, and on illustration of some mathematical tricks.

We treat an N-spin Ising chain with periodic boundary conditions, with $\beta = 1/k_\mathrm{B}T$, $K = \beta J$, and $\sigma_N \equiv \sigma_0$,

$$-\beta E(\sigma_0, \ldots, \sigma_{N-1}) = K \sum_{i=0}^{N-1} \sigma_i \sigma_{i+1}. \tag{5.8}$$

5.2.1 Single-spin flip

Any type of transition probability, $w(\{\sigma\} \to \{\sigma'\})$, can be expressed as

$$w(\{\sigma\} \to \{\sigma'\}) = 2^{-N} \left[1 + \sum_i W_i(\{\sigma\})\sigma'_i + \sum_{\substack{i,j \\ i \neq j}} W_{i,j}(\{\sigma\})\sigma'_i\sigma'_j \right.$$

$$+ \sum_{\substack{i,j,k \\ i \neq j \neq k \neq i}} W_{i,j,k}(\{\sigma\})\sigma'_i\sigma'_j\sigma'_k + \cdots$$

$$\left. + W_{0,1,2,\cdots,N-1}(\{\sigma\})\sigma'_0\sigma'_1\sigma'_2\cdots\sigma'_{N-1} \right], \tag{5.9}$$

where the normalization condition (5.2) has been used. As a physical quantity, the single-spin relaxation function, $m_i(t)$, is introduced by the definition $m_i(t) = \sum_{\{\sigma\}} \sigma_i P(\{\sigma\}; t)$; it can be called the time-dependent magnetization, because the model system is uniform spatially. Only this $m_i(t)$ is treated in this chapter, and other results for time-dependent correlation functions are given in [30]. From (5.4) and (5.9) the expression for

$$m_i(t+1) = \sum_{\{\sigma\}} W_i(\{\sigma\}) P(\{\sigma\}); t) \tag{5.10}$$

is derived. Here the $W_i(\{\sigma\})$ can be again expanded in terms of Ising spins. If the expansion only has terms first-order in the spins, i.e., it is of the form

$$W_i(\{\sigma\}) = \sum_{j=0}^{N_1} t_{ij} \sigma_j, \tag{5.11}$$

then (5.10) becomes

$$m_i(t+1) = \sum_{j=0}^{N-1} t_{ij} m_j(t) \qquad (i = 0, 1, 2, \ldots, N-1), \tag{5.12}$$

or equivalently $\hat{q}(t+1) = \hat{T}\hat{q}(t)$ where the ith element of $\hat{q}(t)$ is $m_i(t)$ and the (i,j)th element of $\hat{T}$ is t_{ij}. Now the relaxation time of the single-spin relaxation function is expressed by the maximum eigenvalue of $\hat{T}$, $\lambda_{\max}$, as $\xi_\tau = 1/\log(1/\lambda_{\max})$. For transition probabilities that do not satisfy the assumption (5.11) the temporal evolution of single-spin functions is tangled with multi-spin functions, and therefore a linear system of the order of 2^N has to be treated. Thus the linearity assumption reduces the eigenvalue problem to an $N \times N$ matrix [31].

Glauber found that the heat-bath type transition probability has the above structure in 1D and leads to an analytic solution. The transition probability of the spin σ_i is

$$w^{\mathrm{HB}}(\sigma_i \to \sigma_i'; \sigma_{i-1}, \sigma_{i+1}) = \frac{1}{2}[1 + g\sigma_i'(\sigma_{i-1} + \sigma_{i+1})], \tag{5.13}$$

where $g = (\tanh 2K)/2$. It is easy to check that it satisfies the detailed balance condition.

Let us first review the random selection (or asynchronous) dynamics. The transition probability for random selection is

$$w(\{\sigma\} \to \{\sigma'\}) = N^{-1} \sum_{i=0}^{N-1} \left[w^{\mathrm{HB}}(\sigma_i \to \sigma_i'; \sigma_{i-1}, \sigma_{i+1}) \prod_{k \neq i} \frac{1 + \sigma_k \sigma_k'}{2} \right]. \tag{5.14}$$

The time scale is determined by defining that N spin flips are tried in unit time. This choice satisfies the assumption (5.11) and $W_i^{\text{random}} = [g\sigma_{i-1} + (N-1)\sigma_i + g\sigma_{i+1}]/N$. So the matrix $\hat{T}^{\text{random}}$ is tridiagonal, and its eigenvalues are easily obtained. We get the relaxation time

$$\xi_\tau^{\text{random}} = -1/\{N\log[1 - (1-2g)/N]\}. \qquad (5.15)$$

This is the solution of the discrete-time version of the Glauber model [32]. In the limit of $N \to \infty$, this relaxation time reproduces the solution of the continuous-time Glauber model [5],

$$\xi_\tau^{\text{HB}} = 1/(1-2g) \sim e^{4K}/2 \qquad \text{(for } K \to \infty\text{)}, \qquad (5.16)$$

apart from the time-scale constant. Zero temperature is regarded as the critical point of the Ising chain, and it corresponds to $K = \infty$. The spatial correlation length is

$$\xi = -1/[\log(\tanh K)] \sim e^{2K}/2 \qquad \text{(for } K \to \infty\text{)}. \qquad (5.17)$$

From (5.16) and (5.17), $z = 2$ for the random-updating kinetic Ising chain.

Let us now discuss sequential selection. 'Sequential' means to update first σ_0, then σ_1, and so on. When we reach σ_{N-1}, time interval one is considered to have passed. This procedure is relevant for computer simulations. The transition probability is written as

$$
\begin{aligned}
w(\{\sigma\} \to \{\sigma'\}) =\ & w^{\text{HB}}(\sigma_0 \to \sigma_0'; \sigma_1, \sigma_{N-1}) \\
& \times \left[\prod_{k=1}^{N-2} w^{\text{HB}}(\sigma_k \to \sigma_k'; \sigma_{k-1}', \sigma_{k+1})\right] \\
& \times w^{\text{HB}}(\sigma_{N-1} \to \sigma_k'; \sigma_{N-2}', \sigma_0').
\end{aligned}
\qquad (5.18)
$$

The assumption (5.11) is true for this w. The matrix $\hat{T}^{\text{seq}}$ is a bit complicated and rather dense. But one can show that the eigenvalues, λ, are solutions of

$$(\lambda - gB_N)(A_N - gA_{N-1}) - g(A_N + 1)(g - B_N + gB_{N-1}) = 0, \qquad (5.19)$$

where $A_j = (\mu_-^{j-1} - \mu_+^{j-1})/(\mu_- - \mu_+)$ and $B_j = (\mu_+^{j-1}\mu_- - \mu_+\mu_-^{j-1})/(\mu_- - \mu_+)$, with $\mu_\pm = (\lambda \pm \sqrt{\lambda^2 - 4\lambda g^2})/2g$.

From the solution of this equation that becomes 1 at zero temperature ($g = 1/2$), the relaxation time near the zero temperature is estimated to be [31]

$$\xi_\tau^{\text{seq}} = \frac{e^{4K}}{4} + \left(\frac{N-2}{2N}\right)^2 + O(e^{-4K}). \qquad (5.20)$$

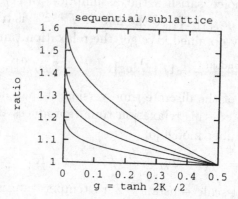

Fig. 5.1. Correlation times of $(M > 2)$-sublattice dynamics. The horizontal axis shows temperature as $g = (\tanh 2K)/2$, which is zero for infinite temperature and 0.5 for zero temperature. The vertical axis is the ratio of the correlation time for M-sublattice dynamics to that for the two-sublattice dynamics at the same temperature. The four curves correspond to $M = 10, 5, 4$ and 3, respectively, from top to bottom. It is observed that the two-sublattice dynamics is the fastest one although the difference is only few per cent in the critical region, i.e., near zero temperature.

This implies that in the critical region, the relaxation time is the same for any length N of chain, and that $z = 2$. At higher temperatures, the relaxation time is longer for longer chains.

Another flip sequence used in simulations is of sublattice type. The M-sublattice dynamics updates spins σ_i for $i \equiv 0 \pmod{m}$ first, then $i \equiv 1 \pmod{m}$, and so on. The chain length N is assumed to be a multiple of $M \, (> 1)$. When the transition probability of each spin is of the form w^{HB}, such a dynamics also satisfies the assumption (5.11) and can be treated within the above formulation. It transpires that the relaxation time of the M-sublattice dynamics is the same as that of the sequential-flip dynamics for chain length M [31]. So the critical behavior is described by replacing N by M in (5.20). The two-sublattice dynamics turns out to be the fastest flip sequence. In this case, the correlation time (see Fig. 5.1) is

$$\xi_\tau^{M=2} = -1/[2\log(\tanh 2K)]. \tag{5.21}$$

5.2.2 Mapping to two-dimensional Ising model

Another aspect of the Glauber model appears when it is mapped to a two-dimensional static model. The similarity of the Glauber model and the two-dimensional Ising model was pointed out when Felderhof solved the former model completely using spin operators [33]. The kinetic Ising model was first shown to be transformable to the static quantum XY chain [34]. In time, this equivalence and its usefulness have become clearer and more direct [31,35-38].

The Glauber transition probability (5.13) can be expressed as an interaction, via its Boltzmann weight,

$$w^{\mathrm{HB}}(\sigma_i \to \sigma_i'; \sigma_{i-1}, \sigma_{i+1}) = \frac{1}{2\sqrt{\cosh 2K}} e^{K\sigma_i'(\sigma_{i-1}+\sigma_{i+1})+L\sigma_{i-1}\sigma_{i+1}}, \quad (5.22)$$

where L is determined by $\tanh L = -(\tanh K)^2$. The r.h.s. of (5.22) corresponds to an anisotropic Ising model on the triangular lattice, apart from a constant. For example, the two-sublattice type dynamics discussed in the previous subsection is mapped to the two-dimensional Ising model on the triangular lattice shown in Fig. 5.2. The relaxation time (5.21) corresponds to the spin correlation length to the t direction in this model. The dynamics with more than two sublattices are mapped to the triangular Ising model with the same coupling constants with skewed boundary conditions [31]. This boundary skew becomes larger for more sublattices, and the larger skew allows a somewhat better propagation of correlations in the temporal direction. This picture gives a physical interpretation of why the two-sublattice relaxation is the fastest. The argument should work in the higher dimensional kinetic ferromagnetic Ising models, and a dynamics using the smallest number of sublattices is expected to be the fastest. Furthermore, this will be true for most transition probabilities, not only the heat-bath type.

We have elucidated the single-spin-flip Glauber dynamics in sufficient detail. Another interesting kinetic Ising chain is that with a multi-spin-flip dynamics, i.e., a dynamics that updates more than one spin at each trial. Two kinds of multi-spin-flip dynamics have been solved. One is the dynamics that flips n successive spins simultaneously [34,39], that is, the updated state of n spins, $\{\sigma_1, \sigma_2, \ldots, \sigma_n\}$ is selected from $\{\sigma_1, \sigma_2, \ldots, \sigma_n\}$ or $\{-\sigma_1, -\sigma_2, \ldots, -\sigma_n\}$ with heat-bath type transition probability. These models were solved using a spin-operator algebra. The other kind is the dynamics that updates n successive spins in such a way that the updated state is selected, from all possible 2^n configurations, with heat-bath type transition probability [38]. They are solved by mapping to a square-lattice Ising model with 1×2 unit-cell structure in the coupling constants.

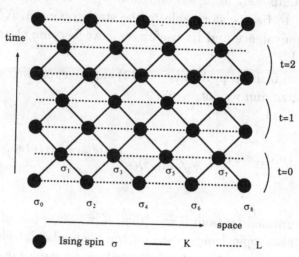

Fig. 5.2. Triangular Ising model, equivalent to the two-sublattice single-spin-flip dynamics with heat-bath transition probability.

5.3 Square and cubic lattices

The analytic studies of kinetic Ising chains reviewed in the previous section yield some insights for higher-dimensional models, for instance, the effects of flip sequences and flip ranges. No analytic solution is, however, known for a higher-dimensional lattice. The physically interesting lattice dimensionalities lie between the lower and upper critical dimensions which are respectively 1 and 4 for the Ising model. So the values of relaxation times and dynamical critical exponents have to be studied and estimated perturbatively or numerically. In this section, recent critical-dynamics studies of the kinetic Ising model on square and cubic lattices with local dynamics are reviewed.

Thus far the results have been restricted to the heat-bath transition probability. Many others are possible that satisfy the detailed balance and ergodicity conditions. We remark here that the Metropolis type has a shorter relaxation time than the heat-bath type in the critical region [26].

When one makes a simulational study of the equilibrium state, the values of the relaxation time and dynamical exponent are relevant. The equilibrium relaxation time is necessary to estimate the MC steps required to get independent samples. The nonequilibrium one is needed to determine the initial relaxation steps required to make the system independent of the initially given configurations. In the critical region, the relaxation time becomes sensitive to the system size and its dependence is characterized by the value of z. Although an accuracy of $z \approx 2$ is usually sufficient for simulational purposes, efforts to get a better determination of z have been central to the statistical mechanical study of dynamical behavior.

The equilibrium relaxation time is estimated from the behavior of the two-time autocorrelation functions. One naive way to get it is to observe the asymptotic decay rate of a correlation $C(t)$ that is identified to be the relaxation time ξ_τ, that is, $C(t) \sim \exp(-t/\xi_\tau)$ for large t. But near the critical region, multi-exponential decay appears and the separation of the longest decay time is not accurate. This difficulty introduces a systematic error in statistical estimation. Empirically, its bias is several per cent for the z estimation.

A well-defined estimator for the relaxation time is the 'integrated correlation time' defined by $\xi_\tau^{\mathrm{int}} = [\int_0^\infty C(t)\mathrm{d}t]/C(0)$. This quantity is a convenient substitute for the asymptotic relaxation time and also facilitates analytic treatment [40].

Another well-defined estimator is introduced from the analysis of the statistical degree of freedom in the variance when correlated data are averaged. If a temporal sequence, $q(t)$, is not correlated in equilibrium, the variance V of a quantity averaged over successive n, $V(Q_n)$, is equal to $V(q)/n$. Here Q_n denotes the stochastic variable defined by $Q_n = \sum_{i=1}^n q(t+i)/n$. But when $q(t)$ is correlated, $V(Q_n)$ is larger than $V(q)/n$. The ratio of these quantities is the 'statistical dependence time' (SDT)[41,42], that is, it is defined by $\xi_\tau^{\mathrm{SDT}} = \frac{1}{2}nV(Q_n)/V(q)$, where $V(Q_n)$ and $V(q)$ are estimated from independent simulations. This is similar to the integrated correlation time, and the estimation of the statistical dependence time is straightforward because it does not require the estimation of $C(t)$. Consequently, any systematic errors are clearly visible in the estimation procedure from the original data. Furthermore the statistical dependence time is a good parameter for which to get an unbiased estimator of cumulants [42,43].

The nonequilibrium relaxation has turned out to be useful for estimating the value of the exponent z. As observed in (5.7), the relaxation process of magnetization $m(t)$, for example, obeys

$$m(t) \sim t^{-\lambda}, \qquad \lambda = \beta/z\nu \qquad (1 \ll t \ll L^z). \qquad (5.23)$$

A convenient way to estimate this asymptotic power λ is to study the behavior of $R(t; m, \tau) = (t/\tau)\{[m(t-\tau)/m(t)] - 1\}$, which approaches λ when $1/t \to 0$ [44-46].

Table 5.1.

The latest results to date on the values of z for square and cubic lattices are shown. NER, HTE, and SDT in the method column denote analyses using nonequilibrium relaxation, high-temperature expansion, and statistical dependence time. In the cubic NER result, the factor ν/β is shown explicitly. The meaning of the error-bar ranges is discussed in the text

		Method	z	Year	Ref.
square	a	NER	2.165 ± 0.010	1993	[43]
	b	HTE	2.183 ± 0.005	1993	[48]
	c	HTE	2.165 ± 0.015	1996	[49]
	d	SDT	2.173 ± 0.016	1994	[42]
cubic	e	NER	$(\beta/\nu)/(0.250 \pm 0.002)$	1993	[45]
	f	SDT	2.03 ± 0.01	1993	[41]

The latest estimates to date of z values for the square- and cubic-lattice kinetic Ising models are shown in Table 5.1 [47]. The error bars in items a and d show 2σ or a more reliable range. The error bars in items e and f include MC statistical errors and systematic errors originating from the numerically known value of the critical point, and again the significance levels are 2σ. The estimates for the square lattice are plotted in Fig. 5.3. It is observed from this figure that they are consistent within 0.5% error. One unresolved and relevant question regarding this z value is the possibility of a logarithmic behavior. This hypothesis came from the study of the equivalent $(d + 1)$-dimensional static correlation length in the time direction [50] and the relation $z\nu = 2 + \alpha$ was conjectured. The coincidence of estimates from several methods, however, justifies the statement that a power-law behavior with $z = 2.165$ describes the phenomena correctly to within 0.5% error in this exponent.

For the cubic lattice, some earlier estimates should be added: 2.04 ± 0.03 [51], 2.10 ± 0.04 [44], and 2.10 ± 0.01 [52]. Systematic errors of unknown origin seem to preclude reaching an accuracy of order 0.5%. When we take the value of $\beta/\nu = 0.515$ as assumed in the NER analysis of [44], the estimate given in item e implies $z = 2.060 \pm 0.016$. To reconcile the results e and f, β/ν should be $(2.03 \pm 0.01) \times (0.250 \pm 0.002) = 0.5075 \pm 0.0046$. So, for the cubic lattice the problem of estimating the dynamic exponent now involves also the estimation of static exponents and of the critical point.

In conclusion, recent results for the critical behavior of the Glauber dynamics have been reviewed in this chapter. It has been observed that this

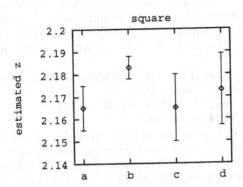

Fig. 5.3. The estimates of the z value for the square lattice.

model has become relevant not only to real physical phenomena but also to simulational and mathematical problems. The chain case has been worked out analytically to yield better understanding of the nature of the dynamics. The methodology reviewed here may be helpful for the study of not only the Glauber-type kinetic Ising model but also other kinds of 1D dynamical model. The numerical results for the square lattice are now accurate enough to allow the conclusion that our understanding and knowledge of the critical dynamics has reached a satisfactory level. In the case of the cubic lattice, some discrepancies remain and further studies are called for. We can expect that more efficient and larger-scale simulations will allow a better clarification of the dynamical behavior.

References

[1] W. Lenz, *Z. Physik* **21**, 613 (1920); E. Ising, *Z. Physik* **21**, 253 (1925); S. G. Brush, *Rev. Mod. Phys.* **39**, 883 (1967) reviews the early developments in the Ising model with a historical description.
[2] R. Kubo, *J. Phys. Soc. Jpn* **12**, 570 (1957).
[3] N. Metropolis, A. W. Rosenbluth, M. N. Rosenbluth, A. H. Teller and E. Teller, *J. Chem. Phys.* **21**, 1087 (1953). This treats the hard-core particle system. A very early Ising simulation is found in L. Fosdick, *Bull. Am. Phys. Soc.* **2**, 239 (1957) and *Phys. Rev.* **116**, 565 (1959), treating binary alloys.
[4] B. I. Halperin and P. C. Hohenberg, *Phys. Rev.* **177**, 952 (1969).
[5] R. J. Glauber, *J. Math. Phys.* **4**, 1 (1963).
[6] K. Kawasaki, *Phys. Rev.* **145**, 224 (1966).
[7] R. H. Swendsen and J.-S. Wang, *Phys. Rev. Lett.* **58**, 86 (1987). A concise review of cluster dynamics: J.-S. Wang and R. H. Swendsen, *Physica* **A167**, 565 (1990). See also X.-J. Li and A. D. Sokal, *Phys. Rev. Lett.* **21**, 827 (1989); N. Ito and G. A. Kohring, *Physica* **201**, 574 (1993).

[8] U. Wolff, *Phys. Rev. Lett.* **60**, 1461 (1988).

[9] N. Ito and G. A. Kohring, *Int. J. Mod. Phys.* C5, 1 (1994).

[10] D. Kandel, E. Domany, D. Ron, A. Brandt and E. Loh, Jr., *Phys. Rev. Lett.* **60**, 1591 (1988); J. Goodman and A. D. Sokal, *Phys. Rev.* D40, 2035 (1989).

[11] M. Creutz, *Phys. Rev. Lett.* **50**, 1411 (1983); H. J. Herrmann, *J. Stat. Phys.* **45**, 145 (1986).

[12] K. Kawasaki, in *Phase Transitions and Critical Phenomena*, Vol. 2, C. Domb and M. S. Green, eds. (Academic Press, New York, 1972).

[13] P. C. Hohenberg, *Rev. Mod. Phys.* **49**, 435 (1977).

[14] H. Falk, *Physica* A104, 459 (1980).

[15] The proof of this statement can be found in textbooks on stochastic processes, e.g., E. Parzen, *Stochastic Processes* (Holdenay, San Francisco, 1960); D. R. Cox and H. D. Miller, *The Theory of Stochastic Processes* (Methuen, London, 1965); K. L. Chung, *Markov Chain With Stationary Transition Probability* (Springer, 1967).

[16] $I(\{\sigma\} \to \{\sigma\})$ may be called the 'cycle length set' of $\{\sigma\}$. Under the connectedness conditions, the largest common divisor of $I(\{\sigma\} \to \{\sigma\})$ is the same for all configurations.

[17] S. Miyashita and H. Takano, *Prog. Theor. Phys.* **73**, 1122 (1985).

[18] For the continuous master equation, (5.3), the formal solution becomes $P(\{\sigma\}; t) = p_0(0)P_{eq}(\{\sigma\}) + \sum_{\ell \neq 0} \exp[-\alpha(1 - \lambda_\ell)t]p_\ell(0)P_\ell(\{\sigma\})$.

[19] M. Suzuki and R. Kubo, *J. Phys. Soc. Jpn* **24**, 51 (1968).

[20] M. Suzuki, *Int. J. Mag.* **1**, 123 (1971).

[21] A. D. Sokal, *Monte Carlo Methods in Statistical Mechanics: Foundations and New Algorithms* (Cours de Troisième Cycle de la Physique en Suisse Romande, Lausanne, 1989).

[22] The Kawasaki dynamics has a larger z, and this is the reason why the Glauber dynamics is better for simulational study of the equilibrium state.

[23] A hypothesis to characterize the value of z based on the cluster shape and its relevance to the dynamics was proposed by Z. Alexandrowicz, *Physica* A167, 322 (1990); A189, 148 (1992).

[24] B. I. Halperin and P. C. Hohenberg, *Phys. Rev. Lett.* **19**, 700 (1967). See also M. Suzuki, *Prog. Theor. Phys.* **51**, 1992 (1974).

[25] R. B. Pearson, J. L. Richardson and D. Toussaint, *Phys. Rev.* B31, 4472 (1985).

[26] N. Ito, M. Taiji and M. Suzuki, *J. Physique* **8**, 1397 (1988).

[27] M. Suzuki, *Prog. Theor. Phys.* **58**, 1142 (1977).

[28] M. Droz, J. Kamphorst Leal Da Silva and A. Malaspinas, *Phys. Lett.* A115, 448 (1986).

[29] F. Haake and K. Thol, *Z. Physik* B40, 219 (1980). See also J. C. Kimball, *J. Stat. Phys.* **21**, 289 (1979).

[30] V. Privman, *J. Stat. Phys.* **69**, 629 (1992). See also Chs. 7 and 8 of this book.

[31] N. Ito, *Prog. Theor. Phys.* **83**, 682 (1990).

[32] H. Falk, *Physica* A117, 561 (1983).

[33] B. U. Felderhof, *Rep. Math. Phys.* **1**, 215 (1971); **2**, 151 (1971).

[34] B. U. Felderhof and M. Suzuki, *Physica* **56**, 43 (1971).

[35] E. D. Siggia, *Phys. Rev.* B16, 2319 (1977).

[36] I. G. Enting, *J. Phys.* A11, 2001 (1978).

[37] I. Peschel and V. J. Emery, *Z. Phys.* B43, 241 (1981).

[38] N. Ito and T. Chikyu, *Physica* A166, 193 (1990).

[39] H. H. Hilhorst, M. Suzuki and B. U. Felderhof, *Physica* **60**, 199 (1972).

[40] M. Suzuki, *Prog. Theor. Phys.* **43**, 882 (1970).
[41] M. Kikuchi and N. Ito, *J. Phys. Soc. Jpn* **63**, 3052 (1993); N. Ito and M. Kikuchi, *Proc. 6th Int. Conf. Physics Computing*, p. 171, R. Gruber and M. Tomassini, eds. (Lugano, 1994).
[42] M. Kikuchi, N. Ito and Y. Okabe, *Proc. Computer Simulation Studies in Condensed-Matter Physics*, Vol. 7, p. 44, D. P. Landau, K. K. Mon and H.-B. Schüttler, eds. (Springer, 1994).
[43] H. Müller-Krumbhaar and K. Binder, *J. Stat. Phys.* **8**, 1 (1973); A. M. Ferrenberg, D. P. Landau and K. Binder, *J. Stat. Phys.* **63**, 867 (1991).
[44] D. Stauffer, *Physica* **A186**, 197 (1992).
[45] N. Ito, *Physica* **A192**, 604 (1993).
[46] N. Ito, *Physica* **A196**, 591 (1993).
[47] There is a long history of such studies for the square lattice, reviewed, e.g., in reference [46].
[48] B. Dammann and J. D. Reger, *Europhys. Lett.* **21**, 157 (1993).
[49] J. Adler, *Annual Rev. Comput. Phys.* **4**, 241 (1996).
[50] E. Domany, *Phys. Rev. Lett.* **52**, 871 (1984).
[51] S. Wansleben and D. P. Landau, *Phys. Rev.* **B43**, 6006 (1991).
[52] H.-O. Heuer, *J. Phys.* **A25**, L567 (1992).

6

1D kinetic Ising models at low temperatures—critical dynamics, domain growth, and freezing

Stephen J. Cornell

6.1 Introduction

One-dimensional (1D) kinetic Ising models are arguably the simplest stochastic systems that display collective behavior. Their simplicity permits detailed calculations of dynamical behavior both at and away from equilibrium, and they are therefore ideal testbeds for theories and approximation schemes that may be applied to more complex systems. Moreover, they are useful as models of relaxation in real 1D systems, such as biopolymers [1].

This chapter reviews the behavior of 1D kinetic Ising models at low, but not necessarily constant, temperatures. We shall concentrate on systems whose steady states correspond to thermodynamic equilibrium, and in particular on Glauber and Kawasaki dynamics. The case of nonequilibrium competition between these two kind of dynamics is covered in Ch. 4. We have also limited the discussion to the case of nearest-neighbor interactions, and zero applied magnetic field. The unifying factor in our approach is a consideration of the effect of microscopic processes on behavior at slow time scales and long length scales. It is appropriate to consider separately the cases of constant temperature, instantaneous cooling, and slow cooling, corresponding respectively to the phenomena of critical dynamics, domain growth, and freezing.

As zero temperature is approached, the phenomenon of critical dynamics ('critical slowing-down') [2] is observed in 1D Ising models. In the exactly solvable cases of uniform chains with Glauber or Kawasaki dynamics, the critical properties are simply related both to the internal microscopic processes and to the conventional Van Hove theory of critical dynamics. It was therefore somewhat surprising when modifications of Glauber and Kawasaki dynamics [3,4], or implementations on slightly different systems [5,6], were found to give rise to nonuniversal values of the dynamic critical exponent.

111

The origin of this nonuniversality is now well understood—it is due to the presence of nonuniversal Arrhenius factors in the dynamics, which may be expressed as powers of the correlation length because the critical temperature is zero. Nevertheless, there remain cases where the dynamic critical exponent is not so simply explained, and the correct scaling description of these systems is still the subject of controversy.

If a 1D kinetic Ising model is subject to a sudden drop in temperature, then the characteristic domain size is much smaller than the new equilibrium correlation length. As these domains grow to restore the equilibrium configuration, there is a wide regime in time where their distribution obeys a dynamic scaling law reminiscent of the behavior of higher-dimensional systems quenched through a phase transition [7]. The behavior in this far-from-equilibrium regime can be related to the critical dynamics. Nevertheless, the exponents describing the temporal scaling of the domain size are universal, and assume the same values as in higher dimensions.

If a kinetic Ising model is cooled slowly, however, the departure from equilibrium is much more gradual, and under some conditions a far-from-equilibrium structure can be made to freeze in. The behavior has many features in common with that which is observed in glasses [8], and the amenability of 1D kinetic Ising models to analytical calculation leads to new insight into the phenomenon of freezing in general.

This chapter is organized in the following way. After a brief summary of the static properties of the 1D Ising model, and a definition of the dynamics, we describe the body of work on dynamic critical phenomena. This includes discussions of the behavior of periodic, quasiperiodic, and randomly modulated systems. The far-from-equilibrium dynamical scaling during domain growth is discussed next. There follows a section on the glass transition in slowly cooled kinetic Ising models with activated dynamics. Among the concluding remarks, we discuss possibilities for further work important to this field.

6.2 The 1D Ising model

We give a brief summary of the equilibrium properties of the 1D Ising model, and an introduction to the dynamics that may be applied to it. For a more detailed discussion of the statics and dynamics of 1D Ising models, see the reviews in [9] and [10] respectively.

6.2.1 Statics

The Ising model in 1D consists of a lattice of N spins $\{S_i\}$, taking values ± 1 and labeled by their site i, with exchange coupling J_i between the ith pair of spins. The Hamiltonian describing this system is

$$\mathcal{H} = -\sum_{i=1} J_i S_i S_{i+1}. \tag{6.1}$$

We assume here (and throughout this chapter) that interactions occur between nearest-neighbors only, and there is no external magnetic field. We shall also implicitly assume that the interactions are ferromagnetic ($J_i > 0$), unless otherwise stated. We impose periodic boundary conditions, $S_{N+1} \equiv S_1$.

The (canonical) partition function of this system is

$$\mathcal{Z} = \sum_{\{S_j = \pm 1\}} \exp\left(-\frac{\mathcal{H}}{k_B T}\right). \tag{6.2}$$

$\mathcal{Z}$ is easily evaluated by the transfer matrix technique (see, e.g., [9]). The result is

$$\mathcal{Z} = 2^N \prod_{i=1}^{N} \cosh\left(\frac{J_i}{k_B T}\right). \tag{6.3}$$

The two-point correlation function $\langle S_i S_j \rangle$ may be obtained by noting that $S_i^2 \equiv 1$, so $S_i S_j = \prod_{k=i}^{j-1}(S_k S_{k+1})$. We therefore have

$$\langle S_i S_j \rangle = \frac{1}{\mathcal{Z}}\left\{\prod_{k=i}^{j-1} \frac{\partial}{\partial(J_k/k_B T)}\right\} \mathcal{Z} = \prod_{k=i}^{j-1} \tanh\left(\frac{J_k}{k_B T}\right). \tag{6.4}$$

Other correlation functions of even order may be obtained in a similar way. All odd correlation functions are zero by symmetry.

This model does not exhibit a phase transition as such, but as zero temperature is approached it displays critical behavior. For the uniform case, $J_i = J$, the two-point correlation function is

$$\langle S_i S_j \rangle = \tanh^{|i-j|}\left(\frac{J}{k_B T}\right) = \exp\left(-\frac{|i-j|}{\xi}\right), \tag{6.5}$$

where the correlation length is

$$\xi = \left\{\log\left[\tanh\left(\frac{J}{k_B T}\right)\right]\right\}^{-1} \to \frac{1}{2}\exp\left(\frac{2J}{k_B T}\right) \quad \text{as } T \to 0. \tag{6.6}$$

The correlation length has an essential singularity at zero temperature, as do other thermodynamic quantities such as the specific heat, so the traditional definition leads to all critical exponents being zero or infinite. However, it is possible to define critical exponents for all thermodynamic quantities in terms of their divergence with respect to ξ.

For the case where the coupling constants assume two values, $J_i = J_1$ for i odd and $J_i = J_2 (< J_1)$ for i even, the correlation length is

$$\xi = 2 \left\{ \log \left[\tanh \left(\frac{J_1}{k_B T} \right) \tanh \left(\frac{J_2}{k_B T} \right) \right] \right\}^{-1} \to \exp \left(\frac{2 J_2}{k_B T} \right) \text{ as } T \to 0.$$
(6.7)

For a more general periodic set of J_i, one finds similarly $\xi \sim \exp[2 J_m / (k_B T)]$ as $T \to 0$, where $J_m = \min(\{J_i\})$. This represents the fact that, at low temperatures, almost all nearest-neighbor pairs of spins are parallel, with a small number of antiparallel pairs (kinks). These kinks tend to lie on the weakest bonds, so as to have the lowest energy. However, although the thermodynamic quantities depend only on the weakest bond, we shall find that the dynamics depends upon the full distribution of coupling constants.

6.2.2 Dynamics

Unlike most statistical mechanical systems defined by a Hamiltonian, the Ising model possesses no intrinsic dynamics. The reason is that all operators in the Hamiltonian commute with one another, and so the Heisenberg equation of motion for them predicts that they remain constant. It is therefore necessary to consider the model as coupled to a heat bath at temperature T, in such a way that the Hamiltonian describes the statics of the problem and also the coupling of the heat bath to the model, inducing changes of state in the model, e.g., spin flips. This *extrinsic* dynamics is described by a 'master equation', the equation of motion of the probability $P(\{S_i\}; t)$ that the system will be at state $\{S_i\}$ at time t, of the following form:

$$\frac{dP(\{S_i\}; t)}{dt} = \sum_{\{S'_i\}} w(\{S'_i\} : \{S_i\}) P(S'_i; t) - w(\{S_i\} : \{S'_i\}) P(S_i; t), \quad (6.8)$$

where $w(\{S'_i\} : \{S_i\})$ is the probability per unit time that a system initially in state $\{S'_i\}$ will change to state $\{S_i\}$.

The equation of motion of the magnetization $\langle S_k \rangle$ at site k, defined by $\langle S_k(t) \rangle \equiv \sum_{\{S_i = \pm 1\}} S_k P(\{S_i\}; t)$, is obtained by multiplying (6.8) by S_k and summing over all configurations.

It is essential that the equilibrium probability distribution,

$$P_{eq} = \frac{1}{Z} \exp \left[-\frac{\mathcal{H}(\{S_i\})}{k_B T} \right], \quad (6.9)$$

be a time-independent solution of the dynamics. To ensure that this is the case, we introduce the constraint of *detailed balance* on the transition rates:

$$w(\{S_i\} : \{S'_i\})P_{eq}(\{S_i\}; t) = w(\{S'_i\} : \{S_i\})P_{eq}(\{S'_i\}; t). \tag{6.10}$$

The detailed balance condition is not strong enough to determine completely the form of w. One of the most natural choices satisfying (6.10) is

$$w(\{S_i\} : \{S'_i\}) = \alpha\Big(1 - \tanh\big\{\left[\mathcal{H}(\{S'_i\}) - \mathcal{H}(\{S_i\})\right]/(2k_BT)\big\}\Big), \tag{6.11}$$

where α is a constant. This choice has the desirable feature that the time scale at which energy-conserving transitions occur does not diverge or vanish as $T \to 0$.

Glauber was the first to attempt to introduce dynamics of this kind into the Ising model [11]. In this approach, only single spin flips are permitted, i.e., $\{S'_i\}$ and $\{S_i\}$ differ in one spin only. The master equation is often written in the form

$$\frac{dP(\{S_i\}; t)}{dt} = -\sum_j (1 - \hat{p}_j)w_j(\{S_i\})P(\{S_i\}; t), \tag{6.12}$$

where $\hat{p}_j$ is the jth spin-flip operator, defined by

$$\hat{p}_j f(S_1, \ldots, S_{j-1}, S_j, S_{j+1} \ldots) = f(S_1, \ldots, S_{j-1}, -S_j, S_{j+1} \ldots); \tag{6.13}$$

$w_j(\{S_i\})$ is now the rate at which a system leaves state $\{S_i\}$ by flipping of the jth spin.

Glauber's particular choice for w_i in fact corresponds to (6.11). When generalized to the case of nonuniform J_i, his form for w_j is

$$w_j(\{S_i\}) = \alpha\left[1 - S_j(\gamma_j^+ S_{j+1} + \gamma_j^- S_{j-1})\right],$$

$$\gamma_j^\pm \equiv \frac{1}{2}\left[\tanh\left(\frac{J_j + J_{j-1}}{k_BT}\right) \pm \tanh\left(\frac{J_j - J_{j-1}}{k_BT}\right)\right]. \tag{6.14}$$

We shall refer to these kinetics as the 'Glauber model', even though Glauber originally treated the case of uniform J only. The equation of motion of the magnetization $m_j \equiv \langle S_j \rangle$ is then

$$\frac{dm_j}{dt} = -2\alpha\left(m_j - \gamma_j^+ m_{j+1} - \gamma_j^- m_{j-1}\right), \tag{6.15}$$

while that for the two-spin correlation function is

$$\frac{d\langle S_i S_j \rangle}{dt} = -2\alpha \Big[2\langle S_i S_j \rangle - \gamma_i^+ \langle S_{i+1} S_j \rangle - \gamma_i^- \langle S_{i-1} S_j \rangle$$
$$- \gamma_j^+ \langle S_i S_{j+1} \rangle - \gamma_j^- \langle S_i S_{j-1} \rangle \Big], \qquad (6.16)$$

for $i \neq j$. If we assume translational invariance, and uniform J, we can write $\langle S_i S_j \rangle = g(j - i, t)$, and (6.16) becomes

$$\frac{\partial g(x, t)}{\partial t} = -4\alpha \left[g(x, t) - \gamma g(x - 1, t) - \gamma g(x + 1, t) \right], \qquad (6.17)$$

with $\gamma = \frac{1}{2}\tanh(2J/k_B T)$. Equations (6.15) and (6.17) may be solved by generating-function methods [11]. A more general solution for all correlation functions, using techniques similar to those used in the solution of the statics of the 2D Ising model, may be found in [12].

Another commonly used class of kinetics for Ising models is spin-exchange, or 'Kawasaki', dynamics [13]. In this case, two nearest-neighbor spins are permitted simultaneously to flip, provided they are antiparallel. The master equation may be written in the form

$$\frac{dP(\{S_i\}; t)}{dt} = -\sum_j \frac{1}{2}(1 - S_j S_{j+1})(1 - \hat{p}_j \hat{p}_{j+1})\omega_j(\{S_i\})P(\{S_i\}; t), \quad (6.18)$$

where the factor $\frac{1}{2}(1 - S_i S_{i+1})$ ensures that only nearest-neighbor antiparallel pairs of spins may flip. The total magnetization is conserved in this model. The form for ω_i corresponding to (6.11) is

$$\omega_i = \alpha \left(1 - S_{i-1} S_i t_i^- - S_{i+1} S_{i+2} t_i^+ \right),$$

$$t_i^\pm \equiv \frac{1}{2} \left[\tanh \left(\frac{J_{i-1} + J_{i+1}}{k_B T} \right) \pm \tanh \left(\frac{J_{i+1} - J_{i-1}}{k_B T} \right) \right]. \qquad (6.19)$$

On writing the equation of motion for a k-spin correlation function, we find that it contains terms that depend on the $(k + 2)$-spin correlation functions. The system is described by an unclosed hierarchy of equations, for which no exact solutions exist.

Generalizations of these dynamics, involving higher-order spin flips or long-range exchange, have also been considered, and will be discussed in Sec. 6.3.4.

6.3 Constant temperature—critical dynamics

In the vicinity of a second-order phase transition, the correlation length ξ of a system diverges. In order for the system to reach equilibrium, correlations have to be established over a range of order ξ, and so the relaxation time for the system also diverges. According to the dynamic scaling hypothesis [2], the relaxation time τ_q for a mode of wavevector q takes the form

$$\tau_q = \xi^z f(q\xi) , \quad \text{for } \xi \to \infty, \tag{6.20}$$

where z is the *dynamic critical exponent*. The conventional, or Van Hove [14], theory of dynamic critical phenomena relates the value of z to the static exponents by $z = 2 - \eta$ ($z = 4 - \eta$ for nonconserved order parameter). η is the exponent describing the spatial decay at the critical point of the two-point correlation function $C(r) \sim r^{2-d-\eta}$. Real systems typically do not obey the conventional theory (except in the case of conserved order parameter [2]), owing to the presence of fluctuations near the critical point. As is the case with static critical phenomena, the renormalization group (RG) permits a classification of dynamical critical phenomena [2]. According to this classification, z should be universal, depending not only upon the usual properties that determine the universality class for statics (spatial dimension, dimension of the order parameter) but also on dynamical conservation laws.

In the case of 1D Ising models, as $T = 0$ is approached the correlation length diverges, and the relaxation times also diverge. The phenomenon of critical dynamics in these models is easily understood, since the diverging length scale corresponds to the distance between domain walls. One finds, however, that the dynamic critical exponent defined in (6.20) has fewer universal properties than would be expected from the classification of Hohenberg and Halperin [2]. The value of z is found generally to depend upon the distributions and ratios of values of the coupling constants in the Hamiltonian, as well as upon choices for the transition rates.

In the simplest case (the 1D Glauber model), it is possible to see explicitly that this nonuniversality comes from an Arrhenius factor associated with an energy barrier Δ that renormalizes the bare spin-flip rate. This energy barrier can either be intrinsic to the choice of coupling constants, or can be added by hand to the transition rates. We may write the relaxation time in the form

$$\tau_q(\xi) = \tilde{\tau}_q(\xi) \exp\left(\frac{\Delta}{k_B T}\right) = \xi^{\Delta/(2J)} \tilde{\tau}_q, \tag{6.21}$$

where $\tilde{\tau}_q$ is the relaxation time for such a system without the energy barrier (if it is possible to construct such a system). The dynamic critical exponents for the two systems are therefore related by $z = \tilde{z} + \Delta/(2J)$.

Some authors [15] argue that because the relaxation time may be separated into universal and nonuniversal contributions this phenomenon is not properly thought of as nonuniversality. However, more complex cases exist where this separation may not be straightforwardly made. If the coupling constants have an anomalous distribution, this nonuniversality can arise from a collective energy barrier. Also, controversy still exists for the case of the spatially modulated Kawasaki model, where analytical techniques disagree with simple physical arguments and simulations.

In this section, we shall first review the techniques that exist for the study of critical dynamics in 1D Ising models, together with the results of these techniques when applied to the uniform case. We shall then discuss separately the results for Glauber and Kawasaki dynamics with periodic, quasiperiodic, or random distributions of coupling constants, and other choices of dynamics.

6.3.1 Techniques

1. Exact solution

As noted in Sec. 6.2.2, the equations of motion for the 1D Glauber model with uniform coupling constants may be solved exactly. Writing (6.15) in the continuum limit (valid for $\xi \to \infty$), we find

$$\frac{\partial m(x,t)}{\partial(2\alpha t)} = \frac{\partial^2}{\partial x^2} m(x,t) - \frac{1}{\xi^2} m(x,t). \qquad (6.22)$$

The equation of motion of the Fourier transform of the magnetization $\tilde{m}(q,t) \equiv \int dx\, m(x,t)e^{iqx}$ is $\frac{\partial \tilde{m}}{\partial t} = -2\alpha \left(q^2 + \xi^{-2}\right)\tilde{m}$, from which we extract

$$\tau_q = \frac{1}{2\alpha}\xi^2 \left[1 + (q\xi)^2\right]^{-1}, \qquad (6.23)$$

and so $z = 2$. For comparison, the conventional value of this exponent is $2 - \eta = 1$; the deviation from this value is due to the vanishing of the kinetic coefficient $\langle \omega_i \rangle$ as ξ^{-1}, when $\xi \to \infty$.

The unclosed hierarchy of equations of motion for the correlation functions in the Kawasaki model does not admit any explicit solution. However, Zwerger [16] has shown that it is possible to solve for the relaxation times in the linear response regime. Adding time-dependent dimensionless fields to the Hamiltonian, i.e., a term of the form $-\sum_j h_j(t)S_j$, the spin exchange rate becomes, in linear response, $\tilde{\omega}_j(t) \equiv \omega_j + \delta_j(t)$, where $\delta_j = -\omega_j (h_j S_j + h_{j+1} S_{j+1})$. The *magnetization current* c_j, defined as the flow of magnetization from site j to site $j+1$, is related directly to the rate at which the pair $S_j S_{j+1}$ exchanges by

$$c_j = \left\langle \tfrac{1}{2}(1 - S_j S_{j+1}) \tilde{\omega}_j(t) S_j \right\rangle = \tfrac{1}{2} \langle \omega_j \rangle (h_j - h_{j+1}). \tag{6.24}$$

The continuity equation for the magnetization is, from the definition of the currents,

$$\frac{dm_j}{dt} = c_j - c_{j+1}. \tag{6.25}$$

Defining the spatial transform of the magnetization, $\tilde{m}_q = N^{-1/2} \sum_j e^{ijq} m_j$, in the limit of low frequency we may relate $\tilde{m}(q)$ to $h(q)$, the Fourier transform of the field, by $\tilde{m}(q) = \chi(q) h(q)$, where $\chi(q)$ is the static susceptibility $\chi(q) = \tfrac{1}{4} \sum_j \langle S_0 S_j \rangle e^{iqj}$. Combining (6.24) and (6.25), we find that

$$\frac{\partial \tilde{m}(q, t)}{\partial t} = -\tau^{-1}(q) \tilde{m}(q, t), \tag{6.26}$$

where the relaxation time is

$$\tau(q) = \frac{2}{\alpha} \xi^5 \frac{1}{(q\xi)^2 \left[1 + (q\xi)^2\right]}. \tag{6.27}$$

This leads to the final result $z = 5$.

According to the RG theory of dynamical critical phenomena, the dynamic critical exponent for conserved-order-parameter dynamics ('model B' of [2]) assumes its conventional value $z = 4 - \eta$. This would predict a value of $z = 3$ for the 1D Kawasaki model. The reason for this discrepancy is that the spin conductance ($\tfrac{1}{2} \langle \omega_j \rangle$ from (6.24)) vanishes as ξ^{-2}, when $T \to 0$, hence giving a further contribution of ξ^2 to the relaxation time.

It is possible to reformulate the master equation as a Schrödinger equation. In this approach, the flipping operators are written in the form of spin operators, and so the problem is mapped onto a quantum spin chain [17]. This is a powerful technique, allowing to apply the Bethe ansatz and related methods, and, in principle, capable of extending the class of exactly solvable problems.

2. Domain-wall arguments

The use of simple physical arguments to explain the exact values of the critical exponents was first used by Cordery et al. [18]. In this approach, one considers the elemental domain-wall-motion processes that lead to relaxation to equilibrium. Estimating the typical time scale of such processes leads to a form for the relaxation time in terms of the correlation length, and hence the critical exponent.

As mentioned above, a typical state of the 1D Ising model at low temperatures consists of long domains of parallel spins, separated by domain walls. The system relaxes by the death of domains, i.e., the mutual annihilation of two neighboring domain walls.

For the case of Glauber dynamics, the dominant process leading to a decay of the domain-wall density is the diffusion of domain walls. If a spin has one neighbor that is parallel to it and another that is antiparallel, i.e., the spin is at a domain wall, then it can flip with no change in its energy, so the rate for such a process is independent of temperature. The result of such a flip is that the domain wall has moved by one lattice site. Domain walls therefore perform random walks, and will annihilate when two walls meet. The number of time steps necessary for two walls separated by a distance ξ to meet is of order ξ^2, so we deduce $\tau = \xi^2$, whence $z = 2$.

For the case of Kawasaki dynamics, domain walls cannot move independently. However, a single spin (i.e., a spin with two antiparallel neighbors) not neighboring a domain wall can perform free diffusion; at each time step, the spin exchanges with an antiparallel neighbor, with no change of energy. At a domain wall, a single spin can split off ('evaporate'), costing energy $4J$, and then will perform a random walk until it meets a domain wall again. The probability that the spin will travel a distance L before returning to its birthplace is $\sim 1/L$ [18]. If it reaches the next domain wall a distance ξ away, it will be absorbed, and the net result is for the domain to have shifted by one lattice unit. The time scale for this process is therefore $\xi \exp(4J/k_B T) = \xi^3$. For the domain to shift a distance ξ, it requires a number ξ^2 such processes. This leads to a characteristic time ξ^5, and hence $z = 5$.

These arguments therefore faithfully reproduce the values of z in the two exactly solved cases above. Cordery *et al.* [18] argued that this approach should always produce an upper bound to the critical exponent, since it is conceivable that more complex, faster modes might exist. However, we have only processes that reduce the number of domains, whereas, close to equilibrium, there most also be processes that tend to create new domains. These two competing types of process might be expected to interact nonlinearly, so a different behavior might be expected. The domain-wall argument therefore gives a reliable description only of the far-from-equilibrium situations described in Secs. 6.4 and 6.5.

3. Renormalization group

The RG is very well suited to studying critical phenomena, because it provides a natural way of extracting scaling behavior without necessarily performing an explicit solution of the model.

While the momentum-space RG gives a systematic approach to calculating dynamical critical behavior [2], the fact that it usually involves an expansion about spatial dimension 4 renders it less useful for 1D problems. However, real-space RG schemes are available to treat the dynamics. These schemes solve the statics of the 1D Ising model exactly, although the dynamics sometimes has to be treated approximately.

The 'decimation' approach to real-space RG calculations of the statics of the 1D Ising model consists of writing the Boltzmann factor for a system of N spins, then performing a partial trace over every second spin to produce a new Boltzmann factor for a system of $N/2$ spins. Consider a reduced Hamiltonian $H = \mathcal{H}/k_B T = -C - \sum K_j S_j S_{j+1}$, where $K_j = J_j/k_B T$. Writing the Boltzmann factor in the form

$$\exp(-H) = e^C \prod_j \cosh K_j (1 + S_j S_{j+1} \tanh K_j), \qquad (6.28)$$

the Boltzmann factor of the decimated system is obtained by tracing over spins at even sites:

$$\exp(-H') = e^{C'} \prod_{j \text{ odd}} \cosh K'_j (1 + S_j S_{j+2} \tanh K'_j), \qquad (6.29)$$

where $\tanh K'_j = \tanh K_j \tanh K_{j+1}$. Shrinking the system by a factor 2 to restore the lattice constant leads to the recursion relation $\xi' = \xi/2$ for the correlation length. Analysis of the fixed points of these recursion relations leads to critical exponents in the usual way. This method is easily generalized to different values b of the rescaling parameter.

For the case of the 1D Glauber model, it is possible to implement a decimation scheme directly upon the equation of motion of the magnetization [19]. The Laplace transform in time of (6.15) for the uniform case gives $(\Omega + 1)m_j = \frac{1}{2}t(m_{j-1} + m_{j+1})$, where $t = \tanh(2J/k_B T)$ (for the uniform case) and $2\alpha\Omega$ is the variable conjugate to t. The recurrence relation $\tanh(K') = \tanh^2(K)$ may be written in the form $t' = 2t^2/(2 - t^2)$. Writing equations of motion for m_{j+1} and m_{j-1}, and eliminating, one finds that $\Omega' + 1 = \frac{1}{2}t'(m_{j-2} + m_{j+2})$, with $\Omega' = [2\Omega(2 + \Omega)]/(2 - t^2)$. Linearizing about the critical fixed point $(t, \Omega) = (1, 0)$, one finds $\delta\Omega' = 2^2 \delta\Omega$, giving $z = 2$ [19].

A more widely applicable scheme of RG, due to Achiam and Kosterlitz [20], performs the decimation on the master equation itself. It is necessary to define a parameter space for the RG procedure, so some kind of ansatz needs to be used for P. The form most commonly used is

$$P(\{S_i\}; t) = P_{eq}(\{S_i\})\left[1 + \sum_j h_j(t)S_j\right]. \quad (6.30)$$

The use of this method for the Glauber model produces the same results as the approach in the previous paragraph. To see why this must be the case, consider the quantity $\Gamma_k \equiv (1 - \hat{p}_k)\omega_k(\{S_i\})P_{eq}(\{S_i\})\sum_j h_j S_j$. By virtue of (6.10), we have

$$\Gamma_k = 2h_k S_k \omega_k P_{eq}(\{S_i\}) = 2h_k\alpha(S_k - \gamma_k^- S_{k-1} - \gamma_k^+ S_{k+1})P_{eq}(\{S_i\}). \quad (6.31)$$

The r.h.s. of (6.31) contains only terms of the form $S_l P_{eq}(\{S_i\})$, so the space of functions of the form (6.30) is an invariant subspace under Glauber dynamics. The Achiam-Kosterlitz ansatz is therefore exact for 1D Glauber dynamics.

Although no direct schemes exist for renormalization of the equations of motion of the Kawasaki model, it is possible to apply the Achiam-Kosterlitz scheme [21]. In this case, however, the ansatz (6.30) no longer forms an invariant subspace under the dynamics, and so it is an approximation. The assumption of small deviations linear in the spins is reminiscent of linear response theory. However, it is possible that, at long times, more complex correlations build up that are not describable in this way. Nevertheless, the application of this RG scheme to the 1D Kawasaki model [21] reproduces the exact result $z = 5$ [16]. Reference [21] also contains a RG argument that the critical dynamics of a system possessing both Glauber and Kawasaki dynamics is determined by the Glauber dynamics.

4. Monte Carlo methods

The use of Monte Carlo (MC) techniques to study Ising models is well documented elsewhere [22]. When one is looking at low-temperature behavior, however, the use of a standard 'Metropolis' algorithm is likely to be very inefficient, because if the system finds itself in a metastable state requiring energy Δ to relax, it will be necessary to perform around $\exp(\Delta/k_B T)$ attempts before it will do so.

A much more efficient method for simulating systems with energy barriers is the 'minimum process' or 'Gillespie' algorithm [23]. Here, if a system is faced with a set of N possible transitions, each having transition rate ω_i ($1 \leq i \leq N$), then the probability that the first transition that takes place is the jth is $\omega_j/\sum_i \omega_i$, and the time before this transition takes place is exponentially distributed with average value $(\sum_i \omega_i)^{-1}$. Instead of performing a large number of acceptance attempts to perform a set of unlikely transitions, it is more efficient to use these probabilities to decide directly which process will be the first to occur, and when it occurs.

5. Other techniques

There are a number of techniques that have been used to obtain inequalities for the dynamic critical exponent. One of these, due to Halperin [24], uses a rigorous inequality to show that the general relaxation time of a system is always larger than the initial relaxation time. It is therefore possible to obtain a lower bound for z from the decay rate at time $t = 0$.

Variational techniques have been used by Haake and Thol [3] to find both upper and lower bounds on z for generalizations of the Glauber and Kawasaki models. These results will be discussed in Sec. 6.3.4. Series expansion methods have been used to study the form of relaxation functions in 1D kinetic Ising models [25] but the results have not been directly applied to critical dynamics.

6.3.2 Glauber dynamics

1. Periodic modulation

The case of Glauber dynamics for a chain whose couplings take alternate values $J_i = J_1$ (i odd) and $J_i = J_2 (< J_1)$ (i even) was first studied by Droz et al. [4] who solved exactly the equation of motion for the zero-wavevector mode of the magnetization. Equation (6.15) for the case $m_j = m$ gives

$$\frac{dm(t)}{dt} = 2\alpha \left[1 - \tanh\left(\frac{J_1 + J_2}{k_B T}\right) \right] m(t), \tag{6.32}$$

so the relaxation time diverges as $\exp[2(J_1 + J_2)/k_B T] \sim \xi^{1+(J_1/J_2)}$. This result has also been obtained by RG techniques, both directly on the equation of motion [26,32] and using the Achiam-Kosterlitz scheme [28].

If the magnetization is allowed to be nonuniform, the reduced translational invariance of the lattice leads to both optical- and acoustic-branch modes [15,29]. The relaxation time for the acoustic mode of wavevector q becomes, in the limit of low temperatures [30],

$$\tau_q = \frac{\alpha}{2} \left\{ \exp\left[\frac{2(J_1 - J_2)}{k_B T} \right] \right\} \frac{1}{q^2 + \xi^{-2}} \tag{6.33}$$

$$\propto \xi^{1+(J_1/J_2)} \left[1 + (q\xi)^2 \right]^{-1} \tag{6.34}$$

(cf. (6.23)). It is evident that the nonuniversality of the critical exponent comes from the Arrhenius prefactor in (6.33), which may be expressed as a power of the correlation length because the critical point is at $T = 0$.

The physical origin of this Arrhenius factor was explained by Droz *et al.*, who showed that the domain-wall argument gave the same value for z [4]. The dominant relaxation process is again the diffusion of domain walls, but this is no longer an energy-conserving process. When the domain wall lies on a strong bond, the diffusion process is exothermic, whereas when it lies on a weak bond it requires an energy $2(J_1 - J_2)$ for the domain wall to move. The wall again needs to move a distance $\sim \xi$, needing $\sim \xi^2$ steps, recovering the above result for the relaxation time at $q = 0$.

The fact that the results for the uniform model are obtained by simply rescaling the bare time scale by an Arrhenius factor has been discussed by Cornell *et al.* [31]. This leads sometimes to 'unusual' forms for the scaling functions describing relaxation in this system. For example, in the limit $\xi \to \infty$, $q\xi \to \infty$, where we expect from the dynamic scaling hypothesis the result $\tau_q = q^{-z} f(q\xi)$. Comparison with (6.34) then leads to $f(x) \sim x^{z-2}$. Achiam and Southern [15] have argued that since the nonuniversal Arrhenius prefactor in (6.33) has a short-range origin it should be treated as a constant, temperature-dependent factor in f, leading to $z = 2$ and $f(x) \to$ constant. Since the microscopic origins of the nonuniversal factor are well understood, preference between these two descriptions is arguably only a matter of taste. However, in the case of more complex systems, for which no exact solutions exist, there may be several different energy barriers that play a role, so the identification of a single Arrhenius prefactor to the dynamics is much less straightforward. In such cases, the scaling description that uses a nonuniversal dynamic critical exponent may be the only unambiguous formalism available.

Results for periodic distributions of bond values with larger unit cells have been obtained by the use of RG techniques [28,32] and directly from the equations of motion [33]. The relaxation time is found again to diverge as $\xi^2 e^{\Delta/(k_B T)}$, where the energy barrier $\Delta = 2(J_M - J_m)$ corresponds to the activation energy needed for a domain wall to cross a unit cell (J_M and J_m are respectively the largest and smallest couplings in the cell). Again, the dynamic critical exponent can be thought of as having the nonuniversal value $z = 1 + (J_M/J_m)$.

2. Quasiperiodic chain

If two coupling strengths J_1 and J_2 $(< J_1)$ are distributed on a Fibonacci lattice (i.e., one created by iterating the procedure $A \to B$, $B \to AB$), the lack of periodicity renders an explicit solution of the equations of motion intractable. However, a real-space RG approach is possible, where the decimation scheme is the inverse of the iteration scheme used to construct the lattice. Ashraff and Stinchcombe [26] showed that the relaxation time diverges as $\xi^{1+(J_1/J_2)}$, as in the periodic case. Southern and Achiam [27] performed a more detailed calculation, and found that the distribution of relaxation times was described by a width proportional to the Arrhenius factor $\exp[-2(J_1 - J_2)/k_B T]$, but with universal spectral exponent. As for the periodic case, they argued that the correct description of this system is one where the dynamic critical exponent is universal.

3. Random chain

The statics for an Ising chain may be solved for a general set of (nearest-neighbor) couplings. Since the two-point correlation function depends only upon the total number of bonds of each type between the two sites, the correlation length is a well-defined quantity (provided no couplings are zero), and is again proportional to $\exp(2J_m/k_B T)$, where J_m is the smallest coupling.

To study the critical dynamics of a (quenched) random chain with Glauber dynamics, it is sufficient to consider the case where all couplings are ferromagnetic, because the case with antiferromagnetic couplings may be transformed by a gauge change $J_i \rightarrow J_i' = |J_i|$, $S_i \rightarrow S_i' = S_i \prod_{j=0}^{i} \text{sign}(J_i)$. The dynamics remains unchanged under this transformation.

For the case of a discrete distribution of couplings, with largest and smallest couplings J_m and J_M, Lage [28] used RG arguments to show that the dynamic critical exponent is $z = 1 + (J_M/J_m)$, just as was found in the periodic and quasiperiodic cases. If the distribution is discrete, but $J_m = 0$ or $J_M = \infty$, the chain breaks up into clusters between which no information can be passed, so the correlation length can no longer diverge and a critical dynamics no longer occurs.

Nunes da Silva and Lage [34] have performed numerical iteration of the RG transformations for two cases of anomalous distributions of couplings: $P(J) = (1-a)J^{-a}$, $0 < J \leq 1$ (case A) and $P(J) = (1-a)J^{-2+a}$, $1 \leq J < \infty$ (case B). For case A, the correlation length has the anomalous form $\xi^{-1} = \overline{\log[\tanh(J/k_B T)]} \sim T^{(1-a)}$. The numerical results suggested $\log \tau \sim T^{-1/(1-a)}$, which was in qualitative agreement with a reformulation of the domain-wall argument. This implies a temperature-dependent value for the dynamic critical exponent. For case B, the correlation length is determined by the smallest value of J, $\xi \sim \exp(1/k_B T)$. The relaxation time is found to have the form $\log \tau \sim \xi^{1/(1-a)} \log \xi$, which was not reproducible by domain-wall-motion arguments. The resulting form $z \sim A + B \log \xi$ for the critical exponent is similar to the form violating dynamic scaling found in dilute magnets near the percolation threshold [35].

The related case where the couplings are not disordered, but the bare spin-flip rate α in (6.14) has a random, anomalous distribution of the form $P(\alpha) = (1 - a)\alpha^{-a}$, has been treated by the same authors [36], using a variety of analytic and heuristic methods. All these approaches predict the value $z = (2 - a)/(1 - a)$ for the critical exponent. This agrees with results for random walks with anomalous distributions of waiting times [37].

Other studies of Glauber dynamics for random chains have been motivated from the direction of spin glasses, and have focused on metastability and relaxation at zero temperature [38] rather than dynamic critical behavior.

6.3.3 Kawasaki dynamics

The exact approach of Zwerger for the uniform case [16] may not be extended to the case of a Kawasaki model with spatially modulated couplings. Accordingly, there has been some controversy as to the values of z for these systems, even in the simplest case, that of alternating bond strengths.

It is necessary here to distinguish between the cases of ferromagnetic and antiferromagnetic exchange couplings, because the gauge transformation that transforms all couplings to ferromagnetic interactions does not leave the dynamics invariant. Specifically, if two spins are connected by an antiferromagnetic coupling, then after a gauge transformation to ferromagnetic coupling the spin-exchange dynamics is transformed into one where the two spins will flip simultaneously if and only if they are parallel. The dynamics is then in the nonconserved-order-parameter class [30].

The domain-wall argument [18] may be extended to the case of alternating bond strengths J_1 and J_2 ($< J_1$) [4]. The argument is similar in that spin diffusion in a large domain still involves no energy change. However, to nucleate a spin at a domain wall requires activation energy $4J_1$ if the wall is on a weak (J_2) bond, and $4J_2$ if the wall is on a strong (J_1) bond. For a domain wall to move, it needs to cross as many weak bonds as strong bonds, so the relaxation time is $\xi^3 \exp(4J_1/k_BT)$, predicting $z = 3 + 2(J_1/J_2)$.

Luscombe found instead the value $z = 4 + (J_1/J_2)$, by using truncation techniques [30]. This value is supported by an Achiam-Kosterlitz RG calculation [39]. It must be remembered that complex correlations might build up at long times, which would be neglected by the Achiam-Kosterlitz ansatz.

This result is what would be expected by neglecting long-time effects in the dynamics, and it coincides with the lower bound provided by the initial response [24]. This corresponds physically to the fact that, if the two spins joined by a strong bond flip, the change in energy is the same as if all couplings had the lower value J_2. The short-time response of the system is then due to processes at these strong bonds, which take place at the normal rate ξ^{-5}. The fraction of domain walls that lie on strong bonds is $\sim \exp[-2(J_1 - J_2)/k_B T]$, so the initial response rate is $\xi^{4+(J_1/J_2)}$. However, for complete relaxation to take place it is necessary for some of the domain walls on weak bonds to move, which requires the higher activation energy $4J_1$. It is quite possible that the approximate analytical techniques that have been used are picking up on a short-term response that has nothing to do with the long-time relaxation to equilibrium.

This controversy has yet to be resolved in the published literature. MC simulation [40], using a technique first developed for the 1D Blume-Emery-Griffiths model [41], finds results in agreement with the domain-wall arguments. However, this technique assumes a relationship between the scaling behaviors near and far from equilibrium, which might not be valid for this model.

The RG treatment has also been applied to other periodic distributions of coupling constants [39], producing a variety of different nonuniversal exponents.

The cases of nonperiodic distributions of bond strengths has received surprisingly little attention in the literature. MC simulations [40] for a Fibonacci chain produce results that are in agreement with domain-wall arguments, assuming that the relaxation is dominated by the largest energy barrier for spin evaporation. This appears also to be the case for random, ferromagnetic chains. For a spin-glass chain, the presence of a very wide distribution of rates makes it very difficult to extract information from MC simulations.

6.3.4 Other dynamics

The Glauber transition rate is by no means the only possible choice, for single-spin flip dynamics, consistent with detailed balance, although it is the only one that admits explicit solution. Detailed balance only constrains the ratio of the rate of one type of process to its inverse, leaving unspecified the relationship between transition rates for independent processes.

For single-spin-flip dynamics, with uniform coupling constants, there are two classes of process: energy-conserving processes of the type $\uparrow\uparrow\downarrow\leftrightarrow\uparrow\downarrow\downarrow$ and endo/exothermic processes of the type $\uparrow\downarrow\uparrow\leftrightarrow\uparrow\uparrow\uparrow$ (in an obvious notation). By allowing the rate of the first type of processes to vanish like $\exp(-\Delta/k_B T)$ as $T \to 0$, the domain-wall argument would predict a relaxation time of the form $\xi^2 \exp(\Delta/k_B T)$, and hence a critical exponent $z = 2 + (\Delta/2J)$. One particular case was treated independently by Deker and Haake [42] and by Kimball [43]. Provided the rate for energy-conserving flips does not vanish at zero temperature, the universal critical exponent $z = 2$ is found [3].

Another source of nonuniversality is the introduction of retarded interactions in the dynamics, by using flip rates of the form

$$\omega_j(\{S_i\}, t) = \alpha \left\{ 1 - S_j(t)[\gamma_i^+ S_{j+1}(t - \tau_{i-1}) + \gamma_j^- S_{j-1}(t - \tau_{i+1})] \right\}. \quad (6.35)$$

Here the dynamics of the spin S_j at time t depends upon its neighbors' states at times $t - \tau_{i\pm 1}$ respectively. The equation of motion of the magnetization becomes

$$\frac{dm_i(t)}{dt} = -2\alpha \left[m_i(t) - \gamma_i^+ m_{i+1}(t - \tau_{i+1}) - \gamma_i^- S_{i-1}(t - \tau_{i-1}) \right]. \quad (6.36)$$

Such interactions can give rise to complex, chaotic behavior [44]. If there is no disorder in the values of J_i and τ_i, however, the conventional result $z = 2$ is recovered [45]. For randomly distributed J_i, and an exponential distribution of τ_i, the results were found numerically to be the same as for no retardation. However, if the J_i are uniform, but the τ_i have an anomalous distribution of the form $P(\tau_i) = (1 - a)\tau_i^{-2+a}$, numerical iterations of the RG transformations suggest that $z = (2 - a)/(1 - a)$, the same result as was found for an anomalous distribution of the bare spin-flip rates [36].

If double spin flips are allowed, the same domain-wall-motion argument can be applied as for single spin flips. The same criteria for universality apply: if the energy-conserving processes take place at constant rate, the universal exponent $z = 2$ is again found [3]. If the coupling strengths take alternating values, then a simultaneous flip of two parallel spins next to a domain wall is energy conserving, removing the energy barrier to domain-wall diffusion and hence restoring the universal value $z = 2$.

Generalizations of the Kawasaki model can also produce changes in the critical exponent. For nearest-neighbor exchanges, Haake and Thol [3] have proved the inequality $z \geq 5$ for the critical exponent. If exchanges of longer, but finite, range may take place, the critical exponent is still found to be $z = 5$, whereas if the range is infinite (or at least allowed to increase like ξ) the exponent $z = 3$ is found [46,47], in accordance with the relation $z_{\text{global}} = z_{\text{local}} - 2$ proposed by Tamayo and Klein for dynamics with global and local conservation laws [48].

6.3.5 Other models

There are a number of models related to the Ising model that may be studied in similar ways. The Blume-Emery-Griffiths model is a spin-1 generalization of the Ising model, introduced initially as a phenomenological model to explain the behavior in ^{3}He-^{4}He mixtures [49]. Even in 1D, this model has a rich phase diagram, including a tricritical point [50]. For Glauber dynamics, RG calculations predict a universal dynamic critical exponent $z = 2$ [51]. However, domain-wall arguments predict a nonuniversal value for z [6], owing to the presence of an energy barrier to domain-wall motion. This result is supported by MC simulations of the relaxation of the magnetization [52].

When Kawasaki dynamics are implemented in the Blume-Emery-Griffiths model, domain-wall arguments again predict a nonuniversal value for z [41]. This was tested against MC simulations by using a scaling argument to relate the domain growth far from equilibrium to the critical exponent. The critical exponent is effectively obtained by simultaneously measuring the domain-growth exponent and the energy barrier to domain-motion processes. This approach gave good agreement with the predictions of domain-wall-motion arguments.

Another model closely related to the Ising model is the Potts model [53]. In this model, each spin may be in one of q states, with the case $q = 2$ corresponding to the Ising model. In 1D, two different choices for the flip rate, both reducing to Glauber's choice for $q = 2$, were found to give rise to different values for z, for $q \neq 2$ [54]. Domain-wall arguments show this discrepancy to be due to the vanishing of the domain-wall diffusion rate in one of the models, and the universal exponent $z = 2$ is restored if this rate remains finite at $T = 0$ [5,55]. These arguments also predict $z = 5$ for Kawasaki-like dynamics, independently of q [5].

Domain-wall-motion arguments have been applied to a 1D Glauber model with both nearest-neighbor and next-nearest-neighbor interactions [57]. The universal value $z = 2$ is again found, since there is no energy barrier to domain-wall diffusion.

6.4 Rapid cooling—domain growth

The dynamic evolution of a system cooled rapidly ('quenched') through a phase transition is a subject that has received considerable attention [7]. Symmetry is broken locally, and domains of the different phases appear and begin to grow. After a transient period, the two-point correlation function

$C(r, t)$ assumes a dynamic scaling form of the type

$$C(\mathbf{r}, t) = \tilde{C}\left(\frac{\mathbf{r}}{L(t)}\right). \tag{6.37}$$

The characteristic length scale is found to increase with time as $L \sim t^x$, where the exponent takes the values $x = 1/2$ and $x = 1/3$ for respectively conserved and nonconserved scalar order parameter. This behavior is explained by the dynamics of the interfaces between the different domains [58].

In 1D, there is no phase transition at finite temperature, and, although domain walls exist, their morphology is qualitatively very different from those for higher dimensions. Nevertheless, dynamic scaling of the type (6.37) has been found for 1D Ising models submitted to an instantaneous quench, with the same growth exponent as for higher dimensions. We discuss separately the cases of Glauber and Kawasaki dynamics.

6.4.1 Glauber model

The equation of motion for the two-point correlation function, (6.17), for an instantaneous quench to low temperatures was solved using Laplace and Fourier transforms by Bray [59]. The solution for zero final temperature, in the limit $\alpha t \gg 1$, is

$$g(x, t) = \int_0^1 \frac{dy}{\pi} \frac{\exp(-x^2/8\alpha t y)}{[y(1-y)]^{1/2}}, \tag{6.38}$$

which is of the form (6.37) with $L \sim t^{1/2}$. The same domain-growth exponent was found by Schilling [60]. For nonzero final temperature, characterized by equilibrium correlation length ξ, (6.38) is found for an intermediate time regime $1 \ll \alpha t \ll \xi^2$, whereas for $t \sim \xi^2$ an exponential decay to equilibrium is found.

Bray also calculated the autocorrelation function $C(t', t) \equiv \langle S_i(t) S_i(t') \rangle$, finding, in the scaling regime $t' \gg t$, $C(t, t') \to \frac{2}{\pi} (2t/t')^{1/2}$.

The dynamical scaling in the domain-growth regime can, in fact, be expressed in terms of the same dynamical scaling theory that describes the critical dynamics regime [31]. This is so because the correlation function obeys exactly a linear, diffusion-like equation of motion, independently of whether it is near to or far from equilibrium.

Calculations of the correlation function after a rapid quench to low temperatures have also been carried out for the Glauber model with alternating bond strengths J_1 and J_2 [31,61]. The temperature cannot be reduced to zero here, because the domain-diffusion dynamics would be completely arrested owing to the additional energy barrier. Although the dynamic critical exponent z is nonuniversal, the domain-growth exponent has the universal value $x = 1/2$ independently of the bond strengths. In the scaling limit of low temperatures and long times, the results reduce exactly to those of the uniform case by the substitution $t \to \theta = t \exp[-2(J_1 - J_2)/k_B T]$. Thus, although the time scales have a temperature-dependent prefactor, the physics is otherwise the same.

Although the growth exponent is independent of dimension, the physical origins of the domain growth are very different. In higher dimensions, the domain walls are driven by their curvature. In 1D, the growth law is explained by the domain-wall motion arguments: a domain of size L has a lifetime $\propto L^2$, and so the equation of motion of the density of domain walls $K \equiv 1/L$ is

$$\frac{1}{K}\frac{dK}{dt} \propto -K^2, \tag{6.39}$$

giving $L \sim t^{1/2}$ at long times.

6.4.2 Kawasaki dynamics

Since the equations of motion for the Kawasaki model do not admit analytic solution, the problem of phase separation after a rapid quench can only be approached from approximations. The domain-wall-motion arguments originally introduced to explain the critical dynamics [18] in fact describe the system in the regime where the domains are much smaller than the correlation length [62,63], which is just the scaling regime of interest. The arguments in the second part of Sec. 6.3.1 show that a domain of length L has a lifetime $L^3 \exp(4J/k_B T)$. The equation of motion of the domain-wall density $K = 1/L$ is then

$$\frac{1}{K}\frac{dK}{dt} \sim -K^3 K, \tag{6.40}$$

leading to $L \sim t^{1/3}$. The full dynamic scaling form for the two-point correlation function has been verified by MC simulation [63].

6.5 Slow cooling—freezing

The dynamics of a system which requires an activation energy for relaxation will be arrested at zero temperature. If the system is cooled slowly, then although at high temperatures it will remain close to equilibrium there will come a point at which it departs from equilibrium. A nonequilibrium state will then freeze if the cooling continues to an arbitrarily low temperature.

This kind of 'laboratory glass transition' has been observed in many different kinds of system [8]. The nonequilibrium structures so formed differ from the equilibrium states at any temperature. One feature common to many glassy states is the appearance of stretched exponential response functions [64], representing a broad spectrum of relaxation times [65].

One 'toy' model that has been used to investigate generic features of activated systems under cooling is a system with only two states, with energies differing by ϵ and with an additional activation energy V [66]. The master equation governing this system can be solved exactly, and studies have been performed of the relation between frozen energy and cooling rate, as well as entropy and hysteresis effects. It is found for several different cooling schedules that the frozen energy E is related to the cooling rate r by a universal power law $E \sim r^a$, with $a = \epsilon/V$, but with possible nonuniversal logarithmic corrections. However, the relaxation at any constant, finite temperature is always exponential.

This section is concerned with studies of simple kinetic Ising models with activated dynamics in the context of continuous cooling. As well as giving different relationships between residual energy and cooling rate, the frozen structures so formed also have been observed to relax nonexponentially.

6.5.1 Glauber dynamics

Apart from its interest as a statistical system in its own right, studies of the 1D Glauber system in connection with the laboratory glass transition have been motivated by a mapping from a chain of particles with anharmonic double-well interactions [67]. In this case, an energy barrier Δ is introduced to the dynamics in an *ad hoc* way, by letting the bare spin-flip rate α in (6.14) depend upon temperature as follows: $\alpha = \alpha_0 \exp(-\Delta/k_B T)$ [60,68]. The energy barrier can be introduced at a more microscopic level by considering instead the Glauber model with alternating couplings J_1 and J_2 [69]; in the low-temperature regime of interest, these two models become equivalent, with $\Delta = 2(J_1 - J_2)$.

If we assume that the heat bath equilibrates much more rapidly than the characteristic Glauber flip rates, we can vary the temperature of the heat bath at will, so that the extrinsic dynamics of the Ising system is just that of the instantaneous temperature of the bath. The normal equation of motion may then be used, the coefficients α and γ being given functions of time. The equation of motion may still be solved by standard techniques.

As an illustration, we follow the approach of [69], where a low-temperature continuum approximation is used, and the equation of motion becomes

$$\frac{\partial}{\partial t} g(x, t) = \alpha \left(\frac{\partial^2}{\partial x^2} - \frac{1}{\xi^2} \right) g(x, t),$$ (6.41)

where $\xi = \exp(2J/k_B T)$ and $\alpha = \exp(\Delta/k_B T)$. Introducing an effective time $u = \int_0^t \alpha(t') \, dt'$, and an integrating factor $J(u) = \int_0^t \alpha(t') \xi^{-2}(t') \, dt' = \int_0^u \xi^{-2}(u') \, du'$, the equation is transformed into a simple diffusion equation

$$\frac{\partial}{\partial u} f(x, u) = \frac{\partial^2}{\partial x^2} f(x, u),$$ (6.42)

where $f(x, u) \equiv g \exp[J(u)]$. This equation is to be solved with initial condition $f(x, 0) = \exp(-x/\xi_0)$ (corresponding to thermal equilibrium at temperature T_0) and boundary condition $f(0, u) = \exp[J(u)]$ (corresponding to the constraint $g(0, t) = 1$). The function g will be able to relax completely if the total effective time $u(\infty)$ for the cooling is infinite. Therefore, the condition for freezing to occur is that $u(\infty)$ be finite. Other authors have treated the discrete-space difference equations [60,70], but the results are the same in the limit of slow cooling. The case of rapid cooling has been treated by Reiss [68].

The cooling rate r is defined unambiguously by using cooling schedules with one characteristic time scale, of the form

$$\alpha = \alpha_0 \phi(rt).$$ (6.43)

The relationship between the residual energy E and the cooling rate is found to depend upon the type of cooling program used [69,71]. For cooling schedules where the inverse temperature varies linearly in time, i.e., $\phi = \exp(-rt)$, it is found [60,69,70] that $E \sim r^{1/z}$, where $z = 2 + (\Delta/2J)$ is the dynamic critical exponent. The same result is found for exponentially decreasing temperature in the limit of small r. However, for a logarithmic cooling program, where $\phi = (1 + rt)^{-b}$, one finds instead [71] $E \sim r^{-1/\zeta}$, where $\zeta = 2 + \Delta(b-1)/(2bJ)$. It is, in fact, possible to construct cooling schedules of the form (6.43) that give rise to any desired relationship between E and r [71].

A naive argument would lead one to expect that the system would freeze when the cooling rate r was of the order of the characteristic equilibration rate ξ^{-z}, leading to $E \sim r^{1/z}$ since the energy density is proportional to the domain-wall density. To be more precise, the domain-wall density equilibrates at a rate ξ^{-z}, whereas the equilibrium domain-wall density is evolving at a characteristic rate $d\ln(1/\xi)/dt$. The frozen domain-wall density may then be estimated by equating these two rates, leading to the correct exponents for all types of cooling schedule [71]. A similar approach is to define a dimensionless 'time remaining' τ for the cooling program in units of the average linear relaxation time: $\tau = \int_t^\infty \xi^{-z} dt$. The system will freeze at the point where $\tau \sim 1$ [70].

The response of the 1D Glauber model to a change in temperature may also be calculated exactly [29,72]. Although the relaxation for $t \sim \xi^z$ is asymptotically exponential, the relaxation at intermediate times may be fitted to a stretched exponential. This arises from the fact that if one has a function of the form $\phi(x) = \exp(-\lambda x^\mu)$, then $\ln |\ln \phi| = \ln \lambda + \mu \ln x$. Consider now a general function $\psi(x)$, and define $\Psi(\ln x) = \ln |\ln \psi(x)|$. If one performs a Taylor expansion about $y = y_0$, defined as the point where $\partial^2 \Psi(y)/\partial y^2 = 0$ (if such a point exists), $\Psi(y) = \Psi(y_0) + (y - y_0)\Psi'(y_0) + \mathcal{O}((y - y_0)^3)$. In the vicinity of $y = y_0$, the function ψ may therefore be considered to be of the form $\psi \approx \exp(-\lambda x^\mu)$, where $\lambda \sim \Psi(y_0)$ and $\mu = \Psi'(y_0)$. Brey and Prados found that the response of the energy to a change in temperature could be fitted to a stretched exponential with $\mu = 1/2$ for more than a decade in time [72]. Stretched exponential relaxation has also been found in other single-spin-flip models [73].

A detailed account of other, related, calculations, including hysteresis for heating and cooling cycles, may be found in [70].

6.5.2 Kawasaki dynamics

Since the 1D Kawasaki model has an internal activation energy, it is not necessary to introduce an extrinsic energy barrier in order for the system to freeze at zero temperatures. Exact calculations of this behavior cannot, however, be performed.

Although the domain-wall arguments used to describe domain growth in the 1D Kawasaki model are strictly only applicable far from equilibrium, where only domain-destruction processes occur, it is possible to extend them to describe domain-creation processes too [63]. Domains are created by the fusion of two diffusing free spins. For this to occur, a single spin must evaporate from a domain wall (a process that requires energy $4J$), and then a second spin must also evaporate before the first one is reabsorbed by the domain wall. Random walk arguments show that this process leads to domain creation at a rate $\propto \exp(-10J/k_B T)K$, where K is the density of domain walls [63]. The quantity K is therefore described approximately by an equation of the form

$$\frac{1}{K}\frac{dK}{dt} = \omega^{5/2} - \omega K^3, \tag{6.44}$$

where $\omega = \exp(-4J/k_BT)$, and t and K have been rescaled to absorb any numerical prefactors. This equation describes the system both at and away from equilibrium—note that both terms on the r.h.s. are of order ξ^5 at equilibrium ($K \propto 1/\xi$).

Equation (6.44) may be integrated for a general cooling program $\omega(t)$. For the particular case $\omega = \omega_0 \exp(-rt)$, the solution is found to be of the scaling form $K(t) = r^{1/5}F(r\xi^5(t))$, where $F(x) \sim x^{-1/5}$ as $x \to 0$, $F(x) \to$ constant as $x \to \infty$, and $\xi(t)$ is the instantaneous value of the correlation length during the cooling process. This simple dependence upon the instantaneous value of ξ comes from the omission of memory effects in (6.44). Nevertheless, the resulting prediction that the frozen domain-wall density is $K_F \propto r^{1/5}$ is reasonable. MC simulations gave $K_F \sim r^{0.18\pm0.02}$ [63].

At zero temperature, energy-conserving spin diffusion processes can still occur. These processes give rise to a residual stretched-exponential relaxation of the form $\exp(-At^{1/3})$. The residual, exponential, relaxation has also been considered for the case where energy-conserving processes are not permitted and only exothermic processes are allowed [63,74].

6.6 Concluding remarks

We now have a very full understanding of low-temperature dynamic phenomena in 1D Ising models. The microscopic domain-wall arguments agree extremely well with the exact calculations of the Glauber model for various types of lattice symmetry. For the Kawasaki model, there are many fewer cases where exact calculations may be made, but the microscopic arguments nevertheless agree well with data from MC simulations.

The critical dynamics of the Glauber model has been investigated with periodic, quasiperiodic, and random lattice symmetry, and there are no obvious gaps in our knowledge. However, for the Kawasaki model with periodic modulations the only analytic techniques that exist do not yield agreement with the domain-wall arguments. What is needed is either a new analytic way to probe the dynamic critical behavior—perhaps quantum chains [17] will be the answer—or MC simulations that probe dynamic behavior at equilibrium, for example by measuring the autocorrelation function. This would appear to be within the capabilities of modern computing power. It would also be interesting to investigate Kawasaki dynamics in a spin glass, to see whether a description in terms of simple microscopic processes may be formulated.

It is worthwhile making a few more comments on the question of whether 1D critical dynamics should really be considered nonuniversal, or whether it is more physical to consider the critical dynamics as universal with an Arrhenius prefactor to all time scales. In the former approach, all temperature dependence is expressed in terms of ξ, and one is led to an unambiguous exponent and scaling function. However, the scaling function so obtained sometimes takes an 'unusual' form, with nonuniversal leading algebraic dependence [31]. In the latter approach, the value of the critical exponent depends upon the choice of the Arrhenius prefactor that is factorized out.

In the case of the alternating-bond Glauber chain, the leading q-dependence at constant temperature of τ_q in the limit $q\xi \to \infty$ is q^{-2} independently of the lattice modulation (see (6.34)). Achiam and Southern [15] therefore concluded that the correct value of the critical exponent is $z = 2$, and identified the correct energy barrier $2(J_1 - J_2)$. This alleviates the need to use 'unusual' forms for the scaling functions.

However, if one considers the uniform Kawasaki model in the same limit, one finds that $\tau_q \sim q^{-4}$ (see (6.27)). In order to maintain a value $z = 4$, one would be forced to conclude that the energy barrier is $2J$, whereas microscopic considerations show that the only energy barrier that is present in this dynamics is $4J$. The natural, physical way to separate the relaxation time into a universal long-range contribution and an energy barrier is to use the conventional value $z = 4 - \eta = 3$ and a microscopic energy barrier $4J$. It would appear that the Kawasaki model is a case where 'unusual' forms for scaling functions are obligatory.

Although the physics of domain growth differs greatly between one dimension and higher dimensions, the equality of the scaling exponents is an interesting corollary to universality. The domain-wall arguments originally used to describe critical behavior are in fact properly applicable in the domain-scaling regime. In the case of the Kawasaki model, certain processes such as spin-spin coalescence may be ignored, and so it should be possible to formulate a simplified dynamics valid in the scaling regime. It would be interesting to see whether the scaling function may be calculated in this way.

The 1D Ising models subject to slow cooling display several features reminiscent of the behavior of true glasses. They display a significant advance over the simple two-level system in that they possess truer collective behavior, and so can have nonexponential relaxation. However, they fall short in describing true glasses in one important respect: in true glasses, relaxation times diverge according to a Vogel-Tamman-Fulcher law [75],

$\tau \sim \exp[A/(T - T_0)]$. In the studies covered in this chapter, all relaxation times may be expressed in an Arrhenius form. This is due to the simplicity of the models considered—to obtain a more glassy behavior it is necessary to have a more complex dynamics that mimics the geometric frustration effects in true glasses [76].

The qualitative arguments that have been used to explain these phenomena are peculiar to 1D, so it is difficult to see how the results described in this chapter could be extended to higher dimensions (an exception would be the case of quasi-1D fractals [55,62,77]). Nevertheless, the body of research on 1D kinetic Ising models has contributed significantly to our understanding of dynamical processes in general.

The author is indebted to Michel Droz for comments on this chapter. He is also grateful to the NSERC for financial support and would like to thank the following people, who have contributed directly to his understanding of the subject matter of this chapter: Alan Bray, Michel Droz, Josef Jäckle, Jafferson Kamphorst Leal da Silva, Kimmo Kaski, Eduardo Lage, Jose Fernando Mendes, Jose Miguel Nunes da Silva, and Robin Stinchcombe.

References

[1] M. Schwartz and D. Poland, *J. Chem. Phys.* **65**, 2620 (1976); R. H. Lascombe, *J. Macromol. Sci. Phys.* **B18**, 697 (1980).

[2] P. C. Hohenberg and B. I. Halperin, *Rev. Mod. Phys.* **49**, 435 (1977).

[3] F. Haake and K. Thol, *Z. Phys.* **B40**, 219 (1980).

[4] M. Droz, J. Kamphorst Leal da Silva and A. Malaspinas, *Phys. Lett.* **A115**, 448 (1986).

[5] M. Droz, J. Kamphorst Leal da Silva, A. Malaspinas and J. Yeomans, *J. Phys.* **A19**, 2671 (1986).

[6] E. J. S. Lage, *Phys. Lett.* **A127**, 9 (1988).

[7] J. D. Gunton, M. San Miguel and P. S. Sahni, in *Phase Transitions and Critical Phenomena*, Vol. **8**, C. Domb and J. L. Lebowitz, eds. (Academic Press, London, 1983); A. J. Bray, *Adv. Phys.* **43**, 357 (1994).

[8] J. Jäckle, *Rep. Prog. Phys.* **49**, 171 (1986).

[9] C. J. Thompson, in *Phase Transitions and Critical Phenomena*, Vol. **1**, C. Domb and M. S. Green, eds. (Academic Press, London, 1972).

[10] K. Kawasaki, in *Phase Transitions and Critical Phenomena*, Vol. **2**, C. Domb and M. S. Green, eds. (Academic Press, London, 1972).

[11] R. Glauber, *J. Math. Phys.* **4**, 294 (1963).

[12] B. U. Felderhof, *Rep. Math. Phys.* **1**, 215 (1970).

[13] K. Kawasaki, *Phys. Rev.* **145**, 224 (1966).

[14] J. van Hove, *Phys. Rev.* **93**, 1374 (1954).

[15] Y. Achiam and B. W. Southern, *J. Phys.* **A25**, L769 (1992).

[16] W. Zwerger, *Phys. Lett.* **A84**, 269 (1981).

[17] F. C. Alcaraz, M. Droz, M. Henkel and V. Rittenberg, *Ann. Phys. (NY)* **230**, 250 (1994).

[18] R. Cordery, S. Sarkhar and J. Tobochnik, *Phys. Rev.* **B24**, 5402 (1981).

[19] Y. Achiam, *J. Phys.* **A11**, 975 (1978).

[20] Y. Achiam and J. M. Kosterlitz, *Phys. Rev. Lett.* **41**, 128 (1978).

[21] J. M. Nunes da Silva and E. J. S. Lage, *J. Phys.* **C21**, 2225 (1988).

[22] K. Binder and D. W. Heermann, *Monte-Carlo Simulation in Statistical Physics—an Introduction* (Springer-Verlag, Berlin, 1988).

[23] D. T. Gillespie, *J. Comput. Phys.* **28**, 395 (1978).

[24] B. I. Halperin, *Phys. Rev.* **B8**, 4437 (1973).

[25] D. Poland, *J. Stat. Phys.* **59**, 935 (1990).

[26] J. A. Ashraff and R. B. Stinchcombe, *Phys. Rev.* **B40**, 2278 (1989).

[27] B. W. Southern and Y. Achiam, *J. Phys.* **A25**, 2519 (1993).

[28] E. J. S. Lage, *J. Phys.* **C20**, 3969 (1987).

[29] S. J. Cornell, D. Phil. thesis (Oxford University, 1990).

[30] J. H. Luscombe, *Phys. Rev.* **B36**, 501 (1987).

[31] S. J. Cornell, M. Droz and N. Menyhárd, *J. Phys.* **A24**, L201 (1991).

[32] B. W. Southern and Y. Achiam, *J. Phys.* **A25**, 2505 (1993).

[33] J. C. Angles d'Auriac and R. Rammal, *J. Phys.* **A21**, 763 (1988).

[34] J. M. Nunes da Silva and E. J. S. Lage, *J. Stat. Phys.* **58**, 115 (1990).

[35] C. L. Henley, *Phys. Rev. Lett.* **54**, 2030 (1985); C. K. Harris and R. B. Stinchcombe, *Phys. Rev. Lett.* **56**, 869 (1986).

[36] J. M. Nunes da Silva and E. J. S. Lage, *Phys. Lett.* **A135**, 17 (1989).

[37] S. Havlin and D. Ben-Avraham, *Adv. Phys.* **6**, 695 (1987).

[38] J. F. Fernandez and R. Medina, *Phys. Rev.* **B19**, 3561 (1979); T. Li, *Phys. Rev.* **B24**, 6579 (1981); B. Derrida and E. Gardner, *J. Physique* **48**, 959 (1986); P. L. Krapivsky, *J. Physique* I1, 1013 (1991).

[39] J. M. Nunes da Silva and E. J. S. Lage, *Phys. Rev.* **A40**, 4682 (1989).

[40] S. J. Cornell, unpublished.

[41] J. F. F. Mendes, S. J. Cornell, M. Droz and E. J. S. Lage, *J. Phys.* **A25**, 73 (1992).

[42] U. Deker and F. Haake, *Z. Phys.* **B35**, 281 (1979).

[43] J. C. Kimball, *J. Stat. Phys.* **21**, 289 (1979).

[44] M. Y. Choi and B. A. Huberman, *Phys. Rev.* **B31**, 2862 (1985).

[45] J. M. Nunes da Silva, *Phys. Lett.* **A142**, 471 (1989).

[46] Y. Achiam, *J. Phys.* **A13**, 1825 (1980).

[47] J. Kamphorst Leal da Silva and F. C. Sá Barreto, *Phys.* **A44**, 2727 (1991); A. G. Moreira, J. Kamphorst Leal da Silva and F. C. Sá Barreto, *Phys. Rev.* **B48**, 289 (1993).

[48] P. Tamayo and W. Klein, *Phys. Rev. Lett.* **63**, 2757 (1989).

[49] M. Blume, V. J. Emery and R. B. Griffiths, *Phys. Rev.* **A4**, 1071 (1971).

[50] S. Krinsky and D. Furman, *Phys. Rev.* **B11**, 2602 (1975).

[51] Y. Achiam, *Phys. Rev.* **B31**, 260 (1985); P. O. Weir and J. M. Kosterlitz, *Phys. Rev.* **B33**, 622 (1986); Y. Achiam, *Phys. Rev.* **B33**, 623 (1986).

[52] J. F. F. Mendes and E. J. S. Lage, *Phys. Lett.* **A159**, 13 (1991).

[53] F. Y. Wu, *Rev. Mod. Phys.* **54**, 235 (1982).

[54] G. Forgacs, S. T. Chiu and H. L. Frisch, *Phys. Rev.* **B22**, 415 (1980); P. O. Weir and J. M. Kosterlitz, *Phys. Rev.* **B33**, 391 (1986); E. J. S. Lage, *J. Phys.* **A18**, 2289; 2411 (1985); M. A. Zaluska-Kotor and L. A. Turski, *J. Phys.* **A22**, 413(1989).

[55] F. Leyvraz and N. Jan, *J. Phys.* **A19**, 603 (1986).

[56] P. O. Weir, J. M. Kosterlitz and S. H. Adachi, *J. Phys.* **A19**, L757 (1986).

[57] Z. R. Yang, *Phys. Rev.* **B46**, 11578 (1992).

[58] S. Allen and J. Cahn, *Acta Metall.* **27**, 1017; 1095 (1979); I. Lifshitz and V. Slyozov, *J. Phys. Chem. Solids* **19**, 35 (1961).

[59] A. J. Bray, *J. Phys.* **A22**, L67 (1989).

[60] R. Schilling, *J. Stat. Phys.* **53**, 1227 (1988).

[61] Z. Lin, X. Wang and R. Tao, *Phys. Rev.* **B45**, 8131 (1992).

[62] F. Leyvraz and N. Jan, *J. Phys.* **A20**, 1303 (1987).

[63] S. J. Cornell, K. Kaski and R. B. Stinchcombe, *Phys. Rev.* **B44**, 12263 (1991).

[64] R. Kohlrausch, *Pogg. Ann. Phys.* **1**, 56; 179 (1854); G. Williams and D. Watts, *Trans. Faraday Soc.* **66**, 80 (1970).

[65] R. Piazza, T. Bellini, V. Degiorgio, R. Goldstein, S. Leibler and R. Lipowsky, *Phys. Rev.* **B38**, 7223 (1988).

[66] W. A. Philips, *J. Low Temp. Phys.* **7**, 351 (1970); D. A. Huse and D. S. Fisher, *Phys. Rev. Lett.* **57**, 2203 (1986); S. A. Langer, A. T. Dorsey and J. P. Sethna, *Phys. Rev.* **B40**, 345 (1990); J. J. Brey and A. Prados, *Phys. Rev.* **B43**, 8350 (1991).

[67] O. Reichert and R. Schilling, *Phys. Rev.* **B32**, 5731 (1985); W. Kob and R. Schilling, *Z. Phys.* **B68**, 245 (1987); W. Tschöp and R. Schilling, *Phys. Rev.* **E48**, 4221 (1993).

[68] H. Reiss, *Chem. Phys.* **47**, 15 (1980).

[69] J. Jäckle, R. B. Stinchcombe and S. J. Cornell, *J. Stat. Phys.* **62**, 425 (1991).

[70] J. J. Brey and A. Prados, *Phys. Rev.* **B49**, 984 (1994).

[71] S. J. Cornell, K. Kaski and R. B. Stinchcombe, *J. Phys.* **A24**, L865 (1991).

[72] J. J. Brey and A. Prados, *Physica* **A197**, 569 (1993).

[73] J. Skinner, *J. Chem. Phys.* **79**, 1955 (1983); H.-U. Bauer, K. Schulten and W. Nadler, *Phys. Rev.* **B38**, 445 (1988).

[74] V. Privman, *Phys. Rev. Lett.* **69**, 3686 (1992); *Mod. Phys. Lett.* **B8**, 143 (1994).

[75] H. Vogel, *Z. Phys.* **22**, 645 (1922); G. Fulcher, *J. Am. Ceram. Soc.* **8**, 339 (1925); G. Tamman and H. Hesse, *Z. Organ. Chem.* **156**, 245 (1926).

[76] R. Palmer, D. Stein, A. Abrahams and P. W. Anderson, *Phys. Rev. Lett.* **53**, 958 (1984).

[77] J. H. Luscombe and R. C. Desai, *Phys. Rev.* **B32**, 488 (1985); Y. Achiam, *Phys. Rev.* **B32**, 1896 (1985); J. H. Luscombe, *J. Phys.* **A20**, 1299 (1987).

Part III: Ordering, Coagulation, Phase Separation

Editor's note

Nucleation, phase separation, cluster growth and coarsening, ordering, and spinodal decomposition are interrelated topics of great practical importance. While most experimental realizations of these phenomena are in three (bulk) and two (surface) dimensions, there has been much interest in lattice and continuum (off-lattice) 1D stochastic dynamical systems modeling these irreversible processes.

The main applications of 1D models have been in testing various scaling theories such as cluster-size-distribution scaling and scaling forms of order-parameter correlation functions. Exact solutions are particularly useful in this regard, and the focus of all three chapters in this Part is on exactly solvable models. Additional literary sources are cited in the chapters, including general review articles as well as other studies in 1D.

Chapter 7 reviews exact solutions of three different models of phase-ordering dynamics, including results based on the Glauber-Ising model introduced in Part II. Chapter 8 reviews a model with synchronous (cellular-automaton) dynamics and relations to reactions (Part I). In both chapters exact results for scaling of the two-point correlation function are obtained. Finally, Ch. 9 describes models of coagulating particles and associated results for cluster-size-distribution scaling.

7

Phase-ordering dynamics in one dimension

Alan J. Bray

Exact solutions for the phase-ordering dynamics of three one-dimensional models are reviewed in this chapter. These are the lattice Ising model with Glauber dynamics, a nonconserved scalar field governed by time-dependent Ginzburg-Landau (TDGL) dynamics, and a nonconserved $O(2)$ model (or XY model) with TDGL dynamics. The first two models satisfy conventional dynamic scaling. The scaling functions are derived, together with the (in general nontrivial) exponent describing the decay of autocorrelations. The $O(2)$ model has an unconventional scaling behavior associated with the existence of two characteristic length scales—the 'phase coherence length' and the 'phase winding length'.

7.1 Introduction

The theory of phase-ordering dynamics, or 'domain coarsening', following a temperature quench from a homogeneous phase to a two-phase region has a history going back more than three decades to the pioneering work of Lifshitz [1], Lifshitz and Slyozov [2], and Wagner [3]. The current status of the field has been recently reviewed [4].

The simplest scenario can be illustrated using the ferromagnetic Ising model. Consider a temperature quench, at time $t = 0$, from an initial temperature T_I, which is above the critical temperature T_C, to a final temperature T_F, which is below T_C. At T_F there are two equilibrium phases, with magnetization $\pm M_0$. Immediately after the quench, however, the system is in an unstable disordered state corresponding to equilibrium at temperature T_I. The theory of phase-ordering kinetics is concerned with the dynamical evolution of the system from the initial disordered state to the final equilibrium state.

Part of the fascination of this field is that in the thermodynamic limit final equilibrium is never achieved. The reason is that the longest relaxation time diverges with system size in the ordered phase, reflecting the broken ergodicity. Instead, a network of domains of the equilibrium phases develops, and the typical length scale $L(t)$ associated with these domains increases with time t. Furthermore, there is ample experimental and simulational evidence for *dynamical scaling*, which in its strongest form asserts that the domain morphology is (in a statistical sense) independent of time, if all lengths are measured in units of the characteristic scale $L(t)$. Testing this hypothesis is one of the motivations for studying one-dimensional (1D) models.

In this chapter we present exact results for the ordering kinetics of a number of models in 1D. These include the Ising model with Glauber dynamics, and time-dependent Ginzburg-Landau models with scalar and two-component vector order parameters. Because 1D systems with only short-range interactions do not exhibit equilibrium phase transitions, we will restrict our discussions to temperature $T = 0$, since for $T > 0$ the equilibrium correlation length $\xi(T)$ sets a limit to the coarsening process, i.e., when the domain size reaches $\xi(T)$ the system has achieved equilibrium. For systems that *do* exhibit an equilibrium phase transition, however, e.g., Ising models in two or more dimensions, it is expected that all final temperatures $T_F < T_C$ lead to the *same* asymptotic scaling behavior, because the coarsening is controlled by a 'strong-coupling' renormalization group (RG) fixed point [5].

Conservation laws play an important role in the coarsening process. For the simple Glauber dynamics [6], the order parameter (i.e., the magnetization) is a nonconserved quantity, since the basic process involves the flipping of a single spin. For Kawasaki dynamics [7], the basic step is the exchange of a nearest-neighbor pair of oppositely oriented spins, a magnetization-conserving process. The conserved dynamics are relevant to, for example, phase separation in binary alloys, where the up and down states of an Ising spin can be used to represent the two-alloy species. In general, conservation laws make analytic solution more difficult, and we are not aware of any exact solutions for late-stage coarsening with conserved dynamics in 1D, although compelling plausibility arguments for the growth of $L(t)$ can be advanced [8,9].

The chapter is organized as follows. In the following section we describe the dynamical models that we shall be considering, and we state the dynamical scaling hypothesis and some of its consequences. Section 7.3 contains the main results of this chapter: exact results for the dynamics of the Glauber model, the scalar TDGL equation, and the two-component TDGL equation, all in 1D. The first two of these exhibit conventional dynamical scaling, while the last exhibits a novel form of scaling, involving two characteristic lengths [10,11]. We also discuss the concept of two-time scaling, introduced in Sec. 7.2, and calculate the new, in general nontrivial, exponent needed to describe the correlations when the two times are well separated. The chapter concludes with a discussion and a summary.

7.2 Dynamical models

In general, two different levels of description are used for systems that undergo phase transitions. The 'microscopic' description starts from spins (or other degrees of freedom) on a lattice, with prescribed interactions between the spins, typically between nearest neighbors only. The prototype of all such models is the ferromagnetic Ising model, with Hamiltonian

$$H = -J \sum_{<i,j>} S_i S_j, \qquad S_i = \pm 1, \tag{7.1}$$

where $< i,j >$ indicates that the summation is only over nearest-neighbor pairs.

One of the key ideas in phase-transition theory is the concept of 'universality', according to which microscopic details, such as the lattice structure, or possible inclusion of, say, next-nearest-neighbor interactions, are 'irrelevant' to the large-scale behavior, i.e., they do not change the scaling behavior (e.g., the values of the critical exponents) associated with the critical point. Accordingly, the 'universality classes' are determined by broad features of the model such as the spatial dimension and the symmetries of the Hamiltonian. Such expectations are borne out within explicit calculation using the renormalization group (RG) approach pioneered by Wilson.

Accordingly it is conventional in RG (or other) studies to start from an equivalent (in the sense of universality classes) coarse-grained description in terms of an order-parameter field (e.g., the 'magnetization density') $\phi(\mathbf{x})$. The quantity that plays the role of the Hamiltonian is the Ginzburg-Landau-Wilson free-energy functional. For the Ising model, the order-parameter will be a scalar quantity, and the functional will be invariant under $\phi \to -\phi$, reflecting the symmetry of (7.1) under $S_i \to -S_i$, for all i. The simplest such form is

$$F[\phi] = \int d^d x \left[\frac{1}{2} (\nabla \phi)^2 + V(\phi) \right], \tag{7.2}$$

with $V(-\phi) = V(\phi)$. For $T > T_C$, the potential $V(\phi)$ has a single minimum at $\phi = 0$, reflecting the vanishing long-range order of the high-temperature phase. For $T < T_C$, it has a double minimum at $\phi = \pm\phi_0$, reflecting the spontaneous symmetry-breaking for $T < T_C$. The squared gradient term in (7.2) represents the free-energy cost of spatial gradients in ϕ, e.g., in creating a domain wall between regions of the two ordered phases.

The conventional view is that, at least as far as equilibrium critical phenomena are concerned, models (7.1) and (7.2), and their dynamical generalizations described below, belong to the *same* universality class. We expect this equivalence to hold also in the nonequilibrium phase-ordering context, as long as the coarse-grained model captures the essence of the large-scale physics. In general dimensions this large-scale behavior corresponds (for a scalar order parameter) to curvature-driven growth, in which the domain walls shed energy by reducing their curvature. In 1D, however, the domain walls are points, and the growth mechanism is then quite different for the Ising and Ginzburg-Landau models. This will become clear in Sec. 7.3. Before proceeding to the solvable 1D models, however, we discuss in general terms the Ising and Ginzburg-Landau dynamics.

7.2.1 The Glauber model

The Glauber model [6] is the simplest extension of the Ising model with explicit dynamics. The 2^N configurations of the N Ising spins are given by $\{S_1, S_2, \ldots, S_N\}$, where $S_i = \pm 1$. Let $P(\{S_i\}; t)$ be the probability for the system to be in the configuration $\{S_i\}$ at time t. If we represent a given configuration by the single index α ($\alpha = 1, \ldots, 2^N$), the Glauber equation takes the form of a master equation,

$$\frac{dP(\alpha; t)}{dt} = \sum_\gamma [(W_{\gamma\alpha} P(\gamma; t) - W_{\alpha\gamma} P(\alpha; t)], \tag{7.3}$$

where $W_{\alpha\gamma}$ is the transition rate from the state α to the state γ. The basic transition consists of the flip of a single spin, $S_i \to -S_i$. The condition of detailed balance, $W_{\alpha\gamma}/W_{\gamma\alpha} = \exp[\beta(E_\alpha - E_\gamma)]$, where $\beta = 1/k_B T$ and E_α is the energy of state α, ensures that the system approaches equilibrium at infinite time. There are many ways to implement this condition. In the Glauber model it is implemented as follows.

Consider the process $S_i \to -S_i$. The energy change associated with this process is $\Delta E = 2h_i S_i$, where $h_i = J\sum_{<j>} S_j$ is the field felt by the spin at site i due to its neighbors. The condition of detailed balance requires that the rate $W(S_i \to -S_i)$ for the process $S_i \to -S_i$ satisfies

$$\frac{W(S_i \to -S_i)}{W(-S_i \to S_i)} = \exp(-2\beta h_i S_i). \tag{7.4}$$

In the Glauber model, this is imposed through the choice

$$W(S_i \to -S_i) = (1 - S_i \tanh \beta h_i)/(2\tau), \tag{7.5}$$

where τ^{-1} is a rate constant, which we set equal to unity. The final form of the Glauber equation then reads

$$\frac{dP(\{S_i\}; t)}{dt} = -\frac{1}{2} \sum_i P(S_1, \ldots, S_i, \ldots, S_N; t) (1 - S_i \tanh \beta h_i)$$

$$+\frac{1}{2} \sum_i P(S_1, \ldots, -S_i, \ldots, S_N; t) (1 + S_i \tanh \beta h_i). \tag{7.6}$$

This has to be supplemented with an initial condition $P(\{S_i\}; 0)$.

7.2.2 Coarse-grained models

For $T = 0$, the potential $V(\phi)$ in (7.2) has a double-well structure, e.g., $V(\phi) = (1 - \phi^2)^2$. The precise form of $V(\phi)$ is not, however, important. We will take the minima of $V(\phi)$ to occur at $\phi = \pm 1$, and adopt the convention that $V(\pm 1) = 0$. The two minima of V correspond to the two equilibrium states.

In the case where the order parameter is not conserved, an appropriate equation for the time evolution of the field ϕ is

$$\partial\phi/\partial t = -\delta F/\delta\phi = \nabla^2 \phi - V'(\phi), \tag{7.7}$$

where $V'(\phi) \equiv dV/d\phi$. A kinetic coefficient Γ, which conventionally multiplies the r.h.s. of (7.7), has been absorbed into the time scale. Equation (7.7), a simple 'reaction-diffusion' equation, corresponds to simple gradient descent, i.e., the rate of change of ϕ is proportional to the gradient of the free-energy functional in function space. This equation provides a suitable coarse-grained description of the Ising model, as well as of alloys that undergo an order-disorder transition on cooling through T_C, rather than a phase separation. Such alloys form a two-sublattice structure, with each sublattice occupied predominantly by atoms of one type. In Ising model language, this corresponds to antiferromagnetic ordering. The magnetization is no longer the order parameter, but a 'fast mode', whose conservation does not significantly impede the dynamics of the important 'slow modes'.

When the order parameter is conserved, as in phase separation, a different dynamics is required. In the alloy system, for example, it is clear physically that A and B atoms can only exchange locally (not over large distances), leading to diffusive transport of the order parameter, and an equation of motion of the form

$$\partial\phi/\partial t = \nabla^2 \delta F/\delta\phi = -\nabla^2 [\nabla^2 \phi - V'(\phi)], \qquad (7.8)$$

which can be written in the form of a continuity equation, $\partial_t \phi = -\nabla \cdot \mathbf{j}$, with current $\mathbf{j} = -\lambda\nabla(\delta F/\delta\phi)$. In (7.8), we have absorbed the transport coefficient λ into the time scale.

Equations (7.7) and (7.8) are called the time-dependent Ginzburg-Landau (TDGL) equation and the Cahn-Hilliard equation respectively. A more detailed discussion of them in the present context can be found in [12]. The same equations with additional Langevin noise terms on the r.h.s. are familiar from the theory of critical dynamics, where they are 'model A' and 'model B' respectively in the classification of [13].

The absence of thermal noise terms in (7.7) and (7.8) indicates that we are effectively working at $T = 0$. The RG flow diagram for T has stable fixed points at $T = 0$ and $T = \infty$, separated by an unstable fixed point at T_C. Under coarse-graining, temperatures above T_C flow to infinity, while those below T_C flow to zero. We therefore expect the final temperature T_F to be an irrelevant variable (in the scaling regime) for quenches into the ordered phase. This can be shown explicitly for systems with conserved order parameter [5]. For this case the thermal fluctuations at T_F simply renormalize the bulk order parameter and the surface tension of the domain walls: when the characteristic scale of the domain pattern is large compared to the domain-wall thickness (i.e., the bulk correlation length in equilibrium), the system behaves *as if* it were at $T = 0$, the temperature dependence entering through T-dependent model parameters.

In a similar way, any short-range correlations present at T_I should be irrelevant in the scaling regime, i.e., all initial temperatures are equivalent to $T_I = \infty$. Therefore we will take the *initial conditions* to represent a completely disordered state. For example, one could choose the 'white noise' form $\langle\phi(\mathbf{x}, 0)\,\phi(\mathbf{x}', 0)\rangle = \Delta\,\delta(\mathbf{x} - \mathbf{x}')$, where $\langle\cdots\rangle$ represents an average over an ensemble of initial conditions, and Δ controls the size of the initial fluctuations in ϕ. The above discussion, however, indicates that the precise form of the initial conditions should not be important, as long as only short-range spatial correlations are present.

The challenge of understanding phase-ordering dynamics, therefore, can be posed as that of finding the nature of the late-time solutions of deterministic differential equations like (7.7) and (7.8), subject to random initial conditions. A physical approach to this formal mathematical problem is based on studying the structure and dynamics of the domain walls. This is the approach that we will adopt.

7.2.3 The scaling hypothesis

Although originally motivated by experimental and simulation results for the structure factor and pair correlation function [14-16], for ease of presentation it is convenient to introduce the scaling hypothesis first, and then discuss its implications for growth laws and scaling functions. Briefly, the scaling hypothesis states that there exists, at long times, a single characteristic length scale $L(t)$ such that the domain structure is (in a statistical sense) independent of time when lengths are scaled by $L(t)$. It should be stressed that scaling has not been rigorously proved, except in some simple models such as the 1D Glauber model [17] (see Sec. 7.3) and the n-vector model with $n = \infty$ [18]. However, the evidence in its favor is compelling.

Two commonly used probes of the domain structure are the equal-time pair correlation function

$$C(\mathbf{r}, t) = \langle \phi(\mathbf{x} + \mathbf{r}, t)\, \phi(\mathbf{x}, t) \rangle, \tag{7.9}$$

and its Fourier transform, the equal-time structure factor,

$$S(\mathbf{k}, t) = \langle \phi_{\mathbf{k}}(t)\, \phi_{-\mathbf{k}}(t) \rangle. \tag{7.10}$$

Here angle brackets indicate an average over initial conditions. The structure factor can, of course, be measured in scattering experiments. The existence of a single characteristic length scale, according to the scaling hypothesis, implies that the pair correlation function and the structure factor have the scaling forms

$$C(\mathbf{r}, t) = f(r/L), \qquad S(\mathbf{k}, t) = L^d\, g(kL), \tag{7.11}$$

where d is the spatial dimensionality, and $g(y)$ is the d-dimensional Fourier transform of $f(x)$. Note that $f(0) = 1$, since (at $T = 0$) there is perfect order within a domain.

At general temperatures $T < T_C$, $f(0) = M^2$, where M is the equilibrium value of the order parameter. (Note that the *scaling limit* is defined by $r \gg \xi$, $L \gg \xi$, with r/L arbitrary, where ξ is the equilibrium correlation length of the ordered phase). Alternatively, we can extract the factor M^2 explicitly by writing $C(\mathbf{r}, t) = M^2 f(r/L)$. The statement that T is irrelevant then amounts to asserting that any remaining temperature dependence can be absorbed into the domain scale L, in such a way that the function $f(x)$ is independent of T. The scaling forms (7.11) are well supported by simulation data and experiment.

For future reference, we note that the different-time correlation function, defined by $C(\mathbf{r}, t, t') = \langle \phi(\mathbf{x} + \mathbf{r}, t)\, \phi(\mathbf{x}, t') \rangle$, can also be written in scaling form. A simple generalization of (7.11) gives [19]

$$C(\mathbf{r}, t, t') = f(r/L, r/L'), \tag{7.12}$$

where L, L' stand for $L(t)$ and $L(t')$. Especially interesting is the limit $L \gg L'$, when (7.12) takes the form

$$C(\mathbf{r}, t, t') \to (L'/L)^{\bar{\lambda}} h(r/L), \qquad L \gg L', \tag{7.13}$$

where the exponent $\bar{\lambda}$, first introduced [20] in the context of nonequilibrium relaxation in spin glasses, is a nontrivial exponent associated with phase ordering kinetics [21,22]. It has recently been measured in an experiment on twisted nematic liquid-crystal films [23]. The *autocorrelation* function, $A(t) = C(\mathbf{0}, t, t')$ is therefore a function only of the ratio L'/L, with $A(t) \sim (L'/L)^{\bar{\lambda}}$ for $L \gg L'$.

7.2.4 Domain walls

It is instructive to look at the properties of a flat equilibrium domain wall. From (7.7) the wall profile is the solution of

$$d^2\phi/dg^2 = V'(\phi), \tag{7.14}$$

with boundary conditions $\phi(\pm\infty) = \pm 1$, where g is a coordinate normal to the wall. We can fix the 'center' of the wall (defined by $\phi = 0$) to be at $g = 0$ by the extra condition $\phi(0) = 0$. Integrating (7.14) once, and imposing the boundary conditions, gives $(d\phi/dg)^2 = 2V(\phi)$. This result can be used in (7.2) to give the energy per unit area of wall, i.e., the surface tension, as

$$\sigma = \int_{-\infty}^{\infty} dg\, (d\phi/dg)^2 = \int_{-1}^{1} d\phi\, \sqrt{2V(\phi)}. \tag{7.15}$$

Note that, for scalar fields, the two terms in (7.2) contribute equally to the wall energy.

The profile function $\phi(g)$ has a sigmoid shape. For $g \to \pm\infty$, linearizing (7.14) around $\phi = \pm 1$ gives

$$1 \mp \phi \sim \exp\left\{ -[V''(\pm 1)]^{1/2} |g| \right\}, \qquad g \to \pm\infty, \tag{7.16}$$

i.e., away from the walls the order parameter saturates exponentially fast. It follows that the excess energy is localized in the domain walls. For spatial dimensions $d > 1$, therefore, the driving force for domain growth is the wall curvature, since the system energy can only decrease through a reduction in the total wall area. For $d = 1$, by contrast, domain walls only interact through the exponential tails of the wall profile function. This leads to logarithmically slow growth [24], instead of the $t^{1/2}$ growth obtained for $d > 1$.

7.3 Solvable models in one dimension

There are few exactly solved models of phase-ordering dynamics and, unfortunately, these models are quite far from describing systems of physical interest. However, the models are not without interest, as some of their qualitative features survive in more physically relevant models. In particular, such models are the only cases in which the hypothesized scaling property has been explicitly established.

Apart from the 1D models considered in this chapter, the ordering dynamics of a vector field is also exactly solvable in the limit that the number of vector components of the field, n, tends to infinity. This limit has been studied, mostly for nonconserved fields, by a large number of authors [18,21,22,25-29]. In principle, the solution is the starting point for a systematic treatment in powers of $1/n$. In practice, the calculation of the $O(1/n)$ terms is technically difficult [21,22]. Moreover, some important physics is lost in this limit. In particular, there are no topological defects, since clearly $n > d + 1$ for any d as $n \to \infty$.

7.3.1 The Glauber model

There are (at least) two different ways to approach the Glauber dynamics of the $d = 1$ Ising model at $T = 0$. One is simply to use the Glauber equation (7.6) for the spin probability function $P(S_1, \ldots, S_N; t)$ to derive equations for the pair correlation functions. It is a special feature of $d = 1$ that these equations are both *closed* (i.e., involve no higher-order correlation functions) and *linear*. They may therefore be solved very simply. This is the approach we will adopt [17].

The second approach is to focus directly on the motion of the domain walls. First note that the Glauber dynamics is equivalent to the following algorithm: (i) choose a spin at random, (ii) calculate the energy change associated with flipping the spin, (iii) flip the spin with probability $[1 -$ $\tanh(\beta \Delta E/2)]/2$, (iv) repeat from the beginning. For $T = 0$, therefore, a move is accepted with probability 1 if $\Delta E < 0$, probability $1/2$ if $\Delta E = 0$, and probability 0 if $\Delta E > 0$. Only spins next to a domain wall can flip, and these flip with probability $1/2$ (unless the spin is a domain of length 1, when it always flips). This means that the domain walls perform independent random walks, and annihilate when they meet. This process is equivalent to the annihilation reaction $A + A \rightarrow 0$, discussed in Part I, and the techniques used to study that process give the same results as those derived below directly from the Glauber equation [30]. Alternatively, one can use synchronous instead of random sequential dynamics for the domain walls. This gives the same scaling function [31].

Starting from (7.6), it is straightforward to derive the equation of motion for the equal-time pair correlation function, $C_{i,j}(t) = \langle S_i(t)S_j(t) \rangle$, where the brackets indicate an average over the distribution P, i.e., $C_{i,j}(t) = \text{Tr}\{P(S_1, \ldots, S_N; t)S_iS_j\}$. Multiplying (7.6) by S_iS_j and performing the trace gives

$$(d/dt)C_{i,j} = -2C_{i,j} + \langle S_i \tanh \beta h_j \rangle + \langle S_j \tanh \beta h_i \rangle. \qquad (7.17)$$

So far this holds in general dimension d. For $d = 1$, however, the equation simplifies greatly. Using $h_i = J(S_{i-1} + S_{i+1})$ gives, exploiting the fact that the S_i take on only two values,

$$\tanh \beta h_i = \tanh[\beta J(S_{i-1} + S_{i+1})] = (\tanh 2K)[(S_{i-1} + S_{i+1})/2], \quad (7.18)$$

where $K = \beta J$. Inserting this into (7.17) and taking the limit $K \rightarrow \infty$, corresponding to $T = 0$, gives for $i \neq j$

$$dC_{i,j}/dt = -2C_{i,j} + (C_{i,j+1} + C_{i,j-1} + C_{i-1,j} + C_{i+1,j})/2, \quad i \neq j, \quad (7.19)$$

a *linear* equation. For $i = j$ one has simply $C_{ii} = 1$.

These results are valid for general initial conditions. Further simplification is achieved by choosing a translationally invariant ensemble of initial conditions. For example, choosing $P(\{S_i\}; 0) = 2^{-N}$ corresponds to completely random initial conditions. Then $C_{i,j}(t) = \tilde{C}(|i - j|, t)$, and (7.19) becomes, suppressing the tilde,

$$dC(r,t)/dt = -2C(r,t) + C(r+1,t) + C(r-1,t), \quad r \neq 0; \quad C(0,t) = 1. \qquad (7.20)$$

The quickest way to obtain the scaling solution to (7.20) is to take the continuum limit, giving the partial differential equation

$$\frac{\partial C}{\partial t} = \frac{\partial^2 C}{\partial r^2},$$ (7.21)

with constraint $C(0, t) = 1$. A scaling solution obviously requires that the characteristic length scale (e.g., the mean domain length) grows as $L(t) \propto t^{1/2}$. Inserting the scaling form $C(r, t) = f(r/t^{1/2})$ into (7.21) gives $f'' = -(x/2)f'$, which can be integrated with boundary conditions $f(0) = 1$, $f(\infty) = 0$ to give

$$f(x) = \text{erfc}\,(x/2),$$ (7.22)

where erfc is the complementary error function. Imposing the obvious symmetry under $r \to -r$ [i.e., noting that the solution of $f'' = -(x/2)f'$ with boundary conditions $f(0) = 1$, $f(-\infty) = 0$ is $f(x) = \text{erfc}\,(-x/2)$] gives the scaling solution

$$C(r, t) = \text{erfc}\,(|r|/2t^{1/2}).$$ (7.23)

It may easily be demonstrated, from the complete solution of (7.20), that the scaling solution (7.23) is an attractor of the dynamics [17]. (In [17], the final result was left in the form of an integral. This reduces to (7.23) after a change of variable.)

So far we have considered only equal-time correlations. It is straightforward to generalize the results to the two-time correlation function $C_{ij}(t, t') = \langle S_i(t) S_j(t') \rangle$. Taking $t > t'$ without loss of generality one finds, instead of (7.19),

$$(d/dt)C_{ij} = -C_{ij} + (C_{i+1,j} + C_{i-1,j})/2, \qquad t > t'.$$ (7.24)

Imposing again a translationally invariant initial distribution, so that $C_{ij}(t, t') = \tilde{C}(r, t, t')$, gives (suppressing the tilde)

$$\frac{\partial C}{\partial t} = \frac{1}{2}\frac{\partial^2 C}{\partial r^2}, \quad t > t'; \qquad C(r, t', t') = \text{erfc}\left[\frac{|r|}{2(t')^{1/2}}\right],$$ (7.25)

the equal-time result (7.23) serving as an initial condition for the partial differential equation. The solution of (7.25) is, for $t > t'$,

$$C(r, t, t') = \int_{-\infty}^{\infty} \frac{dr'}{\sqrt{2\pi(t - t')}}\, \exp\left[-\frac{(r - r')^2}{2(t - t')}\right]\, \text{erfc}\left(\frac{|r'|}{2\sqrt{t'}}\right).$$ (7.26)

Of special interest is the autocorrelation function $A(t, t') = C(0, t, t')$. Evaluating the integral in (7.26) for this case gives

$$A(t, t') = \frac{2}{\pi} \sin^{-1} \left(\sqrt{\frac{2t'}{t + t'}} \right), \qquad t > t'. \tag{7.27}$$

In the limit $t \gg t'$, one obtains

$$A(t, t') \to \frac{2\sqrt{2}}{\pi} \left(\frac{t'}{t} \right)^{1/2} = \frac{2\sqrt{2}}{\pi} \frac{L(t')}{L(t)}, \tag{7.28}$$

since $L(t) \propto t^{1/2}$. Comparing this result with (7.13) gives $\bar{\lambda} = 1$ for the Glauber model in 1D. In general this exponent is expected to be nontrivial (the following subsection gives an explicit example), although for $d = 2$ the value $\bar{\lambda} = 5/4$ has been conjectured [20].

7.3.2 The TDGL equation

The scalar TDGL equation (7.7) is also solvable in 1D, in the sense that the scaling functions can be exactly calculated [32], as can the exponent $\bar{\lambda}$ that describes the decay of autocorrelations [33]. In the scaling limit, when the mean domain size L is much greater than the wall thickness ξ, neighboring domain walls interact only weakly, through the exponential tails (7.16) of the domain-wall profile function. This leads to an exponentially small force, of order $\exp(-r/\xi)$, between domain walls separated by $r \gg \xi$. Moreover, in the limit $L/\xi \to \infty$, the closest pair of domain walls interact strongly compared to other pairs, so that the other walls can be treated as stationary while the closest pair annihilate. This enables one to derive a simple recursion relation for the domain-size distribution, with a scaling solution. Note that this process is deterministic, in contrast to the stochastic domain-wall dynamics of the Glauber model. After time t, the smallest domain has length of order $\xi \ln t$, so the characteristic length scale grows as $L(t) \sim \ln t$ [24].

We shall now derive the scaling form for the distribution of domain sizes. This can be done using either a continuum description, where the domain lengths are real numbers [24], or a discrete one where the lengths are integers [34]. Here we will use the latter approach. It is straightforward to calculate, at the same time, the autocorrelation exponent $\bar{\lambda}$ [33]. We shall find that it has a nontrivial value.

Consider random intervals on the line, representing the domains. To compute $\bar{\lambda}$, we keep track not only of the lengths of the intervals, but also the overlap q of the spins in each interval with their initial configuration. Each interval I is characterized, therefore, by its length $l(I)$ and by the length of its overlap $q(I)$ with its initial condition (initially $q(I) = l(I)$ for all I). At each iteration step, the smallest interval $I_{\min}$ is removed (i.e., the field ϕ is replaced by $-\phi$ in this interval). So, three intervals (the smallest interval $I_{\min}$ and its two neighbors I_1 and I_2) are replaced by a single interval I. The length and the overlap of the new interval I are given by

$$l(I) = l(I_1) + l(I_{\min}) + l(I_2), \quad q(I) = q(I_1) + q(I_2) - q(I_{\min}). \quad (7.29)$$

Then the average length L of domains and the autocorrelation function A are given by $L = \sum_I l(I) / \sum_I 1$, $A = \sum_I q(I) / \sum_I l(I)$, where the sums are over all the intervals I present in the system.

It is easy to see that no correlations between interval lengths develop if none are present initially [24,34]. We take the lengths of the intervals to be integers and i_0 to be the smallest length in the system. We also assume that the total number N of intervals is very large. Let n_i be the number of intervals of length i and q_i the average overlap of an interval of length i. At the beginning, $q_i = i$. We denote with a prime the values of these quantities after all the n_{i_0} intervals of length i_0 have been eliminated, so that the minimal length has become $i_0 + 1$. Then the time evolution of N, n_i and q_i is given by the recurrence relations

$$N' = N - 2n_{i_0}, \quad n'_i = n_i \left(1 - \frac{2n_{i_0}}{N}\right) + n_{i_0} \sum_{j=i_0}^{i-2i_0} \frac{n_j}{N} \left(\frac{n_{i-j-i_0}}{N}\right),$$

$$n'_i q'_i = n_i q_i \left(1 - \frac{2n_{i_0}}{N}\right) + n_{i_0} \sum_{j=i_0}^{i-2i_0} \frac{n_j}{N} \left(\frac{n_{i-j-i_0}}{N}\right) (q_j + q_{i-j-i_0} - q_{i_0}). \quad (7.30)$$

These are only valid under the condition that $n_{i_0} \ll N$ which is indeed the case when i_0 becomes large, provided the system consists of a large number of intervals.

We assume that after many iterations, i.e. when i_0 becomes large, a scaling limit is reached where

$$n_i = \frac{N}{i_0} f\left(\frac{i}{i_0}\right), \quad n_i q_i = N(i_0)^{\lambda-1} g\left(\frac{i}{i_0}\right), \quad (7.31)$$

where $\lambda = 1 - \bar{\lambda}$. Because i_0 is so large, we can consider $x = i/i_0$ as a continuous variable (with the smallest domains having $x = 1$). This gives

$$n'_i = \frac{N'}{i_0+1} f\left(\frac{i}{i_0+1}\right) = \frac{N}{i_0}\left[f(x) - \frac{2}{i_0}f(1)f(x) - \frac{1}{i_0}f(x) - \frac{1}{i_0}xf'(x)\right],$$

$$n_i'q_i' = N'(i_0 + 1)^{\lambda-1} g\left(\frac{i}{i_0 + 1}\right)$$

$$= Ni_0^{\lambda-1}\left[g(x) - \frac{2}{i_0}f(1)g(x) + \frac{\lambda-1}{i_0}g(x) - \frac{1}{i_0}xg'(x)\right]. \quad (7.32)$$

Inserting these expressions into the time evolution equations (7.30) gives

$$i_0\frac{\partial f}{\partial i_0} = f(x) + xf'(x) + \theta(x - 3)f(1)\int_1^{x-2} dy \; f(y)f(x - y - 1), \quad (7.33)$$

$$i_0\frac{\partial g}{\partial i_0} = (1 - \lambda)g(x) + xg'(x) + 2\theta(x - 3)f(1)\int_1^{x-2} dy \; g(y)f(x - y - 1)$$

$$-g(1)\theta(x - 3)\int_1^{x-2} dy \; f(y)f(x - y - 1). \quad (7.34)$$

In (7.31), both n_i and n_iq_i are functions of $x = i/i_0$ and of i_0, and the partial derivatives in (7.33)-(7.34) mean the derivative with respect to i_0, keeping x fixed. Demanding that the system is self-similar, i.e., that the functions $f(x)$ and $g(x)$ do not change with time (and so replacing the l.h.s. of each of (7.33), (7.34) by zero), one finds that the Laplace transforms

$$\phi(p) = \int_1^\infty e^{-px} f(x) \; dx, \qquad \psi(p) = \int_1^\infty e^{-px} g(x) \; dx, \quad (7.35)$$

satisfy the following equations (where primes now indicate derivatives)

$$- f(1)e^{-p} - p\phi'(p) + f(1)e^{-p}\phi^2(p) = 0, \quad (7.36)$$

$$- \lambda\psi(p) - g(1)e^{-p} - p\psi'(p) + 2f(1)e^{-p}\phi(p)\psi(p) - g(1)e^{-p}\phi^2(p) = 0. \quad (7.37)$$

Defining the function $h(p)$ by

$$h(p) = 2f(1)\int_p^\infty (e^{-t}/t)dt, \quad (7.38)$$

the solutions of the above equations are

$$\phi(p) = \tanh[h(p)/2], \quad (7.39)$$

$$\psi(p) = g(1)\int_p^\infty \left[1 + \phi^2(q)\right]\frac{1 - \phi^2(p)}{1 - \phi^2(q)}\left(\frac{q^{\lambda-1}}{p^\lambda}\right) e^{-q}dq. \quad (7.40)$$

The constants of integration implied by these forms were fixed by the requirement that both ϕ and ψ sufficiently decay fast for large p, as is clear from the definitions (7.35). So far the parameters $f(1)$, $g(1)$, and λ are arbitrary. We shall see that they are fixed by physical considerations.

The inverse Laplace transform of $\phi(p)$ gives the domain-size distribution $f(x)$ [24,32,34]. It is instructive to consider the small-p behavior of $\phi(p)$. To this end we introduce the expansion

$$\int_p^\infty \frac{e^{-q}}{q} dq = -\ln p - \gamma - \sum_{n=1}^\infty \frac{(-p)^n}{n\,n!}, \qquad (7.41)$$

where $\gamma = -\int_0^\infty dt\, e^{-t} \ln t = 0.5772156\ldots$ is Euler's constant. Using this in (7.38), we see that $h(p)$ diverges for $p \to 0$. Using the large-argument expansion $\tanh x = 1 - 2\exp(-2x) + \ldots$ in (7.39) gives

$$\phi(p) = 1 - 2e^{2f(1)\gamma} p^{2f(1)} + \cdots \qquad (7.42)$$

for small p. Compare this with the small-p expansion of $\phi(p)$ from (7.35):

$$\phi(p) = \int_1^\infty dx\, f(x)[1 - px + O(p^2)] = 1 - p\langle x \rangle + O(p^2), \qquad (7.43)$$

where $\langle x \rangle$ is the mean domain size (in units of the minimum domain size). Comparing (7.42) and (7.43) shows that if $\langle x \rangle < \infty$ then $f(1) = 1/2$. This is the generic case. More generally, the obvious condition $\langle x \rangle > 0$ implies $f(1) \le 1/2$, but the cases $f(1) < 1/2$ correspond to distributions with an infinite first moment, and one can show that this cannot happen if the initial distribution has a finite first moment. From now on, therefore, we consider $f(1) = 1/2$ only. Then (7.42) and (7.43) give the first moment as $\langle x \rangle = 2e^\gamma = 3.56214\ldots$.

The full scaling distribution $f(x)$ cannot be written down in closed form. By expanding the hyperbolic tangent function in (7.39), and performing the inverse Laplace transform term by term, we obtain [24]

$$f(x) = \sum_{n=0}^{(x-1)/2} A_{2n+1} \int_1^\infty \prod_{m=1}^{2n+1} \frac{dt_m}{t_m} \delta\left(x - \sum_{m=1}^{2n+1} t_m\right), \qquad (7.44)$$

where the first sum is over all n such that $x \ge 2n + 1$, and $2^n A_n$ are the Taylor expansion coefficients of $\tanh(x)$. The distribution is piecewise analytic, with discontinuities at $x = 2n + 1$ in the $2n$th derivative for all $n > 0$ [24]. The distribution is simple for small x,

$$f(x) = \begin{cases} 0, & x \in [0,1), \\ \frac{1}{2}x^{-1}, & x \in [1,3], \end{cases} \qquad (7.45)$$

but gets more complicated in every successive interval of length 2.

The structure-factor scaling function $g(q)$, (7.11), is easily computed [35], by exploiting the fact that the interval lengths are uncorrelated. We start from the expression for the derivative of the order parameter,

$$\phi'(x) = 2\sum_n (-1)^n \delta(x - x_n), \tag{7.46}$$

where the x_n are the locations of the walls, and we have approximated the wall profile by a step function in the limit where the wall width is negligible compared with the mean domain size. Fourier transforming gives

$$ik\phi_k = (2/\sqrt{L_s})\sum_n (-1)^n \exp(-ikx_n), \tag{7.47}$$

where L_s is the system size. This gives the structure factor

$$S(k) = \frac{4}{L_s k^2}\sum_{n,m}(-1)^{n+m}\langle\exp[ik(x_m - x_n)]\rangle. \tag{7.48}$$

Since the intervals lengths are independent we have

$$\langle\exp[ik(x_m - x_n)]\rangle = \begin{cases} \phi(-ik)^{m-n}, & m \geq n, \\ \phi(ik)^{n-m}, & m \leq n. \end{cases} \tag{7.49}$$

Putting it all together, and evaluating the sums over n and $r \equiv n - m$ gives

$$S(k) = \frac{4}{Lk^2}\frac{1 - |\tanh[h(ik)/2]|^2}{|1 + \tanh[h(ik)/2]|^2}, \tag{7.50}$$

where $L = 2e^\gamma$ is the mean domain size, and we have used the minimum domain size as the unit of length.

We turn now to the parameter λ, related via $\lambda = 1 - \bar{\lambda}$ to the exponent $\bar{\lambda}$ that describes the decay of autocorrelations. Equation (7.40) for ψ determines λ. We first write it in the more convenient form

$$\psi(p) = 2g(1)\int_p^\infty \frac{e^{h(q)} + e^{-h(q)}}{e^{h(p)} + 2 + e^{-h(p)}}\left(\frac{q^{\lambda-1}}{p^\lambda}\right)e^{-q}dq. \tag{7.51}$$

Defining the function $r(p)$ by

$$r(p) = h(p) + \ln p = \int_p^\infty \frac{e^{-q}}{q}dq + \ln p, \tag{7.52}$$

one obtains, using (7.51),

$$\psi(p) = 2g(1)\int_p^\infty \frac{e^{r(q)} + q^2 e^{-r(q)}}{e^{r(p)} + 2p + p^2 e^{-r(p)}}\left(\frac{q^{\lambda-2}}{p^{\lambda-1}}\right)e^{-q}dq. \tag{7.53}$$

Now $r(p)$ can be expanded in powers of p, using (7.41), and so this last form makes it easier to analyze the singular behavior of $\psi(p)$ at $p = 0$. One finds that, for small p,

$$\psi(p) = A + Bp^{1-\lambda} + O(p), \qquad (7.54)$$

where $A = 2g(1)/(1 - \lambda)$ and

$$
\begin{aligned}
B &= 2g(1)e^{-r(0)}\left\{\int_0^\infty \frac{q^{\lambda-1}e^{-q}}{1-\lambda}\left[r'(q) - 1\right]e^{r(q)}dq + \int_0^\infty q^\lambda e^{-q}e^{-r(q)}dq\right\} \\
&= 2g(1)e^\gamma(1-\lambda)^{-1}\int_0^\infty q^{\lambda-2}e^{-q}\left[(1 - q - e^{-q})e^{r(q)}\right. \\
&\qquad\qquad \left. + q^2(1 - \lambda)e^{-r(q)}\right]dq. \qquad (7.55)
\end{aligned}
$$

Now compare (7.54) with a direct expansion of (7.35), namely $\psi(p) = \int_1^\infty dx g(x)[1 - px + O(p^2)]$. If the function $g(x)$ is to have a finite first moment then we must have $B = 0$ in (7.54). This condition determines λ as $0.399\,383\,5\ldots$. This result is in good agreement with the simulation data of Majumdar and Huse, when corrections to scaling are taken into account [9].

As in the discussion of the result $f(1) = 1/2$, one can show that $B \neq 0$ would correspond to a power-law decay in $g(x)$ and that such a power law cannot be produced if it is not present in the initial condition. Note that $g(1)$ cannot be determined as we can always multiply all the q_i by a constant without changing our results.

Recently, a new nontrivial exponent associated with ordering kinetics has been introduced [34,36]. For the stochastic Glauber dynamics, one studies the fraction of spins $f(t)$ that have never been flipped up to time t. In 1D, this fraction is found to decay as $f(t) \sim t^{-\theta}$, with $\theta \sim 0.37$ [34]. Very recently, in a remarkable paper, Derrida et al. [37] have shown that $\theta = 3/8$ exactly. Furthermore, they have generalized the result to the q-state Potts model, obtaining

$$\theta(q) = -\frac{1}{8} + \frac{2}{\pi^2}\left[\cos^{-1}\left(\frac{2-q}{\sqrt{2}q}\right)\right]^2. \qquad (7.56)$$

For the deterministic TDGL equation, θ can also be calculated exactly [34]. It is convenient, however, to define θ using the average domain length L, rather than time, as the independent variable. For Glauber dynamics, one has $L \propto t^{1/2}$, so we define $f(t) \sim L^{-2\theta}$. For TDGL dynamics, $f(t)$ is the fraction of the line that has never been traversed by a wall, which we will call the 'dry' part (imagining that the walls 'wet' the line as they move). If $d(I)$ is the dry length of interval I, then the recurrence relation for $d(I)$, analogous to (7.29) for $q(I)$, is simply $d'(I) = d(I_1) + d(I_2)$, because the smallest interval I_{min} becomes completely wet when its surrounding domain walls are eliminated. Proceeding as in the derivation of $\bar{\lambda}$, one finds $2\theta = 0.1750758\ldots$. In contrast with the stochastic model, however, the generalization to Potts symmetry is not straightforward, since the interval lengths are no longer independent random variables [38].

7.3.3 The one-dimensional XY model

As our final example of a solvable model, we consider the $O(2)$ model with nonconserved order parameter. The solution, first given by Newman *et al.* [10], is interesting for the 'anomalous' growth law obtained, $L(t) \sim t^{1/4}$. Here we shall give a more detailed discussion than appears in [10], emphasizing the scaling violations exhibited by, in particular, the two-time correlation function. Recently Bray and Rutenberg [39] have presented a general technique for determining growth laws in phase-ordering systems. The scaling form (7.12) plays an important role in the derivation. For the 1D $O(2)$ model, however, the method fails to predict the correct $t^{1/4}$ growth. The reason is precisely the unconventional form (i.e., different from (7.12)) of the two-time correlation function for this system.

It is simplest to work with 'fixed length' fields, i.e., $\phi^2 = 1$, with Hamiltonian $F = \frac{1}{2} \int dx (\partial\phi/\partial x)^2$. The constraint can be eliminated by the representation $\phi = (\cos\theta, \sin\theta)$, where θ is the phase angle, to give $F = \frac{1}{2} \int dx (\partial\theta/\partial x)^2$. The 'model A' equation of motion, $\partial\theta/\partial t = -\delta F/\delta\theta$, becomes

$$\frac{\partial\theta}{\partial t} = \frac{\partial^2\theta}{\partial x^2}, \tag{7.57}$$

i.e., a simple diffusion equation for the phase. In general dimensions, it is difficult to include vortices, which are singularities in the phase field, in a simple way. Such singularities, however, are absent for $d = 1$.

Equation (7.57) has to be supplemented by suitable initial conditions. It is convenient to choose the probability distribution for $\theta(r, 0)$ to be Gaussian: in Fourier space

$$P(\{\theta_k(0)\}) \propto \exp\left[-\frac{1}{2}\sum_k \beta_k \theta_k(0)\theta_{-k}(0)\right]. \tag{7.58}$$

Then the real-space correlation function at $t = 0$ is readily evaluated using the Gaussian property of the $\{\theta_k(0)\}$:

$$C(r,0) = \langle\cos[\theta(r,0) - \theta(0,0)]\rangle = \exp\{-\tfrac{1}{2}\langle[\theta(r,0) - \theta(0,0)]^2\rangle\}$$
$$= \exp\{-\sum_k (1 - \cos kr)/\beta_k\}. \tag{7.59}$$

The choice $\beta_k = (\xi/2)k^2$ yields $C(r,0) = \exp(-|r|/\xi)$, appropriate to a quench from a disordered state with correlation length ξ.

The general two-time correlation function can be calculated in the same way [4,40]. Using $\beta_k = (\xi/2)k^2$, and the solution $\theta_k(t) = \theta_k(0)\exp(-k^2 t)$ of (7.57), gives

$$C(r,t_1,t_2) = \langle\cos[\theta(r,t_1) - \theta(0,t_2)]\rangle = \exp\{-\tfrac{1}{2}\langle[\theta(r,t_1) - \theta(0,t_2)]^2\rangle\}$$
$$= \exp\left(-\sum_k \frac{1}{\xi k^2}\{[\exp(-k^2 t_1) - \exp(-k^2 t_2)]^2\right.$$
$$\left.+2(1 - \cos kr)\exp[-k^2(t_1 + t_2)]\}\right). \tag{7.60}$$

Since we shall find that r is scaled by $(t_1 + t_2)^{1/4}$, we can take $kr \ll 1$ in the summand for the r values of interest, i.e., we can replace $1 - \cos kr$ by $(kr)^2/2$. Evaluation of the sums then gives

$$C(r,t_1,t_2) = \exp\left\{-\frac{1}{\xi\sqrt{\pi}}\left[\frac{r^2}{2(t_1 + t_2)^{1/2}} + 2(t_1 + t_2)^{1/2}\right.\right.$$
$$\left.\left. - (2t_1)^{1/2} - (2t_2)^{1/2}\right]\right\}. \tag{7.61}$$

For the special case $t_1 = t_2 = t$, (7.61) reduces to

$$C(r,t,t) = \exp\{-r^2/[2\xi(2\pi t)^{1/2}]\}, \tag{7.62}$$

which has the standard scaling form (7.11), with growth law $L(t) \sim t^{1/4}$. This growth law is unusual: the generic form for nonconserved fields is $L(t) \sim t^{1/2}$ [39]. Another, related, feature of (7.61) is the explicit appearance of ξ, the correlation length for the initial condition. The most striking feature of (7.61), however, is the breakdown of the scaling form (7.12) for the two-time correlations. It is this feature that invalidates the derivation of the result $L(t) \sim t^{1/2}$ given in [39]. A similar anomalous scaling is present in the *conserved* $d = 1$ XY model, for which simulation results [40-42] suggest $L(t) \sim t^{1/6}$, instead of the $t^{1/4}$ growth derived assuming simple scaling for two-time correlations [39]. Unfortunately, an exact solution for the conserved case is nontrivial.

The explicit dependence of (7.61) on ξ suggests an unusual sensitivity to the initial conditions in this system. A striking manifestation of this is obtained by choosing initial conditions with a nonexponential decay of correlations. For example, choosing $\beta_k \sim |k|^\alpha$ for small $|k|$ in (7.58) gives, via (7.59), $C(r,0) \sim \exp(-\text{constant} \times |r|^{\alpha-1})$ for large $|r|$, provided $1 < \alpha < 3$. The calculation of the pair correlation function is again straightforward. For example, the equal-time function has a scaling form given by $C(r,t) = \exp[-\text{constant} \times r^2/t^{(3-\alpha)/2}]$, implying a growth law $L(t) \sim t^{(3-\alpha)/4}$, but the two-time correlation still does not scale properly.

An especially interesting case is $\alpha = 1$, which generates power-law spatial correlations in the initial condition. Thus we choose $\beta_k = |k|/\gamma$ for $|k| \le \Lambda$, and $\beta_k = \infty$ for $|k| > \Lambda$, where Λ is an ultraviolet cutoff. Then the initial-condition correlator has the form $C(r,0) \sim (\Lambda r)^{-\gamma/\pi}$ for $\Lambda r \gg 1$. The general two-time correlation function now has the conventional scaling form (7.12), with $L(t) \sim t^{1/2}$. Its form is

$$C(r,t_1,t_2) = f\left(\frac{r}{(t_1+t_2)^{1/2}}\right)\left(\frac{4t_1 t_2}{(t_1+t_2)^2}\right)^{\gamma/4\pi}, \qquad (7.63)$$

where $f(x)$ is the equal-time correlation function. In particular, for $t_2 \gg t_1$ this gives $C(r,t_1,t_2) \simeq f(r/\sqrt{t_2})(4t_1/t_2)^{\gamma/4\pi}$, so the exponent $\bar{\lambda}$ defined by (7.13) is $\gamma/2\pi$ for this model. Also, the large-distance behavior of the equal-time correlation function is $f(x) \sim x^{-\gamma/\pi}$, exhibiting the same power-law decay as the initial condition. These two results are in complete agreement with a general treatment [43] of initial conditions with power-law correlations.

A simple interpretation of the unusual scaling properties of this system comes with the realization that there are in general two characteristic length scales [11]. This is appreciated most simply by introducing the phase gradient, $y = \theta'$. The integral of $y/2\pi$ across the system just gives the total winding number of the phase, which, for periodic boundary conditions, is a conserved quantity. Thus $y/2\pi$ is the local density of a conserved topological charge. Obviously y obeys the same diffusion equation, $\dot{y} = y''$, as the phase angle θ. So one length scale is the 'phase coherence length' $L = t^{1/2}$, which is the typical length scale over which y has a given sign or, equivalently, the typical distance between the zeros of y. This means that the phase winds in a particular sense for a distance of order L, then winds in the opposite direction. The second length scale, the 'winding length' L_w, is the typical distance over which the phase changes by 2π. This plays the dominant role in the order-parameter correlation function C, being the correlation length for $\exp(i\theta)$. Thus $y \sim 1/L_w$. We can therefore estimate L_w from

$$\frac{1}{L_w^2} \sim \langle(\theta')^2\rangle = \int_{-\infty}^{\infty} \frac{dk}{2\pi} \frac{2}{\xi} \exp(-2k^2 t) = \frac{1}{\xi\sqrt{2\pi t}}, \qquad (7.64)$$

i.e., $L_w \sim \sqrt{\xi L}$. On the other hand, derivatives of y are down by a factor $1/L$ per derivative, e.g., $\langle(y')^2\rangle = \langle(\theta'')^2\rangle \sim 1/(\xi L^3)$, i.e., $y' \sim 1/(LL_w)$, etc. This means that the pair correlation $C(r,t) = \langle\exp i[\theta(x+r,t) - \theta(x,t)]\rangle$ is dominated by the first term in the Taylor series expansion of the argument of the exponential, $C(r,t) = \langle\exp[iry(x,t)]\rangle = \exp(-r^2\langle y^2\rangle/2)$, reproducing the previous result. While this approach gives no new results for the non-conserved model, the new insights it provides may be of value in a wider context. For example, the assumption that one has the same separation of length scales in the *conserved* $O(2)$ model leads to the prediction $L_w \sim t^{1/6}$ [11], in good agreement with simulations [40-42].

7.4 Summary

We have discussed three solvable models of coarsening in 1D, each with its own character. The Glauber model and the TDGL model both exhibit conventional scaling, although the underlying physics is very different, as reflected in the different growth laws. In higher dimensions, where the interface curvature drives the growth, one would naively expect these models to belong to the same universality class, although this is far from being established in the general case. Indeed it is possible that lattice models exhibit different behavior above and below the onset temperature for interfacial roughening.

The $O(2)$ model in 1D has a special character, owing to the appearance of two characteristic scales. These lead to unconventional scaling behavior for correlations of the order parameter.

Systems with conserved dynamics present a greater challenge to exact solution, even in 1D, although in some cases their behavior is qualitatively understood [8,9]. One exactly solved model is the Ising model with Kawasaki dynamics at $T = 0$ [44]. In this case, however, the coarsening stops, owing to an activation barrier to spin diffusion. At infinitesimal temperature, heuristic arguments [8] (supported by numerical simulations [8,9]) lead to a $t^{1/3}$ growth as long as the mean domain size is much smaller than the equilibrium correlation length.

Special thanks are due to Claude Godrèche, Bernard Derrida, and Andrew Rutenberg for their contributions to the ideas and results presented in this chapter.

References

[1] I. M. Lifshitz, *Zh. Eksp. Teor. Fiz.* **42**, 1354 (1962); *Sov. Fiz. JETP* **15**, 939 (1962).

[2] I. M. Lifshitz and V. V. Slyozov, *J. Phys. Chem. Solids* **19**, 35 (1961).

[3] C. Wagner, *Z. Elektrochem.* **65**, 581 (1961).

[4] A. J. Bray, *Adv. Phys.* **43**, 357 (1994) and references therein.

[5] A. J. Bray, *Phys. Rev. Lett.* **62**, 2841 (1989); *Phys. Rev.* **B41**, 6724 (1990).

[6] R. Glauber, *J. Math. Phys.* **4**, 294 (1963).

[7] K. Kawasaki, in *Phase Transitions and Critical Phenomena*, Vol. **2**, C. Domb and M. S. Green, eds. (Academic Press, London, 1972).

[8] S. J. Cornell, K. Kaski and R. B. Stinchcombe, *Phys. Rev.* **B44**, 12263 (1991).

[9] S. N. Majumdar and D. A. Huse, *Phys. Rev.* **E52**, 270 (1995).

[10] T. J. Newman, A. J. Bray and M. A. Moore, *Phys. Rev.* **B42**, 4514 (1990).

[11] A. D. Rutenberg and A. J. Bray, *Phys. Rev. Lett.* **74**, 3836 (1995).

[12] J. S. Langer, in *Solids Far from Equilibrium*, C. Godrèche, ed. (Cambridge University Press, 1991).

[13] P. C. Hohenberg and B. I. Halperin, *Rev. Mod. Phys.* **49**, 435 (1977).

[14] K. Binder and D. Stauffer, *Phys. Rev. Lett.* **33**, 1006 (1974).

[15] J. Marro, J. L. Lebowitz and M. H. Kalos, *Phys. Rev. Lett.* **43**, 282 (1979).

[16] H. Furukawa, *Prog. Theor. Phys.* **59**, 1072 (1978); H. Furukawa, *Phys. Rev. Lett.* **43**, 136 (1979).

[17] A. J. Bray, *J. Phys.* **A22**, L67 (1990).

[18] A. Coniglio and M. Zannetti, *Europhys. Lett.* **10**, 575 (1989).

[19] H. Furukawa, *J. Phys. Soc. Jpn* **58**, 216 (1989); *Phys. Rev.* **B40**, 2341 (1989).

[20] D. S. Fisher and D. A. Huse, *Phys. Rev.* **B38**, 373 (1988). Note that λ in this paper corresponds to our exponent $\bar{\lambda}$.

[21] T. J. Newman and A. J. Bray, *J. Phys.* **A23**, 4491 (1990).

[22] J. G. Kissner and A. J. Bray, *J. Phys.* **A26**, 1571 (1993). Note that this paper corrects an error in reference [21].

[23] N. Mason, A. N. Pargellis and B. Yurke, *Phys. Rev. Lett.* **70**, 190 (1993).

[24] A. D. Rutenberg and A. J. Bray, *Phys. Rev.* **E50**, 1900 (1994).

[25] G. F. Mazenko and M. Zannetti, *Phys. Rev. Lett.* **53**, 2106 (1984); *Phys. Rev.* **B32**, 4565 (1985).

[26] M. Zannetti and G. F. Mazenko, *Phys. Rev.* **B35**, 5043 (1987).

[27] F. de Pasquale, in *Nonequilibrium Cooperative Phenomena in Physics and Related Topics*, p. 529, M. G. Velarde, ed. (New York, Plenum, 1984).

[28] F. de Pasquale, D. Feinberg and P. Tartaglia, *Phys. Rev.* **B36**, 2220 (1987).

[29] A. Coniglio, P. Ruggiero and M. Zannetti, *Phys. Rev.* **E50**, 1046 (1994). This paper contains a rather complete discussion of growth kinetics in the large-n limit, for both nonconserved and conserved dynamics.

[30] J. G. Amar and F. Family, *Phys. Rev.* **A41**, 3258 (1990).

[31] V. Privman, *J. Stat. Phys.* **69**, 629 (1992); *Phys. Rev.* **E50**, 50 (1994).

[32] T. Nagai and K. Kawasaki, *Physica* **A134**, 483 (1986).

[33] A. J. Bray and B. Derrida, *Phys. Rev.* **E51**, R1633 (1995).

[34] A. J. Bray, B. Derrida and C. Godrèche, *Europhys. Lett.* **27**, 175 (1994).

[35] K. Kawasaki, A. Ogawa and T. Nagai, *Physica* **B149**, 97 (1988).

[36] B. Derrida, A. J. Bray and C. Godrèche, *J. Phys.* **A27**, L357 (1994).

[37] B. Derrida, V. Hakim and V. Pasquier, *Phys. Rev. Lett.* **75**, 751 (1995).

[38] B. Derrida, C. Godrèche and I. Yekutieli, *Phys. Rev.* **A44**, 6241 (1991); I. Yekutieli, C. Godrèche and B. Derrida, *Physica* **A185**, 240 (1992).

[39] A. J. Bray and A. D. Rutenberg, *Phys. Rev.* **E49**, R27 (1994); A. D. Rutenberg and A. J. Bray, *Phys. Rev.* **E51**, 5499 (1995).

[40] A. D. Rutenberg, A. J. Bray and M. Kay, unpublished.

[41] M. Mondello and N. Goldenfeld, *Phys. Rev.* **E47**, 2384 (1993).

[42] M. Rao and A. Chakrabarti, *Phys. Rev.* **E49**, 3727 (1994).

[43] A. J. Bray, K. Humayun and T. J. Newman, *Phys. Rev.* **B43**, 3699 (1991).

[44] V. Privman, *Phys. Rev. Lett.* **69**, 3686 (1992).

8

Phase separation, cluster growth, and reaction kinetics in models with synchronous dynamics

Vladimir Privman

An exact solution of a lattice spin model of ordering in one dimension is reviewed in this chapter. The model dynamics is synchronous, cellular-automaton-like, and involves interface diffusion and pairwise annihilation as well as spin flips due to an external field that favors one of the phases. At phase coexistence, structure-factor scaling applies, and the scaling function is obtained exactly. For field-driven, off-coexistence ordering, the scaling description breaks down for large enough times. The order parameter and the spin-spin correlation function are derived analytically, and several temporal and spatial scales associated with them analyzed.

8.1 Introduction

Phase separation, nucleation, ordering, and cluster growth are interrelated topics of great practical importance [1-4]. One-dimensional (1D) phase separation, for which exact results can be derived [5,6], is discussed in this chapter. The emphasis will be on dynamical rules that involve *simultaneous* updating of the 1D-lattice 'spin' variables. Such models allow a particularly transparent formulation in terms of equations of motion the linearity of which yields exact solvability.

The results are also related to certain reaction-diffusion models of annihilating particles (see Part I of this book), and to deposition-with-relaxation processes (Part IV). Some of these connections [7] will be reviewed here as well. While certain reaction and deposition processes have experimental realizations in 1D (see Part VII), 1D models of nucleation and cluster growth have been explored mainly as test cases for modern scaling theories of, for instance, structure-factor scaling, which will be reviewed in detail.

The processes of ordering by the formation of ordered domains, e.g., by nucleation, and by their growth due to order-favoring interactions, e.g., ferromagnetic spin-spin coupling and/or due to an applied ordering field, are inherently dynamical, nonequilibrium phenomena. The recent resurgence of interest and research activity in lattice models of ordering has been due to several new emphases. Firstly, the interest in low-dimensional models has been driven by consideration of the systems with non-mean-field fluctuations that are usually obtained in low enough dimensions, d.

Secondly, the use of lattice models, as opposed to the more traditional continuum order-parameter evolution equations, allows a more natural combination of numerical and analytical (in low d) results to be collected and compared. Thirdly, our understanding of scaling theories in general has advanced significantly, mostly owing to developments in equilibrium statistical mechanics. Modern scaling concepts have been applied to nonequilibrium dynamics, although advancement in the nonequilibrium case has been much slower and more difficult than in the equilibrium case.

Finally, as already mentioned, consideration of low-dimensional models has allowed the derivation of new exact results and also provided new insights into relations to other models of nonequilibrium dynamical processes such as reactions and deposition.

The outline of this chapter is as follows. The modeling of ordering in 1D is described in Sec. 8.2. Results for the order parameter are derived in Sec. 8.3. Correlation function scaling, its breakdown for field-driven ordering, etc., are described in parallel to the derivation of exact results for two-point correlations and for certain associated time and length scales, in Secs. 8.4-8.6. Section 8.6 also contains a summarizing discussion.

8.2 Model of the ordering processes

In this section we define the model to be studied [6]. We consider spin variables $\sigma_j(t) = \pm 1$, on a 1D lattice of integer spacing, for discrete times $t = 0, 1, 2, \ldots$. Generally, for cellular automata the value of $\sigma_j(t + 1)$ is determined by the $\sigma_k(t)$ within a certain neighborhood of the site j. Here we consider stochastic evolution rules, with a random decision involved in determining the value of $\sigma_j(t+1)$. The simplest model with a *local* ordering tendency is defined by the rule that if the nearest-neighbor 'parent' sites have spins $\sigma_{j\pm1}(t)$ that are both $+1$ (or both -1), then $\sigma_j(t + 1) = +1$ (or -1, respectively). If the parents are 'undecided', i.e., they have opposite signs, then the value $\sigma_j(t + 1)$ is set to $+1$ or -1 randomly.

An alternative statement of this rule is the 'voter model' formulation [8-10]: each spin takes on randomly one of the neighbor values at the previous time step. We note that this dynamics decouples the even-odd and odd-even space-time sublattices. Specifically, we can then limit our attention to spins at even sites $j = 0, \pm 2, \ldots$ for even times $t = 0, 2, \ldots$, and at odd sites $j = \pm 1, \pm 3, \ldots$ for odd times, $t = 1, 3, \ldots$.

Let us introduce random variables $\zeta_j(t)$ for each time $t > 0$ and location j. They take on values 0 or 1 with equal probabilities, and they are uncorrelated in any way for different times and lattice sites. The dynamical rule just introduced can be written as

$$\sigma_j(t+1) = \zeta_j(t+1)\sigma_{j-1}(t) + [1 - \zeta_j(t+1)]\sigma_{j+1}(t), \qquad (8.1)$$

which applies for $t = 0, 1, \ldots$. The definition of the initial state, $\sigma_i(0)$, will be addressed later on.

The exact solvability (see the next section) of this and other cellular-automaton dynamical rules studied here is directly related to the *linearity* of the r.h.s. of (8.1) in $\sigma_i(t)$. In this respect the present formulation is somewhat more transparent than, for instance, the Glauber-type dynamical rules for continuum-time evolution of Ising models (see Part II) [5,11-13].

Let us discuss in some detail the 'ordering' property of this rule and its relation to the $T \to 0$ limit of the Ising-model dynamics [5,13]. We note that fully ordered domains of $+1$ or -1 spins remain ordered in their interior; see Fig. 8.1. In 1D, interfaces are simply points of contact of opposite-sign spins. In our model, we can formally consider interfaces as located at odd lattice sites at even times and at even lattice sites at odd times, i.e., on the sublattice of opposite parity to that of the spins. An isolated interface has equal probability of hopping to the left or to the right in each time step $t \to t + 1$; see Fig. 8.1. Models with anisotropic hopping can be also considered and solved exactly [14-15].

In the Ising model of spins with ferromagnetic coupling, which favors spin alignment, the fact that ordered domains remain fully ordered internally means that energy-increasing spin flips (which would each create two nearby interfaces) are not allowed; this corresponds to a certain zero-temperature limiting procedure [13]. The hopping of isolated interfaces conserves energy. However, if two neighboring interfaces attempt to 'hop' towards each other they annihilate and the energy is reduced, which is, again, a dynamical move favored in the $T \to 0$ limit.

Interface annihilation occurs when an isolated $+1$ spin flips to align itself with its -1 neighbors, and similarly for an isolated -1 spin. The latter process is illustrated for our model in Fig. 8.2. Thus, this order-

V. Privman

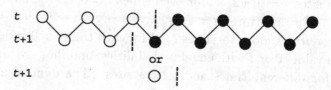

Fig. 8.1. The interiors of fully ordered domains of +1 (full circles) and −1 (open circles) spins at time t remain fully ordered at time $t + 1$. An isolated interface (broken lines) can hop to the left or to the right, with equal probabilities.

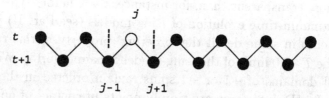

Fig. 8.2. When a −1 spin (open circle) at site j is surrounded by +1 spins (full circles) at $j \pm 2$, there are two interfaces present (broken lines), at $j \pm 1$. The process of interface annihilation corresponds to both sites at $j \pm 1$ assuming values +1. Note that if the interfaces were isolated (cf. Fig. 8.1), these random choices would correspond to both interfaces hopping to j. Annihilation occurs with probability 1/4. Other equally probable processes correspond to interface hopping in different directions (and one or both spins at $j \pm 1$ assuming value −1 at time $t + 1$) and do not result in local annihilation.

ing model also describes the 1D reaction-diffusion two-particle annihilation model $A + A \rightarrow$ inert, of reactant species A diffusing and annihilating pairwise (Part I). Indeed, $T = 0$ Ising models [5,13], two-particle annihilation reactions [7,8,13,14,16–18], the ordering processes reviewed here, and certain deposition-with-relaxation processes [19–24] are all interrelated.

In addition to the local ordering tendency, we would like to model some sort of global 'ordering' field. In order to preserve the linearity of the dynamical rules, we use a particular ±-symmetry-breaking field. Specifically, we assume that the + phase is the stable, 'favored' phase, and we add a new dynamical move: in each time step the outcome of (8.1) is unchanged if it has the stable-phase value $+1$. However, if the value is -1 then it is flipped (to $+1$) with probability p. The physically interesting p values are small, $p \ll 1$, corresponding to small deviations from the symmetric case.

Let us introduce another random variable, $\theta_k(t)$, for all sites k and times $t > 0$. It takes on the value 0 with probability p, and the value 1 with probability $1 - p$. The random variables $\theta_k(t)$ and $\zeta_k(t)$ are uncorrelated. The dynamical rule (8.1) is replaced by

$$\sigma_j(t+1) = 1 - \theta_j(t+1) + \theta_j(t+1)\Big\{ \zeta_j(t+1)\sigma_{j-1}(t)$$

$$+ [1 - \zeta_j(t+1)]\,\sigma_{j+1}(t)\Big\}. \tag{8.2}$$

It is important to notice that the new dynamics still preserves ordered $+$ domains but it no longer preserves $-$ domains. Indeed, for $p > 0$ two 'parent' -1 spins will yield a -1 'offspring' value with probability $1 - p < 1$. Similarly, the probabilities that the opposite-sign parent spins will produce $+1$ or -1 offspring are no longer equal; the $+1$ value is favored. The original symmetric dynamics (8.1) is obtained for $p = 0$.

8.3 Order parameter and initial conditions

Let overbars denote statistical averages. The averaging is over all the random variables ζ and θ, where $\bar{\zeta} = \frac{1}{2}$ and $\bar{\theta} = 1 - p$. However, we would also like to average over the initial conditions (although one could define the same model with more 'deterministic' initial conditions, see below). Thus we assume that the initial values $\sigma_j(0)$ are random and uncorrelated, with $\overline{\sigma_j(0)} = \mu$. Here $-1 \leq \mu \leq 1$ can be viewed as the initial magnetization (order parameter) of the Ising spins. Each $\sigma_j(0)$ takes on values $+1$ with probability $(1 + \mu)/2$, or -1 with probability $(1 - \mu)/2$.

With this definition the statistical averages are translationally invariant. In particular, by averaging (8.2) we obtain the following equation for the time-dependent (but *not* coordinate-dependent) magnetization $m(t) \equiv \overline{\sigma_j(t)}$:

$$m(t+1) = (1-p)m(t) + p. \tag{8.3}$$

The solution of (8.3), with $m(0) = \mu$, is

$$m(t) = 1 - (1 - \mu)(1 - p)^t. \tag{8.4}$$

The density of interfaces, $\rho(t)$, i.e., the probability that two nearby spins are opposite, is in general given by

$$\rho(t) = \tfrac{1}{2}\left[1 - \overline{\sigma_{j-1}(t)\sigma_{j+1}(t)}\right]. \tag{8.5}$$

Since the initial spin values are uncorrelated, we have

$$\rho(0) = \tfrac{1}{2}\left(1 - \mu^2\right). \tag{8.6}$$

One can show, however, that the initial *placement* of interfaces is uncorrelated only in the case $\mu = 0$; see [18] for details. Thus, for instance, in the language of the two-particle annihilation reaction mentioned in the preceding section, the initial state is uncorrelated only for the case of random initial particle or vacant site placement of the reactants, $\rho(0) = \tfrac{1}{2}$. Other methods of solving the two-particle annihilation reactions, which do not utilize the relation to the 1D zero-T Ising or cluster-growth models, have been developed to treat the case $\rho(0) \neq \tfrac{1}{2}$ for uncorrelated placement of reactants (interfaces); see [15] and references therein.

In cases of temporal or spatial variation that is slow compared to the underlying discrete time steps or lattice structure, respectively, one would like to use the *continuum limit* description. The relation for the magnetization, (8.3), is particularly instructive because it contains only one discrete variable. Indeed, it is natural to attempt the approximation

$$m(t + 1) \simeq m(t) + dm(t)/dt. \tag{8.7}$$

The resulting equation $\dot{m} = -pm + p$ is solved by

$$m(t) = 1 - (1 - \mu)e^{-pt}. \tag{8.8}$$

However, we note that (8.8) *does not* provide an accurate large-t approximation to the exact solution (8.4) for all p. Only in the small-p case is the continuum-limit approximation accurate: in the small-p limit the temporal variation of the order parameter is on a scale much larger than the underlying discrete time step, 1.

Finally, we mention that cellular automata related to ours, but with a somewhat different, more $\pm$-symmetric way of favoring one of the spin orientations, have been considered as models of *directed compact percolation* [25-30]. The initial state, however, is no longer translationally invariant. Rather, the spreading of 'wetness', e.g., the $+$ phase, starting from a single $+$ site or a small cluster in the background of the $-$ phase, is considered [25,29], as well as certain surface effects [26] when the system is no longer an infinite line. Further discussion of these models within the cellular-automata nomenclature can be found, e.g., in [27,28,30].

8.4 Spin-spin correlation function and scaling

The two-point correlation function is defined as follows,

$$G_n(t) = \overline{\sigma_j(t)\sigma_{j+n}(t)}. \tag{8.9}$$

Note that we only consider equal-time correlations here. Some results for certain different-time correlation functions in other related models, notably the kinetic Ising model and reaction-diffusion systems, are available in the literature; for recent work, see, e.g., [23,24]. The correlation function in (8.9) depends only upon the relative coordinate n, owing to translational invariance.

The initial and boundary conditions $G_{n>0}(0) = \mu^2$ and $G_0(t) = 1$ follow from the definition. The product of the spins being averaged in (8.9) at time t, is, according to (8.2), expressible in terms of the spins at the earlier time $t - 1$. The exact solvability relates to the fact that only terms quadratic (or lower-order) in the $\sigma_k(t - 1)$ are encountered. More generally, any m-point equal-time correlation function at time t can be calculated in terms of m-point, $(m - 1)$-point, ... correlations at time $t - 1$.

Limiting our consideration to two-point correlations, we note that the above 'closure' observation is further amplified in terms of *connected* correlations. Thus, we define

$$C_n(t) = G_n(t) - m^2(t), \tag{8.10}$$

where the order parameter $m(t)$ was calculated in the preceding section. It then follows from (8.2) and (8.3) that

$$C_n(t + 1) = \tfrac{1}{4}(1 - p)^2 \left[C_{n-2}(t) + 2C_n(t) + C_{n+2}(t) \right]. \tag{8.11}$$

This relation only involves the connected two-point correlation function. The initial and boundary conditions are also easily obtained,

$$C_{n>0}(0) = 0, \tag{8.12}$$

$$C_0(t) = 1 - m^2(t) = 2(1 - \mu)(1 - p)^t - (1 - \mu)^2(1 - p)^{2t}. \tag{8.13}$$

Relation (8.5) for the density of interfaces can be now rewritten as $\rho(t) = \tfrac{1}{2}\left[C_0(t) - C_2(t)\right]$. Interestingly, the density of interfaces, which in an ordering 1D system measures the inverse of the largest characteristic domain sizes, depends on the *short-distance* behavior of the correlation function.

In theories of nucleation and ordering [1-4], there are several implementations of the *scaling* approach. The most straightforward quantity to consider is the cluster-size distribution and, specifically, its evolution with time.

'Cluster-size scaling' is generally quite well understood [1,3]. In $d > 1$, there is always a certain level of ambiguity in the precise definition of the $+$ phase, $-$ phase, etc., clusters of spins, atoms, ... Indeed, the typical Ising $+$ phase at low but finite T is not all $+$ but rather it has $0 < m < 1$.

For voter models in 1D the clusters are, of course, all $+$ or all $-$. For the symmetric case ($p = 0$), certain asymptotic properties of the ordered-interval size distribution are known analytically [8]. Specifically, the distribution scales, with typical length

$$r_{p=0}(t) = 1/\rho_{p=0}(t) \sim \sqrt{t} \tag{8.14}$$

for large times. The $\sqrt{t}$ result will be derived exactly later on; see (8.23). The (scaled) large-distance decay of the cluster-size distribution in space is exponential [8].

The quantity that is probed directly by light-scattering experiments is the structure factor [1-4], i.e., the Fourier transform of the two-point correlation function. Its wavenumber ($\mathbf{q}$) dependence is scaled according to $r(t)\mathbf{q}$. In terms of our spatial distance n, in 1D, the scaling form is

$$C_n(t) \simeq [r(t)]^\phi S[n/r(t)]. \tag{8.15}$$

The exponent ϕ depends on what type of correlations are being considered and is usually determined on dimensional grounds alone. Specifically, for the two-spin correlation function considered here, $\phi = 0$. We also emphasize that the relation (8.14) for the characteristic time-dependent ordering length scale does not apply for $d > 1$. In fact, in $d > 1$ the inverse of the Ising-type 'broken bond' density $\rho(t)$ can only provide a lower bound on $r^d(t)$. Furthermore, the bound will be useful (i.e., diverging with time) only in zero-T type models, where the ordered domains have no internal structure. However, the 'diffusive' power-law behavior $\sim \sqrt{t}$ does apply quite generally for nonconserved order-parameter $d > 1$ dynamics [1-4].

The structure-factor scaling [1-4], i.e., the Fourier-transformed version of (8.15), is well established theoretically, and has been tested experimentally in $d = 3$ and to a lesser extent, in $d = 2$. Furthermore, for 1D zero-T continuous-time Glauber-Ising dynamics, i.e., nonconserved order-parameter dynamics, the scaling relation (8.15) has been obtained exactly [5]. By universality, the same scaling properties should hold for the symmetric voter model ($p = 0$) as is indeed confirmed by our calculations. However, the scaling relation (8.15) breaks down for the nonsymmetric ($p > 0$) ordering process in 1D. These findings and their implications will be discussed in Sec. 8.6.

It is important to comment that symmetric voter models cannot describe all the rich aspects of various nucleation and ordering mechanisms in general d. Specifically, in has been argued [9,10] that the $d = 2$ voter models mimic late-stage spinodal-decomposition type (i.e., no barriers for formation of small ordered domains) growth. However, the applicability of cluster-size scaling to such growth in $d = 2$ has been questioned [9,31-33]. Moreover, in $d = 3$ the symmetric 'voter' dynamics is no longer order inducing. Rather, the system will probably be 'critical'; see [9,31-33] for details. Our discussion here will be limited to $d = 1$.

8.5 Generating functions, temporal and spatial scales

In order to solve (8.11) formally we introduce the generating functions

$$B_n(v) = \sum_{t=0}^{\infty} v^t C_n(t). \tag{8.16}$$

Indeed, these functions satisfy the difference equation

$$4B_n = v(1-p)^2 (B_{n-2} + 2B_n + B_{n+2}), \tag{8.17}$$

for $n = 2, 4, \ldots$, where we have used (8.11) and (8.12). The boundary condition (8.13) yields

$$B_0(v) = \frac{2(1-\mu)}{1-(1-p)v} - \frac{(1-\mu)^2}{1-(1-p)^2 v}. \tag{8.18}$$

The solution for $n > 0$ is now obtained in the exponential form

$$B_n(v) = B_0(v) \left[\Lambda(v)\right]^{n/2}, \tag{8.19}$$

where Λ is selected as that root of the quadratic characteristic equation which yields solutions regular at $v = 0$. The result is

$$\Lambda = \left[1 - \sqrt{1 - (1-p)^2 v}\right]^2 \Big/ \left[v(1-p)^2\right]. \tag{8.20}$$

These formal expressions are not particularly illuminating. More useful information will be obtained in the continuum limit; see the next section. Here we consider the simplest limiting cases of the generating function for the density of interfaces,

$$\sigma(v) = \sum_{i=0}^{\infty} v^t \rho(t) = \tfrac{1}{2} \left[B_0(v) - B_2(v)\right]. \tag{8.21}$$

A more mathematical discussion can be found in [6,14].

For the symmetric case the result is actually quite simple,

$$\sigma_{p=0}(v) = \frac{1-\mu^2}{v}\left(\frac{1}{\sqrt{1-v}} - 1\right). \tag{8.22}$$

The expression for $\rho(t)$ is then obtained in terms of factorials [6,14]. For large times t,

$$\rho_{p=0}(t) \simeq \frac{1-\mu^2}{\sqrt{\pi t}}. \tag{8.23}$$

It is interesting to note that this power-law expression has no characteristic time scale and that its inverse yields the diffusive spatial scale; cf. (8.14).

The general expression for $\sigma_{p>0}(v)$ is more complicated (it is not given here). Inspection of it reveals two singularities in the complex v-plane, one at $v = (1-p)^{-1}$ and another at $v = (1-p)^{-2}$. The large-time behavior is dominated by the former singularity. The leading large-t expression turns out to be

$$\rho_{p>0}(t) \simeq \frac{2(1-\mu)\sqrt{p}}{1+\sqrt{p}}(1-p)^t. \tag{8.24}$$

However, this behavior only sets in at times $t \gg 1/[-\ln(1-p)]$. For small p values this condition becomes $t \gg 1/p$, and the time scale of the exponential time dependence in (8.24) is also of order $1/p$. A similar time dependence was observed for the order parameter; see (8.4) and (8.8).

Consider now small positive p values. The spatial time scale

$$r(t) = 1/\rho(t) \sim e^{pt}/\sqrt{p}, \tag{8.25}$$

for $t \gg 1/p$, is now exponentially large. As emphasized in [6], 'local' (interface diffusion-annihilation) and 'external-field' $(-1 \to +1$ spin flips at a rate proportional to p per site) ordering mechanisms involve different spatial and temporal scales. These will be further discussed in the next section.

8.6 Continuum limit and scaling

As noted in connection with the temporal variation of the order parameter, the continuum limit is valid only in a limited region of the parameter space. Specifically, we expect it to apply when the time and spatial scales of variation of the quantity at hand are large compared to the underlying 'microscopic' time and length parameters of the model. From our earlier discussion this regime was identified as $p \ll 1$.

To be more specific, let us introduce the dimensional coordinate and time, $x = (\Delta x)n$, $\tau = (\Delta\tau)t$, where Δx and $\Delta\tau$ denote the 'microscopic' lattice spacing and time step. Furthermore, we introduce the parameters

$$D = \frac{(\Delta x)^2}{2(\Delta\tau)} \quad \text{and} \quad \lambda = \frac{p}{\Delta\tau}, \tag{8.26}$$

which correspond to the diffusion constant of an isolated interface and flip rate of a single $-$ spin, in the continuum limit.

The continuum-limiting ($p \ll 1$, $x \gg \Delta x$, $\tau \gg \Delta\tau$) form of (8.11) is easily obtained,

$$\frac{\partial C}{\partial \tau} = -2\lambda C + 2D\frac{\partial^2 C}{\partial x^2}, \tag{8.27}$$

the initial and boundary conditions being replaced by

$$C(x > 0, \tau = 0) = 0, \quad C(x = 0, \tau \geq 0) = 1 - m^2(\tau), \tag{8.28}$$

where the magnetization is $m(\tau) = 1 - (1 - \mu)e^{-\lambda\tau}$. We note also that for $\tau \gg \Delta\tau$, $p \ll 1$ the dimensionless interface density per lattice site is obtained as the derivative,

$$\rho(\tau) = -(\Delta x)\left[\frac{\partial C(x, \tau)}{\partial x}\right]_{x=0}. \tag{8.29}$$

The solution of (8.27) with (8.28) is obtained, for instance, by the Laplace transform method [6]. Here we only state the result,

$$C(x, \tau) = (1 - \mu)e^{-\lambda\tau}\left[e^{x\sqrt{\lambda/(2D)}}\,\text{erfc}\left(\frac{x}{\sqrt{8D\tau}} + \sqrt{\lambda\tau}\right)\right.$$

$$+ e^{-x\sqrt{\lambda/(2D)}}\,\text{erfc}\left(\frac{x}{\sqrt{8D\tau}} - \sqrt{\lambda\tau}\right)$$

$$\left. - (1 - \mu)e^{-\lambda\tau}\,\text{erfc}\left(\frac{x}{\sqrt{8D\tau}}\right)\right], \tag{8.30}$$

where $\text{erfc}(z) = (2/\sqrt{\pi})\int_z^\infty e^{-y^2}\,dy$. Thus, we obtain the exact result for the two-spin correlation function in the continuum limit.

For the purpose of discussion, let us now drop the distinction between dimensionless and dimensional quantities, i.e., between p and λ, x and n, t and τ. It is clear that for $p = 0$ ($\lambda = 0$) the result (8.30) is in the scaling form (8.15) with $\phi = 0$. In fact, the erfc-shaped scaling function is consistent with the results for the Glauber-Ising model [5].

Analysis [6] of the explicit result (8.30) shows that the diffusive *scaling combination* $x^2/(Dt)$ enters also for $p > 0$, in the large-x behavior of the correlation function (for fixed t). Thus, the 'tail' of the correlation function involves length scales of order $\sqrt{Dt}$. However, we have already noticed that the short-distance behavior is related to the length scale $1/\rho(t)$, which becomes exponentially large, of order $e^{pt}/\sqrt{p}$, for large enough times, $t \gg 1/p$. Analysis of certain other 'intermediate' length scales [6], defined via moments of the correlation function (8.30), indicates that they remain finite, of order $1/\sqrt{p}$, for $t \gg 1/p$. However, for short times, $t \ll 1/p$, all length scales behave diffusively.

Thus, structure-factor scaling breaks down for field-driven ordering when field-induced alignment (in the case of spins) becomes the dominant ordering mechanism. The latter regime is reached at times for which individual spins have a probability of order 1 of becoming aligned with the field, i.e., for times defined by $pt \gg 1$, where p is proportional to the rate of the local field-induced ordering processes. Note that the other, added mechanism of ordering due to spin interactions (interface annihilation) is always eventually overwhelmed by field-induced ordering. The latter involves multiple time and length scales and no longer obeys structure-factor scaling.

The extension of these ideas to higher dimensions must be done with caution. Indeed, depending on the definition of the ordered clusters, the field-induced mechanism can produce percolation-type ordering, i.e., one spanning domain that actually covers the whole system. Thus, after the 'scaling' diffusion-driven ordering is overtaken by field-induced nonscaling nucleation, the process may actually reach saturation or near saturation at finite times rather than develop the exponential temporal and *spatial* scales observed in the 1D model. The point is that for times larger than those required to form the 'percolating' infinite cluster, the cluster-size distribution will no longer involve large spatial length scales at all. However, the conclusion that scaling breaks down when field-induced ordering dominates should be valid for higher dimensions as well.

References

[1] Review: J. D. Gunton, M. San Miguel and P. S. Sahni, in *Phase Transitions and Critical Phenomena*, Vol. **8**, p. 267, C. Domb and J. L. Lebowitz, eds. (Academic, London, 1983).

[2] A. Sadiq and K. Binder, *J. Stat. Phys.* **35**, 517 (1984).

[3] Review: P. Meakin, in *Phase Transitions and Critical Phenomena*, Vol. **12**, p. 336, C. Domb and J. L. Lebowitz, eds. (Academic, New York, 1988).

[4] Review: O. G. Mouritsen, in *Kinetics of Ordering and Growth at Surfaces*, p. 1, M. G. Lagally, ed. (Plenum, New York, 1990).

[5] A. J. Bray, *J. Phys.* A**23**, L67 (1990).

[6] V. Privman, *J. Stat. Phys.* **69**, 629 (1992).

[7] Review: V. Privman, *Trends Stat. Phys.* **1**, 89 (1994).

[8] M. Bramson and D. Griffeath, *Ann. Prob.* **8**, 183 (1980).

[9] M. Scheucher and H. Spohn, *J. Stat. Phys.* **53**, 279 (1988).

[10] B. Hede and V. Privman, *J. Stat. Phys.* **65**, 379 (1991).

[11] R. J. Glauber, *J. Math. Phys.* **4**, 294 (1963).

[12] Review: A. J. Bray, in *Phase Transitions in Systems with Competing Energy Scales*, T. Riste and D. Sherrington, eds. (Kluwer Academic, Boston, 1993).

[13] Z. Racz, *Phys. Rev. Lett.* **55**, 1707 (1985).

[14] V. Privman, *J. Stat. Phys.* **72**, 845 (1993).

[15] V. Privman, *Phys. Rev.* **E50**, 50 (1994); V. Privman, A. M. R. Cadilhe and M. L. Glasser, *J. Stat. Phys.* **81**, 881 (1995).

[16] D. C. Torney and H. M. McConnell, *J. Phys. Chem.* **87**, 1941 (1983).

[17] M. Bramson and J. L. Lebowitz, *Phys. Rev. Lett.* **61**, 2397 (1988); *J. Stat. Phys.* **62**, 297 (1991).

[18] J. G. Amar and F. Family, *Phys. Rev.* **A41**, 3258 (1990).

[19] P. Nielaba and V. Privman, *Mod. Phys. Lett.* **B6**, 533 (1992).

[20] V. Privman and P. Nielaba *Europhys. Lett.* **18**, 673 (1992).

[21] V. Privman and M. Barma, *J. Chem. Phys.* **97**, 6714 (1992).

[22] B. Bonnier and J. McCabe, *Europhys. Lett.* **25**, 399 (1994).

[23] G. M. Schütz, *J. Phys.* **A28**, 3405 (1995); *Phys. Rev.* **E53**, 1475 (1996).

[24] M. D. Grynberg, T. J. Newman and R. B. Stinchcombe, *Phys. Rev.* **E50**, 957 (1994); M. D. Grynberg and R. B. Stinchcombe, *Phys. Rev.* **E52**, 6013 (1995), *Phys. Rev. Lett.* **76**, 851 (1996).

[25] J. W. Essam, *J. Phys.* **A22**, 4927 (1989).

[26] R. Bidaux and V. Privman, *J. Phys.* **A24**, L839 (1991).

[27] E. Domany and W. Kinzel, *Phys. Rev. Lett.* **53**, 447 (1984).

[28] W. Kinzel, *Z. Phys.* **B58**, 229 (1985).

[29] J. W. Essam and D. Tanlakishani, in *Disorder in Physical Systems*, p. 67, R. G. Grimmet and D. J. A. Welsh, eds. (Oxford University Press, 1990).

[30] R. Dickman and A. Yu. Tretyakov, *Phys. Rev.* **E52**, 3218 (1995).

[31] P. Clifford and A. Sudbury, *Biometrika* **60**, 581 (1973).

[32] R. Holley and T. M. Liggett, *Ann. Prob.* **3**, 643 (1975).

[33] J. T. Cox and D. Griffeath, *Ann. Prob.* **14**, 347 (1986).

9

Stochastic models of aggregation with injection

Misako Takayasu and Hideki Takayasu

A generalized aggregation model of charged particles that diffuse and coalesce randomly in discrete space-time is studied, numerically and analytically. A statistically invariant steady state is established when randomly charged particles are uniformly and continuously injected. The exact steady-state size distribution obeys a power law whose exponent depends on the type of injection. The stability of the power-law size distribution is proved. The spatial correlations of the system are analyzed by a powerful new method, the interval distribution of a level set, and a scaling relation is obtained.

9.1 Introduction

The study of far-from-equilibrium systems has attracted much attention in the last two decades. Though many macroscopic phenomena in nature, such as turbulence, lightning, earthquakes, fracture, erosion, the formation of clouds, aerosols, and interstellar dusts, are typical far-from-equilibrium problems, no unified view has yet been established. The substantial difficulties in studying such systems are the following. First, far-from-equilibrium systems satisfy neither detailed balance nor, at the macroscopic level, the equipartition principle. Second, the system is usually open to an outside source. A common method to describe such systems is by abstracting the macroscopic essential features of the observed system and constructing a model in macroscopic terms irrespective of the microscopic (molecular) dynamics. In other words, we make a far-from-equilibrium model by assuming appropriate irreversible rules for the macroscopic dynamics. One of the early attempts to use this approach was made by Witten and Sander [1], who explained a pattern formed by the cathode deposition of positive metallic ions in a dilute electrolyte solution using a model of random diffusing sticky particles: 'diffusion-limited aggregation' (DLA). Many attempts at modeling aggregating systems by particle dynamics have followed, since aggregation is a typical irreversible macroscopic process [2-5]. In this chapter we discuss a model of an open aggregating-particle system. We do not focus on pattern formation as in the case of DLA but rather concentrate on the size or mass distribution.

Let us consider the aggregation of aerosols. The typical particles in aerosols are of diameter 10Å (the size of a cluster of molecules) to 100μm (the size of a fog water-drop or a particle of fine dust). Aerosols are small enough to be regarded as particles randomly floating in air, which coagulate with a certain probability when they collide. As diffusion and coagulation proceed, fully developed particles form and sedimentation begins. At the same time, small particles are continuously supplied by car fumes and smoke from factories. The effects of aggregation, sedimentation, and the continued supply of small particles balance to realize a steady state. The steady-state size distribution of actual aerosols follows a power law over a wide range, and, due to sedimentation, a fast decay occurs at the largest sizes [6].

The system described above is an example of a *random aggregating system*. In order to see the power-law size distribution more clearly, we neglect the sedimentation process, and consider only coagulation by random collision and injection of small particles. Injection from outside plays an important role in realizing a nontrivial steady state because if such system were closed, and no further particles supplied, then the total number of particles would decay monotonically with time in the irreversible aggregating process. It is known that the decay follows a power law for a model of simplified randomly aggregating particles, in which particles of different sizes diffuse in the same manner [7]. As a result, the steady state will be a trivial configuration in which all particles gather into one large particle. By injecting small particles continuously, a steady size distribution of a fractional power law is realized. It has been shown, in a model to be described in the following section, that this type of power law is stable, namely, the system recovers the same distribution regardless of any perturbation (see Sec. 9.6). A power law is supported by a balance between increasing the number of small particles by injection and decreasing it by aggregation. In other words, if we single out a particle of any size it grows larger and larger by repeated coalescence, but the number of particles of a given size is kept constant by the aggregation of newly injected small particles. Moreover, for a given rate of supply of small particles the aggregation rate is naturally controlled by the injection rate of small particles so that the power law is maintained for any injection rate, although the mean size of particles increases with time. We call this condition, which is characterized by a stable power law, a *statistically steady state* to distinguish it from the usual steady states in which mean values converge as time increases.

The classical understanding of aggregation kinetics is given by the rate equation approach proposed by von Smoluchowski [8]. It has been studied extensively for decades and it can explain the power-law size distributions widely observed in aggregation processes. The exact solutions are known only for special cases [9-11] since the equations are nonlinear. Also, the equations do not account for spatial effects so that the obtained solution corresponds to the mean-field case.

In the next section, we introduce a model for which the size distribution and other statistical properties are rigorously analyzed in one dimension (1D) with spatial effects taken into account. In Sec. 9.3, we give an intuitive explanation of the mathematical results. We discuss the relation to phase transition phenomena in Sec. 9.4, and an application in terms of a stable distribution is introduced in Sec. 9.5. We show the uniqueness and stability of the steady-state power law in Sec. 9.6. In Sec. 9.7, we discuss the spatial correlation of aggregation with injection, and in Sec. 9.8 we introduce a new general method for analyzing the correlation of spatial and temporal data. We then adopt the method to our model and obtain a scaling relation. Concluding remarks are given in Sec. 9.9.

9.2 The model and basic equations

Let us picture a sloping paved road under rain with many pebbles uniformly distributed on the surface. Raindrops fall uniformly and they flow along the slope. When a stream of water encounters one of the pebbles on the surface it is diverted either to the left or to the right randomly. When two such streams happen to meet, they coalesce and continue to glide down the road, randomly diverted by the pebbles without penetration into the ground. Once a path has been created, all later raindrops that come upon it follow the same route. If the surface is large enough, after some time we see a hierarchy of little streams and rivulets.

This picture of rivers is equivalent to a random aggregating particle system in discretized space-time, if we regard the drifting direction as the time axis. In order to avoid complexity, here we consider idealized massive particles (material particles) such that one can ignore their volume in the model. Consider a process in which at every time step massive particles on sites of

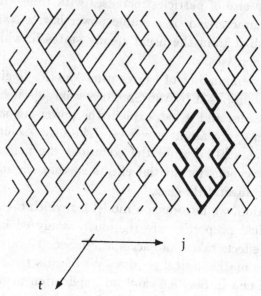

Fig. 9.1. Particle trajectories; see text.

a 1D lattice randomly jump to neighboring sites, while a unit mass particle is added to each site of the lattice. When particles meet at a site they coagulate to form a new particle with mass conserved. Figure 9.1 shows the trajectories of particles in space-time. It involves branched structures that coincide with the river patterns on a road in the rain, introduced in the beginning of this section. The particles at the latest time step are at the bottom of this figure, and each of them has a mass equal to the area of the drainage basin of the corresponding branched structure since we inject a unit mass per time step at each site. The ridges between the drainage basins corresponding to the particles at the latest time step are shown in Fig. 9.2. Thus, we can visualize a configuration of varying sizes of mass.

This model, termed Scheidegger's river model, was originally proposed in 1967 and is of interest in studying river networks. In 1986 it was re-invented as a model of randomly aggregating particle systems by one of the authors and his coworkers, as a basic step towards the statistical physics of systems far from equilibrium [12,13].

The voter model [14,15] is also known to be equivalent to Scheidegger's river model, though the process proceeds oppositely in time. Namely, the voter model is a time-reversed version of Scheidegger's river model. In the voter model the sites of a 1D lattice are occupied by persons who are either in favor of or opposed to some issue. The voters change their opinions

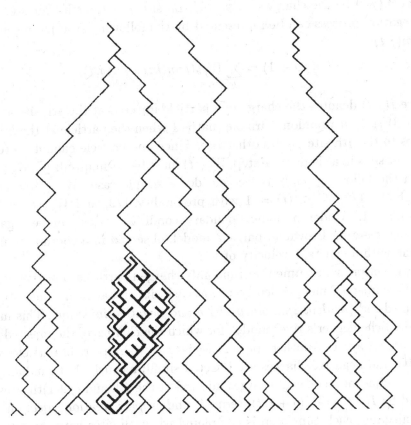

Fig. 9.2. The ridges between the drainage basins; see text.

by being spoken to by their neighbors and also speak frankly about their opinions to the neighbors. This is equivalent to a particle system in which a particle stochastically changes its state to one of the neighboring particle's states at a certain time. After a long time, the particles having the same state start forming clusters on the 1D lattice, and finally all particles take the same state. A space-time pattern of the voter model, which shows who is influenced by whom as time goes on, is identical to a time-reversed version of Scheidegger's model. This clustering process has been analyzed mathematically by Liggett [14] and by Durrett [15].

Scheidegger's river model can be extended to include the case when the dynamical variables have both positive and negative values instead of considering only a positive quantity such as mass. We can interpret this extension as follows: each particle has a positive or negative 'charge'; such charges are added algebraically when two particles aggregate [16].

Let $m(j,t)$ be the charge of a particle on site j at the tth time step. The aggregation process can be represented by the following stochastic equation for $m(j,t)$:

$$m(j, t+1) = \sum_k W_{jk}(t)m(k,t) + I(j,t),\qquad(9.1)$$

where $I(j,t)$ denotes the charge of a particle injected at the jth site at time t, and $W_{jk}(t)$ is a random variable that is 1 when the particle at the kth site jumps to the jth site, and 0 otherwise. Since one particle cannot go to two different sites in a single time step, $W_{jk}(t)$ must be normalized: $\sum_j W_{jk}(t) = 1$. In the following analysis we consider a simple case: $W_{jj}(t) = 1$ with probability 1/2, $W_{jj-1}(t) = 1$ with probability 1/2, and $W_{jk}(t) = 0$ for $k \neq j, j-1$. Under periodic boundary conditions, this can be regarded as an aggregating Brownian-particle model, observed in a coordinate frame moving with a constant velocity of 1/2.

For injection, we assume that randomly charged particles (positive or negative) are injected independently, one at each site and at each time step; this is termed *independent random positive and negative injection*. This model includes Scheidegger's river model, for which the charges of the injected particles are always 1. Another special case is that where pairs of particles with positive and negative charges are injected simultaneously, but randomly, at pairs of adjacent sites, e.g., charges injected at the jth and $(j+1)$th sites are related by $I(j,t) = -I(j+1,t)$, termed *random pair-creation injection*. For pair injection, each time step is completed when all sites have, on average, two injected particles.

The time evolution of the charge distribution can be analyzed by introducing an r-body characteristic function,

$$Z_r(\rho, t) = \left\langle \exp\left[i\rho \sum_{j=1}^r m(j,t)\right] \right\rangle,\qquad(9.2)$$

where $\langle \cdots \rangle$ denotes the average over all realizations of $W_{jk}(t)$ and $I(j,t)$.

We have the following evolution equation for $Z_r(\rho, t)$, from (9.1),

$$Z_r(\rho, t+1) = \left\langle \exp\left[i\rho \sum_{j=1}^r \sum_k W_{jk}\, m(k,t) + i\rho \sum_{j=1}^r I(j,t)\right] \right\rangle.\qquad(9.3)$$

Let us consider the total mass of r successive sites, $M_r(t) = m(k+1,t) + m(k+2,t) + \cdots + m(k+r,t)$, on a 1D lattice. As we follow the time evolution, the difference between $M_r(t)$ and $M_r(t+1)$ is due to the behavior of the two particles, $m(k,t)$ and $m(k+r,t)$, at the edges of the r-site interval. We can consider three different possibilities for the movement of particles at the edges; either $m(k,t)$ or $m(k+r,t)$ (but not both) moves into the r successive sites at $t+1$, with probability $1/4$ respectively, or both particles move into the r successive sites at $t+1$ with probability $1/4$, or neither particle moves into the r successive sites at $t+1$, again with possibility $1/4$. Now some parts of the average, $\langle \cdots \rangle$, in (9.3) can be written explicitly, [13],

$$Z_r(\rho, t+1) = \frac{1}{4} \Big\langle \exp\Big[i\rho \sum_{j=k+1}^{k+r} I(j,t)\Big]\Big\rangle \Big\{ \Big\langle \exp\Big[i\rho \sum_{j=k}^{k+r} m(j,t)\Big]\Big\rangle$$

$$+ \Big\langle \exp\Big[i\rho \sum_{j=k}^{k+r-1} m(j,t)\Big]\Big\rangle + \Big\langle \exp\Big[i\rho \sum_{j=k+1}^{k+r} m(j,t)\Big]\Big\rangle$$

$$+ \Big\langle \exp\Big[i\rho \sum_{j=k+1}^{k+r-1} m(j,t)\Big]\Big\rangle \Big\}. \tag{9.4}$$

Using the translational invariance, we have

$$Z_r(\rho, t) = \tfrac{1}{4}\Phi(\rho)^r \left[Z_{r+1}(\rho, t) + 2Z_r(\rho, t) + Z_{r-1}(\rho, t)\right], \tag{9.5}$$

where $\Phi(\rho) \equiv \langle \exp[i\rho I(j,t)] \rangle$ is the characteristic function for the injection process, which can be expanded as

$$\Phi(\rho) = 1 + i\langle I\rangle\rho - \tfrac{1}{2}\langle I^2\rangle\rho^2 + \cdots. \tag{9.6}$$

In the case of pair-creation injection, positive and negative injections cancel for $j = 2$ to $r-1$ in (9.4). Therefore, the injections that are effective at the r-body level are those of the two particles at the edges of the r successive sites. Then we have the following evolution equation for $r = 1, 2, \ldots$:

$$Z_r(\rho, t) = \tfrac{1}{4}\Phi(\rho)^2 \left[Z_{r+1}(\rho, t) + 2Z_r(\rho, t) + Z_{r-1}(\rho, t)\right]. \tag{9.7}$$

Equations (9.5) and (9.7) give a set of linear equations for $r = 1, 2, \cdots, N-1$. By definition of the r-body characteristic function, $Z_r(\rho, t)$, in (9.2), one of the boundary conditions is given by

$$Z_0(\rho, t) = 1. \tag{9.8}$$

$Z_N(\rho, t)$ is the characteristic function for the total charge of the system, so that the other boundary condition is

$$Z_N(\rho, t+1) = [\Phi(\rho)]^{tN} \, Z_N(\rho, 0). \tag{9.9}$$

The charge distribution in the 1D case with random positive and negative injection can be obtained from the steady-state solution of (9.5). We have the following set of linear equations for $Z_r(\rho)$, $r = 1, 2, 3, \ldots$, with the boundary condition (9.8):

$$Z_{r+1}(\rho) + \left[2 - 4\Phi(\rho)^{-r}\right] Z_r(\rho) + Z_{r-1}(\rho) = 0. \tag{9.10}$$

Dividing both sides of (9.10) by $Z_r(\rho)$, we get a recurrence relation for $Z_r(\rho)/Z_{r-1}(\rho)$ (in what follows we omit the argument ρ in few instances):

$$\frac{Z_r}{Z_{r-1}} = \frac{1}{4\Phi^{-r} - 2 - \dfrac{Z_{r+1}}{Z_r}}. \tag{9.11}$$

Therefore, $Z_1(\rho)$ is given explicitly by the following continued fraction:

$$Z_1 = \cfrac{1}{4\Phi^{-1} - 2 - \cfrac{1}{4\Phi^{-2} - 2 - \cfrac{1}{4\Phi^{-3} - 2 - \cdots}}}. \tag{9.12}$$

Substituting the series expansion for $\Phi(\rho)$, (9.6), into the terms of (9.12) and truncating those terms to linear order leads to a new continued fraction that can be compared with the following continued fraction formula for the ratio of Bessel functions [17]:

$$\frac{J_k(x)}{J_{k-1}(x)} = \cfrac{1}{\dfrac{2k}{x} - \cfrac{1}{\dfrac{2(k+1)}{x} - \cfrac{1}{\dfrac{2(k+2)}{x} - \cdots}}}. \tag{9.13}$$

Comparing (9.12) with (9.13) we then expect that, for $|\rho| \ll 1$,

$$Z_1(\rho) \cong J_{x+1}(x)/J_x(x), \tag{9.14}$$

where $x = 1/\left[-2i\langle I\rangle\rho + \langle I^2\rangle\rho^2\right]$. It can be shown that up to the leading order, the asymptotic behavior of (9.12) is identical to that of (9.14) [18]. From the properties of Bessel functions, as $x \to \infty$, $Z_1(\rho)$ in the vicinity of $\rho = 0$ is given by [18]

$$Z_1(\rho) = 1 - c\langle I\rangle^{1/3} i^{-1/3} |\rho|^{1/3} + \cdots \quad \text{for } \langle I\rangle \neq 0, \tag{9.15}$$

$$Z_1(\rho) = 1 - c\langle I^2\rangle^{1/3} 2^{-1/3} |\rho|^{2/3} + \cdots \quad \text{for } \langle I\rangle = 0, \tag{9.16}$$

where $c = 2\pi \left(\frac{16}{3}\right)^{1/6} / \Gamma\left(\frac{1}{3}\right)^2 = 1.15723\ldots$. The same leading-order result for $Z_1(\rho)$ can be obtained in the continuum limit of (9.11) [18,19].

Since the characteristic function is the Fourier transform of the probability density $p(m)$, we obtain the charge distribution by inversion. In the case where $\langle I \rangle > 0$ (or $\langle I \rangle < 0$), we have

$$p(m) \sim m^{-4/3} \quad \text{for} \quad m \gg \langle I \rangle \quad (\text{or} \ -m \gg -\langle I \rangle), \qquad (9.17)$$

while in the case of $\langle I \rangle = 0$, we have

$$p(m) \sim m^{-5/3} \quad \text{for} \quad m \gg \langle I^2 \rangle^{1/2}. \qquad (9.18)$$

Condition (9.17) shows a one-sided power law whose exponent agrees with that of the steady-state distribution for Scheidegger's river model [12]. Condition (9.18) gives a symmetric power law. Both (9.17) and (9.18) are consistent with simulation results [7].

For pair-creation injection we obtain the following quadratic equation for $Z_1(\rho)$ in the same way as for random injection:

$$1/Z_1(\rho) = 4\Phi(\rho)^{-2} - 2 - Z_1(\rho). \qquad (9.19)$$

Solving (9.19), we get

$$Z_1(\rho) = 1 - 2\langle I^2 \rangle^{1/2} |\rho| + \cdots. \qquad (9.20)$$

The corresponding charge distribution is

$$p(m) \propto m^{-2} \quad \text{for} \quad m \gg \langle I^2 \rangle^{1/2}. \qquad (9.21)$$

This probability density is symmetric and the exponent is the same as that of a Lorentzian. The result agrees with the distribution obtained by a numerical simulation. Also, (9.21) gives the same result as the one for the mean-field case, which is discussed in [7].

9.3 Intuitive prospects

The results can be understood intuitively by contemplating the random walk process. As mentioned in Sec. 9.2, in Scheidegger's model the charge of a particle is equal to the sum of the charges of the particles injected over the corresponding river basin. It is obvious from the evolution rule that the basin's left and right boundaries are formed by simple random walks (see Fig. 9.2) Therefore, the area of the basin is roughly given by $S = hw$, where h and w denote the basin's height and width. Since the boundaries of the basin are random walks, w is proportional to $h^{1/2}$ and the distribution of the height h is identical to the distribution of a Brownian particle's recurrence time [20], $p(h) \propto h^{-3/2}$. Hence, the distribution function for the area of the basin, $S = hw \propto h^{3/2}$, is given by

$$p(S) = p(h) \left(\frac{\partial h}{\partial S} \right) \propto S^{-4/3}. \tag{9.22}$$

By definition, the charge of a particle having a basin of size S is given by the sum of S independent random variables $\{I_j\}$. In the case of $\langle I \rangle \neq 0$ the charge m is obviously proportional to S, hence we have $p(m) \propto m^{-4/3}$. When the average value of I is zero, m becomes proportional to the variance $(\langle I^2 \rangle S)^{1/2}$. Substituting the relation $m \propto S^{1/2}$ into (9.22) we get $p(m) \propto m^{-5/3}$. In the case of pair-creation injection, pair creation inside the area does not contribute to the charge m, because positive and negative charges cancel. The contribution to m comes only from the boundaries, namely, m is given by the sum of $2h$ random variables I_j. The mean value of I_j is automatically zero, so that m is proportional to $(2h)^{1/2}$. Substituting this relation into (9.22), we obtain the charge distribution, $p(m) \propto m^{-2}$. These intuitive results agree perfectly with the analytic results (9.17), (9.18), and (9.21).

9.4 Branching process and phase transition

The time-reversed process of Scheidegger's river model can be regarded as a branching process. Let us recollect the space-time trajectory given in Fig. 9.2. Every site can be regarded as having its origin (or ancestor) at the zero-time layer, which is located at the top of the figure. Reversing the time, we can say that a branching process starting from a bottom site has to have on average one offspring in order to conserve the number of sites at every reversed time step. To be precise, a branch has to have two offspring with probability 1/4, one offspring with probability 1/2, and none with probability 1/4 in one reversed time step.

Let us consider a general simple branching process as shown in Fig. 9.3. A branch has two offspring with probability p and no offspring with probability $1-p$. For the particular branch shown in Fig. 9.3, the probability of existence becomes $p^6(1 - p)^7$. In the general case of the probability of a branch of size s, $N(s)$, it is difficult to count probabilities for all possible shapes. Fortunately, we know that $N(s)$ satisfies the following relation:

$$N(s + 1) = p \sum_{s'=1}^{s} N(s')N(s - s'). \tag{9.23}$$

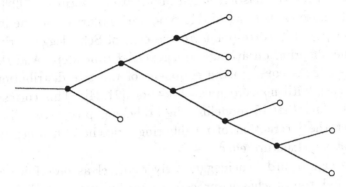

Fig. 9.3. A general simple branching process.

Here we sum the probabilities that two offspring, originating from the same site at some time step, become matured branches of size s and size $s - s'$ after a sufficiently long time. By using the generating function

$$W(x) = \sum_{s=0}^{\infty} x^s W(s),\qquad\qquad (9.24)$$

we obtain, from (9.23), the following result [19]:

$$N(s) \propto s^{-3/2} \exp(-as),\quad a = \log\left[4p(1-p)\right].\qquad (9.25)$$

For $p < 1/2$, the exponential factor in (9.25) becomes dominant, so that $N(s)$ shows an exponential decay and all branches die out in a finite time. For $p = 1/2$, the constant value a becomes zero and $N(s)$ follows a power law with exponent $-3/2$. For $p > 1/2$, the exponential factor dominates again and $N(s)$ shows an exponential decay but in this case there exists a finite value of $N(\infty)$. This picture can be interpreted in terms of a second-order phase transition, as in the percolation problem. In this context, $N(\infty)$ is an order parameter and p is a control parameter and, at the threshold $p_c = 1/2$, a power-law distribution is observed.

Let us now discuss the critical point in more detail. The mean number of offspring is $2p$ because a branch has two offspring with probability p and dies out with probability $1-p$. If $p < 1/2$, the mean number of offspring becomes smaller than 1 so that the number of branches decays exponentially and all of them die out in a finite time. In the opposite case, when $p > 1/2$, the mean number of offspring becomes larger than 1 and the number of branches grows exponentially. From this point of view, $p = 1/2$ is the only case when the number of branches is constant, namely, the critical point is the case when the mean number of offspring in each time step averages to 1. This model is equivalent to the stochastic cellular automaton described in [22], and it is known that phase transitions of this type are similar to the one in directed percolation. $p = 1/2$ corresponds to the case of Scheidegger's river model that has one offspring on average in a reversed time step. And the critical exponent, $-3/2$, is equal to the exponent of the size distribution for the mean-field case with nonzero mean injection [21,24]. This correspondence is due to the fact that in modeling the branching process we did not take into account the interaction of neighboring branches, in other words, we neglected lattice structure effects.

Thus, we can regard Scheidegger's river model as describing a critical condition, and the stochastic process of the model as realizing critical-ity. Namely, Scheidegger's river model is a kind of *self-organized criticality* model [30] that realizes critical-point behavior without parameter tuning.

9.5 View in terms of stable distributions

Throughout this chapter we have focused on the properties of the 'sum of a random number of random variables', $m = \sum I(j,t)$. From the mathematical point of view, for a 'sum of a fixed number of random variables' the central limit theorem can be applied and we have a Gaussian distribution if the 'fixed number' is sufficiently large. However, when the number of variables is not fixed but follows a power-law distribution, as in (9.22), the central limit theorem does not apply. Instead, the sum converges to a power-law distribution.

Let us consider two intervals or sections, denoted section 1 and section 2, consisting respectively of r_1 and r_2 successive particles at time step t. Figure 9.4 shows the ridges between river basins that flow into section 1 and section 2, at time step t. For convenience, let us call the areas of the river basins that contribute to form the total charge of section 1 and section 2, area 1 and area 2, respectively. All particles injected into areas 1 and 2, partitioned by the ridges, flow into the corresponding sections at time t. Let the total charge of r successive sites be m_r. If we assume that m_r is a variable following a stable distribution, then we have the relation

$$m_{r_1+r_2} \stackrel{\mathrm{d}}{=} m_{r_1} + m_{r_2}, \tag{9.26}$$

where $\stackrel{d}{=}$ denotes that variables on both sides have the same distribution function. Also, the distributions for the r-body charge and the 1-body charge are easily shown to satisfy the following relations, by using the same arguments of random walk trajectories as in Sec. 9.3:

$$m_r \stackrel{d}{=} r^3 m_1, \quad \text{for } \langle I \rangle \neq 0 \text{ random injection,} \tag{9.27}$$

$$m_r \stackrel{d}{=} (r^3)^{1/2} m_1, \quad \text{for } \langle I \rangle = 0 \text{ random injection,} \tag{9.28}$$

$$m_r \stackrel{d}{=} (r^2)^{1/2} m_1, \quad \text{for pair-creation injection.} \tag{9.29}$$

From the properties of stable distributions, we know that a variable m satisfying both (9.26) and (9.27) follows a stable distribution with a characteristic exponent equal to 1/3. In that sense we know that the cumulative charge distribution for random injection has a power-law tail with an exponent equal to 3 for large m. In a similar way, we can uniquely determine the charge distribution for the other cases: a two-sided stable distribution with a characteristic exponent 2/3 for the random injection case with $\langle I \rangle = 0$, a two-sided stable distribution with a characteristic exponent 1, i.e., the Lorentzian distribution, for the pair-creation injection case.

9.6 Uniqueness and stability

In the preceding section we have seen that there exists a statistically steady state that is realized by injection. Here we show that the steady state is uniquely determined for a given type of injection, independent of the initial conditions [23]. Let us introduce a perturbation $\tilde{Z}_r(\rho, t)$ around the steady state $Z_r(\rho)$: $Z_r(\rho, t) = Z_r(\rho) + \tilde{Z}_r(\rho, t)$. Due to the linearity of (9.5) and (9.7), the equation for the perturbation in the 1D case is given by

$$\tilde{Z}_r(\rho, t+1) = \tfrac{1}{4}[\Phi(\rho)]^\mu \left[\tilde{Z}_{r+1}(\rho, t) + 2\tilde{Z}_r(\rho, t) + \tilde{Z}_{r-1}(\rho, t) \right], \tag{9.30}$$

where $\mu = r$ for random positive and negative injection, and $\mu = 2$ for pair-creation injection. The boundary conditions are $\tilde{Z}_0(\rho, t) = 0$ and $\tilde{Z}_r(0, t) = 0$.

Taking the absolute value of both sides of (9.30) we obtain the following relation:

$$|\tilde{Z}_r(\rho, t)| \leq |[\Phi(\rho)]^\mu| \max \left\{ |\tilde{Z}_{r'}(\rho, t-1)|, r' = r-1, r, r+1 \right\}. \tag{9.31}$$

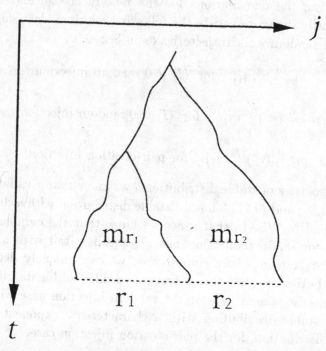

Fig. 9.4. The ridges between river basins that flow into two sections, as discussed in the text.

Iterating (9.31) t times and using the inequality $|\Phi(\rho)| \leq 1$, which is valid for any characteristic function, we get, for $r \neq 0$,

$$|\tilde{Z}_r(\rho, t)| \leq |[\Phi(\rho)]^\mu|^t \max\left\{|\tilde{Z}_{r'}(\rho, 0)|, r' = 1, 2, \ldots\right\}. \qquad (9.32)$$

Therefore, for any ρ that satisfies $|\Phi(\rho)| < 1$ the perturbation of any magnitude goes to zero exponentially with time. This means that for any initial condition the system converges to a steady state that is unique and stable. The stability for the case $|\Phi(\rho)| = 1$ is not trivial, but even in this case, it can be shown that the system relaxes to a unique steady state for any initial perturbation [24]. Thus, the steady-state power-law distribution is very robust. It is uniquely determined by the type of injection and is independent of specific details of injection and of initial conditions. The functional form of the relaxation to the steady state is also determined by the type of injection [21,24].

9.7 Spatial correlations

In the previous sections we analyzed the statistical properties of an aggregation system in 1D using the equation for the many-body characteristic function. We saw that there emerges a nontrivial statistically invariant steady state if we supply particles at a constant rate. The most interesting point here is that the system converges to a steady state even in the absence of a sink or particle breakup. It is easy to confirm from (9.5) that the variance of particle charge increases linearly with time, namely, it diverges in the limit of $t \to \infty$. Usually such divergence means that the system does not have a stationary state, but in this case the charge distribution converges to a power-law distribution in which the mean value and variance diverge. Thus, we call this a quasi-steady state or *statistically steady state*. According to the theory of stable distributions, the divergence of the variance is essential for power-law distributions. The central limit theorem, though, can be generalized to include independent random variables with divergent variances. The limiting distribution of the sum of such variables, if it exists, is a non-Gaussian stable distribution with a power-law tail as we mentioned in Sec. 9.5. In Scheidegger's model, the variance is divergent in the statistically steady state, and a particle's charge is given by a sum of random aggregating variables, $I(j, t)$. If a particle's charge distribution is independent of other particles then we can apply the generalized central limit theorem. But we know that there exists a spatial correlation among particles since exponents of the steady-state distributions of Scheidegger's model are different in the 1D cases and in the mean-field cases, which shows the importance of spatial restriction in 1D (except for the case of pair-creation injection) [16]. Also, by empirical arguments, two large drainage basins are less likely to exist adjacently than a large basin and a small one, see Fig. 9.2, so that the system has some kind of spatial correlation whereby the probability that two large basins will develop nearby one another is reduced. However, we are faced with a conflicting observation that the analysis of stable distributions yields a good estimate of the charge distribution. It seems likely that the charge distribution belongs to the domain of attraction of a non-Gaussian stable distribution and thus has a power-law tail.

Let us see if we can find out anything about correlations from our previous analysis using the characteristic function. From the steady-state equation (9.10), we can show that $Z_r(\rho)$ can be expanded in the vicinity of $\rho = 0$ as $Z_r(\rho) = 1 - c_1 r |\rho|^\beta + \cdots$, where c_1 is a constant and β is either 1/3, 2/3, or 1 depending on the injection type. If the distribution were spatially independent, $Z_r(\rho)$ would be equal to $[Z_1(\rho)]^r$. However, it is easy to prove [25] that these two expressions are identical only up to order ρ^β and that there is a nonvanishing term of order $\rho^{2\beta}$, $[Z_1(\rho)]^r - Z_r(\rho) = \frac{1}{2}r(r-1)(c_1)^2\rho^{2\beta} + \cdots \neq$ 0. Therefore, we can conclude that there really exist spatial correlations. However, it is not easy to detect this subtle correlation numerically because it is of higher order in ρ. Also, another reason that makes it difficult to analyze correlations is that the mass distribution follows a power law for which the moments of the mass distribution diverge. In such a case, it is impossible to calculate the correlation function defined by the second-order moments such as $\langle m(0,t)\, m(j,t) \rangle$. Nevertheless, we can avoid this divergence by modifying the definition of the correlation function: we do this by introducing fractional moments [25]. Using fractional moments has allowed us to detect the existence of spatial correlations but detailed information has not been obtained.

The correlation function of the system is related to the power spectrum by the Wiener-Khinchin theorem. However, in our model, owing to the divergence of the correlation function we cannot determine it by studying the power spectrum either. Namely, the power spectrum of the mass configuration looks the same as in the case of white noise because a site having the largest mass acts as a delta function, and the Fourier transformation gives the same contribution at all wavenumbers. One can also apply the range-of-fluctuations/standard-deviation analysis or the multifractal method, but in both cases the results indicate that the system has no correlations [26].

9.8 Interval distribution of a level set

In order to investigate further the spatial correlations in Scheidegger's river model, we introduce a powerful new approach to systems with the problem of diverging moments [27].

The distribution of gaps (intervals between one incident and the next) has been studied as a way of characterizing fractal properties of Cantor sets and other sets of fractal dimension $D < 1$ [28]. It has been shown that the gap-size distribution for a fractal set follows a power law, $P(r) \propto r^{-D-1}$, where $P(r)$ denotes the probability of finding a gap of size r. Power-law gap-size distributions are found not only in mathematical models but also in real systems such as the error occurrences in data transmission lines [29].

The idea of examining gaps is applicable to fluctuations of a scalar quantity on 1D space by observing the level sets of intersections. A Brownian trajectory in space-time coordinates is one of the examples whose gap-size distribution of the level set on the intersection of $x(t) = c$, where c is a given constant, satisfies $P(r) \propto r^{-D-1}$ with $D = 1/2$ [20]. As for fractional Brownian motion, the exponent D is known to be expressed as $D = 1 - H$ where H is the Hurst exponent, which takes a value between 0 and 1 [28].

The interparticle distribution (the distribution of gaps between successive particles) in a 1D reaction-diffusion model governed by the rule $A + A \to A$ has been examined by ben-Avraham *et al.* [31]. In the case when the process involves the injection of particles at randomly chosen sites, the system reaches a stationary state with a nonzero concentration of particles. At the stationary state, the gap-size distribution shows a distinctive peak at a certain size of gap depending on the rate of injection and the diffusion coefficient.

In this section, we introduce a systematic way of investigating spatial or temporal correlations using the interval distribution of level sets (IDL). First, let us introduce the IDL analysis for 1D spatial or temporal data in general. For given discrete data, we set a threshold height E_c and pay attention only to the data points which have values higher than E_c (see Fig. 9.5). The set of points thus chosen is identified as the level set and E_c is termed the level. The IDL is the size distribution of distances between nearest-neighbor elements of a level set, as illustrated in Fig. 9.5. We observe IDL with different values of E_c and try to scale them on a curve by considering E_c as a parameter. In that sense, the IDL curves obtained with different E_c values are related to one another.

For the purposes of comparison with real data, we list the IDL of an uncorrelated white-noise system. The likelihood of finding an interval of length r, $P_{E_c}(r)$, can be estimated as follows: for large r,

$$P_{E_c}(r) = p_{E_c}^2 (1 - p_{E_c})^{r-1} \propto \exp\left[r \ln(1 - p_{E_c})\right], \qquad (9.33)$$

where p_{E_c} denotes the probability that a chosen site is an element of the level set at level E_c. Equation (9.33) shows that the IDL follows an exponential decay. An exponential probability distribution is fixed uniquely by the mean value of r, $\langle r \rangle_{E_c}$, as

$$P_{E_c}(r) = \frac{1}{\langle r \rangle_{E_c}} \exp\left(-\frac{r}{\langle r \rangle_{E_c}}\right). \qquad (9.34)$$

By comparing (9.33) and (9.34), we see that p_{E_c} is related to $\langle r \rangle_{E_c}$ as follows: for $\langle r \rangle_{E_c} \gg 1$, $p_{E_c} = 1 - \exp\left(-1/\langle r \rangle_{E_c}\right)$. We note that ben-Avraham *et al.* [31] mentioned that in their 1D diffusion reaction model the stationary-state interparticle distribution function becomes exponential when the reaction system includes the reverse process: $A + A \leftarrow A$. This exponential distribution is explained by the maximum entropy principle since the system

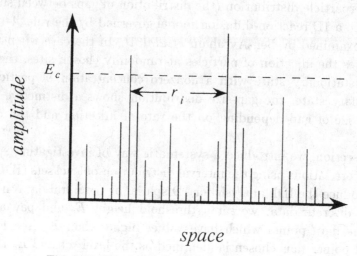

Fig. 9.5. Illustration of the 'level set' concept.

satisfies detailed balance as in the case of thermal equilibrium. That is, in such reversible particle systems, particles are observed to behave independently.

Now, let us apply IDL analysis to the subtle correlations in Scheidegger's model. The IDL for the mass configuration in the 1D model shows a curve with a distinctive peak, the position of which depends on the level, namely, on E_c, i.e., $p_{E_c} \propto E_c^{-\beta}$. So we have

$$\langle r \rangle_{E_c} \propto 1/p_{E_c} \propto E_c^{\beta} \tag{9.35}$$

where $\beta = 1/3$ for Scheidegger's model.

The curves for different E_c ($E_c = 25, 50, 100, 200$, and 400) are confirmed as being congruent. As the value of E_c is doubled, the curve in a double-logarithmic plot shifts constantly in both horizontal and vertical directions. These curves can be scaled on one function by rescaling r and $P_{E_c}(r)$ as follows (see Fig. 9.6):

$$P_{E_c}(r) \propto E_c^{-2/3} f(r E_c^{-1/3}). \tag{9.36}$$

The scaling function, $f(\zeta)$, is estimated to be nearly proportional to ζ for $\zeta < 1$ and it falls off faster than exponentially, $f(\zeta) \propto \exp[-(\zeta - 1)^{-2}]$, for $\zeta > 1$.

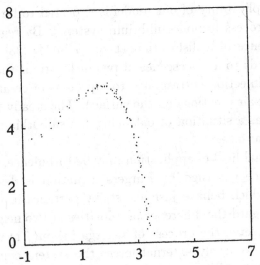

Fig. 9.6. A scaling plot of IDL for Scheidegger's model. Along the vertical axis we have $\log_2[P_{E_c}(r)(E_c/E_0)^{2/3}]$; along the horizontal axis we have $\log_2[r(E_c/E_0)^{-1/3}]$.

The scaling exponents in (9.36) can be obtained theoretically if we make the following scaling assumption [32]:

$$P_{E_c}(r) \propto E_c^{-\theta} f(r E_c^{-z}). \tag{9.37}$$

From the definition, p_{E_c} and $P_{E_c}(r)$ satisfy the following relation:

$$p_{E_c} = \int_1^{\infty} P_{E_c}(r)\, dr \propto E_c^{-\beta}. \tag{9.38}$$

Substituting (9.37) into (9.38) and taking into account that $E_c \gg 1$, we obtain

$$z - \theta = -\beta. \tag{9.39}$$

Also, by (9.37) and (9.38) the mean interval, $\langle r \rangle_{E_c}$, is given as

$$\langle r \rangle_{E_c} \propto \int_1^{\infty} r\, \frac{P_{E_c}(r)}{p_{E_c}}\, dr = E_c^z \int_1^{\infty} \zeta f(\zeta)\, d\zeta \propto E_c^z. \tag{9.40}$$

Comparing (9.40) with (9.35) and using (9.39), we have $z = \beta = 1/3$ and $\theta = 2\beta = 2/3$ for Scheidegger's model, which are consistent with the scaling exponents in (9.36).

9.9 Conclusion

The potential applicability of our model is expected to be wide since it describes a basic process in nonequilibrium systems. By regarding $m(j,t)$ as the height difference of a dislocation at site j in 1D [33], the aggregation process corresponds to a coalescence of two dislocations that forms a larger dislocation, and injection corresponds to a process of creating a new dislocation or bump (pair creation) on the surface. The steady state of this case can be regarded as a situation of balancing the smoothing and roughening processes at the surface.

Another potential field of application may be turbulence. It is well known that 1D turbulence governed by Burgers' equation is described by a set of shock waves, which behave just like sticky particles in the low-viscosity limit [34]. If we regard the difference of velocities at two neighboring sites as $m(j,t)$, we can observe the process of aggregating $m(j,t)$ with momentum conserved. When we apply external forces the system converges to a statistically steady state with a power-law distribution. The exponent is equal to the one in the case of 1D random injection, corresponding to 1/3 in a cumulative distribution [35].

Taguchi and one of the present authors have introduced [36] a modified model of aggregation with injection. In addition to the random aggregating process the model has a rule for the breakup of particles, instead of injection. It is a simple standard model describing general random-transport dynamics, in which a finite portion of a scalar quantity on a site is transported to its neighbor, the rest remaining on the same site for that time step. In this modified model it is shown both numerically and theoretically that the fluctuation of the scalar quantity ('charge' or 'mass') follows neither a Gaussian nor a power law. Gaussian and power-law distributions appear only in the extreme cases, of infinitesimal transport and of the aggregation limit, respectively. In general cases we have non-Gaussian distributions with enhanced tails. Such non-Gaussian distributions are familiar in the study of turbulence and the model gives a simplified explanation of their non-Gaussian nature.

References

[1] T. A. Witten and L. M. Sander, *Phys. Rev. Lett.* **47**, 1400 (1981).
[2] T. Vicsek, *Fractal Growth Phenomena: The Second Version* (World Scientific, Singapore, 1992).
[3] P. Meakin, *Phys. Rev.* **A26**, 1495 (1983).

[4] M. Matsushita, M. Sano, Y. Hayakawa, H. Honjo and Y. Sawada, *Phys. Rev. Lett.* **53**, 286 (1984).

[5] P. Meakin, *J. Colloid Interface Sci.* **105**, 240 (1985).

[6] T. Takahashi, in *Physics of Clouds* (Japanese Tokyo-do Publishing Inc., 1987).

[7] H. Takayasu, M. Takayasu, A. Provata and G. J. Huber, *Stat. Phys.* **65**, 725 (1991).

[8] M. von Smoluchowski, *Z. Phys. Chem.* **92**, 129 (1917).

[9] M. von Smoluchowski, *Z. Phys.* **17**, 557 (1916).

[10] J. B. Mcleod, *Quart. J. Math. Oxford* **13**, 119, 193 (1962).

[11] R. M. Ziff, *J. Stat. Phys.* **23**, 241 (1980).

[12] H. Takayasu and I. Nishikawa, *Proc. 1st Int. Symp. for Science on Form*, p. 15, S. Ishizaka, ed. (KTK Scientific, Tokyo, 1986).

[13] H. Takayasu, I. Nishikawa and H. Tasaki, *Phys. Rev.* **A37**, 3110 (1988).

[14] T. M. Liggett, *Interacting Particle Systems* (Springer-Verlag, New York, 1985).

[15] R. Durrett, *Lecture Notes on Particle Systems and Percolation* (Wadsworth, Belmont, 1988).

[16] H. Takayasu, *Phys. Rev. Lett.* **63**, 2563 (1989).

[17] G. N. Watson, *A Treatise on the Theory of Bessel Functions* (Cambridge University Press, Cambridge, 1944).

[18] G. Huber, *Physica* **A170**, 463 (1991).

[19] C. R. Doering and D. ben-Avraham, *Phys. Rev. Lett.* **62**, 2563 (1989).

[20] W. Feller, *An Introduction to Probability Theory and its Applications* (Wiley, New York, 1966).

[21] M. Takayasu, H. Takayasu and Y.-h. Taguchi, *Int. J. Mod. Phys.* **B8**, 3887 (1994).

[22] E. Domany and W. Kinzel, *Phys. Rev. Lett.* **53**, 311 (1984).

[23] H. Takayasu, A. Provata and M. Takayasu, *Phys. Rev.* **A42**, 7087 (1990).

[24] M. Takayasu, Stochastic models of aggregation with injection, doctoral dissertation (1993).

[25] M. Takayasu, *Phys. Rev.* **A45**, 8965 (1992).

[26] M. Takayasu and H. Takayasu, *Phys. Rev.* **A39**, 4345 (1989).

[27] M. Takayasu, *Physica* **A197**, 371 (1993).

[28] B. B. Mandelbrot, *The Fractal Geometry of Nature* (W. H. Freeman, San Francisco, 1982).

[29] S. M. Sussman, *IEEE, Trans. Commun. System* 213 (1963).

[30] P. Bak, C. Tang and K. Wiesenfeld, *Phys. Rev. Lett.* **59**, 381 (1987).

[31] D. ben-Avraham, M. A. Burschka and C. R. Doering, *J. Stat. Phys.* **60**, 695 (1990).

[32] T. Vicsek, *Fractal Growth Phenomena: The Second Version* (World Scientific, Singapore, 1992).

[33] M. Kardar, G. Parisi and Y.-C. Zhang, *Phys. Rev. Lett.* **56**, 889 (1986).

[34] J. M. Burgers, *The Nonlinear Diffusion Equation* (Reidel, Boston, 1974).

[35] H. Hayakawa, M. Yamamoto and H. Takayasu, *Prog. Theor. Phys.* **78**, 1 (1987).

[36] H. Takayasu and Y.-h. Taguchi, *Phys. Rev. Lett.* **70**, 782 (1993).

Part IV: Random Adsorption and Relaxation Processes

Editor's note

The next three chapters cover the topics of monolayer adsorption and, to a limited extent, multilayer adsorption [1], in those systems where finite particle dimensions provide the main interparticle interaction mechanism. Furthermore, the particles are larger than the unit cells of the underlying lattice (for lattice models). As a result, deposition without relaxation leads to interesting random jammed states where small vacant areas can no longer be covered. This basic process of random sequential adsorption, and its generalizations, are described in Ch. 10.

Added relaxation processes lead to the formation of denser deposits, yielding ordered states (full coverage in 1D). Chapter 11 is devoted to diffusional relaxation. The detachment of recombined particles is another relaxation mechanism, reviewed in Ch. 12. The detachment of originally deposited particles, although modeling an important experimental process, has been studied much less extensively [2].

While several exact results are available, as well as extensive Monte Carlo studies, it is interesting to note that many theoretical advances in deposition models with added relaxation have been derived by exploring relations to other 1D systems. These range from Heisenberg spin models to reaction-diffusion systems (Part I). However, most of these relations are limited to 1D.

Besides their theoretical interest, 1D deposition models find applications in characterizing certain reactions on polymer chains, in modeling traffic flow, and in describing the attachment of small molecules on DNA. The latter application is described in Ch. 22. Of course, experimentally the most important dimensionality for deposition is that of the planar substrate, lattice or continuum. For a review of such two-dimensional deposition models dominated by particle jamming effects with added relaxation processes, see [3] and references therein.

[1] Multilayer deposition will be further discussed in Ch. 15 (Part V).
[2] See, e.g., P. L. Krapivsky and E. Ben-Naim, *J. Chem. Phys.* **100**, 6778 (1994).
[3] V. Privman, *Annual Rev. Comput. Phys.* **3**, 177 (1995).

10

Random and cooperative sequential adsorption: exactly solvable models on 1D lattices, continuum limits, and 2D extensions

James W. Evans

Random sequential adsorption (RSA) and cooperative sequential adsorption (CSA) on 1D lattices provide a remarkably broad class of far-from-equilibrium processes that are amenable to exact analysis. We examine some basic models, discussing both kinetics and spatial correlations. We also examine certain continuum limits obtained by increasing the characteristic size in the model (e.g., the size of the adsorbing species in RSA, or the mean island size in CSA models having a propensity for clustering). We indicate that the analogous 2D processes display similar behavior, although no exact treatment is possible here.

10.1 Introduction

In the most general scenario for chemisorption [1] or epitaxial growth [2] at single crystal surfaces, species adsorb at a periodic array of adsorption sites, hop between adjacent sites, and possibly desorb from the surface. Such processes can be naturally described within a lattice-gas formalism [3]. The microscopic rates for different processes in general depend on the local environment and satisfy detailed-balance constraints [1-3]. The net adsorption rate is determined by the difference in chemical potential between the gas phase and the adsorbed phase [2]. In many cases, thermal desorption can be ignored for a broad range of typical surface temperatures, T. Furthermore, for sufficiently low T, thermally activated surface diffusion is also inoperative, so then species are irreversibly (i.e., permanently) bound at their adsorption sites. Henceforth, we consider the latter regime exclusively. Clearly the resultant adlayer is in a far-from-equilibrium state determined by the kinetics of the adsorption process. In epitaxial growth, the adsorbing species typically occupies a single empty site [2]. However, in the chemisorp-

205

tion of molecular species, often a certain ensemble of more than one empty site is required for adsorption [4], either because the adspecies actually occupy more than one site, or because strong short-range repulsion between adspecies prevents them from occupying, e.g., adjacent or nearest-neighbor (NN) sites. We consider both cases here.

The simplest situation of *random sequential adsorption* (RSA) [5-7] applies where adsorption is typically not thermally activated, and occurs directly from the gas phase. Here adsorption is attempted at a single fixed rate at randomly selected sites or ensembles of sites, and is successful only if the selected site or ensemble of sites is empty. The RSA of 'monomers', at single sites, on an initially empty lattice trivially produces a random adlayer, and the coverage, θ, at time t satisfies $\theta = 1 - \exp(-kt)$; k is the impingement rate. However, the RSA of species requiring adsorption ensembles of more than one site will clearly produce nontrivial adlayers. Here there is particular interest in characterizing the final 'jammed' or 'saturation' state. For example, the RSA of 'dimers', requiring an adjacent pair of empty sites for adsorption, produces a jammed state containing isolated empty sites.

If adsorption is thermally activated, or occurs via some weakly bound precursor state, then in general one expects the adsorption rate to depend on the local environment of the adsorption site or ensemble [1], i.e., distinct rates must be specified for each configuration of this local environment. This more complicated situation is described as *cooperative sequential adsorption* (CSA) [7]. Clearly, if nearby filled sites enhance adsorption rates then cooperativity will produce islanding or clustering, analogous to the effect of attractive interactions in equilibrium adlayers. The case where nearby filled sites inhibit adsorption is analogous to repulsive interactions.

The focus of this chapter (and this book) is on one-dimensional (1D) models. These could reasonably describe low-temperature adsorption on 'grooved' fcc (110) surfaces, and on surfaces with missing-row reconstructions [8]. However, most interest in surface adsorption centers on two-dimensional (2D) systems, so we also comment on the 2D analogues of our 1D models. In fact, much of the impetus for the development of RSA and CSA, and much of the historical interest in these models, was driven by their application to the description of reaction or binding at the pendant groups along polymer chains [9-11]—natural 1D systems! Indeed most of the examples presented in this chapter are directly relevant to such processes. In this context, the state of sites might be better referred to as 'unreacted' or 'reacted', rather than 'empty' or 'filled'. However, we will use the latter language of adsorption for uniformity.

In Sec. 10.2, we first describe the general techniques of analysis, and general properties of these RSA and CSA models, which allow exact analysis in 1D. The key feature is a 'shielding' or Markov property of empty (but not of filled) sites. In Sec. 10.3, we consider the RSA of M-mers on a linear lattice, and the classic continuum 'car parking' limit. CSA on a linear lattice, where contiguous islands of filled sites are produced by a propensity for clustering, is described in Sec. 10.4, together with the continuum 'grain growth' limit. Generalized CSA models producing islands with 'superlattice' structure, and permanent domain boundaries, are discussed in Sec. 10.5. In all cases, we analyze both the adsorption kinetics and the spatial correlations, and comment on the behavior of analogous 2D models. A variety of generalizations of these models are discussed in Sec. 10.6, and some conclusions presented in Sec. 10.7.

10.2 General techniques of analysis and properties of models

For kinetic lattice-gas models of general adsorption processes in finite systems, one can always write down exact master equations describing the evolution of (probabilities of) various configurations for the entire system [12]. In the ideal limit of infinite system size (considered exclusively here), these equations are necessarily recast in the form of an infinite coupled linear hierarchy for probabilities of various subconfigurations of finite sets of sites [12]. For general adsorption processes with adsorption-desorption or hopping, such equations cannot be solved exactly, even for 1D systems. One notable exception is the 1D Glauber model for adsorption-desorption [13] in which there is a special relationship between adsorption and desorption rates. Another exception is the case of a randomly hopping lattice gas (subject to the exclusion of multiple occupancy of sites), where analysis is possible in any dimension [14,15]. In contrast, exact analysis of the hierarchical evolution equations is possible for a remarkably broad class of RSA and CSA processes in 1D (and also for branching media) [7].

Of primary interest is the evolution of the coverage, θ, with time, t. This adsorption kinetics is often characterized in terms of a sticking probability, $S_{ads} \propto d\theta/dt$, normalized to unity at $t = 0$. It is often convenient to regard S_{ads} as a function of θ (rather than t), so then $S_{ads}(\theta)$ decreases from unity when $\theta = 0$ to zero at jamming. However, as indicated above, it is also appropriate to characterize in detail the spatial statistics of the adlayer: ordering, islanding, spatial correlations, etc.

Below we denote empty sites by 'o' and filled sites by '•'. Unless otherwise stated, we consider exclusively infinite systems with translational invariance. The probabilities of various subconfigurations are denoted by $\mathcal{P}[\cdots]$, e.g., $\theta = \mathcal{P}[\bullet]$. Standard 'conservation relations' apply: $\mathcal{P}[\bullet] + \mathcal{P}[\text{o}] = 1$, $\mathcal{P}[\bullet\text{o}] + \mathcal{P}[\text{oo}] = \mathcal{P}[\text{o}]$, $\mathcal{P}[\bullet\bullet] + \mathcal{P}[\bullet\text{o}] + \mathcal{P}[\text{o}\bullet] + \mathcal{P}[\text{oo}] = 1$, etc., which allow one to rewrite the probabilities of arbitrary configurations in terms of those for empty configurations, say. For example, one has $\mathcal{P}[\bullet] = 1 - \mathcal{P}[\text{o}]$, $\mathcal{P}[\bullet\text{o}] = \mathcal{P}[\text{o}] - \mathcal{P}[\text{oo}]$, and $\mathcal{P}[\bullet\bullet] = 1 - 2\mathcal{P}[\text{o}] + \mathcal{P}[\text{oo}]$, as well as more complicated relationships like $\mathcal{P}[\bullet\bullet\bullet] = 1 - 3\mathcal{P}[\text{o}] + 2\mathcal{P}[\text{oo}] + \mathcal{P}[\text{o-o}] - \mathcal{P}[\text{o o o}]$, which relate the probabilities of connected filled clusters to those of generally disconnected empty configurations. Here '-', in $\mathcal{P}[\text{o-o}]$, denotes a site of unspecified state. It is thus sufficient to write down the hierarchical evolution equations only for empty configurations, since these can be recast as a closed set using the above conservation relations. This turns out to be the most appropriate presentation of the evolution equations for RSA and CSA from the perspective of 'solvability', as discussed below. There is another indication of the intrinsic significance for RSA and CSA of the hierarchy for empty configurations. If it is recast as a linear matrix equation, $d\mathbf{P}/dt = \mathbf{AP}$, where $\mathbf{P}$ is (the transpose of) an infinite vector $(\mathcal{P}[\text{o}], \mathcal{P}[\text{oo}], \mathcal{P}[\text{o o o}], \ldots)$ of probabilities of empty configurations of nondecreasing size, then the linear operator $\mathbf{A}$ generating the time evolution is an (infinite) upper triangular matrix (for RSA or CSA). Thus its spectrum can be read off directly from the diagonal entries, immediately elucidating the asymptotic kinetics of adsorption [7].

The universal genesis of 'exact solvability' for 1D RSA and CSA is a Markov-type shielding property for empty sites, which actually holds in all dimensions [16]: *Consider a wall of sites specified empty that separates the lattice into two topologically disconnected regions. Suppose that the wall is sufficiently thick that any filling event is not simultaneously affected by the state of the sites on both sides of the wall. Then such a wall completely shields sites on one side from the influence of those on the other.* This condition is most directly manifested by the equality of certain conditional probabilities, various explicit examples of which will be presented below. Thus for adsorption with finite-range cooperative effects in 1D, a sufficiently long string of consecutive empty sites will 'shield'. Let $P_n = \mathcal{P}[\text{o o} \cdots \text{o o}]$ denote the probability of finding a string of n empty sites in 1D (where adjacent sites could be either empty or filled). This string of n empty sites will also be referred to as an empty n-tuple. Also, let $Q_n = P_{n+1}/P_n$ denote the conditional probability of finding an empty site adjacent to a specified string of n empty sites. Then, if L empty sites are sufficient to shield, one has

$$Q_n = Q_L = Q, \text{ say}, \quad \text{and} \quad P_n = Q^{n-L}P_L, \quad \text{for all } n \geq L. \quad (10.1)$$

An immediate consequence of this relation is that the distribution, $\mathcal{P}[\bullet\circ\circ\cdots\circ\circ\bullet] = P_n - 2P_{n+1} + P_{n+2}$, of the lengths of strings of exactly n empty sites is exactly geometric for $n \geq L$ (i.e., the ratio of probabilities for successive lengths is constant).

It should be emphasized that no analogous shielding property holds for walls of filled sites in RSA or CSA. This asymmetry between filled and empty sites means that exact analysis of these models via hierarchical truncation, as described below, is necessarily formulated in terms of empty-site configurations. The asymmetry is a reflection of the irreversible nature of filling, $\circ \rightarrow \bullet$, in these models. Here one does not expect a 'particle-hole' symmetry, wherein the statistics of the filled sites at coverage θ is identical to that of the empty sites at coverage $1 - \theta$. One should contrast this behavior with the properties of equilibrium lattice-gas models, which do display symmetry between filled and empty sites, at least for pairwise additive interactions. Correspondingly, the spatial statistics reflect a symmetric Markov or shielding property of both empty and filled sites.

Next we discuss the consequences of this shielding relation for solvability in 1D. If the above-mentioned hierarchy for empty configurations contains a closed infinite subset of equations for the P_n, then this infinite subset can be exactly truncated to a finite set of equations using the above relation (10.1). This allows exact solution. Thus the general criterion for exact solution of 1D models is the existence of a closed subhierarchy for the P_n. Note that while the shielding property holds in higher dimensions, it does not lead to exact solution.

Although not appreciated in the early work on 1D models, it is clear that the above determination of P_n does not constitute a complete solution. Additional information is contained in the probabilities of disconnected empty configurations. However, these can also be determined for solvable 1D models by refined application of the shielding condition. This leads directly to pair, triplet, etc., spatial correlations for arbitrary separations, and indirectly to a distribution of the lengths of filled strings of sites (which must be reconstructed from multiply disconnected empty configuration probabilities, as indicated above). This general scheme was first recognized by Plate *et al.* [17], and implemented by Evans and coworkers [18-20]. Specific examples will be presented in the following sections.

Access to spatial pair correlations also allows determination of the 'structure factor' for the adlayer, a quantity probed directly in diffraction studies. Let $P(\mathbf{m})$ denote the probability of finding two filled sites separated by $\mathbf{m}$ (with $P(\mathbf{0}) = \theta$), and let $C(\mathbf{m}) = P(\mathbf{m}) - \theta^2$ denote the corresponding pair correlation. Then the structure factor is given by $S(\mathbf{q}) = \sum_{\mathbf{m}} \exp[i\mathbf{q} \cdot \mathbf{m}] C(\mathbf{m})$. From the spatial correlations or the structure factor one can also extract information about fluctuations in the adlayer. The best-known relation determines the magnitude of fluctuations in the number of adsorbed particles, #, in a region of V sites, as [3]

$$\langle [\# - \langle \# \rangle]^2 \rangle \sim V S(\mathbf{0}), \tag{10.2}$$

in the limit of large V.

Finally, we mention other general techniques of analysis. For general kinetic lattice-gas models in any dimension, it is possible to develop formal short-time Taylor expansions for quantities of interest [7]. One can then eliminate the time, t, in favor of some other preferred variable, e.g., the coverage, θ [21]. This approach can be recast in terms of a creation-annihilation operator formalism, facilitating the determination of coefficients in the expansions [22]. For solvable problems, one can obtain coefficients to all orders, and summing the expansions then recovers the solutions obtained by more direct truncation techniques. It is also possible to formulate general kinetic lattice-gas models as *infinite particle systems* [23], and then to exploit general results from this field to demonstrate, e.g., the super-exponential decay of correlations in RSA and CSA. Of course, one should also note that Monte Carlo simulation of these processes provides a powerful and practical tool for analysis of their properties.

10.3 RSA of M-mers

In this section, we consider the RSA of M-mers, which occupy M consecutive sites on a linear lattice; see Fig. 10.1. We also explore the continuum 'car parking' limit as $M \to \infty$, and generalizations to 2D lattices where the adsorbing 'animals' occupy M sites of various shapes.

10.3.1 RSA of dimers ($M = 2$) in 1D

The RSA of dimers or 'dumbbells', at adjacent pairs of sites on a linear lattice, Fig. 10.1(a), was considered in 1939 by Flory [24] in the context of a polymer cyclization reaction. This constituted the first analysis of a nontrivial problem in RSA. Now, a string of n empty sites can be destroyed by dimer adsorption at $n - 1$ locations completely within the string, or at

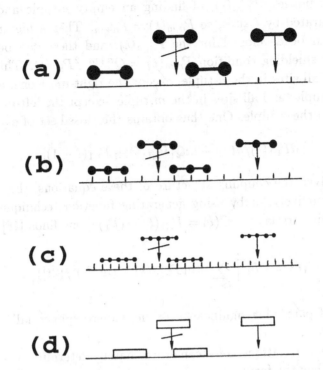

Fig. 10.1. RSA of M-mers on a linear lattice for (a) dimers; (b) trimers; (c) pentamers; and (d) the 1D continuum car parking limit.

two locations partly overlapping each end (which requires the adjacent site also to be empty). Thus, if k denotes the impingement rate at each pair of sites, one has [25]

$$dP_n/dt = -k(n-1)P_n - 2kP_{n+1}, \qquad \text{for all} \quad n \geq 1. \qquad (10.3)$$

From inspection, one can immediately see that the shielding condition, $Q_n = P_{n+1}/P_n = \exp(-kt) = Q$, say, is satisfied for all $n \geq 1$, given an initially empty lattice. Note that for this model, only a single empty site is required for shielding. Using this condition to truncate the hierarchy, one immediately obtains $\theta = 1 - P_1 = 1 - \exp\left(2e^{-kt} - 2\right)$, yielding a jamming coverage of $\theta_J = 1 - e^{-2}$. For the normalized sticking probability, one obtains $S_{\text{ads}}(\theta) = (1 - \theta)[1 + \frac{1}{2}\ln(1 - \theta)]$, which vanishes at θ_J, and also at a 'maximal' coverage of unity, if one considers the analytic extension beyond jamming.

To determine spatial correlations, one naturally considers rate equations for the probabilities, $P_{n,m}(\ell)$, of finding an empty n-tuple and an empty m-tuple separated by ℓ sites, so $P_{n,m}(\ell) = P_{n+m}$. Then a hierarchy of rate equations can be developed for the $P_{n,m}(\ell)$, and these can be truncated by using the shielding condition $P_{n,m}(\ell) = Q^{n+m-2}P_{1,1}(\ell)$. This condition follows since all sites in the n-tuple, except the right-most one, are shielded from the m-tuple, and all sites in the m-tuple, except the left-most one, are shielded from the n-tuple. One thus obtains the closed set of equations [18]

$$dP_{1,1}(\ell)/dt = -2kQ[P_{1,1}(\ell) + P_{1,1}(\ell - 1)], \qquad (10.4)$$

for $\ell \geq 2$. Given the coupling structure of these equations, they can either be solved recursively, or by using generating function techniques. For the associated pair correlation, $C(\ell) = P_{1,1}(\ell) - (P_1)^2$, one finds [18]

$$C(\ell) = -P_1 \left[\sum_{k \geq \ell+1} (\ln P_1)^k/k! + (\ln P_1)^\ell/(2\ell!) \right], \qquad (10.5)$$

for $\ell > 0$. Of particular significance is the 'super-exponential' asymptotic decay for large ℓ.

From these correlations, one can immediately determine the structure factor, which has the form

$$S(q) = (P_1)^{1+\cos q} \sin(|\ln P_1| \sin q)/\tan(q/2), \qquad (10.6)$$

One then finds that the mean-square of the particle number fluctuations per site is given by

$$S(0) = 2|\ln P_1|P_1^2 = 2|\ln(1 - \theta)|(1 - \theta)^2, \qquad (10.7)$$

which increases initially from zero, but then decreases again to a 'small' value of $4e^{-4}$ at jamming, and actually vanishes at the 'maximal' coverage of unity, if one considers the analytic extension beyond jamming. The result for jamming was obtained long ago by MacKenzie [27].

An important consequence of the feature that, for this model, only a single empty site is required for shielding, is that all multiply disconnected empty configuration probabilities can be factorized in terms of P_1, Q, and the $P_{1,1}(\ell)$, e.g., one has $\mathcal{P}[\text{o--oo-o---oo}] = \mathcal{P}[\text{o--o}]\mathcal{P}[\text{o-o}]\mathcal{P}[\text{o---o}]Q^2\{\mathcal{P}[\text{o}]\}^{-2}$, where again '-' denotes sites of unspecified state. Thus all quantities can now be determined, including the probabilities of filled strings that display an asymptotically geometric distribution [19].

10.3.2 RSA of M-mers in 1D for $M > 2$

For these models, one can again generate an infinite closed set of equations for the P_n, analogous to the dimer case. One sees immediately that the generic form of these equations for $n \geq M-1$ is consistent with the shielding property of empty $(M-1)$-tuples. This property can be used to truncate the infinite hierarchy, obtaining a finite closed set of equations for P_n, where $n < M$, together with a conditional probability $Q = P_M/P_{M-1}$. This closed set can be integrated to obtain the exact kinetics and jamming coverages [26]. Of particular interest is the limit, $M \to \infty$, where, after suitable rescaling of time and length, one obtains the continuum 'car parking' problem of the random deposition of nonoverlapping unit intervals on an infinite line [26,27]; see Fig. 10.1. In particular, one can consider the convergence of the jamming coverage [28], $\theta_J(M) = \theta_J(\infty) + 0.216181/M + 0.362559/M^2 + \cdots$, to the car parking value of $\theta_J(\infty) = 0.747597\ldots$ [29].

One can naturally continue to consider the spatial correlations in these models. Following the general prescription of Evans *et al.* [18], and the above treatment of the dimer case, one considers probabilities, $P_{n,m}(\ell)$, for separated empty n-tuples and m-tuples. Now the application of shielding allows one to truncate these to consider just $n < M$ and $m < M$. The coupling structure of these equations is similar to the dimer case, and again solution is possible exploiting generating function techniques. A detailed analysis can be found in Bonnier *et al.* [30], but here we shall only comment further on behavior in the $M \to \infty$ car parking limit; see Sec. 10.3.3.

10.3.3 1D continuum 'car parking' limit

Here one randomly deposits nonoverlapping unit intervals on the line at an impingement rate of k per unit length. Certainly a complete analysis of this problem follows from the solution for RSA of M-mers, after taking the continuum limit as $M \to \infty$. However, a more direct analysis is also possible [29]. In fact, this analysis can be performed in a way entirely analogous to that for the discrete lattice models above [7]. Thus instead of the P_n, one considers the probabilities, $P[x]$, of finding empty or uncovered intervals of the line of length x (which could be part of longer empty intervals). One writes down a set of integro-differential evolution equations for these, as for the RSA of dimers above, simply accounting for all ways in which the empty stretch can be destroyed. Their truncation and solution simply follows from implementation of the shielding property of empty intervals of length unity, which implies that $P[x] = \exp[-k(x-1)t]P[1]$, for $x > 1$.

One thus obtains Rényi's result [29], $\theta = 1 - P[0] = \int_0^{kt} [h(w)]^2 dw$, where $h(w) = \exp[-\int_0^w (1 - e^{-u})/u \, du]$.

More recently, the above analysis of kinetics has been extended to consider spatial correlations [7,30]. Evans [7] showed that the pair correlations can be obtained entirely analogously to the discrete dimer (and M-mer) cases by first considering the evolution of probabilities, $P[x, y; r]$, of finding empty stretches of lengths x and y separated by a stretch of length r, of unspecified state. Application of the shielding property implies that $P[x, y; r] = \exp[-k(x + y - 2)t]P[1, 1; r]$ for $x > 1$ and $y > 1$. This ultimately allows determination of the probabilities, $P[0, 0; r]$, of finding two uncovered points separated by r, and of the associated correlations $C(r) = P[0, 0; r] - P[0]^2$. A complete analysis was performed only recently by Bonnier *et al.* [30] of the related center-to-center pair probability, $g(r)$, where $\theta^2 g(r) dr$ gives the probability for finding the centers of two cars separated by ℓ. Their results are shown in Fig. 10.2. In particular, they note that

$$g(r + 1) - 1 \sim (2/\ln r)^\ell/\Gamma(r + 1), \quad \text{as} \quad r \to \infty, \tag{10.8}$$

revealing an even more dramatic super-exponential decay than in the lattice models. Bonnier *et al.* [30] also determined the mean-square car number fluctuations per unit length as

$$S(0, t) = \theta + 2\theta^2 + 2\theta \left[h^2(t) - 1 \right]$$
$$- 4 \int_0^t dt_1 h^2(t_1) \int_0^{t_1} dt_2 h^2(t_2) \int_0^{t_2} dt_3 \frac{t_3 - 1 - e^{-t_3}}{h^2(t_3) t_3^2} e^{-t_3}. \tag{10.9}$$

This quantity increases initially from zero, but then decreases to a 'small' asymptotic value of about 3.7×10^{-2}.

10.3.4 2D extensions

There have been extensive studies of the RSA of dimers on a square lattice [5,7,31], where the jamming coverage increases to about 0.9068, compared with $1 - e^{-2} \simeq 0.864664$ in 1D. A density or coverage expansion is available for S_{ads}, which satisfies $S_{\text{ads}}(\theta) \sim (\theta_J - \theta)/4$ near jamming [7]. Separating-walls of empty sites of thickness unity exhibit a shielding property (an empty-site Markov field property), which can be exploited to develop approximate, but not exact, hierarchical truncation techniques [31]. Here we just mention a recent simulation study by Bartelt and Evans [32] of the decay of the spatial pair correlations $C(\ell, 0)$, for horizontal separations $l = (\ell, 0)$. At jamming, they found that $C(\ell, 0) = 0.0844, -0.00865$,

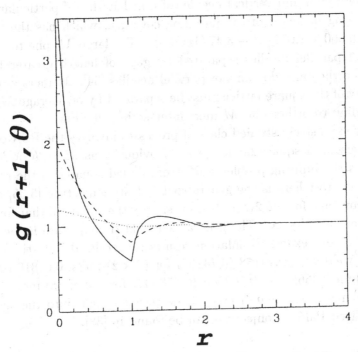

Fig. 10.2. The pair correlation function, $g(r + 1, \theta)$, vs. r, for 1D continuum car parking, with $\theta = 0.25$ (dotted curve) , 0.6 (broken curve), and at jamming (solid curve); from [30].

-0.00240, 0.00288, -0.00106, 0.00028, -0.0006, for $\ell = 0, 1, \ldots, 6$ respectively, which clearly display super-exponential decay. In fact, the decay is much more dramatic than in 1D at jamming where, from (7), one has $C(\ell) = 0.1170$, -0.0183, -0.0183, 0.0268, -0.0183, 0.0088, -0.0033, for $\ell = 0, 1, \ldots, 6$ respectively. This behavior for the square lattice, and the differences from 1D, can be elucidated in an approximate 2D treatment where, for each separation, one ignores correlations corresponding to longer separations [18]. If one applies this approximation for ℓ larger than some selected 'large' value ℓ^*, using $C(\ell^*)$ as input, then one predicts $C(\ell)$, for $\ell > \ell^*$, with reasonable accuracy. However, currently, sophisticated analytic approximations are not available. Finally, we note that the 'chord length' distribution of 1D strings of filled sites should exhibit asymptotic geometric decay [19], but again no detailed analysis is available.

Next we mention the more general problem of linear M-mers depositing 'horizontally' or 'vertically' on a square lattice. In the limit as $M \to \infty$, one obtains a nontrivial limiting problem that can be described roughly as RSA of horizontal and vertical needles of unit length. Of particular interest is the limiting behavior of the jamming coverage, which has the form [33] $\theta_J(M) = 0.660 + 1.071/M - 3.47/M^2 + \cdots$. For large M, one tends to get domains of parallel needles separated by gaps of length greater than M, filled with orthogonal domains of parallel needles [34]. M-mers lying along any 1D row of the square lattice must be separated by on average M 'points' corresponding to orthogonal M-mers intersecting this row.

Another extensively studied class of processes involves the RSA of square $K \times K$-mers on a square lattice [5,7]. Obviously, as $M = K \times K \to \infty$, one obtains the limiting problem: RSA of aligned squares in the plane. Of particular interest here is the generalized Palásti conjecture [5,35] that the jamming coverage for the 2D problem is simply the square of that for the 1D problem, for any $K \leq \infty$. The generalized Palásti conjecture is remarkably accurate, but not exact. Simulation results (generalized Palásti conjecture estimates) yield $\theta_J = 0.74788$ (0.74765) for (2×2)-mers, 0.64818 (0.64625) for (3×3)-mers [36], ..., 0.562009 (0.558902) for $(\infty \times \infty)$-mers (i.e., for aligned squares in the plane) [37]. Some recent insight into the success of the generalized Palásti conjecture can be found in [38].

10.4 CSA of monomers with NN cooperativity

10.4.1 Exact analysis in 1D

In these models, monomers adsorb at single empty sites on a linear lattice at a rate k_i for a site with $i = 0$, 1, or 2 occupied neighbors; see Fig. 10.3. Any choice of rates with $k_0 > 0$ is allowed. Clearly, if $k_0 = k_1 = k_2 = k$, say, then one obtains trivial RSA of monomers, discussed previously. If $k_1 > k_0$, then the adsorption process has a propensity for clustering or island formation, i.e., a propensity for adsorption adjacent to already occupied sites. Then one can think of k_0 as corresponding to island birth, k_1 to island growth, and k_2 to island coalescence. Considering all possible scenarios for filling an empty site, one obtains [39]

$$dP_1/dt = dP[\circ]/dt = -k_0 P[\circ \, \circ \, \circ] - k_1 P[\bullet \, \circ \, \circ] - k_1 P[\circ \, \circ \, \bullet] - k_2 P[\bullet \, \circ \, \bullet]$$

$$= -k_2 P_1 - 2(k_1 - k_2)P_2 - (k_0 - 2k_1 + k_2)P_3, \tag{10.10}$$

using 'conservation relations' for the P's. Similarly, one obtains [39]

$$dP_n/dt = -[(n-2)k_0 + 2k_1]P_n - 2(k_0 - k_1)P_{n+1}, \quad \text{for all } n > 1. \tag{10.11}$$

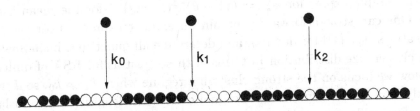

Fig. 10.3. CSA of monomers with NN cooperativity in 1D.

Clearly (10.11) is consistent with the shielding property of empty pairs, $Q_n = P_{n+1}/P_n = \exp(-k_0 t) = Q$, say, for all $n \geq 2$, which can be used to close and solve the hierarchy (10.10)-(10.11). Below we discuss the behavior of the kinetics in specific regimes. It is appropriate to mention that the above analysis also determines the island density, $D = \mathcal{P}[\bullet\circ] = \mathcal{P}[\circ\bullet] = P_1 - P_2$, and thus a measure, s_{av}, of the mean island size (i.e., the mean length of filled strings or the mean number of particles), $s_{av} = \theta/D$.

Just as for the RSA of dimers, spatial correlations can be determined by consideration of the rate equations for the $P_{n,m}(\ell)$ [17,18]. Application of the shielding yields closed coupled sets of equations for $P_{2,2}(\ell)$, $P_{2,1}(\ell)$, and $P_{1,1}(\ell)$ (rather than simply a closed set of equations for $P_{1,1}(\ell)$, as for the RSA of dimers), but these can still be solved, recursively or by generating function techniques [18], and reveal the ubiquitous super-exponential asymptotic decay. Below we discuss the behavior of the pair correlations in specific rate regimes. In particular, we emphasize that in the strong clustering regime the correlations have a 'scaled form' for separations of the order of the correlation length, only crossing over to super-exponential decay for much larger separations. Determination of the filled cluster size distribution is intrinsically more difficult [19,20], as this first requires determination of multiply disconnected empty configurations (in contrast to the RSA of dimers). Here we note only that this cluster-size distribution is asymptotically geometric for $k_2 > 0$, but decays more quickly if $k_2 = 0$ (i.e., in the absence of island coalescence).

Next we discuss a remarkable special case in which the k_i form an arithmetic progression, i.e., $k_1 = k_0(1 + d)$ and $k_2 = k_0(1 + 2d)$. Mityushin [40] first noted that here only a single empty site (rather than a pair) is required for shielding. (Since here $k_0 - 2k_1 + k_2 = 0$, the last term in (10.10) drops out, facilitating this reduction.) This produces the simplified kinetics

$$P_1 = 1 - \theta = \exp[-(1 + 2d)k_0 t + 2d(1 - e^{-k_0 t})], \qquad (10.12)$$

and the simple expression $s_{av} = \theta(1 - \theta)^{-1}(1 - Q)^{-1}$ for the mean island size. One can straightforwardly obtain an explicit expression for the pair correlations (see [41] for details), and determine all quantities, including the filled cluster size distribution in terms of these (just as for RSA of dimers).

Below we focus on the strong clustering regime where $k_1 \gg k_0$, so $d \gg 1$. Clearly here the adsorption kinetics and correlation length (or mean island size before coalescence) depend only weakly on k_2, since coalescence events are relatively rare. Thus these quantities can be most easily determined using the Mityushin rate choice. From (10.12), it follows that the characteristic time, t_c, for adsorption satisfies $k_0 t_c = d^{-1/2}$ and, for $d \gg 1$, one obtains

$$P_1 = 1 - \theta \approx \exp[-(t/t_c)^2], \quad \text{and} \quad s_{av} \approx d^{1/2}\theta(1 - \theta)^{-1}|\ln(1 - \theta)|^{-1/2}.$$
$$(10.13)$$

Thus the characteristic length scales as $d^{1/2}$.

10.4.2 Continuum 'grain growth' limit

For $\alpha = k_1/k_0 = 1 + d \gg 1$, a 'continuum picture' of the nucleation and growth of islands applies. Let a denote the lattice constant. Then one can regard islands as being nucleated randomly at a rate $I = k_0/a$ per unit length, thereafter expanding with a fixed velocity of $V = k_1 a$. This corresponds to a classic Kolmogorov 'grain growth' model [42] in which one neglects fluctuations in island growth (but not in their nucleation). A simple dimensional analysis produces a characteristic time $\tau = (IV)^{-1/2}$ and length $\lambda = (V/I)^{1/2}$, consistent with the analysis in Sec. 10.4.1. For this model, one can determine exactly both the kinetics and spatial correlations [43], by focusing on the probability that a point, or pair of points, is not covered by an island at time t (analogously to focusing on the empty-site statistics in the lattice models). One thus obtains $1 - \theta = \exp(-IVt^2)$, consistent with (10.13), and, for the pair correlations for separation r, one has $C(r) = F(r/\lambda, t/\tau)$, with

$$F(y, s) = \begin{cases} \exp(-2s^2)\{\exp[(2s - y)^2/4] - 1\}, & \text{if } y < 2s, \\ 0, & \text{if } y > 2s. \end{cases} \qquad (10.14)$$

The feature that the pair correlations are identically zero for separations $r > 2\lambda t/\tau$ reflects the deterministic nature of grain growth, and the finite speed of propagation of 'influence' in this model.

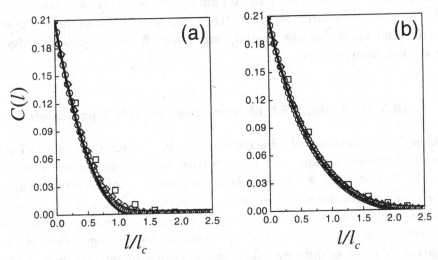

Fig. 10.4. The pair correlation, $C(\ell)$, vs. ℓ/ℓ_c for the CSA of monomers with NN cooperativity and $k_1/k_0 = 10$ (squares), 100 (diamonds), 1000 (circles). We have chosen $k_2 - 2k_1 + k_0 = 0$ (Mityushin rates). Here we show data for (a) $\theta = 0.3$, and (b) 0.7. The solid curve shows the limiting 'grain growth' scaling function; see [41].

Relating the grain growth and CSA models, as described above, shows that the pair correlations in the CSA model of Sec. 10.4.1 have the scaling form [41] $C(\ell) \sim F(\alpha^{-1/2}\ell, \alpha^{1/2}k_0 t)$, for $r \gg 1$. Thus the $C(\ell)$ display a nontrivial nonzero scaling form only for $\ell < 2\alpha^{1/2}k_0 t$. The regime $\ell > 2\alpha^{1/2}k_0 t$, in which the scaling function is identically zero, includes the asymptotic regime in the lattice model where the (very small) $C(\ell)$ exhibit super-exponential decay. Convergence of the lattice $C(\ell)$ to the asymptotic scaling form is shown in Fig. 10.4, where $\ell_c = \alpha^{1/2}$.

10.4.3 2D extensions

We now consider the CSA of monomers on a square lattice with rates k_i, for sites with $i = 0, 1, 2, 3, 4$ occupied neighbors [44]. For $k_1 > k_0$, the model exhibits clustering or island formation. In 2D, there is some flexibility in the choice of rates, and an associated variation in island morphology. For example, when $k_i = \alpha k_0$, for all $i \geq 1$, then the individual islands are Eden clusters. However, when $k_i = \alpha^i k_0$, for all $i \geq 1$, the individual islands tend to be square (before coalescence). Of course, no exact analysis is possible here, but one can still determine the scaling of characteristic times and lengths in the strong clustering regime. Often the limit as $\alpha = k_1/k_0 \to \infty$ corresponds to a continuum 2D grain growth model, where the grain growth velocity is not necessarily constant. Then one can determine the asymptotic kinetics and scaling of the spatial correlations from the exact results for these continuum models.

10.5 RSA and CSA of monomers with NN exclusion

Exclusion of occupancy of adjacent sites leads to the formation of double-spaced islands or domains on a linear lattice, and to 'checkerboard' or 'centered 2×2', denoted $c(2 \times 2)$, islands or domains on a square lattice. In both cases, this superlattice structure leads to domains that can reside on one of two interpenetrating sublattices (denoted '+' and '−'), corresponding to two different 'phases' of the domains (i.e., domains have a degeneracy of two). Adjacent domains of different phase are separated by domain or anti-phase boundaries. RSA with NN exclusion clearly leads to small disordered domains, whereas CSA can lead to the nucleation and growth of large islands with superlattice structure. Just as for equilibrium lattice gases with infinite NN repulsive interactions, the pair correlations alternate in sign. The superlattice structure (or alternation in sign of the $C(\ell)$) produces additional maxima or '1/2-order superlattice peaks' in the structure factor at positions $q^* = \pi$, in 1D, and $\mathbf{q}^* = (\pi, \pi)$, in 2D. Finally we mention a fluctuation-correlation relation, distinct from (10.2), appropriate for these situations. Let $\#_-$ and $\#_+$ denote the number of adsorbed particles on the '−' and '+' sublattices, in a region of V sites. Then one has $\langle (\#_- - \#_+)^2 \rangle \sim V S(\mathbf{q}^*)$, in the limit of large V.

10.5.1 RSA with NN exclusion in 1D

The RSA of monomers on a linear lattice with an impingement rate of k per site, subject to the constraint that sites with occupied neighbors cannot be filled, is actually 'isomorphic' to the RSA of dimers. (Each monomer corresponds to the center of a dimer on the dual lattice.) Thus the behavior of the kinetics, and even the fluctuations in particle number, are obtained from the results for the RSA of dimers after a simple transformation $\theta \to 2\theta$ of the coverage. In particular, the jamming coverage for RSA of monomers with NN exclusion is given by $\theta_J = (1 - e^{-2})/2$. However, this does not provide a complete solution to the problem, which, instead, can be obtained as a special case of the analysis of the CSA of monomers (Sec. 10.4.1) with $k_1 = k_2 = 0$. Since an empty pair, rather than a single empty site, is required for shielding, analysis of the pair correlations is more complicated than for the RSA of dimers. Note that filled-site pair probabilities for the RSA of dimers do not give precise information on dimer location (the filled sites could be either left or right ends of dimers), in contrast to the corresponding quantities for the RSA of monomers with NN exclusion. For the latter problem, one obtains [45]

$$C(\ell) = \tfrac{1}{2}(-1)^\ell (1 - 2\theta) \sum_{k \geq 0} |\ln(1 - 2\theta)|^{\ell + 2k + 1} / (\ell + 2k + 1)!, \qquad (10.15)$$

again revealing super-exponential decay.

From these $C(\ell)$, one can determine the structure factor [45]:

$$S(q) = \tfrac{1}{2}(1 - 2\theta)^{1 + \cos q} \sin(|\ln(1 - 2\theta)| \sin q) / \sin q. \qquad (10.16)$$

Of particular interest is its value at the 1/2-order peak $S(\pi) = \tfrac{1}{2}|\ln(1 - 2\theta)|$, which provides information on mean-square fluctuations in the population difference of the two sublattices. Note that this quantity diverges if analytically extended beyond jamming to the 'maximal' coverage of 1/2. Finally, $S(0) = (1 - 2\theta)^2 S(\pi)$ provides information on (total) particle number fluctuations consistent with the previous analysis of RSA of dimers.

For contrast, we note that the 1D equilibrium lattice gas with infinitely repulsive NN interactions has $C(\ell) = (-1)^\ell \theta^{|\ell|+1}(1 - 2\theta)^{1-|\ell|}$, from which one can readily determine $S(q)$. Here we just mention that $S(\pi) = \theta(1 - \theta)(1 - 2\theta)^{-1}$, and that again $S(0) = (1 - 2\theta)^2 S(\pi)$.

10.5.2 CSA with NN exclusion in 1D and the continuum two-state grain growth limit

Here sequential adsorption of monomers with NN exclusion occurs at a rate k_i for sites with i occupied second-nearest neighbors; see Fig. 10.5. For $\alpha = k_1/k_0 > 1$, the model exhibits a propensity for the formation of double-spaced islands: k_0 corresponds to birth, k_1 to island growth, and k_2 to the coalescence of in-phase islands. When out-of-phase islands meet, a permanent anti-phase or domain boundary, $\cdots \circ \bullet \circ \bullet \circ \circ \circ \bullet \circ \bullet \circ \cdots$, is formed. Here exact solvability follows from the shielding property of empty quartets. For $\alpha \gg 1$, clearly the scaling of the characteristic time t_c and length ℓ_c and the form of the kinetics, will be the same as described for CSA with NN cooperativity in Sec. 10.4.1. As noted above, the pair correlations alternate

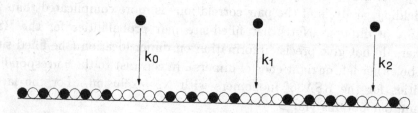

Fig. 10.5. CSA of monomers with NN exclusion and second NN cooperativity in 1D.

in sign, so here one must specify two scaling functions for the upper and lower envelopes of the $C(\ell)$, i.e., $C(\ell) \sim F_{+(-)}(\ell/\ell_c, t/t_c)$, for even (odd) ℓ. Here one has $F_+ > 0$, with $F_+(\ell/\ell_c \to 0) = C(\ell = 0) = \theta(1 - \theta)$, and $F_- < 0$, with $F_-(\ell/\ell_c \to 0) = C(\ell = 1) = -\theta^2$.

The above scaling functions, $F_{+(-)}$, can be determined exactly from analysis of an appropriate 'two-state' grain growth problem [41], where grains of two phases (corresponding to the two phases of double spaced islands) are nucleated randomly at a constant rate, and thereafter expand at constant velocity. When growing in-phase grains impinge, they merge. When growing out-of-phase grains impinge, a permanent grain boundary is formed. Exact determination of spatial correlations in the grain growth model [43] provides the desired $F_{+(-)}$. Here we only note one basic property, which follows directly from probability conservation and symmetry relations: $F_-(t = \infty) = -F_+(t = \infty)$ [41]. The convergence of the $C(\ell)$ in the lattice problem to the scaling functions determined from the grain growth problem is shown in Fig. 10.6.

10.5.3 2D extensions

The RSA (of monomers) with NN exclusion on a square lattice has become the prototypical lattice RSA problem [7]. Here the jamming coverage decreases to about 0.3641, compared with $(1 - e^{-2})/2 \simeq 0.432332$ for the 1D problem. Coverage or density expansions are available for S_{ads} [22,46], which satisfies $S_{ads} \sim \theta_J - \theta$ near jamming [7]. A simulation study reveals the super-exponential decay of the spatial pair correlations. For horizontal separations, $\mathbf{l} = (\ell, 0)$, one finds at jamming: $C(\ell, 0) = 0.2316, -0.1327,$ $0.0519, -0.0142, 0.0027, -0.0003,$ for $\ell = 0, 1, \ldots, 5$. Just as for the RSA of dimers, the decay is much more dramatic than in 1D, where, from (10.15),

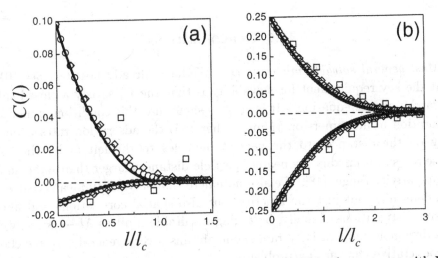

Fig. 10.6. The pair correlation, $C(\ell)$, vs. ℓ/ℓ_c for the CSA of monomers with NN exclusion and second-NN cooperativity; $k_1/k_0 = 10$ (squares), 100 (diamonds), 1000 (circles). We have chosen $k_2 - 2k_1 + k_0 = 0$. Here we show data for (a) $\theta = 0.11$, and (b) at jamming. The solid curve shows the limiting 'grain growth' scaling function; see [41].

one finds at jamming: $C(\ell) = 0.2425, -0.1869, 0.1101, -0.0516, 0.0199, -0.0065$, for $\ell = 0, 1, \ldots, 5$, respectively. The 2D structure factor at the 1/2-order superlattice peak is given by $S(\pi, \pi) = 0, 0.126, 0.366, 0.874, 1.97$ when $\theta/\theta_J = 0, 1/4, 1/2, 3/4, 1$, respectively. One also finds that $S(0,0) = 0, 0.064, 0.058$ when $\theta/\theta_J = 0, 1/2, 1$, respectively. As noted above, these quantities reflect mean-square fluctuations in the difference in sublattice populations, and in the total particle number, respectively, and display qualitatively similar behavior to the 1D case.

It is natural to further consider CSA with NN exclusion, where the co-operativity introduces a propensity for $c(2 \times 2)$ island formation. Now the specific choice of cooperativity affects not just the pre-coalescence morphology and scaling of islands, but also the residual pattern of domain boundaries in the jammed state [47]. This model provides a paradigm for (precursor-mediated) island-forming chemisorption. In the regime of strong clustering, the kinetics and scaling can often be deduced from appropriate 2D two-state grain growth models. The spatial pair correlations still alternate in sign, being positive (negative) for separations connecting sites on the same (different) $c(2 \times 2)$ sublattice [47], and the scaling functions for the upper (F_+) and lower (F_-) envelopes can often be determined exactly from grain growth models.

10.6 Generalizations

Most general solvable model in 1D: We have already noted in Sec. 10.2 that the key requirement for solvability is that the P_n satisfy a closed sub-hierarchy of rate equations. It is easy to show that this condition is met for adsorption of monomers or M-mers where: (i) the adsorption rates depend only on the distances to the nearest particles to the left and right; (ii) these rates are constant for nearest-particle distances larger than some finite 'cooperativity range' [48]. We also note here that solvability means that one can determine not just the kinetics, but also spatial correlations and filled cluster distributions. Finally, by taking a suitable limit as $M \to \infty$, as well as a divergent cooperativity range, one obtains a very general solvable class of cooperative car parking problems.

Simplest nonsolvable model in 1D: The simplest model where the P_n do not satisfy a closed hierarchy, and thus solvability is lost, is the CSA of monomers with the most general choice of range-two cooperative effects. However there is still an exact shielding property of empty quartets that can be incorporated into suitable approximate hierarchy truncation schemes [49].

General initial conditions in 1D: The above presentation has focused exclusively on the case of an initially empty lattice. However, this is quite unnecessary as exact solvability is preserved for a broad range of initial conditions [50]. Suppose, for the model under consideration, that L empty sites are required for shielding (with an initially empty lattice). Then consider initial conditions which satisfy Nth-order spatial Markov statistics (i.e., strings of N sites, each of which is specified as either filled or empty, shield sites on one side from the influence of those on the other). If $N \leq L$, then the initial state clearly satisfies a shielding property for a string of L empty sites, so truncation of the P_n-hierarchy can proceed exactly as for the case of an initially empty lattice. If $N > L$, then the initial conditions satisfy a shielding property for strings of N empty sites, and it is easy to show that this property will be satisfied for all times, and can be used to truncate exactly the P_n-hierarchy [50].

Edge effects and adsorption site inhomogeneity: In semi-infinite or finite systems, obviously one looses the translational invariance utilized in the above studies. Thus instead of a single coverage one must determine a separate (mean) coverage for each site. In general these will vary significantly near the edge, but become uniform in the interior of the system. Again the empty-site shielding property can yield exact results. Loosely speaking, one can use this property to shield the influence of the edge of a semi-infinite system, and to show that certain conditional probabilities reduce to their values for infinite systems. Thus, e.g., for the RSA of dimers, one obtains a closed coupled set of equations for the probabilities of empty sites, labeled by their position, and this set can be solved recursively or by generating function techniques [18,51] (akin to the analysis of pair correlations).

One can also consider systems with a mixture of site types, characterized by different adsorption rates. The site types could be distributed, e.g., periodically or stochastically. The latter case has been analyzed extensively within the context of binding or reaction at the pendant groups of 1D 'random' copolymers [52]. Here an efficient strategy is to consider probabilities for empty strings of sites of specified type (which satisfy standard shielding properties). Then one can reconstruct probabilities for empty strings of sites of unspecified type using the known statistics of the distribution of site types [53].

Competitive adsorption: Epstein [54] first noted that the exact treatment of the RSA of M-mers in 1D easily extends to allow consideration of the RSA of mixtures of M-mers with different sizes. In fact, exact solvability also extends to CSA problems [55].

Transient mobility: Here we consider the scenario where adsorbing species are not immediately accommodated at the site where they impinge. In the case of dimer adsorption, dissociation might occur on impact with the surface, and the constituent atoms might separate a number of lattice spacings before becoming thermally accommodated. This feature was first incorporated into an RSA model to describe the dissociative adsorption of oxygen on Pd(100) [56]. For metal homoepitaxy (where single atoms are deposited), transient mobility between adsorption sites on a clean surface apparently does not occur [57], although deposited adatoms may be drawn to nearby pre-deposited atoms [58] in the regime where thermal diffusion is inoperative. A broad class of these processes can be solved exactly in 1D, again by virtue of a shielding property of empty sites [7,56,59].

Multilayer growth: Unfortunately, on the whole natural multilayer generalizations of the RSA and CSA processes discussed here are not exactly solvable (even for a 1D substrate). This is the case for multilayer RSA of dimers on a 1D substrate, where adsorption in layer M requires a pair of empty layer-M sites, and that the two underlying layer-$(M-1)$ sites are either filled, or have both neighbors filled. This guarantees that the jammed state of each layer includes only isolated empty sites [60]. Clearly the filling of each layer is independent of the filling of higher layers. However, it is not possible to analyze exactly even the filling of the second layer.

While there have been many studies of asymptotic kinetic roughening in multilayer RSA type models [61], few studies have focused on the details of the adsorption kinetics and correlations for each layer during the experimentally important initial stage of growth. Here we just note that exact results are available for these properties for a number of models involving RSA with different adsorption-site geometries [62], as well as for 'single-step' and for various 'restricted solid-on-solid' models [63].

10.7 Conclusions

We have presented the remarkably broad class of solvable lattice models of RSA and CSA processes in 1D, as well as described the solvability of the corresponding continuum limits. These 1D studies are particularly valuable in elucidating RSA and CSA in higher dimensions since, in contrast to equilibrium systems, the behavior of these models is not strongly dimension dependent. Certainly, application to adsorption on single-crystal surfaces provides motivation for further analysis of 2D models that are not exactly solvable. Several obvious questions arise. As regards the adsorption kinetics, does the analytic extension of S_{ads} above jamming vanish at some 'maximal' coverage, as in the above 1D examples? We have found that this is not the case for branching media that mimic 2D systems [64]. Can one develop sophisticated techniques for analysis of super-exponentially decaying spatial correlations, and for other features of the stochastic geometry of the adlayer [7] (chord-length distributions, domain sizes, domain-boundary structure, etc.)? In conclusion, RSA and CSA constitute a particularly rich and important class of far-from-equilibrium processes that continues to provide challenging problems in nonequilibrium statistical mechanics.

The author gratefully acknowledges recent discussions and collaborations with Maria Bartelt on several of the topics presented here. His work was supported by the Division of Chemical Sciences, Office of Basic Energy Sciences, of the US Department of Energy (USDOE). Ames Laboratory is operated for the USDOE by Iowa State University under Contract W-7405-Eng-82.

References

[1] E. S. Hood, B. H. Toby and W. H. Weinberg, *Phys. Rev. Lett.* **55**, 2437 (1985).

[2] J. D. Weeks and G. H. Gilmer, *Adv. Chem. Phys.* **40**, 157 (1979).

[3] H. E. Stanley, *Introduction to Phase Transitions and Critical Phenomena* (Oxford University Press, New York, 1971).

[4] C. T. Campbell, M. T. Paffett and A. F. Voter, *J. Vac. Sci. Tech.* **A4**, 1342 (1986).

[5] H. Solomon and H. Weiner, *Comm. Stat.* **A15**, 2571 (1986).

[6] M. C. Bartelt and V. Privman, *Int. J. Mod. Phys.* **B5**, 2883 (1991).

[7] J. W. Evans, *Rev. Mod. Phys.* **65**, 1281 (1993).

[8] S. Günther, Ph. D. thesis (Universität Ulm, 1995); S. Günther, M. C. Bartelt, J. W. Evans and R. J. Behm, unpublished.

[9] P. Rempp, *Pure Appl. Chem.* **46**, 9 (1976).

[10] E. A. Boucher, *Prog. Polymer Sci.* **6**, 63 (1978).

[11] N. A. Plate and O. V. Noah, *Adv. Polymer Sci.* **31**, 134 (1978).

[12] K. Kawasaki, in *Phase Transitions and Critical Phenomena*, Vol. **2**, C. Domb and M. S. Green, eds. (Academic Press, New York, 1972).

[13] R. J. Glauber, *J. Math. Phys.* **4**, 294 (1963).

[14] R. Kutner, *Phys. Lett.* **A81**, 239 (1981).

[15] J. W. Evans and D. K. Hoffman, *Phys. Rev.* **B30**, 2704 (1984).

[16] J. W. Evans, D. R. Burgess and D. K. Hoffman, *J. Chem. Phys.* **79**, 5011 (1983).

[17] N. A. Plate, A. D. Litmanovich, O. V. Noah, A. L. Toom, and N. B. Vasilyev, *J. Polymer Sci.* **12**, 2165 (1974).

[18] J. W. Evans, D. R. Burgess and D. K. Hoffman, *J. Math. Phys.* **25**, 3051 (1984).

[19] R. S. Nord, D. K. Hoffman and J. W. Evans, *Phys. Rev.* **A31**, 3820 (1985).

[20] J. W. Evans and R. S. Nord, *Phys. Rev.* **A31**, 1759 (1985).

[21] D. K. Hoffman, *J. Chem. Phys.* **65**, 95 (1976).

[22] R. Dickman, J. S. Wang and I. Jensen, *J. Chem. Phys.* **94**, 8252 (1991).

[23] T. M. Liggett, *Interacting Particle Systems* (Springer, Berlin, 1985).

[24] P. J. Flory, *J. Am. Chem. Soc.* **61**, 1518 (1939).

[25] E. R. Cohen and H. Reiss, *J. Chem. Phys.* **38**, 680 (1963).

[26] J. J. Gonzalez, P. C. Hemmer and J. S. Høye, *Chem. Phys.* **3**, 288 (1974).

[27] J. K. MacKenzie, *J. Math. Phys.* **37**, 723 (1962).

[28] M. C. Bartelt, J. W. Evans and M. L. Glasser, *J. Chem. Phys.* **99**, 1438 (1993).

[29] A. Rényi, *Sel. Trans. Math. Stat. Prob.* **4**, 205 (1963).

[30] B. Bonnier, D. Boyer and P. Viot, *J. Phys.* **A27**, 3671 (1994).

[31] R. S. Nord and J. W. Evans, *J. Chem. Phys.* **82**, 2795 (1985).

[32] M. C. Bartelt and J. W. Evans, unpublished.

[33] B. Bonnier, M. Hontebeyrie, Y. Leroyer, C. Meyers, and E. Pommiers, *Phys. Rev.* **E49**, 305 (1994).

[34] Y. Leroyer and E. Pommiers, *Phys. Rev.* **B50**, 2795 (1994).

[35] I. Palásti, *Publ. Math. Inst. Hung. Acad. Sci.* **5**, 353 (1960).

[36] R. S. Nord, *J. Stat. Comput. Sim.* **39**, 231 (1991).

[37] B. J. Brosilow, R. M. Ziff and R. D. Vigil, *Phys. Rev.* **A43**, 631 (1991); V. Privman, J. S. Wang and P. Nielaba, *Phys. Rev.* **B43**, 3336 (1991).

[38] B. Bonnier, M. Hontebeyrie and C. Meyers, *Physica* **A198**, 1 (1993).

[39] J. B. Keller, *J. Chem. Phys.* **38**, 325 (1963).

[40] L. G. Mityushin, *Prob. Peredachi Inf.* **9**, 81 (1973).

[41] M. C. Bartelt and J. W. Evans, *J. Stat. Phys.* **76**, 867 (1994).

[42] A. N. Kolmogorov, *Bull. Acad. Sci. USSR* **3**, 355 (1937).

[43] S. Ohta, T. Ohta and K. Kawasaki, *Physica* A**140**, 478 (1987).

[44] D. E. Sanders and J. W. Evans, *Phys. Rev.* A**38**, 4186 (1988).

[45] J. W. Evans, *Surf. Sci.* **215**, 319 (1989).

[46] A. Baram and D. Kutasov, *J. Phys.* A**22**, L251 (1989); J. W. Evans, *Phys. Rev. Lett.* **62**, 2642 (1989); Y. Fan and J. K. Percus, *ibid.* **67**, 1677 (1991).

[47] J. W. Evans, R. S. Nord and J. A. Rabaey, *Phys. Rev.* B**37**, 8598 (1988).

[48] J. W. Evans, *J. Phys.* A**23**, 2227 (1990).

[49] J. W. Evans and D. R. Burgess, *J. Chem. Phys.* **79**, 5023 (1983).

[50] N. O. Wolf, J. W. Evans and D. K. Hoffman, *J. Math. Phys.* **25**, 2519 (1984).

[51] J. T. Terrell and R. S. Nord, *Phys. Rev.* A**46**, 5260 (1992).

[52] J. J. Gonzalez and P. C. Hemmer, *J. Polymer Sci. (Polymer Lett. Edition)* **14**, 645 (1976).

[53] J. W. Evans and R. S. Nord, *J. Stat. Phys.* **38**, 681 (1985).

[54] I. R. Epstein, *Biopolymers* **18**, 765 (1979).

[55] J. W. Evans, D. K. Hoffman and D. R. Burgess, *J. Chem. Phys.* **80**, 936 (1984).

[56] J. W. Evans, *J. Chem. Phys.* **87**, 3038 (1987).

[57] J. W. Evans, D. E. Sanders, P. A. Thiel and A. E. DePristo, *Phys. Rev.* B**41**, 5410 (1990).

[58] S. C. Wang and G. Ehrlich, *Phys. Rev. Lett.* **71**, 4174 (1993).

[59] E. V. Albano and V. D. Pereyra, *J. Chem. Phys.* **98**, 10044 (1993); V. Privman, *Europhys. Lett.* **23**, 341 (1993).

[60] M. C. Bartelt and V. Privman, *J. Chem. Phys.* **93**, 6820 (1990); P. Nielaba and V. Privman, *Phys. Rev.* A**45**, 6099 (1992).

[61] F. Family and T. Vicsek, eds., *Dynamics of Fractal Surfaces* (World Scientific, Singapore, 1991).

[62] J. W. Evans, *Phys. Rev.* B**43**, 3897 (1991).

[63] J. W. Evans and M. C. Bartelt, *Phys. Rev.* E**49**, 1061 (1994).

[64] J. W. Evans and C. J. Westermeyer, unpublished.

11

Lattice models of irreversible adsorption and diffusion

Peter Nielaba

11.1 Introduction

In many experiments on the adhesion of colloidal particles and proteins on substrates, the relaxation time scales are much longer than the times for the formation of the deposit [1-5]. Owing to its relevance for the theoretical study of such systems, much attention has been devoted to the problem of *irreversible monolayer* particle deposition, termed random sequential adsorption (RSA) or the car parking problem [6-24]; for reviews see [25,26]. In RSA studies the depositing particles (on randomly chosen sites) are represented by hard-core extended objects; they are not allowed to overlap.

In this chapter, numerical Monte Carlo studies and analytical considerations are reported for 1D and 2D models of multilayer adsorption processes. Deposition without screening is investigated; in certain models the density may actually increase away from the substrate. Analytical studies of the RSA late stage coverage behavior show the crossover from exponential time dependence for the lattice case to the power-law behavior in continuum deposition. In 2D, lattice and continuum simulations rule out some 'exact' conjectures for the jamming coverage. For the deposition of dimers on a 1D lattice with diffusional relaxation the limiting coverage (100%) is approached according to the $\sim 1/\sqrt{t}$ power law; this is preceded, for fast diffusion, by the mean-field crossover regime with intermediate, $\sim 1/t$, behavior. In the case of k-mer deposition ($k > 3$) with diffusion the void fraction decreases according to the power law $t^{-1/(k-1)}$. In the case of the RSA of lattice hard-squares in 2D with diffusional relaxation, the approach to full coverage is $\sim t^{-1/2}$. In case of RSA-deposition with diffusion of two-by-two square objects on a 2D square lattice the coverage also approaches 1 according to the power law $t^{-1/2}$, while on a finite periodic lattice the final state is a frozen random regular grid of domain walls connecting single-site defects.

11.2 Irreversible multilayer adsorption

In the monolayer deposition of colloidal particles and macromolecules one can assume that the adhesion process is irreversible [1,3,27]. Experimental studies of multilayer deposition have been reported recently [27-34], as well as theoretical mean-field studies [24,35] and rate-equation approximations [24]. In irreversible deposition, the most profound correlations are due to blocking by the already deposited particles of the available area for deposition of new particles. This infinite-memory effect has been studied in the monolayer case under the term RSA (for reviews see [25,26]). The deposition process stops at a certain jamming coverage which is less than close-packing. The blocking also plays an important role in the higher-layer particle-on-particle deposition [24]. Another effect present only in the multilayer case is the *screening* of lower layers by the particles in higher layers. Models without blocking but with screening allowed, fall in the class of ballistic deposition [36] or diffusion-limited aggregation [37], depending on the mechanism of particle transport to the surface. Since colloidal experiments usually involve not too many layers (up to about 30), the details of the transport mechanism of particles to the surface are less important than in ballistic deposition or diffusion-limited aggregation. Appropriate models have been formulated [24,38-40] to eliminate or suppress the screening of lower layers by particles adhering in higher layers, emphasizing those correlation and dynamics effects that result from the 'jamming' or blocking due to finite particle size and irreversibility of the deposit formation. Generally, deposition dominated by jamming effects will result in an amorphous deposit so that the notion of 'layers' in a true continuum deposition can be employed only as an approximate concept. However, simplified lattice models can reveal many general aspects of the deposition processes as well as new unexpected features.

Let us survey the theoretical studies of lattice models with screening eliminated by disallowing overhangs [38] or by allowing overhangs only over gaps that are small enough [39,40]. We study the deposition of k-mers, of length k, on the linear periodic 1D lattice of unit spacing, and the deposition of square-shaped $(k \times k)$-mers on the periodic square lattice of unit spacing in 2D. The deposition site is chosen at random. The time scale, t, is fixed by having exactly L^d deposition attempts per unit time, where L is the linear lattice size. Systems have been studied with sizes up to $L = 10^5$ in 1D and $L \times L = 1000^2$ in 2D. Various Monte Carlo runs (for $k = 2, 3, 4, 5, 10$ in 1D, and $k = 2, 4$ in 2D) have gone up to $t = 150$. The numerical values are consistent with the exact results [9] for layer $n = 1$ in 1D.

We first focus on a model where the saturation layer coverage decreases with layer height [38]. Only if *all* the lattice segments in the selected landing site are already covered by exactly $n - 1$ layers is the arriving k-mer deposited, increasing the coverage to n $(n \geq 1)$. For small t, the coverage (fraction of the total volume covered by depositing particles) increases according to $\theta_n(t) \propto t^n$, as expected from the mean-field theory. The numerical results suggest that for lattice models the fraction of the occupied area in the nth layer, $\theta_n(t)$, approaches the saturation value exponentially,

$$\theta_n(t) \approx \theta_n(\infty) + B_n \exp(-t/\tau_n). \tag{11.1}$$

A plot of the coverages in the first layers of a 2D lattice as functions of time is shown in Fig. 11.1. In the higher layers, the larger the n value, the more gaps are contained by the jammed state *in the deposition without overhangs*. The growth in the higher layers proceeds more and more via uncorrelated 'towers'. The jamming coverages vary according to a power law, with no length scale, reminiscent of critical phenomena,

$$\theta_n(\infty) - \theta_\infty(\infty) \approx An^{-\phi}. \tag{11.2}$$

Within the limits of the numerical accuracy the values of the exponent ϕ are universal for $k \geq 2$. Based on numerical data analysis one obtains the estimates $\phi(1D) = 0.58 \pm 0.08$ and $\phi(2D) = 0.48 \pm 0.06$. These values are most likely to be exactly $1/2$, as suggested by analytical random walk arguments [41,42].

Next we consider a model for the multilayer deposition of k-mer 'particles' when the saturation layer coverage increases with layer height [24,39,40]. The deposition attempts are 'ballistic'; particles arrive at a uniform rate per site. The group of those k lattice sites that are targeted in each deposition attempt is examined to find the lowest layer $n \geq 1$ such that all the k sites are empty in that layer (and all layers above it). The deposition rules are illustrated in Fig. 11.2. Note that initially the substrate is empty, in all the sites and layers $1, 2, 3, \ldots$. If the targeted group of sites is in the layer $n = 1$, then the arriving particle is deposited: the k sites become occupied. However, if the targeted layer is $n > 1$, then the deposition attempt is accepted only provided no gaps are thereby *partially* covered in layer $n - 1$. Thus, successful deposition of a k-mer in layer $n > 1$ requires that it *fully* covers any gap underneath it in layer $n - 1$. This lower gap must therefore be of size $1, 2, \ldots, k - 1$; see Fig. 11.2. Otherwise the attempt is rejected. In successful deposition, at least one of the lattice sites below one of the end coordinates of the arriving k-mer is already occupied in layer $n - 1$. The other end coordinate has an occupied nearest-neighbor lattice site in layer

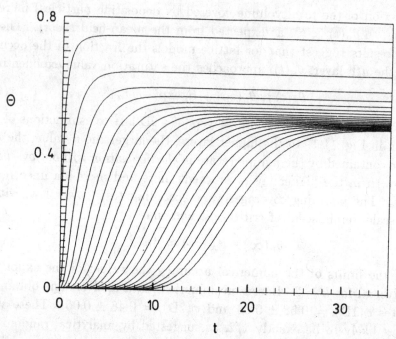

Fig. 11.1. The coverages in layers $n = 1, 2, \ldots, 20$ as functions of the time of deposition, t, of (2×2)-mers on the square lattice. The monolayer coverage is the uppermost curve, and generally $\theta_n(t) < \theta_{n-1}(t)$, for each t. These results were obtained on a 1000×1000 lattice, and represent averages over 140 Monte Carlo runs.

$n - 1$, or is itself occupied (in layer $n - 1$). Thus all overhangs are disallowed that would *partially* block (screen) gaps in lower layers. In particular, gaps that are large enough to accommodate future deposition events (in layers $n - 1$ or lower) are k-site or larger. These gaps will not be blocked until they are filled up in lower layers. The final, large-time configuration in each layer contains gaps of at most $k - 1$ consecutive empty sites. In fact, deposition in lower layers $1, \ldots, N$ is unaffected by deposition in layers $N + 1$ onwards.

For the particular deposition rule considered here the coverage in layer n eventually exceeds that in layer $n - 1$ at larger times. This unexpected behavior was found numerically for all layers $n \leq 55$ and for all k ($k = 2, 3, 4, 5, 10$) studied. Clear evidence was found of the power-law behavior

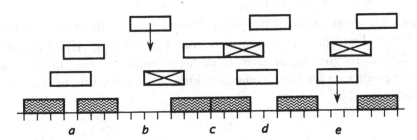

Fig. 11.2. Deposition of trimers on the 1D substrate. The shaded trimers illustrate a possible configuration in the layer $n - 1$. Lower layers $(n - 2, n - 3, \ldots)$ are not shown. Instead, the underlying lattice structure is indicated. Possible locations of trimer arrival in a deposition attempt are illustrated. Gap a of size 1 will be successfully covered by any overlaying trimer (two out of the three possible locations are shown). Gap b of size 4, however, cannot be blocked in any of the possible arrival positions: in four 'partial overhang' cases (one of which is shown) the deposition will be rejected (crossed trimer). Two other configurations (one of which is shown by an arrow-marked trimer) will result in deposition in a layer lower than n. Deposition over gaps of size 0, marked c, is always possible. For gaps of size 2, marked d, the successful deposition configurations are those that *fully* cover this gap. Both of them are shown, as is one of the two disallowed configurations. Finally, the gap of size 3, marked e, cannot be fully covered. Therefore, the deposition will either be rejected (crossed trimer) or occur in a lower layer (arrow-marked). We note that head-on depositions (rightmost trimer), which involve no gaps, are allowed as well by the deposition rules.

(11.2) with $A < 0$, and $\phi = 0.3 \pm 0.15$ was established to be universal for all k studied. When regions sufficiently covered, i.e, containing k-mers or gaps of sizes up to $k - 1$, have formed in layer $n - 1$, then deposition with overhangs beyond those regions will be delayed. Thus, there will be some preference for higher density in layer n especially near the ends of the regions occupied in layer $n-1$. To test the above suggestion, the following monolayer dimer-deposition model was considered. We select randomly $\rho L/2$ dimers and make the ρL sites thus selected unavailable for deposition for times $0 \leq t \leq t_s$. A 'sleeping time' t_s for a fraction ρ of lattice sites (grouped in dimers) in monolayer deposition will model, supposedly, the effect of disallowed overhangs (over gaps of size larger than 1 in the lower layer) on multilayer deposition in layer n, provided we loosely identify $t_s \propto n$. Indeed, the multilayer data suggest that the time needed to build up the nth layer coverage grows linearly with n. For instance, the time $t_{1/2}$ defined via $\theta(t_{1/2}) = (1/2)\theta_n(\infty)$ grows according to $t_{1/2} \simeq \eta n$, where the coefficient η is of order 1. After time t_s all the blocked sites are released and can be occupied in subsequent deposition attempts. The variation of the jamming coverage is $\propto t_s^{-\phi}$, with $\phi \simeq 1/3$.

Another interesting quantity besides the coverage, is the time evolution of the interfacial width. The standard, Kardar-Parisi-Zhang (KPZ) model [43] of the kinetic roughening of growing surfaces (reviewed, e.g., in [44]) yields a scaling prediction for the interfacial width W as a function of time, t, and substrate size, L,

$$W \simeq L^{\zeta} F\left(tL^{-z}\right), \tag{11.3}$$

where for 1D surfaces the exponent values are $\zeta_{\text{KPZ}} = 1/2$ and $z_{\text{KPZ}} = 3/2$. In fact, the value $\zeta = 1/2$ is common to many 1D models of fluctuating interfaces, stationary or growing. However, $z = 3/2$ is characteristic of the KPZ universality class. For instance, for stationary, thermally fluctuating interfaces, $z = 2$. These values have been well established by numerical simulations and are believed to be exact (in 1D). For large lattice sizes $(L \to \infty)$ one should not expect an L-dependence of W; this corresponds to the assumption that $F(y) \propto y^{\zeta/z}$ for small arguments $y = tL^{-z}$. For large arguments, the function F approaches a constant, corresponding to $W \propto L^{\zeta}$. In deposition, the KPZ approach focuses on fluctuations of the growing surface that are determined by the evolving structure of the uppermost deposit layers; these are in the process of being formed due to the arrival and adhesion of new particles according to the rules of the particular model at hand. However, once the advancing surface has passed each fixed height h (measured from the substrate, which is at $h = 0$), and once all the transient rearrangement of particles (if allowed in the model) has run its course, the remaining asymptotic (large-time) 'saturated' deposit density will be a function of h only. In an interesting study, Krug and Meakin [45] argued that to the leading order, the KPZ fluctuations affect the growth rate such that (in 1D) the average deposit density at the height h decreases $\propto h^{-2/3}$ away from the substrate to the limiting large-h value. This prediction has been verified for several ballistic deposition models in 1D and 2D [45,46].

In the model described above the density actually *increases* away from the substrate, according to (11.2) with $A < 0$ and an exponent that is smaller (in absolute value) than the KPZ-contribution exponent. The growth of the interface width W,

$$W = \left\langle \left[\frac{1}{L}\sum_{j=1}^{L} h_j^2 - \left(\frac{1}{L}\sum_{j=1}^{L} h_j \right)^2 \right]^{1/2} \right\rangle, \qquad (11.4)$$

where h_j is the height of the deposit at site j, has been studied recently [40]. The results place the model within the KPZ universality class. In fact, the exponent values and scaling form, (11.3), are confirmed quite accurately as $\zeta/z = 0.34 \pm 0.04$ and $\zeta = 0.49 \pm 0.03$. Thus the suppression of screening apparently does not change the universality class of the growing interface fluctuations even though it can dominate other aspects of the deposit structure such as variation of the saturated deposit density with height. In the latter, the KPZ fluctuation effects are probably present as a subleading correction.

We note that several numerical studies of the KPZ and other growth-universality classes have been done by scaling data collapse and tests of universality of quantities derived from scaling functions similar to $F(t)$ [47-49]. In a Monte Carlo study of ballistic and other deposition processes, Ko and Seno [50] question the KPZ universality class identification for ordinary ballistic deposition in 1D.

11.3 Continuum limit in RSA

For continuum deposition RSA (monolayer) models, the long-time behavior is generally given by the power law

$$\theta(t) = \theta(\infty) - \text{constant} \times (\ln t)^q/t^p, \qquad (11.5)$$

where $q = 0$ in most cases. In 1D one has $p = 1$, $q = 0$ [9]. Analytical arguments [12,13] support the numerical conjecture [11] that $p = 1/d$ and $q = 0$ for deposition of spherical objects in d dimensions. However, there are analytical [13] and numerical [21,23,51] indications that the precise form of the convergence law depends on the shape and orientational freedom of the depositing objects. Numerical lattice simulations [17-19,38] are easier to perform than continuum deposition simulations. However, the approach to jamming coverage is asymptotically exponential in lattice deposition models; see (11.1) with $n = 1$. For the continuum version of the deposition of hypercubic objects of fixed orientation in d dimensions, Swendsen [13] proposed an analytical argument for the asymptotic law (11.5) with $q = d - 1$ and $p = 1$.

Table 11.1.

Jamming coverage estimates for the $d = 2$ lattice deposition of $k \times k$ oriented squares. The error bars shown are statistical. The data were averaged over several Monte Carlo runs the number of which is given in the 'Runs' column

k	$\theta_k(\infty)$	Runs
2	0.74793 ± 0.00001	30000
3	0.67961 ± 0.00001	30000
4	0.64793 ± 0.00001	15000
5	0.62968 ± 0.00001	17000
10	0.59476 ± 0.00004	4400
20	0.57807 ± 0.00005	600
50	0.56841 ± 0.00010	150
100	0.56516 ± 0.00010	199

The crossover from the characteristic lattice behavior to the continuum asymptotic form was analyzed in [52]. In order to study the continuum limit in RSA, consider deposition of objects of size l on a 1D substrate of size L. R is the rate of random deposition attempts per unit time and volume. The lattice approximation is introduced by choosing the cubic mesh size $b = l/k$. The lattice deposition is defined by requiring that objects of size l can only deposit in sites of linear size k lattice units. According to [12,13], the late stage (after time τ), deposition in continuum, can be described as the filling up of voids only large enough to accommodate exactly one depositing object. At this time τ, the density of those small gaps (of various sizes) will be ρ. For lattice models, a similar picture applies for $k \gg Rl\tau$. Typical small gaps can be assumed to be of sizes $k + m$ ($m = 0, 1, \ldots, k - 1$) with density ρ/k at time τ, and will be filled up at a rate $Rb(m + 1)$. We consider $t \gg \tau$ so that no new small gaps are created by the filling up of large gaps. Then the density Ω of each type of small gap will have the time dependence $\Omega(m) = (\rho/k) \exp[-Rb(m+1)(t-\tau)]$. In each deposition event the coverage is increased by l/L. Thus, since depositions occur at a rate $Rb\Omega(m)(m+1)$, we have

$$d\theta/dt \simeq \sum_{m=0}^{k-1} (Rbl\rho/k)(m + 1) \exp[-Rb(m + 1)(t - \tau)]. \qquad (11.6)$$

After integration we get the asymptotic $(t \gg \tau)$ estimate, generalized to d dimensions:

$$\theta_k(t) = \theta_k(\infty) - \frac{\rho l^d}{k^d} \sum_{m_1=0}^{k-1} \cdots \sum_{m_d=0}^{k-1} \exp\left\{-\left(\frac{Rl^d t}{k^d}\right) \prod_{p=1}^{d} (m_p + 1)\right\}. \quad (11.7)$$

We consider some special limits. For k fixed, the 'lattice' long-time behavior sets in for $Rl^d t \gg k^d$. In this limit the $m_j = 0$ term in the sums in (11.7) dominates:

$$\theta(t) \approx \theta(\infty) - \frac{\rho l^d}{k^d} \exp\left[-\left(\frac{Rl^d t}{k^d}\right)\right]. \quad (11.8)$$

Thus, the time decay constant increases as k^d. The continuum limit of (11.7) is obtained for $k^d \gg Rl^d t$. In this limit one can convert the sums to integrals. Recall that all the expressions here apply only for $t \gg \tau$ and $k^d \gg Rl^d \tau$, where $Rl^d \tau$ is a fixed quantity of order 1. Thus, the large-k and large-t conditions are simply $k \gg 1$ and $t \gg 1/(Rl^d)$. The latter condition allows us to evaluate the integrals asymptotically, to the leading order for large t, which yields

$$\theta(t) \approx \theta(\infty) - \frac{\rho \left[\ln\left(Rl^d t\right)\right]^{d-1}}{(d-1)! \, Rt}. \quad (11.9)$$

The asymptotic $(\ln t)^{d-1} t^{-1}$ law was derived by Swendsen [13] for the continuum deposition of cubic objects.

The maximum coverage (jamming coverage) values in 2D have been investigated by several authors [14,17,19,51]. Privman *et al.* [52] evaluated $\theta_k(\infty)$ numerically for system sizes $L/k = 200$. The data suggest a fit of the form $\theta_k(\infty) = \theta_\infty(\infty) + (A_1/k) + (A_2/k^2) + \ldots$. By standard manipulations of the sequence given in Table 11.1 to cancel the leading $1/k$ term, followed by a further extrapolation to $k \to \infty$, they arrived at the estimate $\theta_\infty(\infty) = 0.5620 \pm 0.0002$ ($> 0.5589\ldots = (0.747597\ldots)^2$) [52]. The errors are small enough to rule out the conjecture of Palásti [53], and its generalization for finite k [54], which state that the jamming coverages for 2D ($k \times k$) oriented squares are equal to the *squared* jamming coverages of the corresponding 1D k-mer models. The latter are known exactly [9].

11.4 RSA with diffusional relaxation

Recent experiments on protein adhesion at surfaces [55-57] indicate that in biomolecular systems the effects of surface relaxation, due to diffusional rearrangement of particles, are observable on time scales of the deposition process. The resulting large-time coverage is denser than in irreversible RSA and in fact is experimentally comparable to a fully packed (i.e., locally semicrystalline) particle arrangement. Diffusion affects the lattice deposition process in two ways. Firstly, the coverage reaches full saturation for large times and secondly, a slow, power-law approach to full coverage occurs, owing to the vacancy properties.

Here we survey studies of the effects of diffusion on RSA in 1D and 2D. In the deposition of k-mers on a 1D linear lattice, holes of $k-1$ sites or less cannot be reached. Diffusion of the deposited objects can combine small holes to form larger landing sites accessible to further deposition attempts leading to a fully covered lattice at large times [58-61]. For large times the holes are predominantly single-site vacancies that hop due to k-mer diffusion. They must be brought together in groups of k to be covered by a depositing k-mer. If the deposition rate is small, the k-site holes may be broken up again by diffusion before a successful deposition attempt. Thus the process of k-mer deposition with diffusion will reach its asymptotic large-time behavior when most of the empty space is in single-site vacancies. The approach of the coverage to 1 for large times will then be related to the reaction $k\mathcal{A} \to$ inert, with partial reaction probability on each encounter of k diffusing particles $\mathcal{A}$. Such reactions were studied extensively in 1D for $k = 2$ [62-64] including the partial reaction case [64-67]; see Part I for more recent literature. Scaling arguments [68] indicate that the particle density for $k \geq 3$ will follow the mean-field law $\sim t^{-1/(k-1)}$ for large times, with possible logarithmic corrections for $k = 3$ (borderline). This corresponds to $1 - \theta(t\text{-large}) \propto t^{-1/(k-1)}$ in deposition.

Consider now the effect of diffusional relaxation in 1D dimer deposition [58]. At each Monte Carlo step a pair of adjacent sites on a linear lattice ($L = 2000$) is chosen at random. Deposition is attempted with probability p, or otherwise diffusion, with equal probabilities of moving one lattice spacing to the left or right. The time step $\Delta T = 1$ corresponds to L deposition- or diffusion-attempt Monte Carlo steps and the time variable t is defined as $t = pT$. Monte Carlo results were obtained for $p = 0.9, 0.8, 0.5, 0.2$. The coverage increases monotonically with $(1-p)/p$ at fixed pT. For $p < 1$ we obtain $\theta(\infty) = 1$, whereas for $p \equiv 1$,

$$\theta(T) = 1 - \exp\left\{-2\left[1 - \exp(-T)\right]\right\} \to 1 - e^{-2} < 1; \qquad (11.10)$$

see [9]. The convergence of $1 - \theta$ to the limiting value at $t = pT = \infty$ is exponential without diffusion. Small diffusional rates lead to an asymptotic $\sim t^{-1/2}$ convergence to $\theta(p < 1, t = \infty) = 1$. For faster diffusion, the onset of the limiting behavior is preceded by a region of $\sim t^{-1}$ behavior followed by a crossover to $\sim t^{-1/2}$ for larger times. In the cases $k = 3, 4$ the large-time results [59] are roughly consistent with the mean-field relation. For all p values studied, the void area is dominated by the single vacancies precisely in the regime where the mean-field law sets in. For $k = 2$ the single-site vacancies take over for $t \gtrsim 2$. For fast diffusion there follows a long crossover region from the initial mean-field behavior to the asymptotic fluctuation-dominated behavior.

Studies of RSA with diffusional relaxation by analytical means encounter several difficulties associated with possible collective effects in hard-core particle systems at high densities (such as, for instance, phase separation) and with the possibility, in certain lattice models, of locally 'gridlocked' vacant sites. The latter effect may actually prevent full coverage in some models. Neither difficulty is present in 1D; there are no equilibrium phase transitions, traces of which might manifest themselves as collective effects in $d > 1$ deposition with diffusion, and furthermore in 1D diffusional relaxation leads to a simple hopping-diffusion interpretation of the motion of vacant sites, which recombine to form larger open voids accessible to deposition attempts. In a study of $d > 1$ models [69], a low-density-expansion approximation scheme has been applied to the off-lattice deposition of circles on a plane accompanied by diffusional relaxation.

Wang et al. [70] recently reported simulation results for RSA with diffusion on the $d = 2$ square lattice with objects occupying two by two squares of four sites. The distinctive feature of this model is the existence of locally frozen single-site defects. RSA with diffusion on a periodic lattice then leads to frozen states with domains of four different phases. The corresponding equilibrium ground states are highly degenerate [71-73]. With probability $p \ (= 0.1$ here), particle deposition is attempted at site (i, j) on a periodic square lattice of size $L \times L$ (L is even), and with probability $1 - p$ diffusion occurs. The deposition attempt will be successful if the sites (i, j), $(i + 1, j)$, $(i, j + 1)$, $(i + 1, j + 1)$ are all empty. A diffusion attempt is made by trying to move the square in one of four directions (up, down, left, right) chosen at random, by one lattice spacing. Diffusion is successful if the move is not blocked by other squares. Owing to diffusional relaxation, the coverage θ almost reaches the fully crystalline value 1 for large times, $t \to \infty$; however, the final configuration on a finite-size lattice typically has frozen defects that are single-site and serve as points of origin of domain walls separating four different sublattice arrangements ('phases'). One such configuration is illustrated in Fig. 11.3. The defects are locally 'gridlocked', and any state that contains all the empty area, i.e., the lattice sites not covered by

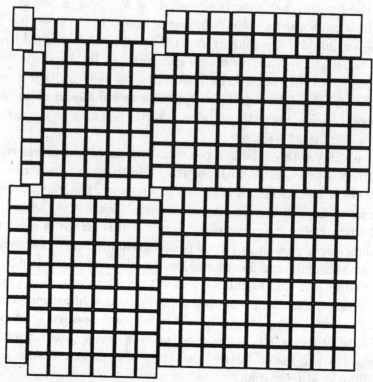

Fig. 11.3. A 'frozen' configuration with four defect sites, on a 32×32 lattice. The 2×2 deposited objects are shown as square outlines while the defects are shown as four 'holes', half the linear size of the outlined objects. The periodic lattice was cut in such a way as to keep the 2×2 objects whole.

the deposited objects, as such single-site isolated defects no longer evolves dynamically. The defect line structure is essentially a random rectangular grid. This frozen-in rectangular grid was also observed [72] for a related equilibrium model by slow cooling from a disordered configuration to zero temperature. The single-site defect density $\varrho_d(t)$ in the limit of large t decreases at least as $1/L$, thus $1 - \theta(\infty) \sim 1/L$, on a periodic lattice. The domain linear size squared grows as $\sim L^{1.6}$. This seems to suggest that the domain size, and possibly shape, distributions are nontrivial. Furthermore, $\varrho_d(t)$ has a peak around $t = 10$ and saturates at $\sim 1/L$. The infinite-L envelope, obtained for $t \gtrsim 50$, fits a power law, $\sim t^{-0.573 \pm 0.004}$. The empty-area fraction $1 - \theta(t)$ is proportional to $t^{-\alpha}$, where the effective exponent value α decreases from 0.61 to 0.53 as t increases. Prior to saturation, i.e., asymptotically for an infinite lattice, the domain growth is of power-law type with a growth exponent near, or possibly somewhat smaller than, $\frac{1}{2}$.

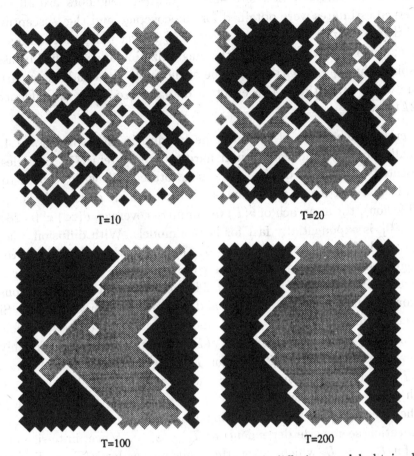

Fig. 11.4. Configurations of hard-square RSA-with-diffusion model obtained with the deposition rate parameter $p = 0.1$, at times $T = 10, 20, 100$, and 200. Particles centered on the even sublattice are shown in grey, while those on the odd sublattice are shown in black. The lattice size was 32×32.

Wang *et al.* [74] studied collective effects in the RSA of diffusing hard squares on a square lattice. This model has been well studied for its equilibrium phase transition [71]. In it the only gridlocked (locally frozen) vacancies are parts of domain walls. As a result the coverage reaches the full crystalline limit at large times, by a diffusional domain-wall motion leading to cluster growth reminiscent of quenched binary alloys and fluids at low temperatures [75-77]. In each Monte Carlo trial [78] of the simulation [74] on a $L \times L$ square lattice, a site is chosen at random with deposition probability p. Only if the chosen site and its four nearest-neighbor sites are all empty is the deposition performed. A diffusion move by one lattice spacing is made if the targeted new site and its nearest neighbors are all empty. Numerical estimates were obtained for the coverage and the 'susceptibility' $\chi = L^2[\langle m^2 \rangle - \langle |m| \rangle^2]$, where the average $\langle \ \rangle$ is over independent runs. The order parameter m was defined as the magnetization and specified by assigning 'spin' values $+1$ to particles on one of the sublattices and -1 to particles on the other sublattice. The effective domain size, $\ell(T)$, was defined by $\ell = 2L\sqrt{\langle m^2 \rangle}$.

A series of snapshots of the coverage buildup is shown in Fig. 11.4. As is usually done for the equilibrium hard-square system [71], particles are represented by squares of size $\sqrt{2} \times \sqrt{2}$, rotated $45°$ with respect to the original square lattice on which the particle centers are deposited. For $p = 1$ (no diffusion), the approach of $\theta(T)$ to jamming coverage, $\theta(\infty) \simeq 0.728 < 1$, see [79,80], is exponentially fast for lattice models. With diffusion, one can always reach full coverage, $\theta(\infty) = 1$. However, the convergence to full coverage is slow, power-law. Here the coverage-growth mechanism for large times is due to interfacial dynamics. The void space at late times consists of domain walls separating spin-up and spin-down ordered regions. Since a typical domain has area $\sim \ell^2(T)$ and boundary $\sim \ell(T)$, we anticipate that, for large times, $1 - \theta(T) \propto \ell^{-1}(T)$. The numerical data roughly fit the power law $1 - \theta(T) \propto T^{-1/2}$, for $T > 10^3$. Thus, the RSA quantity $1 - \theta(T)$ behaves analogously to the energy excess in equilibrium domain-growth problems. The 'susceptibility' χ for a given finite size L has a peak and then decreases to zero, indicating long-range order for large T. The peak location seems size dependent, at $T_{\text{peak}} \propto L^2$. Since finite-size effects set in for $\ell(T) \sim L$, which, given the 'bulk' power law $\ell(T) \sim T^{1/2}$ leads precisely to the criterion $T \sim L^2$, this maximum in fluctuations is expected to be a manifestation of the ordering process at high densities.

In summary, in this chapter an overview of studies of certain 1D and 2D models of multilayer adsorption without screening has been presented. We have also surveyed models of monolayer adsorption processes with diffusional relaxation. In particular the coverage and the interfacial width at large times have been analyzed, as well as the crossover from deposition on a lattice to continuum deposition.

It is a pleasure to thank V. Privman and J.-S. Wang for the good cooperation over many years on which the results discussed in this chapter are based. The author thanks K. Binder for discussions and the DFG for support (Heisenberg foundation) as well as the SFB 262 of the DFG.

References

[1] J. Feder and I. Giaever, *J. Colloid Int. Sci.* **78**, 144 (1980).

[2] A. Schmitt, R. Varoqui, S. Uniyal, J. L. Brash and C. Pusiner, *J. Colloid Int. Sci.* **92**, 25 (1983).

[3] G. Y. Onoda and E. G. Liniger, *Phys. Rev.* **A33**, 715 (1986).

[4] N. Kallay, M. Tomić, B. Biškup, I. Kunjašić and E. Matijević, *Colloids Surf.* **28**, 185 (1987).

[5] J. D. Aptel, J. C. Voegel and A. Schmitt, *Colloids Surf.* **29**, 359 (1988).

[6] P. J. Flory, *J. Am. Chem. Soc.* **61**, 1518 (1939).

[7] A. Rényi, *Publ. Math. Inst. Hung. Acad. Sci.* **3**, 109 (1958).

[8] B. Widom, *J. Chem. Phys.* **44**, 3888 (1966).

[9] J. J. Gonzalez, P. C. Hemmer and J. S. Høye, *Chem. Phys.* **3**, 228 (1974).

[10] E. R. Cohen and H. Reiss, *J. Chem. Phys.* **38**, 680 (1963).

[11] J. Feder, *J. Theor. Biology* **87**, 237 (1980).

[12] Y. Pomeau, *J. Phys.* **A13**, L193 (1980).

[13] R. H. Swendsen, *Phys. Rev.* **A24**, 504 (1981).

[14] R. S. Nord and J. W. Evans, *J. Chem. Phys.* **82**, 2795 (1985).

[15] J. W. Evans and R. S. Nord, *J. Stat. Phys.* **38**, 681 (1985).

[16] L. A. Rosen, N. A. Seaton and E. D. Glandt, *J. Chem. Phys.* **85**, 7359 (1986).

[17] M. Nakamura, *J. Phys.* **A19**, 2345 (1986).

[18] M. Nakamura, *Phys. Rev.* **A36**, 2384 (1987).

[19] G. C. Barker and M. J. Grimson, *Molecular Phys.* **63**, 145 (1988).

[20] P. Schaaf, J. Talbot, H. M. Rabeony and H. Reiss, *J. Phys. Chem.* **92**, 4826 (1988).

[21] R. D. Vigil and R. M. Ziff, *J. Chem. Phys.* **91**, 2599 (1989).

[22] P. Schaaf and J. Talbot, *Phys. Rev. Lett.* **62**, 175 (1989).

[23] J. Talbot, G. Tarjus and P. Schaaf, *Phys. Rev.* **A40**, 4808 (1989).

[24] M. C. Bartelt and V. Privman, *J. Chem. Phys.* **93**, 6820 (1990).

[25] M. C. Bartelt and V. Privman, *Int. J. Mod. Phys.* **B5**, 2883 (1991).

[26] J. W. Evans, *Rev. Mod. Phys.* **65**, 1281 (1993).

[27] N. Ryde, N. Kallay and E. Matijević, *J. Chem. Soc. Faraday Trans.* **87**, 1377 (1991).

[28] N. Ryde, H. Kihira and E. Matijević, *J. Colloid Interface Sci.* **151**, 421 (1992).

[29] M. F. Haque, N. Kallay, V. Privman and E. Matijević, *J. Adhesion Sci. Technol.* **4**, 205 (1990).

[30] V. Privman, N. Kallay, M. F. Haque and E. Matijević, *J. Adhesion Sci. Technol.* **4**, 221 (1990).

[31] M. Elimelech and C. R. O'Melia, *Environ. Sci. Technol.* **24**, 1528 (1990).

[32] J. E. Tobiason and C. R. O'Melia, *J. Am. Water Works Assoc.* **80**, 54 (1988).

[33] M. T. Habibian and C. R. O'Melia, *J. Environ. Eng. Div. (Am. Soc. Civ. Eng.)* **101**, 567 (1975).

[34] K. M. Yao, M. T. Habibian and C. R. O'Melia, *Environ. Sci. Technol.* **5**, 1105 (1971).

[35] V. Privman, H. L. Frisch, N. Ryde and E. Matijević, *J. Chem. Soc. Faraday Trans.* **87**, 1371 (1991).

[36] F. Family and T. Vicsek, *J. Phys.* **A18**, L75 (1985).

[37] P. Meakin and F. Family, *Phys. Rev.* **A34**, 2558 (1986).

[38] P. Nielaba, V. Privman and J.-S. Wang, *J. Phys.* **A23**, L1187 (1990).

[39] P. Nielaba and V. Privman, *Phys. Rev.* **A45**, 6099 (1992).

[40] P. Nielaba and V. Privman, *Phys. Rev.* **E51**, 2022 (1995).

[41] R. Hilfer and J.-S. Wang, *J. Phys.* **A24**, L389 (1991).

[42] V. Privman and J.-S. Wang, *Phys. Rev.* **A45**, R2155 (1992).

[43] M. Kardar, G. Parisi and Y. C. Zhang, *Phys. Rev. Lett.* **56**, 889 (1986).

[44] F. Family and T. Vicsek, eds., *Dynamics of Fractal Surfaces* (World Scientific, Singapore, 1991); J. Krug and H. Spohn, in *Solids Far from Equilibrium: Growth, Morphology, Defects*, C. Godrèche, ed. (Cambridge University Press, 1991).

[45] J. Krug and P. Meakin, *J. Phys.* **A23**, L987 (1990).

[46] B. D. Lubachevsky, V. Privman and S. C. Roy, *Phys. Rev.* **E47**, 48 (1993).

[47] J. Krug, P. Meakin and T. Halpin-Healy, *Phys. Rev.* **A45**, 638 (1992).

[48] M. Schroeder, M. Siegert, D. E. Wolf, J. D. Shore and M. Plischke, *Europhys. Lett.* **24**, 563 (1993).

[49] M. Siegert and M. Plischke, *J. Physique* I3, 1371 (1993).

[50] D. Y. K. Ko and F. Seno, *Phys. Rev.* **E50**, R1741 (1994).

[51] B. J. Brosilow, R. M. Ziff and R. D. Vigil, *Phys. Rev.* **A43**, 631 (1991).

[52] V. Privman, J.-S. Wang and P. Nielaba, *Phys. Rev.* **B43**, 3366 (1991).

[53] I. Palásti, *Publ. Math. Inst. Hung. Acad. Sci.* **5**, 353 (1960).

[54] E. M. Tory, W. S. Jodrey and D. K. Pickard, *J. Theor. Biology* **102**, 439 (1983).

[55] J. J. Ramsden, *J. Phys. Chem.* **96**, 3388 (1992).

[56] J. J. Ramsden, *Phys. Rev. Lett.* **71**, 295 (1993).

[57] J. J. Ramsden, *J. Stat. Phys.* **73**, 853 (1993).

[58] V. Privman and P. Nielaba, *Europhys. Lett.* **18**, 673 (1992).

[59] P. Nielaba and V. Privman, *Mod. Phys. Lett.* **B6**, 533 (1992).

[60] V. Privman and M. Barma, *J. Chem. Phys.* **97**, 6714 (1992).

[61] M. D. Grynberg and R. B. Stinchcombe, *Phys. Rev. Lett.* **74**, 1242 (1995).

[62] D. ben-Avraham, M. A. Burschka and C. R. Doering, *J. Stat. Phys.* **60**, 695 (1990).

[63] D. C. Torney and H. M. McConnell, *Proc. Roy. Soc. Lond.* **A387**, 147 (1983).

[64] V. Kuzovkov and E. Kotomin, *Rep. Prog. Phys.* **51**, 1479 (1988).

[65] K. Kang and S. Redner, *Phys. Rev.* **A32**, 435 (1985).

[66] S. Redner and K. Kang, *Phys. Rev.* **A30**, 3362 (1984).

[67] L. Braunstein, H. O. Martin, M. Grynberg and H. E. Roman, *J. Phys.* **A25**, L255 (1992).

[68] K. Kang, P. Meakin, J. H. Oh and S. Redner, *J. Phys.* **A17**, L665 (1984).

[69] G. Tarjus, P. Schaaf and J. Talbot, *J. Chem. Phys.* **93**, 8352 (1990).

[70] J.-S. Wang, P. Nielaba and V. Privman, *Physica* **A199**, 527 (1993).

[71] L. K. Runnels, in *Phase Transitions and Critical Phenomena*, Vol. **2**, p. 305, C. Domb and M. S. Green, eds. (Academic, London, 1972).

[72] K. Binder and D. P. Landau, *Phys. Rev.* **B21**, 1941 (1980).

[73] W. Kinzel and M. Schick, *Phys. Rev.* **B24**, 324 (1981).

[74] J.-S. Wang, P. Nielaba and V. Privman, *Mod. Phys. Lett.* **B7**, 189 (1993).

[75] J. D. Gunton, M. San Miguel and P. S. Sahni, in *Phase Transitions and Critical Phenomena*, Vol. **8**, p. 267, C. Domb and J. L. Lebowitz, eds. (Academic, London, 1983).

[76.] O. G. Mouritsen, in *Kinetics and Ordering and Growth at Surfaces*, p. 1, M. G. Lagally, ed. (Plenum, NY, 1990).

[77] A. Sadiq and K. Binder, *J. Stat. Phys.* **35**, 517 (1984).

[78] K. Binder, in *Monte Carlo Methods in Statistical Physics*, 2nd edition, p. 1, K. Binder, ed. (Springer-Verlag, Berlin, 1986).

[79] P. Meakin, J. L. Cardy, E. Loh and D. J. Scalapino, *J. Chem. Phys.* **86**, 2380 (1987).

[80] R. Dickman, J.-S. Wang and I. Jensen, *J. Chem. Phys.* **94**, 8252 (1991).

12

Deposition-evaporation dynamics: jamming, conservation laws, and dynamical diversity

Mustansir Barma

The dynamics of the deposition and evaporation of k adjacent particles at a time on a linear chain is studied. For the case $k = 2$ (reconstituting dimers), a mapping to the spin-$\frac{1}{2}$ Heisenberg model leads to an exact evaluation of the autocorrelation function $C(t)$. For $k \geq 3$, the dynamics is more complex. The phase space decomposes into many dynamically disconnected sectors, the number of sectors growing exponentially with size. Each sector is labeled by an irreducible string (IS), which is obtained from a configuration by a nonlocal deletion algorithm. The IS is shown to be a shorthand way of encoding an infinite number of conserved quantities. The large-t behavior of $C(t)$ is very different from one sector to another. The asymptotic behavior in most sectors can be understood in terms of the diffusive, noncrossing movement of individual elements of the IS. Finally, a number of related models, including several that are many-sector decomposable, are discussed.

12.1 Introduction

Problems related to random sequential adsorption (RSA), initially studied several decades ago [1], have aroused renewed interest over the past few years [2-4]. The reason for this is the growing realization that the basic process of deposition of extended objects, which is modeled by RSA, has diverse physical applications. In turn, this has led to the examination of a number of extensions, including the effect of interactions between atoms on adjacent sites, and the diffusion and desorption of single atoms.

In this chapter, we study a simple generalization of RSA that brings a rich and complex dynamical behavior into play [5]. The elementary stochastic kinetic steps involve the deposition of k atoms at a time onto adjacent unoccupied sites of a substrate lattice, and the evaporation (detachment) of any

247

k adjacent atoms from the lattice. The k atoms that leave need not be the same k atoms that fell in together—in other words, k-mers can reconstitute freely. Reconstitution is a crucial feature of deposition-evaporation (DE) dynamics, and is responsible for the interesting temporal properties of the system.

We study the process on a one-dimensional (1D) lattice containing L sites. It turns out that there is a fundamental difference between the case $k = 2$ (dimers), and cases with higher k (≥ 3), the latter showing complex behavior characterized by jamming, strongly broken ergodicity, and long-time properties that depend on the initial conditions. These are the characteristics of a class of systems that we term as many-sector decomposable (MSD). In the DE system, these properties may be understood in terms of conservation laws of the stochastic operator that governs the time evolution of the system. In the remainder of this section, we give an overview of the principal results, before turning to a more detailed exposition in subsequent sections, so that the reader has a bird's-eye view of the terrain.

Dimers: The case $k = 2$ corresponds to the deposition and evaporation of reconstituting dimers. The phase space is divided into $L + 1$ sectors, an outcome of a single conservation law. If the deposition and evaporation rates are equal, the evolution operator can be mapped to the Hamiltonian of the spin-$\frac{1}{2}$ Heisenberg model. Then the steady-state site-occupation autocorrelation function $C(t)$ can be found in closed form; at long times, $C(t) \sim t^{-1/2}$ [5,6].

Many-sector decomposition and jamming: For $k \geq 3$, the phase space is partitioned into very many dynamically disjoint sectors; configurations belonging to different sectors are not connected by DE dynamics. The number of sectors grows exponentially with L. Some sectors consist of completely jammed configurations that do not evolve at all, but the majority of sectors are dynamically interesting. A hint about the nature of the conservation laws responsible for many-sector decomposability comes from a study of the motion and interactions of unjammed patches in a jammed background [5,6].

The irreducible string: The definition of the irreducible string (IS) is the key step leading to a simple and powerful method of categorizing sectors, and ultimately even the long-time behavior in different sectors. The IS is obtained by repeatedly applying a deletion algorithm to a configuration until no sets of k adjacent occupied or unoccupied sites remain. It provides a label for sectors and explains their multiplicity and sizes. The IS is a compact and useful way of encoding an infinite number of conservation laws [7,8].

Dynamical diversity: The long-time behavior of $C(t)$ varies from one sector to another, one of the unusual and interesting features of DE dynamics. Typically, $C(t)$ decays as a power law, with a sector-dependent exponent. In some sectors $C(t)$ follows a stretched exponential decay. This dynamical diversity can be understood in most sectors in terms of the IS, by following the noncrossing, diffusive movement of its elements [9].

In the subsequent sections, we give a more detailed account of each of these features. The concluding section has a review of related work, and a discussion of some open problems.

12.2 Dimers

The behavior of reconstituting dimers ($k = 2$) can be worked out in detail. At the outset, we note the existence of a conservation law: divide the lattice into two sublattices A and B and let M_A and M_B denote the total number of particles on each sublattice. Then $M_A - M_B$ is conserved as the elementary step of deposition or evaporation increments or decrements the occupation of each sublattice equally. Since $M_A - M_B$ can run over $L+1$ distinct values, the phase space is divided into as many sectors.

Let us associate a pseudospin $S_i = \pm 1$ ($|\uparrow\rangle$ or $|\downarrow\rangle$) with site i, the value $+1$ corresponding to an occupied site and -1 corresponding to an empty site. Consider the action of the dynamics at adjacent sites. The deposition of a dimer corresponds to $|\downarrow\downarrow\rangle \to |\uparrow\uparrow\rangle$ and occurs with probability ϵdt in a small time dt while evaporation corresponds to the transition $|\uparrow\uparrow\rangle \to |\downarrow\downarrow\rangle$, which occurs with probability $\epsilon' dt$. In an infinitesimal time dt, pairs of antiparallel spins $|\uparrow\downarrow\rangle$ or $|\downarrow\uparrow\rangle$ cannot evolve at all.

If $\epsilon = \epsilon'$, the action of DE dynamics can be represented by the spin operator

$$h_{ij} = \tfrac{1}{2}\epsilon \left[1 + \sigma_i^z \sigma_j^z - \tfrac{1}{2}(\sigma_i^+ \sigma_j^+ + \sigma_i^- \sigma_j^-) \right], \tag{12.1}$$

where the σ's are Pauli operators. The $\sigma_i^+ \sigma_j^+$ and $\sigma_i^- \sigma_j^-$ terms evidently represent the deposition and evaporation processes, while the $\sigma_i^z \sigma_j^z$ terms arise from the fact that the stay-put probability depends on the configuration of the pair. The full configuration $|\{S_i\}\rangle$ then obeys the evolution equation

$$\frac{d|\{S_i\}\rangle}{dt} = -\mathcal{H}|\{S_i\}\rangle, \tag{12.2}$$

where $\mathcal{H} = \sum_i h_{i,i+1}$, and the aim is to find the autocorrelation function

$$C(t) = \langle \sigma_i^z(0)\sigma_i^z(t)\rangle - \langle \sigma_i^z\rangle^2. \tag{12.3}$$

On performing a spin rotation $\sigma^x \rightarrow \sigma^x$, $\sigma^y \rightarrow -\sigma^y$, $\sigma^z \rightarrow -\sigma^z$ on all spins on the B sublattice, we see that $\mathcal{H} \rightarrow \mathcal{H}'$, the Hamiltonian of the spin-$\frac{1}{2}$ isotropic Heisenberg ferromagnet,

$$\mathcal{H}' = \frac{\epsilon}{2} \sum_i (1 - \sigma_i \cdot \sigma_{i+1}). \tag{12.4}$$

The rotational invariance of $\mathcal{H}'$ has some interesting consequences. The conservation of $\mathcal{S}^z = \frac{1}{2} \sum_i \sigma_i^z$ is a re-expression of the conservation of $M_A - M_B$, evident in the stochastic description. The meaning of the conservation of the total spin operator $\mathcal{S}^2 = (\mathcal{S}^x)^2 + (\mathcal{S}^y)^2 + (\mathcal{S}^z)^2$ of the Heisenberg model is less directly transparent in terms of the stochastic model. Nevertheless, it is a useful invariance that allows explicit evaluation of the autocorrelation function in the steady state, as sketched below. In the sector with r down spins ($\mathcal{S}^z = L/2 - r$) the site-occupation autocorrelation function is given by $\langle \tilde{G}|\sigma_i^z \exp(-\mathcal{H}'t)\sigma_i^z|\tilde{G}\rangle - |\langle \tilde{G}|\sigma_i^z|\tilde{G}\rangle|^2$ where $|\tilde{G}\rangle$, the steady state of the stochastic model, corresponds to a rotated version of the all-spins-up ground state of the Hamiltonian, viz., $(\mathcal{S}^-)^r| \uparrow\uparrow \cdots \uparrow \rangle$. Making a spectral decomposition of $\mathcal{H}'$, the first term becomes $\sum_k \exp[-E(k)t]\langle \tilde{G}|\sigma_i^z|k\rangle\langle k|\sigma_i^z|\tilde{G}\rangle$, where k runs over all eigenstates of $\mathcal{H}'$ and $E(k)$ is the corresponding energy eigenvalue. Since the total spin $\mathcal{S}$ is a good quantum number and $\mathcal{S} = L/2$ in the state $|\tilde{G}\rangle$, an angular momentum selection rule implies that the matrix element vanishes unless $|k\rangle$ is a state with spin $L/2$ or $L/2 - 1$. Spin-$(L/2)$ states are (rotated) ground states, while spin-$(L/2 - 1)$ states are obtained from single spin-wave states by rotation. This observation facilitates the calculation, and the result [5] is $C(t) = (1 - m^2) \exp(-2\epsilon t)I_0(2\epsilon t)$, where $m = 2\mathcal{S}^z/L$ and I_0 is a Bessel function of imaginary argument. At long times, there is a slow diffusive decay, $C(t)/C(0) \approx 1/\sqrt{4\pi\epsilon t}$.

The long-time diffusive behavior of $C(t)$ is an outcome of the conservation of $M_A - M_B$. By making a particle-hole transformation on the B sublattice (equivalent to the spin rotation discussed above), DE dynamics maps onto the problem of diffusing particles with a hard-core interaction. $M_A - M_B$ maps onto the total number of particles, which is thus conserved. The elementary processes of deposition or evaporation of single dimers map onto particle-hole interchanges. If $\epsilon = \epsilon'$, DE dynamics thus maps onto the symmetric exclusion process [10], in which the long-time properties of density fluctuations are described by the diffusion equation.

This mapping also sheds light on the long-time behavior of $C(t)$, if $\epsilon \neq \epsilon'$. In this case, we find that the equivalent problem is one of particles with hard-core constraints diffusing on a lattice with a staggered external potential. Since the potential is flat on a coarse-grained scale, the long-time behavior of $C(t)$ is diffusive again, $C(t)/C(0) \approx 1/\sqrt{Dt}$, where the diffusion constant D can be evaluated explicitly [6].

12.3 Many-sector decomposition and jamming

The deposition-evaporation dynamics of k-mers with $k \geq 3$ differs qualitatively from the case of dimers ($k = 2$). One of the principal differences is that the number of sectors in phase space is enormously larger and grows exponentially with L, in contrast to the linear growth with L for $k = 2$. The case $k = 3$ is typical of all larger k, and in the remainder of this chapter we confine ourselves to trimers ($k = 3$).

What is the reason for the decomposition into so many sectors? Ultimately, it can be traced to an infinity of conservation laws, which are encoded in the irreducible string to be discussed in Sec. 12.4. In this section, we discuss the question at a more qualitative level and try to understand the phenomenon in terms of jamming [5,6].

A completely jammed configuration is one that cannot evolve at all under DE dynamics, as it has no triplets of adjacent parallel spins. Such a configuration is dynamically uninteresting, and constitutes an isolated 1×1 sector in phase space. There are many such configurations. In a linear chain with free boundary conditions, their number can be shown to be $2F_L$, where F_L is the Lth Fibonacci number [5,6]. As $L \to \infty$, the number of totally jammed configurations thus grows exponentially, as μ^L, where $\mu = (1 + \sqrt{5})/2$ is the golden mean.

Now consider creating a local patch of activity in an otherwise completely jammed background. For instance, starting from the jammed configuration $\uparrow\downarrow\uparrow\downarrow\uparrow\downarrow\uparrow\downarrow \cdots$, such a local patch could be created by flipping a single spin, thereby forming a triplet of parallel spins at which some activity is possible. When the triplet flips under DE dynamics, the size of the parallel-spin patch increases to 5. Depending on exactly which three of the five spins are flipped next, the size of the parallel-spin patch becomes 3 or 4. It is then not difficult to see that under stochastic evolution with DE rules, the size of the patch remains between 3 and 5, and that it performs a random walk through the jammed background (Fig. 12.1). This sort of random walk of a single active patch can occur in a variety of jammed backgrounds, and has been studied in [6].

The next question that arises naturally is: what happens if there is more than one active patch? The answer, from Monte Carlo simulation in several jammed backgrounds, is that each patch performs a diffusive motion. When

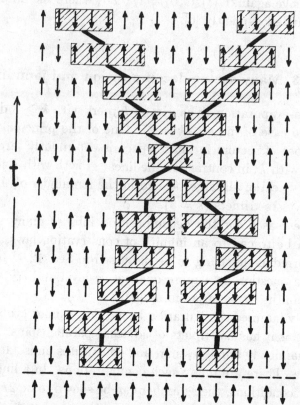

Fig. 12.1. Dynamics of two unjammed patches. Each patch is formed by flipping a spin of the jammed configuration shown at the bottom. Under DE dynamics, each patch performs a random walk, while its size undulates between 3 and 5. The number of patches is conserved except when they collide and interact. At that instant, the number drops momentarily from two to one.

patches collide, there is an effective attractive interaction, but eventually, with probability one, they separate. It is instructive to look at a typical collision process in some detail, as in Fig. 12.1. If we associate a set of three, four, or five adjacent parallel spins with one walker, and two such sets with two walkers, then, as the figure shows, the patches diffuse and collide. During the collision, the number of patches seems to change from two to one, momentarily. But the number of walkers, before and after a collision, is conserved. Thus we begin to get an inkling of the missing quantum numbers that account for the large number of sectors in phase space, namely, the number of active-patch walkers, plus a description of the jammed background in which the walks occur. Since the number of such jammed backgrounds grows exponentially with L, so does the number of partially jammed sectors—resulting in strongly broken ergodicity.

Nevertheless, we are faced with a piquant situation. Given a configuration such as the one with a single active patch in Fig. 12.1, how can one tell that there are really two walkers? From the past and the future evolution (Fig. 12.1), we know that the number of walkers is two—yet the configuration in question has only one parallel spin patch. Evidently, it would be very useful to have an algorithm to count the total number of possible walkers at *all* instants, without having to evolve the configuration in time. A method of doing this is described in Sec. 12.4.

12.4 The irreducible string

A number of properties of DE systems can be characterized in a precise way in terms of a nonlocal construct, the *irreducible string* (IS). The IS is conserved by DE dynamics. It provides a label for sectors in phase space, and enables the number of sectors and their sizes to be found. The length of the IS quantifies the notion of jamming. Finally, the dynamics of the IS sheds considerable light on the the strong sector-wise variation of autocorrelation functions.

The IS is defined as follows [7,8]. Consider a spin configuration $C \equiv \{S_1, S_2, \ldots, S_L\}$ on a 1D lattice with free boundary conditions. Starting (say) from the left, delete triplets of successive parallel spins, reducing the size of the spin string by three in each such reduction. After completing a pass through the full string, re-start from the left and repeat the procedure, again and again. This 'deletion do loop' comes to a halt only when further reduction is impossible. The resulting irreducible string $\{\alpha_1, \alpha_2, \ldots, \alpha_\ell\}$ evidently has no parallel triplets of spins.

The IS is a constant of the motion and uniquely labels each sector of phase space. The elementary step of deposition or evaporation changes three successive ↑ spins to three successive ↓ spins, or vice versa; evidently, the IS is invariant under this. Also, if two configurations have the same IS, they are necessarily reachable from each other by DE dynamics. This follows from the observation [8] that every configuration is connected by DE dynamics to a reference configuration $\{S_1 = \alpha_1, S_2 = \alpha_2, \ldots, S_\ell = \alpha_\ell, S_{\ell+1} = S_{\ell+2} = \cdots = +1\}$.

Given this labeling scheme for sectors, it is possible to prove the following facts about the number of sectors and their sizes [7,8]:

(i) There are many configurations for which the length ℓ of the IS is the length L of the lattice. These are the completely jammed configurations discussed in Sec. 12.3, which each constitute a 1×1 sector in phase space. The number of such sectors is $2F_L$, where F_L is the Lth Fibonacci number, which approaches μ^L as $L \to \infty$.

(ii) The total number of sectors N_L obeys the recursion relation $N_L = N_{L-3} + 2F_L$, since each sector is either completely irreducible or not irreducible. Consequently, $N_L \approx (1 + \mu)\mu^L$.

(iii) The size $D(L, I)$ of a sector is the number of configurations in it. Here I is the IS that labels it. It turns out that D depends on I only through its length $\ell(I)$. An explicit formula can be obtained for the generating function $g(I) \equiv \sum_{L=\ell}^{\infty} x^L D(L, I)$. It is then possible to find the asymptotic growth law for $D(L, I)$, for given I.

(iv) The set of all fully reducible configurations $\{\mathcal{C}\}_\phi'$ constitutes the sector labeled by the null irreducible string $I \equiv \phi$. For this sector, $D_\phi \sim (27/4)^{L/3}$.

(v) Consider picking a configuration $\mathcal{C}_{\mathrm{ran}}$ totally at random from amongst the 2^L possible configurations—an important case, since most configurations are of this type. We may ask for the likely length ℓ_{ran} of the corresponding IS. We find that $\ell_{\mathrm{ran}} \approx L(1 - 3\mu^{-2}/2)$.

The irreducible string encapsulates an infinite number of conservation laws, a fact which explains its success in providing a complete labeling scheme for sectors. We now sketch a procedure for enumerating these conservation laws. With an arbitrary configuration of spins, say $\{\uparrow\downarrow\uparrow\downarrow\uparrow\uparrow\uparrow\downarrow\downarrow\uparrow\downarrow\cdots\}$ associate an operator product of the form $\Pi = \cdots QPQQPPPQPQP$, where we associate operator P with an up-spin and operator Q with a down-spin, and we write the operator product from right to left as we go down the configuration from left to right. If the operators satisfy the condition $P^3 = Q^3 = 1$, and do not commute with each other, the operator product Π for the full configuration is evidently the same as that associated with the corresponding IS. Examples of P and Q represented as 2×2 matrices are given in [7]. Suppose P and Q are functions of a parameter x. The operator product is then a matrix $\Pi(x)$ whose elements are functions of x. We may expand a typical element $f(x)$ in a power series $\sum f_m x^m$. The coefficients f_m constitute an infinite, commuting set of conserved quantities, each a function of the spins $\{\sigma_i^z\}$. For instance, f_1 is linear in the σ_i^z, and its real and imaginary parts imply the conservation of $M_A - M_B$ and $M_B - M_C$, where A, B, C denote sublattices, and the M's the respective occupation numbers. The higher-order f_m involve sums over complexes of spins. Since the IS determines the order of the P and Q operators in the product, it carries information about $f(x)$, and therefore all the f_m as well. Thus, the IS encapsulates an infinite number of conservation laws in a compact form.

12.5 Dynamical diversity

We have seen that the DE process partitions phase space into an exponentially large number of sectors, each labeled by a distinct IS. Monte Carlo studies show that the system exhibits dynamical diversity—there is a strong variation in the long-time behavior of the autocorrelation function $C(t)$ from one sector to another [9]. For instance, in the sector with IS $= (\uparrow\uparrow\downarrow)^{fL}$, i.e., the string $\uparrow\uparrow\downarrow\uparrow\uparrow\downarrow\cdots$ formed by repeating $\uparrow\uparrow\downarrow$ fL times, with $f < 1/3$, we find $C(t) \sim 1/t^{1/2}$. If we start with a totally random initial configuration, we find $C(t) \sim 1/t^{1/4}$. In the sector with IS $= (\uparrow\downarrow)^{fL}$ with $f < 1/2$, we find $C(t) \sim \exp(-bt^{1/2})$. In the sector where the IS is a periodic string $(\uparrow\uparrow\downarrow\downarrow\uparrow\downarrow)^{fL}$ with $f < 1/6$, the autocorrelation function $C(t)$ depends on which sublattice the site is on (in a three-sublattice decomposition of the lattice): $C(t) \sim 1/t^{1/2}$ when i is on two of the three sublattices, while $C(t) \sim \exp(-bt^{1/2})$ when i is on the third sublattice.

An explanation of dynamical diversity can be found in terms of the IS, for, besides being a label on sectors, it has a dynamical meaning as discussed below. Recall that the elements of the IS $\{\alpha_1, \alpha_2, \ldots, \alpha_\ell\}$ are themselves a subset of the full configuration of spins $\mathcal{C} \equiv \{S_1, S_2, \ldots, S_L\}$, namely those that are left undeleted at the end of repeated application of the triplet reduction procedure (the 'deletion do loop' referred to above). Under time evolution, the configuration evolves, and as a result, the α's move on the lattice, keeping their relative ordering intact, since the IS is a constant of the motion.

In any sector in which the density ℓ/L of the IS is finite, the diffusive motion of the set of conserved spins $\{\alpha_k\}$ constitutes the slow modes of the system. We conjecture that the long-time properties in such a sector are the same as those of a simpler system, a set of hard-core random walkers with conserved spin (HCRWCS), whose jump rules are identical to the symmetric exclusion process. A direct numerical test of this conjecture is possible. In the HCRWCS system, the mean square fluctuation in the number of particles between two distant points in space, at two different times 0 and t, grows as $t^{1/2}$. In the DE system, this translates into a similar growth law for fluctuations in the length of the IS between two distant space points. This is verified by Monte Carlo simulations, which reveal that the $t^{1/2}$ growth law very well describes IS length fluctuations in all sectors except those in which $\ell/L \to 0$.

Quite a lot can be said about HCRWCS dynamics, as a good deal is known already about the diffusion of particles on a lattice, with mutual hard-core interactions (the exclusion process) [10]. The arrangement of spins on the particles in no way affects the particle dynamics, but is of course crucial in determining the spin autocorrelation function, which is what we need. The latter involves a convolution over the spin pattern $\{\alpha_k\}$ (which in turn identifies a particular sector of the deposition-evaporation model), and the conditional probability in the exclusion process that particle m is at a particular site at $t = 0$, and particle $m + \Delta m$ is at the same site at time t. Not surprisingly, the nature of the spin pattern has a profound effect on the result. The upshot is that the HCRWCS model successfully explains the full variety of long-time behaviors found numerically in different sectors in which the density of walkers ℓ/L is finite, as detailed in the first paragraph of this section.

In the sector consisting of the set of all fully reducible configurations, the dynamical behavior is not determined by HCRWCS diffusion, as there is no IS when the full configuration from 0 to L is examined. Nevertheless, fluctuations lead to a time-varying length of irreducible string between two fixed space points near the center of the chain. Monte Carlo simulations [9] show that the mean square difference between fluctuations grows as $t^{2\beta}$ with $\beta \simeq 0.19$. The autocorrelation function $C(r) \sim 1/t^\theta$ with $\theta \simeq 0.59$. Exact diagonalization of the transfer matrix for finite systems [11] yields consistent estimates of critical exponents. The estimated value of the dynamical exponent is $z \simeq 2.5$, quite different from the value for diffusive cases, $z = 2$.

12.6 Related models

We have seen that the DE system in 1D has several interesting features, including slow, nonexponential time-dependent decays, and (for $k \geq 3$) strongly broken ergodicity and sector-dependent dynamics, i.e., the system is many-sector decomposable (MSD). In this section we briefly discuss related models, and comment on similarities and differences of behavior.

Potts antiferromagnet: Consider the time-dependent properties of a 1D antiferromagnetically coupled three-state Potts model in which unlike particles on neighboring sites can exchange (a generalized Kawasaki dynamics). At $T = 0$, the system is confined to the manifold of ground states in which no like Potts spins are nearest neighbors, and the dynamics is constrained to act within that subspace. There is a mapping between the dynamics of this system and DE dynamics [12]: with every nearest-neighbor pair of Potts spins, associate a spin that is ↑ if the pair is ab, bc, or ca, while the spin is ↓ if the pair is ba, cb, or ac. It is then easy to check that every allowed move of $T = 0$ Potts exchange dynamics corresponds to the deposition or evaporation of a trimer, e.g., $ababcabac \to abacbabac$ corresponds to $\cdots \uparrow\downarrow\uparrow\uparrow\uparrow\uparrow\downarrow\downarrow \to \cdots \uparrow\downarrow\uparrow\uparrow\uparrow\uparrow\downarrow\downarrow$. Thus the Potts antiferromagnet is another example of an MSD system, and has the same phase space structure as the DE system.

Higher dimensions: Most of the results of Sec. 12.2 for dimer deposition and evaporation, including exact evaluation of $C(t)$ if $\epsilon = \epsilon'$, can be generalized to any bipartite lattice. $C(t)$ continues to show a diffusive decay $\sim 1/t^{d/2}$, where d is the spatial dimension. On nonbipartite lattices, there is no mapping to a diffusion problem. In fact, Monte Carlo simulations of dimers on a triangular lattice indicate that $C(t)$ decays exponentially [13]. Simulations of DE dynamics with triangular trimers on the triangular lattice, on the other hand, show that $C(t)$ decays as a power law [13]. A systematic investigation of phase space decomposition in these cases remains an interesting open problem.

Single-particle diffusion in the dimer DE problem: In addition to the dimer deposition and evaporation moves with rates ϵ and ϵ', suppose that a particle is allowed to hop to a nearest-neighbor site with rate h rightward and h' leftward provided that the site in question is unoccupied. Building up the stochastic evolution operator as in Sec. 12.2, the effective Hamiltonian is found to be anisotropic [6], so that there is a gap in its spectrum, in general, leading to exponential decay of the autocorrelation functions. The full parameter space was studied in [14]. An especially interesting case is $\epsilon' \to 0, h + h' = \epsilon$ (RSA plus single-particle diffusion, with related rates), where the gap vanishes and the autocorrelation function starting from an empty lattice varies as $1/t$ [15]. Another problem that has been investigated is that of a single defect (a single bond at which the rates differ from the rest of the lattice), with the result that formation of a bound state can strongly affect the slowest relaxation rate [16].

Multicolored dimers: In [17] a three-species (aa, bb, or cc) dimer model on a line was studied in which the basic moves are $aa \to bb$ or cc, etc. This model exhibits the many-sector decomposition property—there are $\sim 2^L$ sectors in 3^L-dimensional phase space, and a deletion algorithm generates the IS that labels sectors. The variation of $C(t)$ from sector to sector can be accounted for in terms of the IS formed by recursively deleting adjacent pairs of like atoms. The model has properties very similar to the k-mer

DE model, but it has some attractive features, e.g., pairwise as opposed to triplet interactions. Besides, it has a number of extra symmetries, which allow some eigenvectors of the stochastic matrix to be found exactly.

Dimer diffusion: This model allows a pair of atoms (a dimer) to move one step to the right or left, provided there is a vacancy, $\uparrow\uparrow\downarrow \leftrightarrow \downarrow\uparrow\uparrow$ [18]. Interestingly, there is an exponentially large number of sectors labeled by an IS formed by deleting only $\uparrow\uparrow$ pairs. Substantial progress can be made on observing a connection between this model and the exclusion process with two types of hole. Thus far, it seems that dimer diffusion is the simplest model in the class of MSD systems. By virtue of its connection to the exclusion process, and thus to the Heisenberg model, it may provide some insights into the relationships between the dynamics in different sectors.

In conclusion, the deposition-evaporation system of reconstituting k-mers has interesting dynamical properties. For $k = 2$, the mapping to the Heisenberg model leads to exact solvability. For $k \geq 3$, the DE system is a paradigm for the dynamical behavior of the MSD class—systems with an exponentially large number of sectors, and sector-dependent dynamics. The irreducible string, which embodies an infinite number of conservation laws, is central to the theoretical description. It yields a complete description of the manner in which phase space is divided, and also elucidates the sector dependence of long-time dynamics in almost all sectors. The notion of the IS seems to be of more general applicability, and has proved useful in analyzing several other many-sector decomposable systems whose phase space is very finely divided.

It is a pleasure to acknowledge fruitful collaboration on deposition-evaporation and related systems with Robin Stinchcombe, Marcelo Grynberg, Deepak Dhar and Gautam Menon.

References

[1] P. J. Flory, *J. Am. Chem. Soc.* **61**, 1518 (1939).
[2] M. C. Bartelt and V. Privman, *Int. J. Mod. Phys.* **B5**, 2883 (1991).
[3] J. W. Evans, *Rev. Mod. Phys.* **65**, 1281 (1993).
[4] J. J. Ramsden, *J. Stat. Phys.* **79**, 491 (1995).
[5] M. Barma, M. D. Grynberg and R. B. Stinchcombe, *Phys. Rev. Lett.* **70**, 1033 (1993).
[6] R. B. Stinchcombe, M. D. Grynberg and M. Barma, *Phys. Rev.* **E47**, 4018 (1993).
[7] D. Dhar and M. Barma, *Pramana - J. Phys.* **41**, L193 (1993).
[8] M. Barma and D. Dhar, in *Proc. Int. Colloq. Mod. Quantum Field Theory II*,

p. 123, S. R. Das, G. Mandal, S. Mukhi and S. R. Wadia, eds. (World Scientific, Singapore, 1995).

[9] M. Barma and D. Dhar, *Phys. Rev. Lett.* **73**, 2135 (1994).

[10] T. M. Liggett, *Interacting Particle Systems* (Springer, New York, 1985).

[11] P. B. Thomas, M. K. Hari Menon and D. Dhar, *J. Phys.* **A27**, L831 (1994).

[12] D. Dhar, *J. Ind. Inst. Sci.* **75**, 297 (1995).

[13] N. N. Chen, M. D. Grynberg and R. B. Stinchcombe. *J. Stat. Phys.* **78**, 971 (1995).

[14] M. D. Grynberg, T. J. Newman and R. B. Stinchcombe, *Phys. Rev.* **E50**, 957 (1994).

[15] M. D. Grynberg and R. B. Stinchcombe, *Phys. Rev. Lett.* **74**, 1242 (1995).

[16] N.-N. Chen and R. B. Stinchcombe, *Phys. Rev.* **E49**, 2784 (1994).

[17] M. K. Hari Menon and D. Dhar, *J. Phys.* A, to be published.

[18] G. I. Menon, M. Barma and D. Dhar, report TIFR/TH/95-58, to be published.

Part V: Fluctuations in Particle and Surface Systems

Editor's note

Much interest has been devoted recently to various systems described in the continuum limit by variants of nonlinear diffusion equations. These include versions of the KPZ equation [1], Burgers' equation [2], etc. Chapter 13 surveys nonlinear effects associated with shock formation in hard-core particle systems. Exact solution methods and results for such systems are then presented in Ch. 14.

Selected nonlinear effects in surface growth are reviewed in Ch. 15. Their relation to kinetic Ising models and a survey of some results were also presented in Ch. 4 (Sec. 4.6). This is a vast field with many recent results; see [3] (and Chs. 4, 15) for review-type literature. Some surface-growth effects were also reviewed in Ch. 11.

The nonequilibrium 1D systems covered in this book are effectively (1+1)-dimensional, where the second 'dimension' is time. For stochastic dynamics, the latter is frequently viewed as 'Euclidean time' in the field-theory nomenclature. Certain directed-walk models of surface fluctuations associated with wetting transitions, etc., as well as related models of polymer adsorption at surfaces, are effectively (0 + 1)-dimensional in this classification, where the spatial dimension along the surface is effectively the Euclidean-time dimension. This property is shared by 1D quantum mechanics, to which the solution of many surface models reduces in the continuum limit.

These models share simplicity, the availability of exact solutions, and the importance of fluctuations with the (1+1)-dimensional systems. A short review of recent results with emphasis on surveying the literature is presented in Ch. 16. We also note that the (0 + 1)-dimensional surface-fluctuation models are among the few systems where the effects of randomness can be studied by analytical means; for a review see [4].

[1] The KPZ equation was introduced in Sec. 4.6. Various aspects of the KPZ theory were addressed in Ch. 3, 4, 11.

[2] Burgers' equation has already been encountered, in connection with nonlinear effects in other systems, in Chs. 1, 9.

[3] H. Spohn, *Large Scale Dynamics of Interacting Particles* (Springer, New York, 1991); J. Krug and H. Spohn, in *Solids Far from Equilibrium*, C. Godrèche, ed. (Cambridge University Press, 1991); T. Halpin-Healy and Y.C. Zhang, *Phys. Rep.* **254**, 215 (1995); J. Krug, in *Scale Invariance, Interfaces, and Non-Equilibrium Dynamics*, A. McKane, M. Droz, J. Vannimenus and D. Wolf, eds. (Plenum, New York, 1995).

[4] G. Forgacs, R. Lipowsky and Th. M. Nieuwenhuizen, in *Phase Transitions and Critical Phenomena*, Vol. **14**, p. 135, C. Domb and J. L. Lebowitz, eds. (Academic, New York, 1991).

13

Microscopic models of macroscopic shocks

Steven A. Janowsky and Joel L. Lebowitz

We present some rigorous and computer-simulation results for a simple microscopic model, the asymmetric simple exclusion process, as it relates to the structure of shocks.

13.1 Introduction

In this chapter our concern is the underlying microscopic structure of hydrodynamic fields, such as the density, velocity and temperature of a fluid, that are evolving according to some deterministic autonomous equations, e.g., the Euler or Navier-Stokes equations. When the macroscopic fields described by these generally nonlinear equations are smooth we can assume that on the microscopic level the system is essentially in local thermodynamic equilibrium [1]. What is less clear, however, and is of particular interest, both theoretical and practical, is the case where the evolution is not smooth—as in the occurrence of shocks. Looked at from the point of view of the hydrodynamical equations these correspond to mathematical singularities—at least at the compressible Euler level—possibly smoothed out a bit by the viscosity, at the Navier-Stokes level. But what about the microscopic structure of these shocks? Is there really a discontinuity, or at least a dramatic change in the density, at the microscopic scale or does it look smooth at that scale?

It is clear that this question cannot be answered by the macroscopic equations. Also, both the traditional methods of deriving these equations from the microscopic dynamics, which use (uncontrolled) Chapmann-Enskog type expansions [2], and the more recent mathematical methods, which use ergodic properties of the microscopic dynamics and the large separation between microscopic and macroscopic space-time scales, appear at first glance

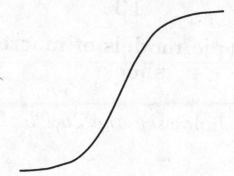

Fig. 13.1. Smooth initial data for Burgers' equation.

to require that the hydrodynamic fields vary smoothly on the macroscopic scale [1,3,4]. In fact until recently all rigorous derivations of hydrodynamical equations were restricted to macroscopic times when their solutions are smooth. This is in particular the case for the derivation of the inviscid Burgers' equation [5],

$$\frac{\partial u(x,t)}{\partial t} + c\frac{\partial}{\partial x}\left\{u(x,t)\left[1 - u(x,t)\right]\right\} = 0, \qquad (13.1)$$

from various microscopic model systems. Burgers' equation can produce shocks: smooth initial data of the form shown in Fig. 13.1 quickly evolve into a discontinuous profile. After a long time the macroscopic profile takes the form sketched in Fig. 13.2. The validity of the derivations of (13.1) stops, however, just at the development of the discontinuity. It is therefore quite remarkable that recently F. Rezakhanlou was able to extend previous results and prove the validity of (13.1) in describing the time evolution of a microscopic model system *even after the formation of a shock* [6].

13.2 Microscopic model

The microscopic model that Rezakhanlou considered is the so-called asymmetric simple exclusion process or ASEP (some generalizations are possible). In this model the microscopic system consists of particles on some lattice (typically Z^d) in which there is hard-core exclusion preventing more than one particle from occupying a given lattice site at any one time. The configuration of the system is specified by $\eta = \{\eta_i : i \in Z^d, \eta_i = 0, 1\}$. The time evolution proceeds by allowing every particle independently to attempt to jump to one of its neighboring sites at a rate τ^{-1}. Once it decides to jump it chooses a direction $\mathbf{e}_\alpha \in \{\pm\mathbf{e}_1, \pm\mathbf{e}_2, \ldots, \pm\mathbf{e}_d\}$ with probability $p_{\mathbf{e}_\alpha}$ (the

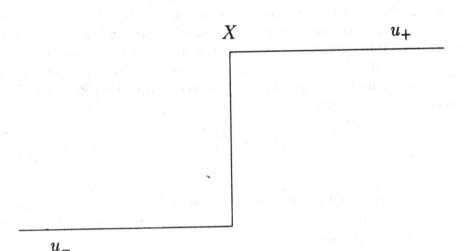

Fig. 13.2. For large times, the shock is located at a position X and moves with velocity $v = c(1 - u_+ - u_-)$.

e_α form a basis for the lattice). The jump is actually performed only if the target site is empty—otherwise the particle does not go anywhere. The dynamics is symmetric if $p_{e_\alpha} = p_{-e_\alpha}$ for all α; otherwise it is asymmetric.

In all cases the only stationary translation-invariant states are product measures ν_ρ, $\rho \in [0, 1]$, where ρ is the average density at any given site: $\langle \eta_i \rangle_{\nu_\rho} = \rho$, $\langle \eta_i \eta_j \rangle_{\nu_\rho} = \rho^2$ $(i \neq j)$, $\langle \eta_i \eta_j \eta_k \rangle_{\nu_\rho} = \rho^3$ $(i, j, k$ distinct), etc. These states can be thought of as the equilibrium state of a lattice gas with on-site hard-core interaction. Note, however, that the dynamics obeys detailed balance with respect to these stationary states only for the symmetric case. (There is also, in the asymmetric case, a stationary non-translation-invariant state in which the density approaches unity along the direction of the asymmetry, e.g., along the e_1-direction if $p_{e_1} > p_{-e_1}$ and $p_{e_\alpha} = p_{-e_\alpha}$ for $\alpha > 1$. The dynamics satisfies detailed balance with respect to such 'blocked' states, which can be thought of as equilibrium states in the presence of a constant external 'electric' field [7]. These states, however, will not be relevant in our further consideration.)

To study the microscopic structure of our model system, whose macroscopic evolution is described by (13.1), we start the system in a nonuniform state that converges under rescaling to a piecewise-smooth density $u_0(x)$. For simplicity we will state results only for the one-dimensional (1D) case, but most of the results carry over to higher dimensions. A simple initial state corresponds to having each site occupied independently with probability $\langle \eta_i \rangle_0 = u_0(\epsilon i)$, where $u_0(x)$ takes values in $[0, 1]$. The parameter ϵ measures the ratio of the microscopic to macroscopic length scales. For a typical system, deterministic, autonomous, macroscopic evolution is achieved in the limit $\epsilon \to 0$. More precisely, let us count the number of particles in a macroscopic region of length Δ, containing Δ/ϵ sites, located at a point x at a macroscopic time t, corresponding to microscopic time t/ϵ ('Euler scaling') and divide by the number of sites, to obtain

$$N^\epsilon(x, t, \Delta) = \frac{\epsilon}{\Delta} \sum_{i=[x/\epsilon]}^{[(x+\Delta)/\epsilon]} \eta_i(t\epsilon^{-1}). \tag{13.2}$$

A theorem then says that as $\epsilon \to 0$ the random variable $N^\epsilon(x, t, \Delta)$ converges almost surely, i.e., for almost every microscopic configuration, to a deterministic quantity

$$N^\epsilon(x, t, \Delta) \to \frac{1}{\Delta} \int_x^{x+\Delta} u(x, t)\, dx, \tag{13.3}$$

where $u(x, t)$ satisfies (13.1) with initial condition $u(x, 0) = u_0(x)$. For simplicity we consider $\tau = 1$, which gives $c = 2p - 1$.

13.2.1 Shock tracking

Given the existence of singularities in the hydrodynamic equation we return to our original question: what about the microscopic structure of these shocks? Actually, we first must answer a more basic question: how does one locate the shock? If $u_- = 0$ this is an easy question; the first particle determines the shock position. No such simple tool will work if $u_- > 0$; however, there is a device that has been used with great success: the *second-class particle* [8].

The second-class particle is a special particle added to the system whose position is updated via a modified dynamical rule: basically the second-class particle behaves like a particle when attempting to jump to other sites, but it behaves like a hole when other particles (now called first-class particles) attempt to jump to its site. A second-class particle in the region of density ρ will have a mean velocity equal to $(2p - 1)(1 - 2\rho)$. This gives the second-class particle a drift towards the location of any shock, and in 1D its position may be used as the microscopic definition of the shock position. (In more than 1D one can still use second-class particles to track the shock, but one cannot necessarily define the position in this way.)

13.3 Detailed shock structure

Using the second-class particle to define the shock position, it has been shown that the shock front in the 1D ASEP remains sharp even on the microscopic level [8-12]. When $p = 1$ there is additional information about the structure of the microscopic system at the location of the shock, i.e., about the 'shape' of the shock as seen from the point of view of the second-class particle. This amounts to determining the time-invariant distribution of first-class particles relative to the second-class particle's location. It is done by first examining what happens when we add many second-class particles to the ASEP, and then studying the nonequilibrium stationary states of this two-species system. One recovers the original shock profile [13,14] via a trick of [11]. For the case $p = 1$, total asymmetry, this correspondence is particularly easy to describe: in a system with a density ρ_1 of first-class particles and ρ_2 of second-class particles, if we take the point of view of a specific second-class particle then all second-class particles in front of it behave as first-class particles, while all second-class particles behind it behave like holes. Thus this particular particle sees the same microscopic shock profile as a single second-class particle in a system containing density $u_- = \rho_1$ on the left and $u_+ = \rho_1 + \rho_2$ on the right.

The ordinary ASEP has a trivial stationary state in a closed or infinite system, but exhibits interesting behavior when combined with nontrivial boundary conditions. This model was solved exactly in [15,16], by embodying the ASEP dynamics into an algebra. An extension of this algebra serves to solve the two-species model (on a ring or in infinite volume); these results are presented in [13,14,17-19]. The primary fact needed to derive all the other results is that the probability that a given configuration occurs in the stationary distribution $P(\eta_1, \eta_2, \ldots, \eta_N)$ is proportional to $\text{Tr}[X_1 X_2 \cdots X_N]$ where $X_i = \mathcal{O}_k$ if $\eta_i = k$ (k takes values 0, 1, 2) and the operators $\mathcal{O}_k$ satisfy the algebra

$$\mathcal{O}_2 = [\mathcal{O}_1, \mathcal{O}_0], \qquad \mathcal{O}_1 \mathcal{O}_0 = \mathcal{O}_1 + \mathcal{O}_0. \tag{13.4}$$

A specific representation of the operators $\mathcal{O}_k$ is then used to determine various properties of the two-species stationary state. A convenient representation is $(\mathcal{O}_0)_{ij} = \delta_{i,j} + \delta_{i-1,j}$, $(\mathcal{O}_1)_{ij} = \delta_{i,j} + \delta_{i,j-1}$, $(\mathcal{O}_2)_{ij} = \delta_{i,1} \delta_{j,1}$, where $i, j \in \{1, 2, 3, \ldots\}$. If one now considers a grand-canonical ensemble in which the particles and holes have fugacities z_0, z_1, and z_2, then the partition function and other expectations can be expressed in terms of the traces of powers of $G = z_0 \mathcal{O}_0 + z_1 \mathcal{O}_1 + z_2 \mathcal{O}_2$. This allows many results to be formulated in terms of known results for random walks since G is the (unnormalized) transition matrix for a biased random walk restricted to the nonnegative half-line.

The solution illustrates several novel features of the stationary state, such as the properties that two second-class particles will form a state where the particles are bound together yet the expectation of their distance is infinite and that certain correlations between particles are unexpectedly zero— neither property was predicted prior to the development of the exact solution. It also describes the full microscopic structure of a shock. Defining the density about the position of the second-class particle as $\delta_i = \langle \eta_{X+i} \rangle$, where X is the position of the second-class particle, one has

$$\delta_i = \begin{cases} u_+ + u_-(1-u_+) - g_i, & i > 0, \\ u_- u_+ + g_{-i}, & i < 0, \end{cases} \qquad (13.5)$$

where

$$g_i = \sum_{n=1}^{i-1} \sum_{m=0}^{n-1} \frac{1}{m+1} \binom{n}{m} \binom{n-1}{m} (u_- u_+)^{m+1} [(1-u_-)(1-u_+)]^{n-m}. \quad (13.6)$$

This shows in particular that there is an exponential approach to the product measure with density u_+ or u_- as we move away from the second-class particle. The correlation length diverges as $(u_+ - u_-)^{-2}$ when $u_+ \to u_-$.

An interesting feature of the exact solution is that the shock profile is *not* monotonic. The profile for $u_- = 1/4$, $u_+ = 3/4$ is plotted in Fig. 13.3.

It is not at all clear whether this structure is the 'true' nature of the shock or whether it is an artifact resulting from determining the shock position via the use of the second-class particle. This is an important question—of course one is interested in the shock's own inherent structure, but it appears that the only reliable tool for studying the shock has so much structure of its own that the shock itself may become obscured. In particular, when the asymptotic densities on either side of the shock are the same, and we would expect there not to be any shock at all, the second-class particle still produces a nontrivial density profile.

To overcome this problem one can attempt to separate the portion of the probability measure resulting from the motion of the second-class particle from that due to the inherent shock structure. This is not easy, however— although the macroscopic definition of a shock is clear, there is no one simple and consistent way to define the exact location of a shock on the microscopic scale. A multitude of definitions exists, and the microscopic structure may depend on the definition that is chosen.

One can attempt to eliminate the second-class particle altogether. The second-class particle has significant theoretical advantages, but other types

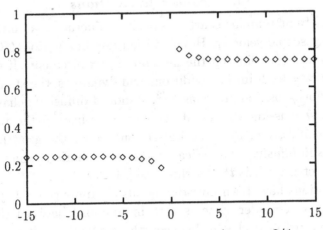

Fig. 13.3. The profile for $u_- = 1/4$, $u_+ = 3/4$.

of shock 'tracers' may more accurately (or more intuitively) identify the instantaneous shock location. If the tracer particle need only follow the shock, rather than also obey the ASEP dynamics as the second-class particle does, it may possibly avoid the artifacts seen in the shock as defined by the second-class particle. One can then study the shock structure using these other measurements, both theoretically and numerically, and see how it compares with results determined through the use of second-class particles. One possibility is the use of the instantaneous ASEP configuration as a background potential field for a random walker, which will have a distribution centered around the shock position. All such indicators examined so far produce a nonmonotone profile—which seems intrinsic to using a probe sensitive to local variations in the environment for determining the instantaneous position of a global change superimposed on many local fluctuations.

One can also study particle models other than the ASEP and examine their shock structure. After all, one is not so much interested in the detailed behavior of the ASEP as the structure of shocks in particle systems in general. If the behavior in the ASEP turns out to be generic then it is a very useful model; if it is very different from that in other models then it is simply an obscure special case that has an exact solution. One possible avenue of study is the Boghosian-Levermore model [20], which has the same hydrodynamic behavior [20,21] and shock position fluctuations [21,22] as the

ASEP but where preliminary numerical studies indicate that a variant of the second-class particle behaves more intuitively than in the ASEP.

13.3.1 Shock fluctuations

Although the front remains sharp, its position fluctuates around the deterministic macroscopic velocity. How big are these fluctuations? One obtains different results depending on the method of measurement. If we compute expectations over both initial conditions and dynamics, the standard deviation of the shock position grows as $t^{1/2}$, standard diffusive behavior [1]. On the other hand it is known that if one computes instead the variance just over the dynamics, i.e., fixes the initial conditions, the growth in time is $o(t^{1/2})$ [12], subdiffusive. Numerical studies and heuristic analysis indicate that the correct growth is $t^{1/3}$ in this case [21-23].

These questions have been investigated also in more general systems. For example one can consider systems with interactions between the particles in which the rates of exchange between the contents of sites i and j in a configuration η, $c(i,j;\eta)$, leading to a new configuration $\eta^{i,j}$ has the form

$$c(i,j;\eta) = c^{(0)}(i,j;\eta)e^{-\beta E(i-j)(\eta_i - \eta_j)}, \tag{13.7}$$

where $c^{(0)}(i,j;\eta)$ satisfies detailed balance for an equilibrium Gibbs distribution with some Hamiltonian $H_0(\eta)$, i.e.,

$$c^{(0)}(i,j;\eta)/c^{(0)}(i,j;\eta^{ij}) = e^{-\beta[H_0(\eta^{i,j})-H_0(\eta)]}. \tag{13.8}$$

For a review of the behavior of these driven diffusive systems see [24]. Then the translation-invariant stationary distributions are generally not product measures, but can still be classified by their density ρ, i.e., there are, at least at high temperatures, measures μ_ρ such that $\langle \eta_i \rangle_{\mu_\rho} = \rho$, $\langle \eta_i \eta_j \rangle - \rho^2 \to 0$ as $|i-j| \to \infty$.

Burgers' equation (13.1) needs to be generalized to fit driven diffusive systems. Based on the assumption (which can be proven in many cases) that on the appropriate spatial and temporal scale the probability distribution is locally that of a stationary state with local density $u(x,t)$, we can evaluate the current $J(u(x,t))$ entering the conservation law,

$$\frac{\partial u(x,t)}{\partial t} + \frac{\partial}{\partial x}J(u(x,t)) = 0. \tag{13.9}$$

In the ASEP we have

$$\begin{aligned} J(u) &= [p\langle \eta_i(1-\eta_{i+1})\rangle_{\nu_u} - (1-p)\langle \eta_{i+1}(1-\eta_i)\rangle_{\nu_u}]/\tau \\ &= (2p-1)u(1-u)/\tau, \end{aligned} \tag{13.10}$$

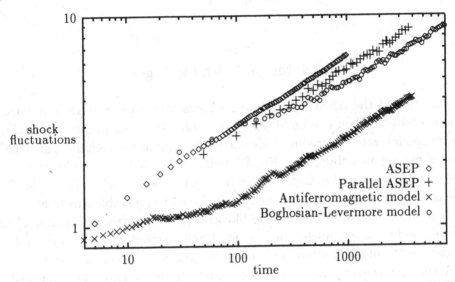

Fig. 13.4. Numerical evaluation of shock fluctuation exponents for ASEP and other models.

which gives equation (13.1) with $c = (2p - 1)/\tau$.

One expects that the scaling of the fluctuations in these models should be universal, for example the exponents should not depend on the details of the model but only on gross features such as the dimension, symmetry properties, etc. An extensive series of computer simulations of the motion of shock fronts in a variety of 1D stochastic lattice models with parallel and serial dynamics, infinite and finite temperatures, and ferromagnetic and antiferromagnetic particle interactions, was conducted [22], and this was found to be the case: all the models have the same shock fluctuation exponents, as illustrated by the slopes in Fig. 13.4. This strengthens the rationale for using the exceptionally uncomplicated ASEP in the first place.

In addition, it was shown [22,25] that this exponent could be changed if the dynamics were tuned to a special critical point: when one considers the current $J(\rho)$ as a function of the density ρ in a uniform system, if the system is at a point where $\partial^2 J/\partial \rho^2 = 0$ then the fluctuations will be reduced from the universal value of $O(t^{1/3})$. In the ordinary ASEP we have $J = (2p - 1)\rho(1 - \rho)$ and $\partial^2 J/\partial \rho^2 = 2 - 4p \neq 0$, and there is no critical point. However, a variant with non-nearest-neighbor jumps allowed has a current $J = \rho(1 - \rho)[1/2 + \rho^3 + (1 - \rho)^3]$ and therefore at $\rho = 1/2$ one has

$\partial^2 J/\partial\rho^2 = \partial^3 J/\partial\rho^3 = 0$, where one finds that the standard deviation of the shock position grows as $t^{1/4}$ [22,25].

13.4 Models with blockages

A variation of the ASEP can produce shocks that do not move: one introduces a 'blockage' on one bond (or hypercolumn of bonds in more than 1D); the transition rates are reduced for particles jumping between one particular pair of nearest-neighbor sites [26]. Particles pile up behind the blockage and the density is reduced in front of it; if one uses periodic boundary conditions the transition from low to high density will result in a stable shock localized at a position that depends on the blockage and the average density. Now initial conditions are irrelevant and one would expect the previous scaling of the growth of fluctuations in time to be converted into a spatial scaling.

Actually, however, there are two components to the shock fluctuations—the aforementioned 'dynamical randomness' producing the $L^{1/3}$ behavior, and a 'blockage randomness', since the timing of jumps across the blockage is random. The variance of the shock position due to the blockage randomness is proportional to the expected net number of excitations in the system and thus provides no contribution when the system is half-filled, i.e., when the number of particles is the same as the number of holes; and indeed a half-filled system of size L has shock fluctuations of order $L^{1/3}$, which confirms the behavior observed earlier. In general the blockage randomness produces fluctuations that scale as $L^{1/2}$. The validity of this picture can be seen by changing the dynamics to reduce artificially the noise resulting from the blockage jumps while keeping the rest of the system (relatively) unchanged. This new system has fluctuations that scale as $L^{1/3}$ for *all* densities, further reinforcing the belief in the exponent 1/3.

Of course the model with a blockage provides two subjects worthy of study: the region around the blockage as well as the region around the shock. The motivation for this model came from studying the shock, but the blockage has proved to be very interesting as well, and is perhaps of more general interest, serving as a model for long-range correlations and phase transitions in nonequilibrium systems: it exhibits the sensitivity to changes in parameters that are typical of driven diffusive systems; a local change in the transition rates (i.e., the jump rate across the blockage) has a global effect [27] in that the system segregates into high- and low-density phases.

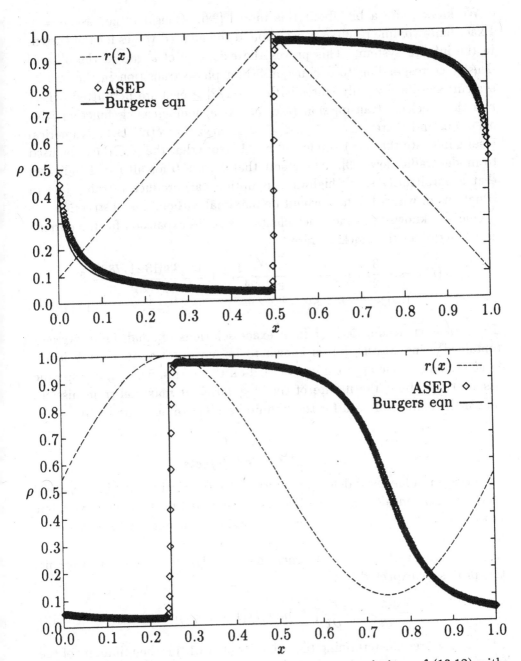

Fig. 13.5. Two examples of rate functions: stationary solutions of (13.12) with average value 1/2, and the stationary density of the ASEP with 500 sites and the same rate function.

We know quite a bit about this model [28]. Consider the case $p = 1$. Exact finite volume results serve to bound the allowed values for the current in the infinite system. This proves the existence of a gap in the allowed density corresponding to a nonequilibrium phase transition in the infinite system: specifically, only states with $|\rho - 1/2| \geq \delta(r)$ are permitted, where r is the blockage transmission rate, $\delta(\cdot)$ is a nonincreasing function with $\delta(1) = 0$, and there exists r^*, $0 < r^* < 1$, such that $\delta(r^*) > 0$. Numerical results indicate that $\delta(r) > 0$ for all $r < 1$. Note that the ASEP can be used to model traffic flow [29]. The result that $\delta(r) > 0$ for all $r < 1$ indicates that an arbitrarily small highway disruption can produce a traffic jam in situations in which the flow would be maximal without the obstruction.

One also knows the exact coefficients of a series expansion for the current as a function of the blockage rate:

$$J(r) = r - \frac{3}{2}r^2 + \frac{19}{2^4}r^3 - \frac{5 \times 59 \times 73}{2^{10} \times 3^3}r^4 + \frac{13 \times 33613 \times 177883}{2^{26} \times 3^7}r^5$$
$$-0.3278724755(1)r^6 + O(r^7). \tag{13.11}$$

The series expansion, derived from exact solutions of small finite systems (obtained using MAPLE V), is known to be asymptotic for all sufficiently large systems. Padé approximants based on this series, which make specific assumptions about the nature of the singularity at maximal transmission, match the numerical data for the 'infinite' system to one part in 10^4.

13.4.1 Extended defects

A blockage is a localized defect in an otherwise translation-invariant system. One can also consider extended defects. One model for this is an exclusion process where the rates vary over the system. In keeping with a hydrodynamic picture, we consider a smooth rate function r on the unit circle, i.e., the jump rate at site i is $r(i/L)$. Then the hydrodynamic equation governing this process is expected to be

$$\frac{\partial u(x,t)}{\partial t} + \frac{\partial}{\partial x}\left\{ r(x)u(x,t)\left[1 - u(x,t)\right]\right\} = 0. \tag{13.12}$$

The key feature in determining the behavior of (13.12) is the minimum of the function r. Consider stationary viscosity solutions $u(x)$ of (13.12) (obtained by considering the limit of adding a vanishingly small viscosity term). Let $u_\pm(x) = [1 \pm \sqrt{1 - r_{min}/r(x)}]/2$, assuming that r_{min} is the unique minimum of r located at x_{min}. Then we have three types of solutions: if $\int u(x)\,dx \geq \int u_+(x)\,dx$ the solution is of the form $u(x) = [1 + \sqrt{1 - R/r(x)}]/2$ with

$R \leq r_{\min}$; if $\int u(x)\,dx \leq \int u_-(x)\,dx$ the solution is of the form $u(x) = [1 - \sqrt{1 - R/r(x)}\,]/2$ with $R \leq r_{\min}$; otherwise the solution is discontinuous, with $u(x) = u_+(x)$ to the left of $x_{\min}$ and $u(x) = u_-(x)$ to the right of $x_{\min}$. This results in a solution continuous at $x_{\min}$ and discontinuous at some arbitrary other point.

Does this have anything to do with the particle system? In Fig. 13.5 we see two examples of rate functions: stationary solutions of (13.12) with average value 1/2, and the stationary density of the ASEP with 500 sites and the same rate function. Thus we see that shocks are not caused by discontinuities in the jump rates, but can occur and remain stable even when the rates vary smoothly.

If r has multiple minima, the picture is much less clear; certainly numerous weak stationary solutions of (13.12) exist but it is not clear if they have anything to do with the particle system and whether they exist as the limit of vanishing viscosity.

We would like to thank Francis Alexander and Eugene Speer for helpful discussions. Our work was supported in part by AFOSR Grant 4-26435. See also the added note [30].

References

[1] H. Spohn, *Large Scale Dynamics of Interacting Particles* (Springer-Verlag, New York, 1991), and references therein.

[2] C. Cercignani, *The Boltzmann Equation and Its Applications* (Springer-Verlag, New York, 1988).

[3] A. de Masi and E. Presutti, *Mathematical Methods for Hydrodynamic Limits. Lecture Notes in Mathematics*, Vol. **1501** (Springer-Verlag, 1991), and references therein.

[4] J. Lebowitz, E. Presutti and H. Spohn, *J. Stat. Phys.* **51**, 841 (1988).

[5] J. M. Burgers, *Adv. Appl. Mechanics* **1**, 171 (1948).

[6] F. Rezakhanlou, *Comm. Math. Phys.* **165**, 1 (1994).

[7] P. A. Ferrari, S. Goldstein and J. L. Lebowitz, in *Statistical Physics and Dynamical Systems: Rigorous Results*, J. Fritz, A. Jaffe and D. Szaszm, eds. (Birkhäuser, Boston, 1985).

[8] E. D. Andjel, M. Bramson and T. M. Liggett, *Prob. Theory Rel. Fields* **78**, 231 (1988).

[9] P. Ferrari, *Ann. Prob.* **14**, 1277 (1986).

[10] A. De Masi, C. Kipnis, E. Presutti and E. Saada, *Stochastics Rep.* **27**, 151 (1989).

[11] P. Ferrari, C. Kipnis and E. Saada, *Ann. Prob.* **19**, 226 (1991).

[12] P. Ferrari, *Prob. Th. Rel. Fields* **91**, 81 (1992).

[13] B. Derrida, S. A. Janowsky, J. L. Lebowitz and E. R. Speer, *Europhys. Lett.* **22**, 651 (1993).

[14] B. Derrida, S. A. Janowsky, J. L. Lebowitz and E. R. Speer, J. Stat. Phys. **73**, 813 (1993).

[15] B. Derrida, E. Domany and D. Mukamel, J. Stat. Phys. **69**, 667 (1992).

[16] B. Derrida, M. R. Evans, V. Hakim and V. Pasquier, J. Physique **A26**, 1493 (1993); G. Schütz and E. Domany, J. Stat. Phys. **72**, 277 (1993).

[17] E. R. Speer, in On Three Levels: Micro-, Meso-, and Macro-Approaches in Physics, M. Fannes, C. Maes and A. Verbeure, eds. (Plenum, New York, 1994).

[18] S. A. Janowsky, REBRAPE (Rev. Brasileira Prob. Estat.) **8**, 85 (1994).

[19] P. A. Ferrari, L. R. G. Fontes and Y. Kohayakawa, J. Stat. Phys. **76**, 1153 (1994).

[20] B. M. Boghosian and C. D. Levermore, Complex Systems **1**, 17 (1987).

[21] Z. Cheng, J. L. Lebowitz and E. R. Speer, Comm. Pure Appl. Math. **XLIV**, 971 (1991).

[22] F. J. Alexander, S. A. Janowsky, J. L. Lebowitz and H. van Beijeren, Phys. Rev. **E47**, 403 (1993).

[23] H. van Beijeren, J. Stat. Phys. **63**, 47 (1991).

[24] B. Schmittmann and R. K. P. Zia, Statistical Mechanics of Driven Diffusive Systems. Phase Transitions and Critical Phenomena, Vol. **17**, C. Domb and J. L. Lebowitz, eds. (Academic Press, New York, 1995).

[25] H. van Beijeren, R. Kutner and H. Spohn, Phys. Rev. Lett. **54**, 2026 (1985).

[26] S. A. Janowsky and J. L. Lebowitz, Phys. Rev. **A45**, 618 (1992); see also L.-H. Tang and D. Wolf, Phys. Rev. Lett. **65**, 1591 (1990); G. Schütz, J. Stat. Phys. **71**, 471 (1993).

[27] P. L. Garrido, J. L. Lebowitz, C. Maes and H. Spohn, Phys. Rev. **A42**, 1954 (1990).

[28] S. A. Janowsky and J. L. Lebowitz, J. Stat. Phys. **77**, 35 (1994).

[29] K. Nagel and M. Schreckenberg, J. Physique **I2**, 2221 (1992).

[30] Note added in proofs. B. Derrida, J. L. Lebowitz and E. Speer have recently obtained results about the structure of the shock for the case when the asymmetry is not total, $\frac{1}{2} < p < 1$. They find in particular that for any asymptotic densities ρ_+ and ρ_-, there exists a p^* such that for $p \leq p^*$ the local density profile, as seen from the second class particle, is monotone. For $\rho_+ = 1$, $\rho_- = 0$, $p^* = 1$. O. Costin, J. L. Lebowitz and E. Speer have found additional evidence (but no proof) that the behavior of the current $J(r)$ is, in the case of a single blockage, as described in Sec. 13.4.

14

The asymmetric exclusion model: exact results through a matrix approach

Bernard Derrida and Martin R. Evans

14.1 Introduction

We present here a series of recent exact results for a simple one-dimensional model of hopping particles, the fully asymmetric exclusion model [1-3]. These results, which include various exact expressions of steady-state properties, have been obtained using a matrix method, the description of which constitutes the main part of what follows.

14.1.1 Definition of the model and evolution of correlation functions

Let us start by defining the stochastic dynamics of the asymmetric exclusion model. Each site of a 1D lattice is either occupied by one particle or empty. A configuration of the system is characterized by binary variables $\{\tau_i\}$ where $\tau_i = 1$ if site i is occupied by a particle and $\tau_i = 0$ if site i is empty. During an infinitesimal time interval dt, each particle hops with probability dt to its right if this site is empty (and does not move otherwise).

From this stochastic dynamical rule, one can easily derive the equations that govern the time evolution of any correlation function. For example, if one considers the occupation of site i (for the moment we consider a non-boundary site to avoid choosing any particular boundary conditions) one can write

$$\tau_i(t + dt) = \begin{cases} \tau_i(t), & \text{with prob. } 1 - 2dt, \\ \tau_i(t) + [1 - \tau_i(t)]\,\tau_{i-1}(t), & \text{with prob. } dt, \\ \tau_i(t)\tau_{i+1}(t), & \text{with prob. } dt. \end{cases} \tag{14.1}$$

The first line comes from the fact that with probability $1 - 2dt$ neither of the sites $i - 1$ or i is updated and therefore τ_i remains unchanged. The second and the third lines correspond to updating sites $i - 1$ and i respectively.

If one averages (14.1) over the events that may occur between t and $t + dt$ and over all histories up to time t one obtains

$$\frac{d\langle \tau_i \rangle}{dt} = \langle \tau_{i-1}(1 - \tau_i) \rangle - \langle \tau_i(1 - \tau_{i+1}) \rangle = \langle \tau_{i-1} \rangle - \langle \tau_i \rangle - \langle \tau_{i-1}\tau_i \rangle + \langle \tau_i\tau_{i+1} \rangle.$$

(14.2)

The same kind of reasoning allows one to write down an equation for the evolution of $\langle \tau_i\tau_{i+1} \rangle$,

$$\frac{d\langle \tau_i\tau_{i+1} \rangle}{dt} = \langle \tau_{i-1}(1 - \tau_i)\tau_{i+1} \rangle - \langle \tau_i\tau_{i+1}(1 - \tau_{i+2}) \rangle$$

$$= \langle \tau_{i-1}\tau_{i+1} \rangle - \langle \tau_i\tau_{i+1} \rangle - \langle \tau_{i-1}\tau_i\tau_{i+1} \rangle + \langle \tau_i\tau_{i+1}\tau_{i+2} \rangle \quad (14.3)$$

or for any other correlation function $\langle \tau_i\tau_j \cdots \rangle$. These equations are exact and give in principle the time evolution of any correlation function. However, the evolution (14.2) of $\langle \tau_i \rangle$ requires the knowledge of $\langle \tau_i\tau_{i+1} \rangle$, which itself (14.3) requires the knowledge of $\langle \tau_{i-1}\tau_{i+1} \rangle$ and $\langle \tau_{i-1}\tau_i\tau_{i+1} \rangle$, so that the problem is intrinsically an N-body problem in the sense that the calculation of any correlation function requires the knowledge of all the others. This is a situation quite common in equilibrium statistical mechanics where, although one can write relationships between different correlation functions, there is an infinite hierarchy of equations, which in general makes the problem intractable.

In the following, different variants of the model will be considered corresponding to different boundary conditions, as follows.

- An infinite 1D system, where (14.2), (14.3) are valid everywhere.
- A finite system of N sites with periodic boundary conditions $\tau_{i+N} = \tau_i$ (i.e., a ring of N sites). In this case the number of particles $M = \sum_i \tau_i$ is fixed. As for the infinite system, (14.2), (14.3) are valid everywhere [4-7].
- A finite system of N sites with open boundary conditions, where in time dt a particle may enter the lattice at site 1 with probability αdt (if this site is empty) and a particle at site N may leave the lattice with probability βdt (if this site is occupied). In this case the number of particles in the system is not conserved [8-11]. The evolution equations (14.2), (14.3) are then valid everywhere in the bulk but are modified at the boundary sites. For example (14.2) becomes

$$\frac{d\langle \tau_1 \rangle}{dt} = \alpha\langle (1 - \tau_1) \rangle - \langle \tau_1(1 - \tau_2) \rangle, \quad \frac{d\langle \tau_N \rangle}{dt} = \langle \tau_{N-1}(1 - \tau_N) \rangle - \beta\langle \tau_N \rangle. \quad (14.4)$$

Our main purpose is to describe how a matrix approach can be used to calculate exactly various steady-state properties of such systems: steady-state current, profiles, correlations for a system with open boundary conditions;

phase transitions induced by boundaries; diffusion constants in a finite system with periodic boundary conditions; infinite or finite systems with several species of particles; shocks in an infinite system.

14.1.2 *Simple known results*

Let us first mention some simple known results:

• For an infinite system, a Bernoulli measure, which is a product measure where all sites are occupied with the same density ρ (i.e., the τ_i are independent and take the value 0 with probability $1 - \rho$ and the value 1 with probability ρ), is a steady state [2]. Clearly, taking $\langle \tau_i \rangle = \rho$, $\langle \tau_i \tau_j \rangle = \rho^2$, and so on, gives 0 for the r.h.s.s of (14.2), (14.3).

• For an infinite system, there are also traveling steady states (shocks) [12-16]. If one considers an initial situation where the measure is a product measure of density ρ_+ at the right of the origin and a product measure of density ρ_- at the left of the origin, with $\rho_+ > \rho_-$, the dynamics produces in the long-time limit a shock with a steady shape that moves with a velocity v given by $v = 1 - \rho_- - \rho_+$.

A simple way of understanding the latter result is to consider a window of size $2L$, large enough to contain the shock and such that outside the window the measure is almost exactly a Bernoulli measure with density ρ_- to the left of the window and ρ_+ to the right of the window. Then from (14.2), one has

$$\frac{d}{dt} \sum_{i=-L}^{L} \langle \tau_i \rangle = \langle \tau_{-L-1}(1 - \tau_{-L}) \rangle - \langle \tau_L(1 - \tau_{L+1}) \rangle$$

$$= \rho_-(1 - \rho_-) - \rho_+(1 - \rho_+). \tag{14.5}$$

If one assumes that the shape of the shock does not evolve in time, this implies that during the time dt, a region of length vdt changes its density from ρ_+ to ρ_-, i.e., $(d/dt) \sum_{i=-L}^{L} \langle \tau_i \rangle = v(\rho_- - \rho_+)$, and together with (14.5) this gives $v = 1 - \rho_- - \rho_+$.

• On a ring the steady state is such that all configurations (with a given number M of particles) are equally likely. This can easily be checked since if all configurations have equal weight the rate at which the system leaves a given configuration (which is equal to the number of clusters of particles in that configuration: the first particle in each cluster can hop forward) is equal to the rate at which the system enters that configuration (by the move of the last particle of each cluster) [4].

• For open boundary conditions on a finite lattice there is a line $\alpha + \beta = 1$ where the steady state becomes a product measure (that is, each n-point correlation function factorizes into a product of n 1-point correlation functions) with $\langle \tau_i \rangle = \alpha = 1 - \beta$ for all i . We will see in Sec. 14.2 that the line $\alpha + \beta = 1$ appears as a special case of the exact solution valid for all α and β. At this stage, it is easy to check that this condition gives 0 for the r.h.s.s of (14.2)-(14.4) when correlations are absent, i.e., $\langle \tau_i \tau_{i+1} \rangle = \langle \tau_i \rangle \langle \tau_{i+1} \rangle$, $\langle \tau_i \tau_{i+1} \tau_{i+2} \rangle = \langle \tau_i \rangle \langle \tau_{i+1} \rangle \langle \tau_{i+2} \rangle$ and so on.

Remark: Partially asymmetric exclusion. The fully asymmetric exclusion model described above can easily be generalized to the case of partial asymmetry where each particle has during each time dt a probability qdt of hopping to its left neighbor and pdt of hopping to its right neighbor (provided that the target is empty) [10,17,18]. The same reasoning that led to (14.2), (14.3) gives in this case

$$\frac{d\langle \tau_i \rangle}{dt} = p\langle \tau_{i-1} \rangle + q\langle \tau_{i+1} \rangle - (p+q)\langle \tau_i \rangle + (p-q)\left[\langle \tau_i \tau_{i+1} \rangle - \langle \tau_i \tau_{i-1} \rangle\right], \quad (14.6)$$

and

$$\frac{d\langle \tau_i \tau_{i+1} \rangle}{dt} = p\langle \tau_{i-1} \tau_{i+1} \rangle + q\langle \tau_i \tau_{i+2} \rangle - (p+q)\langle \tau_i \tau_{i+1} \rangle$$
$$+ (p-q)\left[\langle \tau_i \tau_{i+1} \tau_{i+2} \rangle - \langle \tau_{i-1} \tau_i \tau_{i+1} \rangle\right]. \quad (14.7)$$

Of course, for $p = 1$ and $q = 0$, one recovers (14.2), (14.3). One also sees quite clearly that in the symmetric case, $p = q$, the hierarchy decouples and that the calculation of a given correlation function does not require the knowledge of higher-order correlations [1,3].

Remark: Short-time expansions. Equations (14.2)-(14.4) and their generalizations to arbitrary correlations are valid for arbitrary initial conditions. If one takes successive time derivatives of any of these equations, the l.h.s. has always one more derivative than the r.h.s. In this way, one can relate the successive time derivatives of any correlation function at $t = 0$ to other correlation functions at $t = 0$. For example (14.2), (14.3) imply that

$$\frac{d^2\langle \tau_i \rangle}{dt^2} = \frac{d}{dt}\left[\langle \tau_{i-1}(1 - \tau_i)\rangle - \langle \tau_i(1 - \tau_{i+1})\rangle\right] = \langle \tau_{i-2} \rangle - 2\langle \tau_{i-1} \rangle + \langle \tau_i \rangle$$
$$- \langle \tau_{i-2}\tau_{i-1}\rangle + 3\langle \tau_{i-1}\tau_i\rangle - 2\langle \tau_i\tau_{i+1}\rangle - \langle \tau_{i-2}\tau_i\rangle + \langle \tau_{i-1}\tau_{i+1}\rangle$$
$$+ \langle \tau_{i-2}\tau_{i-1}\tau_i\rangle - 2\langle \tau_{i-1}\tau_i\tau_{i+1}\rangle + \langle \tau_i\tau_{i+1}\tau_{i+2}\rangle. \quad (14.8)$$

This allows one, in principle, to obtain the short-time expansion of any correlation function, for arbitrary initial conditions.

14.2 Exact steady state for open boundaries

14.2.1 Steady-state equations

In the steady state, all the correlation functions are time independent. Therefore, the time evolution equations (14.2)-(14.4) become the following steady-state equations:

$$\langle \tau_{i-1} \rangle - \langle \tau_i \rangle - \langle \tau_{i-1}\tau_i \rangle + \langle \tau_i\tau_{i+1} \rangle = 0, \tag{14.9}$$

$$\langle \tau_{i-1}\tau_{i+1} \rangle - \langle \tau_i\tau_{i+1} \rangle - \langle \tau_{i-1}\tau_i\tau_{i+1} \rangle + \langle \tau_i\tau_{i+1}\tau_{i+2} \rangle = 0, \tag{14.10}$$

$$\alpha\langle(1 - \tau_1)\rangle - \langle \tau_1(1 - \tau_2)\rangle = 0, \tag{14.11}$$

$$\langle \tau_{N-1}(1 - \tau_N)\rangle - \beta\langle \tau_N \rangle = 0. \tag{14.12}$$

Some of these steady-state equations have a simple physical meaning: for example, (14.9), (14.11), (14.12) rewritten as

$$\alpha\langle(1 - \tau_1)\rangle = \langle \tau_1(1 - \tau_2)\rangle = \cdots = \langle \tau_{i-1}(1 - \tau_i)\rangle$$
$$= \cdots = \langle \tau_{N-1}(1 - \tau_N)\rangle = \beta\langle \tau_N \rangle, \tag{14.13}$$

simply express that the current of particles is conserved in the steady state.

14.2.2 The matrix formulation

In collaboration with V. Hakim and V. Pasquier, we have developed a matrix method to calculate the steady-state properties in the case of open boundary conditions [10]. This type of approach, initially introduced in connection with the Bethe ansatz [19,20], has been used to solve several problems in statistical mechanics (directed lattice animals [21] and quantum antiferromagnetic spin chains [22-24]).

The main idea is to try to write the weights $f_N(\tau_1, \ldots, \tau_N)$ of the configurations in the steady state as

$$f_N(\tau_1, \ldots, \tau_N) = \langle W| \prod_{i=1}^{N} [\tau_i D + (1 - \tau_i)E] |V\rangle, \tag{14.14}$$

where D, E are matrices, $\langle W|, |V\rangle$ are vectors (we use the standard 'bra ket' notation of quantum mechanics) and τ_i are the occupation variables. Thus, in the product (14.14) we use matrix D whenever $\tau_i = 1$ and E whenever $\tau_i = 0$. Since, as we shall see, the matrices D and E in general do not commute, the weights $f_N(\tau_1, \ldots, \tau_N)$ are complicated functions of the configuration $\{\tau_1, \ldots, \tau_N\}$. As the weights $f_N(\tau_1, \ldots, \tau_N)$ given by (14.14) are not necessarily normalized, the probability $p_N(\tau_1, \ldots, \tau_N)$ of a configuration $\{\tau_1, \ldots, \tau_N\}$ in the steady state is

$$p_N(\tau_1, \ldots, \tau_N) = f_N(\tau_1, \ldots, \tau_N) \Big/ \sum_{\tau_1=1,0} \cdots \sum_{\tau_N=1,0} f_N(\tau_1, \ldots, \tau_N). \quad (14.15)$$

Of course, it is not obvious that such matrices D, E and vectors $\langle W|, |V\rangle$ exist. We are going to show below that if they satisfy

$$D|V\rangle = \frac{1}{\beta}|V\rangle, \quad \langle W|E = \frac{1}{\alpha}\langle W|, \quad (14.16)$$

$$DE = D + E, \quad (14.17)$$

then (14.14) does give the steady state.

Correlation functions

Before presenting the proof that (14.16), (14.17), via (14.14), give the steady state, let us show how the approach leads to a straightforward computation for the correlation functions. If one defines the matrix C by $C = D + E = DE$, it is clear that $\langle \tau_i \rangle_N$ defined by

$$\langle \tau_i \rangle_N = \sum_{\tau_1=1,0} \cdots \sum_{\tau_N=1,0} \tau_i \, f_N(\tau_1, \ldots, \tau_N) \Big/ \sum_{\tau_1=1,0} \cdots \sum_{\tau_N=1,0} f_N(\tau_1, \ldots, \tau_N),$$

$$(14.18)$$

is given by

$$\langle \tau_i \rangle_N = \langle W|C^{i-1}DC^{N-i}|V\rangle \Big/ \langle W|C^N|V\rangle. \quad (14.19)$$

In the same way, any higher correlation function will take a simple form in terms of the matrices C, D, and E. For example, when $i < j$, the two-point function is given by

$$\langle \tau_i \tau_j \rangle_N = \langle W|C^{i-1}DC^{j-i-1}DC^{N-j}|V\rangle \Big/ \langle W|C^N|V\rangle. \quad (14.20)$$

Therefore, all we require in order to be able to calculate arbitrary spatial correlation functions is that the matrix elements of any power of $C = D+E$ have manageable expressions.

Remark. Note that (14.13) and (14.10) are equivalent to

$$\alpha\langle W|EC^{N-1}|V\rangle = \langle W|DEC^{N-2}|V\rangle = \cdots = \langle W|C^{i-2}DEC^{N-i}|V\rangle$$
$$= \cdots = \beta\langle W|C^{N-1}D|V\rangle, \quad (14.21)$$

and

$$\left\langle W \middle| C^{i-2}[DCDC - CD^2C - D^3C + CD^3]C^{N-i-2} \middle| V \right\rangle = 0. \qquad (14.22)$$

One can easily verify that these are satisfied when the vectors $\langle W|, |V\rangle$ and the matrices C, D, E satisfy (14.16), (14.17).

14.2.3 Proof of the matrix formulation

We are now going to prove that the weights (14.14) do indeed satisfy the steady-state conditions. Let us first consider a configuration of the form

$$1^{m_1} \, 0^{n_1} \, 1^{m_2} \, 0^{n_2} \cdots 1^{m_k} \, 0^{n_k}, \qquad (14.23)$$

where the integer exponents satisfy $m_i \geq 1$ and $n_i \geq 1$, and write its weight according to (14.14) as $\langle D^{m_1} E^{n_1} D^{m_2} E^{n_2} \cdots D^{m_k} E^{n_k}\rangle$, where to lighten the notation we have replaced $\langle W|$ by $\langle$ and $|V\rangle$ by $\rangle$. There are k ways of leaving this configuration. Also, one can enter into this configuration from any of the following configurations,

$$\underline{0} \, 1^{m_1-1} \, 0^{n_1} \, 1^{m_2} \, 0^{n_2} \cdots 1^{m_k} \, 0^{n_k},$$
$$1^{m_1} \, 0^{n_1-1} \, \underline{10} \, 1^{m_2-1} \, 0^{n_2} \cdots 1^{m_k} \, 0^{n_k},$$
$$\vdots$$
$$1^{m_1} \, 0^{n_1} \, 1^{m_2} \, 0^{n_2} \cdots 1^{m_k} \, 0^{n_k-1} \, \underline{1},$$

where the underlining indicates where the move occurs. In order to verify that the weights (14.14) do give the steady state, one has to check that the probability of entering a configuration is the same as the probability of leaving that configuration. Let us do it first for a configuration of the type (14.23) when $k = 2$. One requires

$$\begin{aligned}
2\langle D^{m_1} E^{n_1} D^{m_2} E^{n_2}\rangle &= \alpha\langle ED^{m_1-1} E^{n_1} D^{m_2} E^{n_2}\rangle \\
&\quad + \langle D^{m_1} E^{n_1-1} DED^{m_2-1} E^{n_2}\rangle \\
&\quad + \beta\langle D^{m_1} E^{n_1} D^{m_2} E^{n_2-1} D\rangle. \qquad (14.24)
\end{aligned}$$

By replacing, in each term not involving α, β, one product DE by $D + E$, (14.24) becomes

$$\begin{aligned}
&\langle D^{m_1-1} E^{n_1} D^{m_2} E^{n_2}\rangle + \langle D^{m_1} E^{n_1-1} D^{m_2} E^{n_2}\rangle \\
&+ \langle D^{m_1} E^{n_1} D^{m_2-1} E^{n_2}\rangle + \langle D^{m_1} E^{n_1} D^{m_2} E^{n_2-1}\rangle = \\
&\qquad \alpha\langle ED^{m_1-1} E^{n_1} D^{m_2} E^{n_2}\rangle + \langle D^{m_1} E^{n_1-1} D^{m_2} E^{n_2}\rangle \\
&\qquad + \langle D^{m_1} E^{n_1} D^{m_2-1} E^{n_2}\rangle + \beta\langle D^{m_1} E^{n_1} D^{m_2} E^{n_2-1} D\rangle. \qquad (14.25)
\end{aligned}$$

Clearly, the effect is that each DE produces two terms: one where the block before the DE term has one D less and one where the block after the DE term has one E less. One sees that this gives rise to a cancellation of two terms from both sides of (14.25), and one ends up with

$$\langle D^{m_1-1} E^{n_1} D^{m_2} E^{n_2} \rangle + \langle D^{m_1} E^{n_1} D^{m_2} E^{n_2-1} \rangle =$$
$$\alpha \langle E D^{m_1-1} E^{n_1} D^{m_2} E^{n_2} \rangle + \beta \langle D^{m_1} E^{n_1} D^{m_2} E^{n_2-1} D \rangle, \quad (14.26)$$

which is easily satisfied, owing to (14.16).

For all configurations of the type (14.23) when $k \neq 2$, the same calculation using (14.17) leads to a systematic cancellation of $2(k-1)$ terms from both sides of the equation for the conservation of probability, and one always obtains an equation with only four terms as in (14.26). This last equation is always satisfied, owing to (14.16).

The configuration chosen above (14.23) is of the type that starts with a 1 and ends with a 0. However, the above calculation extends without any difficulty to the other three possibilities.

14.2.4 *Properties of the matrices*

We mentioned in the introduction that along the line $\alpha + \beta = 1$, the steady state becomes trivial. This is reflected by the fact that one can choose commuting matrices D and E to solve (14.16), (14.17). If D and E commute one can write $\langle W|D + E|V \rangle = \langle W|DE|V \rangle = \langle W|ED|V \rangle$ and this becomes $(\alpha^{-1} + \beta^{-1}) \langle W|V \rangle = (\alpha\beta)^{-1} \langle W|V \rangle$. As $\langle W|V \rangle \neq 0$, this clearly implies that $\alpha + \beta = 1$. Under this condition it is easy to check that $D = \beta^{-1}$ and $E = \alpha^{-1}$ solves (14.16), (14.17) .

On the other hand, for $\alpha + \beta \neq 1$, the matrices D, E cannot commute and therefore their size must be greater than unity. The next question is whether one can find finite-dimensional matrices that will satisfy (14.16), (14.17). It turns out that one can prove [10] that this is impossible. Indeed, if $DE = D + E$, then there is no vector $|X\rangle$ such that $E|X\rangle = |X\rangle$ (otherwise, we would have $D|X\rangle = DE|X\rangle = D|X\rangle + E|X\rangle = D|X\rangle + |X\rangle$, so that $|X\rangle = 0$). Therefore, if the matrices are finite dimensional, $1 - E$ has an inverse (since E has no eigenvalue 1) and $DE = D + E$ implies that $D = E(1 - E)^{-1}$. As D is a function of E, it is clear that D commutes with E. So the only possibility left for D and E not to commute is that D and E are infinite-dimensional matrices.

In order to perform calculations within the matrix formulation there are basically two approaches one can take. Either one can work with the algebra

(14.16), (14.17) directly, or one can make a particular choice of matrices (see Appendix B) and use it to the full.

Here, instead of using explicit forms for the matrices, we calculate directly matrix elements such as those which appear in (14.19), (14.20) from the commutation rules (14.16), (14.17). For example, one can easily show that $\langle W|C|V\rangle/\langle W|V\rangle = \langle W|D + E|V\rangle/\langle W|V\rangle = \alpha^{-1} + \beta^{-1}$, $\langle W|C^2|V\rangle/\langle W|V\rangle = \langle W|D^2 + ED + E^2 + D + E|V\rangle/\langle W|V\rangle = \alpha^{-2} + (\alpha\beta)^{-1} + \beta^{-2} + \alpha^{-1} + \beta^{-1}$. A general expression of $\langle W|C^N|V\rangle$ (where $C = D + E$) valid for all values of α and β and $N \geq 1$ is [10]

$$\frac{\langle W|C^N|V\rangle}{\langle W|V\rangle} = \sum_{p=1}^{N} \frac{p\,(2N-1-p)!}{N!\,(N-p)!} \frac{\beta^{-p-1} - \alpha^{-p-1}}{\beta^{-1} - \alpha^{-1}}. \qquad (14.27)$$

For large N one finds that

$$\langle W|C^N|V\rangle \sim N^{-3/2}4^N, \quad \text{for } \alpha > \tfrac{1}{2} \text{ and } \beta > \tfrac{1}{2}, \qquad (14.28)$$

$$\langle W|C^N|V\rangle \sim [\alpha(1-\alpha)]^{-N}, \quad \text{for } \alpha < \tfrac{1}{2} \text{ and } \beta > \alpha, \qquad (14.29)$$

$$\langle W|C^N|V\rangle \sim [\beta(1-\beta)]^{-N}, \quad \text{for } \beta < \tfrac{1}{2} \text{ and } \alpha > \beta. \qquad (14.30)$$

The line $\alpha = \beta < 1/2$ is special: one has $\langle W|C^N|V\rangle \sim N\,[\alpha(1-\alpha)]^{-N}$.

14.2.5 Some results

Once the matrix elements of C are known, expressions for several quantities can be derived. For example, in the steady state, the current through the bond $i, i+1$ is simply $J = \langle \tau_i(1 - \tau_{i+1})\rangle$, because during a time dt the probability that a particle jumps from i to $i+1$ is $\tau_i(1 - \tau_{i+1})dt$. Therefore, J is given by

$$J = \langle W|C^{i-1}DEC^{N-i-1}|V\rangle/\langle W|C^N|V\rangle = \langle W|C^{N-1}|V\rangle/\langle W|C^N|V\rangle, \qquad (14.31)$$

where we have used the fact that $DE = C$. This expression is of course independent of i, as expected in the steady state (14.13). From the large-N behavior of the matrix elements $\langle W|C^N|V\rangle$ given by (14.28)-(14.30) one sees that there are three different phases, where the current J is given by

$$J = \tfrac{1}{4}, \qquad \text{for } \alpha \geq \tfrac{1}{2} \text{ and } \beta \geq \tfrac{1}{2}, \qquad (14.32)$$

$$J = \alpha(1-\alpha), \qquad \text{for } \alpha < \tfrac{1}{2} \text{ and } \beta > \alpha, \qquad (14.33)$$

$$J = \beta(1-\beta), \qquad \text{for } \beta < \tfrac{1}{2} \text{ and } \alpha > \beta. \qquad (14.34)$$

From knowledge of the matrix elements $\langle W|C^N|V\rangle$, one can also obtain [10] exact expressions for all equal-time correlation functions. For example, using the fact that

$$C^n E = \sum_{p=0}^{n-1} \frac{(2p)!}{p!\,(p+1)!} C^{n-p} + \sum_{p=2}^{n+1} \frac{(p-1)(2n-p)!}{n!\,(n+1-p)!} E^p , \qquad (14.35)$$

which is a direct consequence of (14.16), (14.17), one finds that the profile $\langle \tau_i \rangle_N$ is given for $i > 1$ by

$$\langle \tau_i \rangle_N = 1 - \sum_{p=0}^{i-2} \frac{(2p)!}{p!\,(p+1)!} \frac{\langle W|C^{N-1-p}|V\rangle}{\langle W|C^N|V\rangle}$$

$$- \frac{\langle W|C^{N-i}|V\rangle}{\langle W|C^N|V\rangle} \sum_{p=2}^{i} \frac{(p-1)(2i-2-p)!}{(i-1)!\,(i-p)!} \alpha^{-p} \qquad (14.36)$$

and $\langle \tau_1 \rangle_N = \langle W|C^{N-1}|V\rangle/(\alpha\langle W|C^N|V\rangle)$, where $\langle W|C^N|V\rangle$ is given by (14.27). Several limiting behaviors (N large, i large) are discussed in [10,11]. In the bulk limit (i.e., for $i \gg 1$ and $N - i \gg 1$) one finds the density $\rho = \langle \tau_i \rangle_N$ is given in the three phases by

$$\rho = \tfrac{1}{2}, \qquad \text{for } \alpha \geq \tfrac{1}{2} \text{ and } \beta \geq \tfrac{1}{2}, \qquad (14.37)$$

$$\rho = \alpha, \qquad \text{for } \alpha < \tfrac{1}{2} \text{ and } \beta > \alpha, \qquad (14.38)$$

$$\rho = (1 - \beta), \quad \text{for } \beta < \tfrac{1}{2} \text{ and } \alpha > \beta. \qquad (14.39)$$

Thus (14.37) is a maximal-current phase, (14.38) is a low-density phase and (14.39) is a high-density phase. We see that the lines $\alpha = 1/2$, $\beta > 1/2$ and $\beta = 1/2$, $\alpha > 1/2$ are second-order transition lines (as ρ is continuous) whereas the line $\alpha = \beta < 1/2$ is a first-order transition line (with a discontinuity of $1 - 2\alpha$ in ρ).

In the case $\alpha = \beta = 1$ one can perform the sums in (14.36) to obtain [9]

$$\langle \tau_i \rangle_N = \frac{1}{2} + \frac{N - 2i + 1}{4} \frac{(2i)!}{(i!)^2} \frac{(N!)^2}{(2N+1)!} \frac{(2N - 2i + 2)!}{[(N - i + 1)!]^2} . \qquad (14.40)$$

In this case the profile near the left boundary, in the limit $N \to \infty$ keeping i fixed is

$$\langle \tau_i \rangle_\infty = \frac{1}{2} + \frac{(2i)!}{(i!)^2} \frac{1}{2^{2i+1}} \simeq \frac{1}{2} + \frac{1}{2} \frac{1}{\sqrt{\pi}} \frac{1}{i^{1/2}} \qquad \text{for large } i. \qquad (14.41)$$

Remark: Comparison with the mean-field theory. We may compare these results with the mean-field theory outlined in Appendix B of this chapter. The mean-field theory predicts the correct phase diagram; however, the profile is incorrect: for example the decay $i^{-1/2}$ of the exact solution (14.41) is replaced by i^{-1} in the mean-field approximation (14.87).

Remark: Shocks. Along the first-order transition line $\alpha = \beta < 1/2$ one finds that for large N and i the profile is given by $\langle \tau_i \rangle_N \simeq \alpha + (1 - 2\alpha)i/N$. This is consistent with the existence of a shock, between regions of density α and $1 - \alpha$, located at an arbitrary position.

14.3 Diffusion constant

So far we have seen that for open boundary conditions, one can calculate all equal-time correlations using matrices. Here we are going to show that at least one simple quantity, the diffusion constant of a tagged particle [25,26], which is related to non-equal-time correlation functions, can also be calculated using the matrix approach [7]. In order to keep the discussion as straightforward as possible, we consider the case of M particles on a ring of size N, for which the steady state is known to be particularly simple (see Sec. 14.1).

14.3.1 Exact expression for the diffusion constant

If we tag one of the particles (without in any way changing its dynamics) and call Y_t the number of hops the particle has made up to time t, one expects that in the long-time limit

$$\lim_{t \to \infty} (\langle Y_t \rangle / t) = v, \qquad \lim_{t \to \infty} \left[\left(\langle Y_t^2 \rangle - \langle Y_t \rangle^2 \right) / t \right] = \Delta. \qquad (14.42)$$

The exact expressions for the velocity v and the diffusion constant Δ are given by [7]

$$v = \frac{N - M}{N - 1}, \qquad \Delta = \frac{(2N - 3)!}{(2M - 1)!\,(2N - 2M - 1)!} \left[\frac{(M - 1)!\,(N - M)!}{(N - 1)!} \right]^2. \qquad (14.43)$$

The expression for v is a simple consequence of the fact that in the steady state all configurations with M particles have equal probability. The expression for Δ, on the other hand, is nontrivial.

An interesting limit of (14.43) is that of a given density ρ of particles in an infinite system ($M = N\rho$ as $N \to \infty$ in (14.43)) where one finds

$$\Delta \simeq \frac{1}{2} \sqrt{\frac{\pi}{N}} \left[\frac{(1 - \rho)^{3/2}}{\rho^{1/2}} \right]. \qquad (14.44)$$

The fact that Δ vanishes for $N \to \infty$ indicates that in an infinite system for fixed initial conditions the fluctuations in the distance traveled by a tagged particle are subdiffusive [27,28].

Remark. There are several other diffusion constants one can define, all simply related to Δ owing to the fact that the particles cannot overtake each other. For example one can consider the fluctuations of the total number of moves made by all the particles (center-of-mass fluctuations), which are given by $M^2 \Delta$, or the fluctuations in the number of particles that have passed through a marked bond, which are given by $(M^2/N^2)\Delta$.

14.3.2 Calculation of the diffusion constant

The proof of (14.43) is somewhat long and we do not give it here. Instead we would just like to point out some of the ideas involved. Consider quantities $\langle Y_t | \mathcal{C} \rangle$, which are the conditional averages of the number of hops made by the tagged particle up to time t, given that the system is in configuration $\mathcal{C}$ at time t. The angle brackets denote an average both on the initial condition at $t = 0$ (with the steady-state weights $p(\mathcal{C})$) and on the process up to time t. One can show [7,29] that in the long-time limit

$$\langle Y_t | \mathcal{C} \rangle \to vt + r(\mathcal{C})/p(\mathcal{C}), \qquad (14.45)$$

where the velocity v is given by $v = \sum_{\mathcal{C}, \mathcal{C}'} M_1(\mathcal{C}, \mathcal{C}')p(\mathcal{C}')$. Here $M_1(\mathcal{C}, \mathcal{C}')$ is equal to 1 if a transition from $\mathcal{C}'$ to $\mathcal{C}$ can occur by the tagged particle's hopping forward, and is equal to zero otherwise. With knowledge of the quantities $r(\mathcal{C})$ one can compute the diffusion constant through the formula [7]

$$\Delta = v + 2 \sum_{\mathcal{C}, \mathcal{C}'} M_1(\mathcal{C}, \mathcal{C}')r(\mathcal{C}'). \qquad (14.46)$$

It turns out [7] that the $r(\mathcal{C})$ may be calculated by writing them in the form

$$r(\mathcal{C}) = R \operatorname{Tr}[BX_1 X_2 \cdots X_{N-1}] + \text{constant}, \qquad (14.47)$$

where the constant is chosen so that $\sum_{\mathcal{C}} r(\mathcal{C}) = 0$, as required by (14.45) to give $\langle Y_t \rangle = \sum_{\mathcal{C}} p(\mathcal{C})\langle Y_t | \mathcal{C} \rangle = vt$. Equation (14.47) involves the trace of a product of matrices: matrix B marks the position of the tagged particle; $X_i = D$ if the site i places in front of the tagged particle is occupied, $X_i = E$ if this site is empty; and R is a normalization. The matrices D, E satisfy the relation $DE = D + E$ and we use representation (14.77) of Appendix A,

$$D = \sum_{i=1}^{\infty} \Big(|i\rangle\langle i| + |i\rangle\langle i+1| \Big), \qquad E = \sum_{i=1}^{\infty} \Big(|i\rangle\langle i| + |i+1\rangle\langle i| \Big). \quad (14.48)$$

The matrix B has a finite number of nonzero elements,

$$B = \sum_{i=1}^{M-1} \Big(|i\rangle\langle i| + |i\rangle\langle i+1| \Big), \qquad (14.49)$$

and R is given by $\binom{N-1}{M-1}^{-2}$. With expression (14.47) for the $r(\mathcal{C})$ one obtains (14.43) from (14.46).

Remark: Open boundary conditions. The calculation of a diffusion constant has been extended to the case of open boundary conditions. In this case if one denotes by Y_t the number of particles that have entered the lattice at site 1 between times 0 and t one can evaluate the current and diffusion constant,

$$\lim_{t \to \infty} (\langle Y_t \rangle / t) = J, \qquad \lim_{t \to \infty} \left[\left(\langle Y_t^2 \rangle - \langle Y_t \rangle^2 \right) / t \right] = \Delta, \qquad (14.50)$$

where J is the current calculated in Sec. 14.2. Using a more complicated matrix technique [29] one can obtain expressions of Δ as a function of N, α, β. The general expression is rather complicated; however in the case $\alpha = \beta = 1$ it simplifies to

$$\Delta = \Big\{ 3(4N+1)! \, [N!(N+2)!]^2 \Big\} \Big/ \Big\{ 2 [(2N+1)!]^3 (2N+3)! \Big\}. \quad (14.51)$$

One should notice that again Δ vanishes as $O(N^{-1/2})$, as $N \to \infty$. However, in the low-density and high-density phases ($\alpha < 1/2$ or $\beta < 1/2$) the diffusion constant remains finite as $N \to \infty$ [29]. Along the line of shocks $\alpha = \beta < 1/2$ the diffusion constant is discontinuous and drops to 2/3 of the value it takes just off this line.

14.4 Several species of particle

14.4.1 Steady state on a ring

An important generalization of the matrix method gives the exact solution of systems with more than one species of particle [16]. For example, one can consider a system containing two species of particle, which we represent by 1 and 2, and holes represented by 0 in which the hopping rates of the two species of particles are

$$1\,0 \;\to\; 0\,1 \text{ with rate } 1, \quad 2\,0 \;\to\; 0\,2 \text{ with rate } \alpha, \quad 1\,2 \;\to\; 2\,1 \text{ with rate } \beta.$$
$$(14.52)$$

Even for the case of periodic boundary conditions the steady state of this model is in general nontrivial. Nevertheless, the steady-state weights may be obtained by writing them in the form [16]

$$\text{Tr}\,(X_1 X_2 \cdots X_N), \tag{14.53}$$

where $X_i = D$ if site i is occupied by a 1-particle, $X_i = A$ if it is occupied by a 2-particle and $X_i = E$ if it is empty. By a proof similar to that of Sec. 14.2, one can show that (14.53) gives the steady state of this system provided that the matrices D, A, and E satisfy the following algebra [16]:

$$DE = D + E; \qquad \beta DA = A; \qquad \alpha AE = A. \tag{14.54}$$

The second and third of these equations are satisfied when A is given by

$$A = |V\rangle\langle W|, \quad \text{with } D|V\rangle = \beta^{-1}|V\rangle \quad \text{and} \quad \langle W|E = \alpha^{-1}\langle W|. \tag{14.55}$$

So one can use any of the matrices D, E presented for the case of open boundary conditions, see appendix equations (14.77), (14.80), and construct matrix A from the vectors $\langle W|, |V\rangle$, see appendix equations (14.76), (14.79).

A case of the two-species problem of particular interest and which has been used to study shocks is that of first- and second-class particles. This corresponds to $\alpha = \beta = 1$ so that all hopping rates are 1 and both first- and second-class particles hop forward when they have a hole to their right, but when a first-class particle has a second-class particle to its right the two particles interchange positions. Thus a second-class particle tends to move backwards in an environment with a high density of first-class particles and tends to move forwards in an environment with a low density of first-class particles.

An interesting result concerns the case of a finite number of second-class particles in an infinite uniform system of first-class particles at density ρ. It can be shown that they form an algebraic bound state, i.e., the probability of finding them a distance r apart decays as a power law in r. For example, in the case of two second-class particles in an infinite system of first-class particles at density ρ, the probability $P(r)$ of finding them a distance r apart is given by [16]

$$P(r) = \sum_{p=0}^{r-1} \rho^{2p+1}\,(1-\rho)^{2r-2p-1}\,\frac{r!(r-1)!}{p!(p+1)!(r-p)!(r-p-1)!}, \tag{14.56}$$

which decays for large r as $P(r) \simeq \left[2\sqrt{\pi\rho(1-\rho)}\right]^{-1} r^{-3/2}$. Thus, the two second-class particles form a bound state although their average distance is infinite. This result was obtained [16] by using the weights (14.53) on a ring of N sites with M first-class particles, two second-class particles and by taking the limit $N \to \infty$ with $M/N = \rho$ fixed.

14.4.2 Extension to open systems

One may wonder for which boundary conditions might the matrices (14.55) be used to describe an open system of two species of particles. For the class of boundary conditions where 1-particles enter at the left and leave at the right and holes enter at the right and leave at the left [30], i.e., *at the left-hand boundary*:

$$0 \to 1 \text{ with rate } \alpha_1, \quad 0 \to 2 \text{ with rate } \alpha_2, \quad 2 \to 1 \text{ with rate } \alpha_3, \tag{14.57}$$

and at *the right-hand boundary*:

$$1 \to 0 \text{ with rate } \beta_1, \quad 1 \to 2 \text{ with rate } \beta_2, \quad 2 \to 0 \text{ with rate } \beta_3, \tag{14.58}$$

one can use (14.55) to write the steady-state weights as $\langle W | \prod_{i=1}^{N} X_i | V \rangle$ (where $X_i = D$ if site i is occupied by a 1-particle, $X_i = A$ if it is occupied by a 2-particle and $X_i = E$ if it is empty) provided that the following conditions hold:

$$\alpha_1 + \alpha_2 = \alpha, \qquad \beta_1 + \beta_2 = \beta, \qquad \frac{\alpha_2}{\alpha\alpha_3} = \frac{\beta_2}{\beta\beta_3}. \tag{14.59}$$

In order to show this, one can use the technique outlined in Sec. 14.2. A simple check is that the current of 1-particles be conserved, which requires

$$\frac{\langle W | [\alpha_1 E + \alpha_3 A] G^{N-1} | V \rangle}{\langle W | G^N | V \rangle} = \frac{\langle W | G^{i-1} D(E + \beta A) G^{N-i-1} | V \rangle}{\langle W | G^N | V \rangle}$$
$$= \frac{\langle W | G^{N-1} D(\beta_1 + \beta_2) | V \rangle}{\langle W | G^N | V \rangle}, \tag{14.60}$$

and that the current of zeros be conserved, which requires

$$\frac{\langle W | (\alpha_1 + \alpha_2) E G^{N-1} | V \rangle}{\langle W | G^N | V \rangle} = \frac{\langle W | G^{i-1} (D + \alpha A) E G^{N-i-1} | V \rangle}{\langle W | G^N | V \rangle}$$
$$= \frac{\langle W | G^{N-1} [\beta_1 D + \beta_3 A] | V \rangle}{\langle W | G^N | V \rangle}, \tag{14.61}$$

where $G = D + E + A$. It is easy to see using (14.55) and the conditions (14.59) that if $\langle W|V \rangle = \frac{\alpha_2}{\alpha\,\alpha_3} = \frac{\beta_2}{\beta\,\beta_3}$, then both currents are given by $\langle W|G^{N-1}|V \rangle / \langle W|G^N|V \rangle$.

14.4.3 Spontaneous symmetry breaking

There are boundary conditions for which the steady state is not yet known exactly. One example that has been studied in [30] is the case of first- and second-class particles with transition rates *at the left-hand boundary*

$$0 \rightarrow 2 \text{ with rate } \delta, \quad 2 \rightarrow 1 \text{ with rate } \gamma, \qquad (14.62)$$

and *at the right-hand boundary*

$$1 \rightarrow 2 \text{ with rate } \delta, \quad 2 \rightarrow 0 \text{ with rate } \gamma \qquad (14.63)$$

(although in [30] a different notation was used). These dynamics and boundary conditions have a symmetry between first-class particles and holes (i.e., the dynamics of first-class particles moving to the right is the same as that of holes moving to the left). Only for $\delta = 1$ does this model fall into a class of boundary conditions for which the steady state is known ($\alpha = \beta = \alpha_2 = \beta_2 = 1$, $\alpha_1 = \beta_1 = 0$, $\alpha_3 = \beta_3 = \gamma$ in (14.58), (14.57)). For other values of δ a mean-field theory calculated in [30] and similar to the one presented in Appendix B reveals that for sufficiently low values of δ there are solutions to the mean-field steady-state equations in which the currents of first-class particles and holes are unequal. Monte Carlo simulations of the original model indicate that a state of this kind with unequal currents of first-class particles and holes persists for a time exponentially long in the system size before flipping to the symmetric state (in which the currents have been interchanged). This has been interpreted as spontaneous symmetry breaking [30].

14.5 Steady states in an infinite system

In all the cases considered so far, we have used the matrix method to find the steady state of particle systems on a finite lattice (open boundary condition, or ring). In this section we show that one can also use matrices to construct steady states directly in an infinite system. We will limit our discussion here to the following two situations.

14.5.1 Shock profiles as seen from a second-class particle

First, consider a moving shock where the density is ρ_- to the left of the shock and ρ_+ to the right of the shock. It is known that if a single second-class particle is introduced into this system it will be trapped by the shock [13,31,16]. So one can try to calculate the steady-state properties as seen from the second-class particle.

It is possible to prove (by generalizing the proof of Sec. 14.2) that one can write the probability of a given environment of the second-class particle as a product of two matrix elements. For example

$$\text{prob}(1\ 1\ 0\ 2\ 1\ 1\ 0\ 1) = \langle w_1|\delta^2\epsilon|v_1\rangle\langle w_2|\delta^2\epsilon\delta|v_2\rangle, \qquad (14.64)$$

where the matrices and the vectors satisfy

$$\delta\epsilon = (1 - \rho_-)(1 - \rho_+)\delta + \rho_-\rho_+\epsilon,$$

$$\langle w_1|(\delta + \epsilon) = \langle w_1|, \qquad \delta|v_1\rangle = \rho_-\rho_+|v_1\rangle,$$

$$\langle w_2|\epsilon = (1 - \rho_-)(1 - \rho_+)\langle w_2|, \qquad (\delta + \epsilon)|v_2\rangle = |v_2\rangle,$$

$$\langle w_1|v_1\rangle = 1, \qquad \langle w_2|v_2\rangle = 1. \qquad (14.65)$$

So for each particle we write a δ, for each hole an ϵ, and for the second-class particle we write the projector $|v_1\rangle\langle w_2|$. This allows one to calculate the profile as seen from the second-class particle: for $i \geq 1$,

$$\langle \tau_i\rangle = \langle w_2|(\delta + \epsilon)^{i-1}\delta|v_2\rangle, \qquad \langle \tau_{-i}\rangle = \langle w_1|\delta(\delta + \epsilon)^{i-1}|v_1\rangle, \qquad (14.66)$$

and the calculation leads to the same expressions as those obtained by working first on a finite system and afterwards taking the thermodynamic limit [16]. For example, from the above algebraic rules it is very easy to show that

$$\langle \tau_1\rangle = \rho_+ + \rho_- - \rho_+\rho_-, \qquad \langle \tau_{-1}\rangle = \rho_+\rho_-,$$

$$\langle \tau_{-2}\rangle = \rho_+\rho_- + \rho_-\rho_+(1 - \rho_-)(1 - \rho_+), \qquad (14.67)$$

and so on, in agreement with [16].

14.5.2 A finite density of first- and second-class particles

One can also consider an homogeneous situation: an infinite system with a nonzero density ρ_1 of first-class particles and a nonzero density ρ_2 of second-class particles. One can here also find three matrices δ, α, and ϵ, to represent a first-class particle, a second-class particle, or a hole, such that any correlation function can be written in terms of these matrices. More precisely, if the matrices δ, α, ϵ satisfy the following algebraic rules,

$$\delta\alpha = \rho_1(\rho_1 + \rho_2)\alpha, \qquad \alpha\epsilon = (1 - \rho_1 - \rho_2)(1 - \rho_1)\alpha,$$

$$\delta\epsilon = \rho_1(\rho_1 + \rho_2)\epsilon + (1 - \rho_1)(1 - \rho_1 - \rho_2)\delta, \qquad (14.68)$$

and there exist two vectors $\langle w|, |v\rangle$ that satisfy $\langle w|\epsilon = (1 - \rho_1 - \rho_2)\langle w|$, $\langle w|(\delta + \alpha + \epsilon) = \langle w|$, and $\delta|v\rangle = \rho_1|v\rangle$, $(\delta + \alpha + \epsilon)|v\rangle = |v\rangle$, one can write the probability of finding any configuration in a finite region (of the infinite system) as a product of matrices. For example, prob(1 1 2 1 0 1 1 2 0 1) = $\langle w|\delta^2\alpha\delta\epsilon\delta^2\alpha\epsilon\delta|v\rangle$. Note that if $\rho_2 = 0$, the measure is Bernoulli, as is seen by taking $\delta = \rho_1$, $\epsilon = 1 - \rho_1$, and $\alpha = 0$. As a consequence of the algebraic rules, one can calculate any correlation function. If one calls $p_{ij}(r)$ the probability of having a particle i at the origin and a particle j at site r, one finds

$$p_{11}(r) = \langle w|\delta\gamma^{r-1}\delta|v\rangle, \quad p_{12}(r) = \langle w|\delta\gamma^{r-1}\alpha|v\rangle, \quad p_{10}(r) = \langle w|\delta\gamma^{r-1}\epsilon|v\rangle,$$
$$p_{21}(r) = \langle w|\alpha\gamma^{r-1}\delta|v\rangle, \quad p_{22}(r) = \langle w|\alpha\gamma^{r-1}\alpha|v\rangle, \quad p_{20}(r) = \langle w|\alpha\gamma^{r-1}\epsilon|v\rangle,$$
$$p_{01}(r) = \langle w|\epsilon\gamma^{r-1}\delta|v\rangle, \quad p_{02}(r) = \langle w|\epsilon\gamma^{r-1}\alpha|v\rangle, \quad p_{00}(r) = \langle w|\epsilon\gamma^{r-1}\epsilon|v\rangle,$$
$$(14.69)$$

where $\gamma = \delta + \epsilon + \alpha$. From the algebraic rules (14.68), several of these correlation functions are very easy to calculate and show an absence of correlation: $p_{11}(r) = \rho_1^2$; $p_{21}(r) = \rho_2\rho_1$; $p_{01}(r) = (1 - \rho_1 - \rho_2)\rho_1$; $p_{02}(r) = (1 - \rho_1 - \rho_2)\rho_2$; $p_{00}(r) = (1 - \rho_1 - \rho_2)^2$. However, the other correlation functions do depend on r. For example $p_{22}(r) = \langle w|\alpha(\delta + \alpha + \epsilon)^{r-1}\alpha|v\rangle$, which gives $p_{22}(1) = \rho_2(1 - \rho_1)(\rho_1 + \rho_2) \neq \rho_2^2$. It is interesting to note that in the limit $\rho_2 \to 0$, the quantity $\langle w|\alpha(\delta + \epsilon)^{r-1}\alpha|v\rangle$, which is the probability that two consecutive second-class particles are a distance r apart, is equal to $\rho_2 P(r)$ where $P(r)$ is given by (14.56).

14.6 Conclusion

We have shown here how the matrix formulation leads to several exact results for the asymmetric exclusion model. Firstly we saw in Sec. 14.2 how to represent the steady-state weights of the model with one species of particle on a finite lattice with open boundary conditions as elements of products of noncommuting matrices. This formalism was generalized to the case of two species of particle in Sec. 14.4 where we saw how nontrivial steady states on a ring may be obtained and how for certain boundary conditions an open system may be treated. We added to the range of steady states that may be described by showing in Sec. 14.5 how to represent the steady states for two species of particles on a finite segment of an infinite system.

In addition to steady-state equal-time correlations we indicated in Sec. 14.3 how diffusion constants may be calculated.

Through most of this chapter we have been concerned with totally asymmetric exclusion although one can also deal with the partially asymmetric situation via the matrix approach (recall that in the partially asymmetric exclusion problem particles can hop either to the right with probability pdt or to the left with probability qdt where $p > q$). In this case one can show [10,17] that replacing (14.17) by $pDE - qED = D + E$ still gives a steady state.

14.6.1 Open problems

We now outline some open problems, which, to our knowledge, have not yet been solved within the matrix (or any other) formulation.

1. An interesting case for which an exact solution is still lacking is the two-species problem with broken symmetry [30] discussed in Sec. 14.4.3.

2. For the diffusion constant we have not been able to generalize the exact expression (14.43) to the partially asymmetric case (i.e., for $p > q > 0$) although the first terms of an expansion in powers of $p - q$ are known [32]. Additionally, the diffusion constant for a single second-class particle on a ring in the presence of an arbitrary number of first-class particles has been calculated exactly [32]; however, we have not succeeded in extending the calculation to more than one second-class particle.

3. Another case [33] that has so far resisted exact solution is the steady state on a ring with a blockage (a blockage is a bond through which particles pass at a reduced rate).

4. Numerical studies indicate that steady states for systems with more than two species of particle should be tractable; work to write their steady state properties in terms of matrices is in progress [34].

5. For the asymmetric exclusion process we have seen that all equal-time steady-state properties are computable using the matrix approach. One may wonder whether the approach could be extended to deal with arbitrary steady-state non-equal-time correlation functions.

6. It would be interesting to know whether non-steady-state properties such as the relaxation to the steady state could be calculated within a matrix approach as suggested by [35].

Let us now review briefly how the results presented above are connected with the existing literature.

Firstly, there are a large number of results on asymmetric exclusion obtained by mathematicians and mathematical physicists: these concern the existence of invariant measures, hydrodynamical limits, the diffusion of tagged particles, and expressions for diffusion constants in infinite systems [2,36,25,37,12,15]. What the matrix approach adds to this body of knowledge is that it allows all kinds of steady-state properties, such as arbitrary correlation functions in the steady state of finite and infinite systems, to be calculated. In particular, it has been used to prove that the invariant measure for a mixture of two species of particles is not Gibbsian [38].

The evolution operator of the partially or fully asymmetric exclusion process may be written as the Hamiltonian of a quantum spin chain. This equivalence has allowed techniques such as the Bethe ansatz, and results on vertex models, to be applied [6,39-43,64].

There is a whole class of systems out of equilibrium that are related to the asymmetric exclusion process: superionic conductors [44,45]; traffic flow models [46,47]; models of polymer dynamics and electrophoresis [48-50]; the kinetics of polymerization [51]; driven diffusive systems [52]; interfacial growth [53-58]; directed polymers in a random medium [53,57,58].

Finally, we outline the equivalence between the asymmetric exclusion model and two other problems of interest in statistical mechanics: growing interfaces and directed polymers in a random medium.

14.6.2 Mapping to interface growth

The asymmetric exclusion model may be mapped exactly onto a model of a growing interface in $1 + 1$ dimensions [4] by associating with each configuration $\{\tau_i\}$ of the particles a configuration of an interface: a particle at a site corresponds to a downwards step of the interface height of one unit whereas a hole corresponds to an upward step of one unit. The heights of the interface are thus defined by

$$h_{i+1} - h_i = 1 - 2\tau_i. \qquad (14.70)$$

The dynamics of the asymmetric exclusion process, in which a particle at site $i-1$ may interchange position with a neighboring hole at site i, corresponds to growth at a site i which corresponds to a minimum of the interface height, i.e., if $h_i(t) = h_{i-1}(t) - 1 = h_{i+1}(t) - 1$ then

$$h_i(t + dt) = \begin{cases} h_i(t) & \text{with prob. } 1 - dt, \\ h_i(t) + 2 & \text{with prob. } dt, \end{cases} \tag{14.71}$$

otherwise $h_{i+1}(t)$ remains unchanged. A growth event turns a minimum of the surface height into a maximum; thus the system of hopping particles maps onto what is known as a single-step growth model [4,54], meaning that the difference in heights of two neighboring positions on the interface is always of magnitude one unit. Since whenever a particle hops forward the interface height increases by two units, the velocity v at which the interface grows is related to J, the current in the asymmetric exclusion process, by $v = 2J$.

The periodic boundary conditions for the particle problem with M particles and $N - M$ holes correspond to the requirement that $h_{i+N} = h_i + N - 2M$ for the interface problem, i.e., to helical boundary conditions with an average slope $1 - 2M/N$. The case of open boundary conditions corresponds to special growth rules at the boundaries. Because of this equivalence, several results obtained for the asymmetric exclusion process can be translated into exactly computable properties of a growing interface [59].

14.6.3 Mapping to a directed polymer in a random medium

It is also well known that this growing-interface model is formally equivalent to the problem of a directed polymer in a random medium. This can be seen by considering the time $T(i, h)$ at which site i of the interface reaches height h. $T(i, h)$ obeys the equation

$$T(i, h) = \max\left[T(i - 1, h - 1),\ T(i + 1, h - 1)\right] + \eta, \tag{14.72}$$

where the random variable η is the waiting time for a growth event to occur at site i once it has become a minimum of the interface height. In the asymmetric exclusion process η is just the waiting time for a particle at site $i - 1$ to hop forward (once site $i - 1$ is occupied and site i is empty) and thus has an exponential distribution

$$p(\eta) = e^{-\eta} \quad \text{for} \quad \eta > 0. \tag{14.73}$$

If we define

$$E(i, h) = -T(i, h), \tag{14.74}$$

(14.72) becomes

$$E(i, h) = \min\left[E(i - 1, h - 1), E(i + 1, h - 1)\right] - \eta. \tag{14.75}$$

Now let us consider a directed walk of h steps on a two-dimensional lattice: at each step of the walk the length h increases by 1 and the coordinate i in the perpendicular direction either increases or decreases by 1. Thus the coordinates of the end of a polymer of length h are (i, h). The random medium refers to random variables $-\eta(i, h)$ (site energies) associated with each lattice site (i, h) and the energy of a directed polymer is the sum of the energies of the sites through which it passes. The ground-state energy $E(i, h)$ of a directed polymer finishing at (i, h) satisfies (14.75): this equation just expresses the fact that $E(i, h)$ will be given by the minimum of the ground-state energies for the possible coordinates of the directed polymer at the previous step, plus $-\eta(i, h)$, the energy at site (i, h). Thus from (14.74) we see that the time for the interface to reach height h is equivalent to the ground-state energy of a directed polymer of length h. For long times we expect $h/t \to v = 2J$, which implies that $T(i, h) \simeq h/(2J)$. For a lattice of $2L$ sites with periodic conditions in the direction i, and which is infinite in the h-direction, we then obtain the exact ground-state energy per unit length for the exponential distribution of site energies: $\lim_{h \to \infty}[E(i, h)/h] = -1/(2J) = -(2L-1)/L$, where $J = M(N-M)/[N(N-1)]$ (owing to the fact that all configurations have equal weight) with $N = 2L$ and $M = L$ to represent an interface perpendicular to the growth direction. So that, for a lattice infinite in both directions, one finds that the exact ground-state energy per unit length is -2 when the distribution of the site energies $-\eta$ is exponential (14.73).

In a similar way one can relate the fluctuations of the ground-state energy of the directed polymer to the fluctuations of the absolute height of the interface [60]. The variance of the absolute height of the interface corresponds to the diffusion constant that was discussed in Sec. 14.3.

For open boundary conditions in the asymmetric exclusion model, the three different phases discussed in Sec. 14.3 correspond to transitions in the problem of a directed polymer confined between two walls with attractive potentials [61]. The high- and low-density phases correspond to phases where the directed polymer is bound to the right wall and to the left wall respectively. The maximal current phase corresponds to a directed polymer that explores the full lateral extension (size N) of the medium.

Most of the results discussed here have been obtained in collaboration with E. Domany, D. P. Foster, C. Godrèche, V. Hakim, S. A. Janowsky, J. L. Lebowitz, K. Mallick, D. Mukamel, V. Pasquier, and E. R. Speer. We thank them as well as V. Rittenberg, G. Schütz, H. Spohn and R. Stinchcombe for interesting discussions. One of the authors (MRE) was a Royal Society University Research Fellow at

Oxford, while the work reported here was being undertaken. See also the added note [65].

Appendix A: Explicit forms of the matrices

We give here some representations of the matrices that satisfy (14.16), (14.17). One particularly simple choice is

$$\langle W| = \left(1, \left(\frac{1-\alpha}{\alpha}\right), \left(\frac{1-\alpha}{\alpha}\right)^2, \cdots\right),$$

$$|V\rangle = \left(1, \left(\frac{1-\beta}{\beta}\right), \left(\frac{1-\beta}{\beta}\right)^2, \cdots\right)^t; \qquad (14.76)$$

$$D = \begin{pmatrix} 1 & 1 & 0 & 0 & \cdots \\ 0 & 1 & 1 & 0 & \\ 0 & 0 & 1 & 1 & \\ 0 & 0 & 0 & 1 & \\ \vdots & & & & \ddots \end{pmatrix}, \qquad E = \begin{pmatrix} 1 & 0 & 0 & 0 & \cdots \\ 1 & 1 & 0 & 0 & \\ 0 & 1 & 1 & 0 & \\ 0 & 0 & 1 & 1 & \\ \vdots & & & & \ddots \end{pmatrix}; \qquad (14.77)$$

here t denotes the transpose. This choice makes the particle-hole symmetry of the problem apparent since the matrices D and E have very similar forms and the boundary conditions α and β only appear in the vectors $\langle W|$ and $|V\rangle$. For this choice (14.77) of D, E, the elements of C^N (where $C = D + E$ and N denotes the Nth power of matrix C) are given by

$$\left(C^N\right)_{nm} = \begin{pmatrix} 2N \\ N+n-m \end{pmatrix} - \begin{pmatrix} 2N \\ N+n+m \end{pmatrix}. \qquad (14.78)$$

This expression can be obtained by noting that $(C^N)_{nm}$ is proportional to the probability that a random walker who starts at site $2m$ of a semi-infinite chain, with an absorbing boundary at the origin, is at site $2n$ after $2N$ steps of a random walk. This probability may be calculated by the method of images.

An apparent disadvantage of the choice (14.77), (14.76) is that, owing to the form of $\langle W|$ and $|V\rangle$, one has to sum geometric series to obtain the correlation functions and these series diverge in some range of α, β, namely $\alpha + \beta \leq 1$. However, at least for finite N, all expressions are rational functions of α, β so that in principle one can obtain results for $\alpha + \beta \leq 1$ by analytic continuation from those for $\alpha + \beta > 1$.

Other choices of matrices and vectors are possible [10], which solve the equations (14.16), (14.17). For example, a possible choice of $D, E, \langle W|, |V\rangle$, that avoids the divergences is

$$\langle \tilde{W}| = (1,0,0,\dots), \qquad |\tilde{V}\rangle = (1,0,0,\dots)^t; \qquad (14.79)$$

$$\tilde{D} = \begin{pmatrix} 1/\beta & a & 0 & 0 & \cdots \\ 0 & 1 & 1 & 0 & \\ 0 & 0 & 1 & 1 & \\ 0 & 0 & 0 & 1 & \\ \vdots & & & & \ddots \end{pmatrix}, \qquad \tilde{E} = \begin{pmatrix} 1/\alpha & 0 & 0 & 0 & \cdots \\ a' & 1 & 0 & 0 & \\ 0 & 1 & 1 & 0 & \\ 0 & 0 & 1 & 1 & \\ \vdots & & & & \ddots \end{pmatrix}. \qquad (14.80)$$

The two constants a and a' must satisfy $aa' = (\alpha+\beta-1)/(\alpha\beta)$ but otherwise are arbitrary. One should note that for $\alpha = \beta = 1$, one can choose $a = a' = 1$ and the choice (14.80), (14.79) coincides with our previous bidiagonal choice (14.77), (14.76). Also, aa' vanishes for $\alpha+\beta = 1$ so that the 1, 1-elements of the matrices $\tilde{D}, \tilde{E}$ decouple from the other elements. This choice of matrices then becomes effectively one dimensional as is sufficient for this special case of α and β.

In Sec. 14.2 we presented some exact results for finite systems; these were obtained directly from the algebraic rules (14.16), (14.17) by combinatorial methods. An alternative approach, well suited to calculating asymptotic behavior, is to diagonalize particular representations of $C = D + E$. This has been done for (14.77) in [10] (see Appendix B therein), and for matrices of form (14.80) in [30] (see Appendix B therein).

For the partially asymmetric exclusion model a choice of matrices that satisfies $pDE - qED = D + E$ is [10] (for $p \geq q$)

$$D = \frac{1}{p-q} \begin{pmatrix} 1-d & a_1 & 0 & 0 & \cdots \\ 0 & 1-d(q/p) & a_2 & 0 & \\ 0 & 0 & 1-d(q/p)^2 & a_3 & \\ 0 & 0 & 0 & 1-d(q/p)^3 & \\ \vdots & & & & \ddots \end{pmatrix}, \qquad (14.81)$$

$$E = \frac{1}{p-q} \begin{pmatrix} 1-e & 0 & 0 & 0 & \cdots \\ a_1' & 1-e(q/p) & 0 & 0 & \\ 0 & a_2' & 1-e(q/p)^2 & 0 & \\ 0 & 0 & a_3' & 1-e(q/p)^3 & \\ \vdots & & & & \ddots \end{pmatrix}, \qquad (14.82)$$

where the off-diagonal terms a_i and a_i' satisfy

$$a_i a_i' = \left[1 - \left(\frac{q}{p}\right)^i\right]\left[1 - de\left(\frac{q}{p}\right)^{i-1}\right]. \tag{14.83}$$

and d and e are arbitrary. Choosing (14.79) for the vectors $\langle W|$, $|V\rangle$ and setting $d = 1 - (p-q)/\beta$, $e = 1 - (p-q)/\alpha$, one satisfies (14.16) and obtains steady-state weights, where particles enter at site 1 with rate α and holes at site N with rate β (but no holes are introduced at site 1 or particles at site N). The matrices (14.81), (14.82) are related to the creation and annihilation operators of the 'q-deformed' harmonic oscillator; see [17].

We saw that for $p = 1$ and $q = 0$ the matrices are one dimensional along the line $\alpha + \beta = 1$. For general p and q it has recently been shown that for some choices of boundary conditions the matrices can be finite dimensional [62].

We note that for $p \to q$ the limiting forms of the matrices (14.81), (14.82) are well defined, and for $p = 1$, $q = 0$ the matrices and vectors reduce to the choice (14.80) that we presented for fully asymmetric exclusion.

Appendix B: Mean-field equations

To illustrate how one can derive a mean-field theory [9,8,51,63] of the fully asymmetric exclusion model, we will consider the steady state of a system of N sites with open boundary conditions.

The spirit of the mean-field approximation is to neglect correlations in the steady-state equations and to replace $\langle \tau_i \tau_j \rangle$ by $\langle \tau_i \rangle \langle \tau_j \rangle$ so that (14.13) becomes

$$\alpha(1 - t_1) = t_1(1 - t_2) = t_2(1 - t_3) = \cdots = t_{N-1}(1 - t_N) = \beta t_N, \tag{14.84}$$

where $t_i = \langle \tau_i \rangle$. Thus, in the mean-field approximation, we have N equations (14.84) for the N unknowns t_i. A simple way of solving these mean-field equations is to use the fact that $t_i(1 - t_{i+1})$ does not depend on i; in fact, $t_i(1 - t_{i+1}) = c$, where c is the current. Then it is easy to check that the solution of (14.84) is

$$t_i = \frac{1}{2}\frac{\alpha(r_1^{i+1} - r_2^{i+1}) - 2c(r_1^i - r_2^i)}{\alpha(r_1^i - r_2^i) - 2c(r_1^{i-1} - r_2^{i-1})}, \tag{14.85}$$

where $r_1 = 1 + \sqrt{1 - 4c}$ and $r_2 = 1 - \sqrt{1 - 4c}$, and the current c is determined by the right-hand boundary condition

$$\frac{r_2^N}{r_1^N} = \frac{(\alpha\beta - c)r_1 + 2c(1 - \alpha - \beta)}{(\alpha\beta - c)r_2 + 2c(1 - \alpha - \beta)}. \tag{14.86}$$

In the limit $N \to \infty$, the only possible solutions of this equation, compatible with the fact that all the t_i are positive, are $c = \alpha(1 - \alpha)$, $c = \beta(1 - \beta)$ and $c = 1/4$. One can then show [9] that the mean-field theory predicts the following phase diagram. For $\alpha \geq 1/2$ and $\beta \geq 1/2$ the current c and the density $\rho = \langle \tau_i \rangle$ for a site i far from the boundaries are given by $c = 1/4$, $\rho = 1/2$. For $\alpha < 1/2$ and $\beta > \alpha$, $c = \alpha(1 - \alpha)$, $\rho = \alpha$. For $\beta < 1/2$ and $\alpha > \beta$, $c = \beta(1 - \beta)$, $\rho = 1 - \beta$.

We see that the phase diagram predicted by the mean-field theory is the same as that of the exact solution discussed in Sec. 14.2. Once the current is known, one can use (14.85) to determine the profiles near the boundaries. For example when $\alpha = \beta = 1$, and for $N \to \infty$, one finds, for i finite,

$$\langle \tau_i \rangle_\infty = \frac{i + 2}{2i + 2}, \tag{14.87}$$

and this mean-field result disagrees with the exact expression (14.41).

Remark. One way of seeing that the mean-field theory is merely an approximation is to write more general steady-state equations than (14.9), (14.11), (14.12). For example (14.10) becomes in the mean-field approximation $t_{i-1}t_{i+1} - t_i t_{i+1} - t_{i-1}t_i t_{i+1} + t_i t_{i+1}t_{i+2} = 0$, and it is easy to check that the mean-field result (14.87) does not satisfy this steady-state equation.

References

[1] F. Spitzer, *Adv. Math.* **5**, 246 (1970).
[2] T. M. Liggett, *Interacting Particle Systems* (Springer-Verlag, New York, 1985).
[3] H. Spohn, *Large Scale Dynamics of Interacting Particles* (Springer-Verlag, New York, 1991).
[4] P. Meakin, P. Ramanlal, L. M. Sander and R. C. Ball, *Phys. Rev.* **A34**, 5091 (1986).
[5] D. Dhar, *Phase Transitions* **9**, 51 (1987).
[6] L. H. Gwa and H. Spohn, *Phys. Rev.* **A46**, 844 (1992).
[7] B. Derrida, M. R. Evans and D. Mukamel, *J. Phys.* **A26**, 4911 (1993).
[8] J. Krug, *Phys. Rev. Lett.* **67**, 1882 (1991).
[9] B. Derrida, E. Domany and D. Mukamel, *J. Stat. Phys.* **69**, 667 (1992).
[10] B. Derrida, M. R. Evans, V. Hakim and V. Pasquier, *J. Phys.* **A26**, 1493 (1993).
[11] G. Schütz and E. Domany, *J. Stat. Phys.* **72**, 277 (1993).
[12] E. D. Andjel, M. Bramson and T. M. Liggett, *Prob. Theory Rel. Fields* **78**, 231 (1988).

[13] P. A. Ferrari, C. Kipnis and E. Saada, *Ann. Prob.* **19**, 226 (1991).

[14] P. A. Ferrari, *Prob. Theory Rel. Fields* **91**, 81 (1992).

[15] P. A. Ferrari and L. R. G. Fontes, *Prob. Theory Rel. Fields* **99**, 305 (1994); *Ann. Prob.* **22**, 820 (1994).

[16] B. Derrida, S. A. Janowsky, J. L. Lebowitz and E. R. Speer, *Europhys. Lett.* **22**, 651 (1993); *J. Stat. Phys.* **73**, 813 (1993).

[17] S. Sandow, *Phys. Rev.* **E50**, 2660 (1994).

[18] S. Sandow and G. M. Schütz, *Europhys. Lett.* **26**, 7 (1994).

[19] L. D. Faddeev, *Sov. Sci. Rev.* **C1**, 107 (1980).

[20] R. J. Baxter, *Exactly Solved Models in Statistical Mechanics* (Academic, New York, 1982).

[21] V. Hakim and J. P. Nadal, *J. Phys.* **A16**, L213 (1983).

[22] A. Klümper, A. Schadschneider and J. Zittartz, *J. Phys.* **A24**, L955 (1991).

[23] M. Fannes, B. Nachtergaele, R. F. Werner, *Comm. Math. Phys.* **144**, 443 (1992).

[24] A. Klümper, A. Schadschneider and J. Zittartz, *Europhys. Lett.* **24**, 293 (1993).

[25] A. De Masi and P. A. Ferrari, *J. Stat. Phys.* **38**, 603 (1985).

[26] R. Kutner and H. van Beijeren, *J. Stat. Phys.* **39**, 317 (1985).

[27] H. van Beijeren, *J. Stat. Phys.* **63**, 47 (1991).

[28] S. N. Majumdar and M. Barma, *Phys. Rev.* **B44**, 5306 (1991).

[29] B. Derrida, M. R. Evans and K. Mallick, *J. Stat. Phys.* **79**, 833 (1995).

[30] M. R. Evans, D. P. Foster, C. Godrèche and D. Mukamel, *Phys. Rev. Lett.* **74**, 208 (1995); *J. Stat. Phys.* **80**, 69 (1995).

[31] C. Boldrighini, G. Cosimi, S. Frigio and M. G. Nunes, *J. Stat. Phys.* **55**, 611 (1989).

[32] B. Derrida and M. R. Evans, Exact steady state properties of the one dimensional asymmetric exclusion model, in *Probability and Phase Transition*, p. 1, G. Grimmett, ed. (Kluwer Academic, Dordrecht, 1994).

[33] S. A. Janowsky and J. L. Lebowitz, *Phys. Rev.* **A45**, 618 (1992).

[34] B. Derrida, K. Mallick and V. Pasquier, work in progress.

[35] R. B. Stinchcombe and G. M. Schütz, *Europhys. Lett.* **29**, 663 (1995); *Phys. Rev. Lett.* **75**, 140 (1995).

[36] A. De Masi and E. Presutti, Mathematical methods for hydrodynamical behavior, in *Lecture Notes in Mathematics* (Springer-Verlag, 1991).

[37] P. A. Ferrari, *Ann. Prob.* **14**, 1277 (1986).

[38] E. R. Speer, The two species totally asymmetric exclusion process, in *Micro, Meso and Macroscopic Approaches in Physics*, M. Fannes, C. Maes and A. Verbeure, eds. (Plenum, 1994).

[39] G. M. Schütz, *J. Stat. Phys.* **71**, 471 (1993).

[40] F. C. Alcaraz, M. Droz, M. Henkel and V. Rittenberg, *Ann. Phys. (NY)* **230**, 250 (1994).

[41] D. Kandel, E. Domany and B. Nienhuis, *J. Phys.* **A23**, L755 (1990).

[42] G. Schütz, *Phys. Rev.* **E47**, 4265 (1993).

[43] I. Peschel, V. Rittenberg and U. Schultze, *Nucl. Phys.* **B430**, 633 (1994).

[44] P. M. Richards, *Phys. Rev.* **B16**, 1393 (1977).

[45] W. Dieterich, P. Fulde and I. Peschel, *Adv. Phys.* **29**, 527 (1980).

[46] A. Schadschneider and M. Schreckenberg, *J. Phys.* **A26**, L679 (1993).

[47] T. Nagatani, *J. Phys.* **A26**, L781 (1993); *J. Phys.* **A26**, 6625 (1993).

[48] M. Rubinstein, *Phys. Rev. Lett.* **59**, 1946 (1987).

[49] T. A. J. Duke, *Phys. Rev. Lett.* **62**, 2877 (1989).

[50] J. M. J. van Leeuwen and A. Kooiman, *Physica* **A184**, 79 (1992).

[51] C. T. MacDonald, J. H. Gibbs and A. C. Pipkin, *Biopolymers* **6**, 1 (1968); C. T. MacDonald, J. H. Gibbs, *Biopolymers* **7**, 707 (1969).

[52] B. Schmittmann and R. K. P. Zia, *Statistical Mechanics of Driven Diffusive Systems. Phase Transitions and Critical Phenomena*, Vol. **17**, C. Domb and J. L. Lebowitz, eds. (Academic Press, New York, 1995).

[53] M. Kardar, G. Parisi and Y.-C. Zhang, *Phys. Rev. Lett.* **56**, 889 (1986).

[54] J. Kertész and D. E. Wolf, *Phys. Rev. Lett.* **62**, 2571 (1989).

[55] D. E. Wolf and L.-H. Tang, *Phys. Rev. Lett.* **65**, 1591 (1990).

[56] D. Kandel and D. Mukamel, *Europhys. Lett.* **20**, 325 (1992).

[57] J. Krug and H. Spohn in *Solids Far from Equilibrium*, C. Godrèche, ed. (Cambridge University Press, 1991).

[58] T. Halpin-Healy and Y.-C. Zhang, *Phys. Rep.* **254**, 215 (1994).

[59] B. Derrida and M. R. Evans, *J. Physique* I3, 311 (1993).

[60] J. Krug, P. Meakin and T. Halpin-Healy, *Phys. Rev.* **A45**, 638 (1992).

[61] J. Krug and L.-H. Tang, *Phys. Rev.* **E50**, 104 (1994).

[62] F. H. L. Essler and V. Rittenberg, *J. Phys.* **A29**, 3375 (1996).

[63] D. Kandel, G. Gershinsky, D. Mukamel and B. Derrida, *Physica Scripta* T49, 622 (1993).

[64] G. M. Schütz, *J. Stat. Phys.* **79**, 243 (1995).

[65] *Note added in proofs.* Several recent developments of the matrix approach have been published, in particular H. Hinrichsen, S. Sandow and I. Peschel, *J. Phys.* **A29**, 2643 (1996); N. Rajewsky, A. Schadschneider and M. Shreckenberg, *J. Phys.* **A29**, L305 (1996); H. Hinrichsen, *J. Phys.* **A29**, 3659 (1996).

15

Nonequilibrium surface dynamics with volume conservation

Joachim Krug

15.1 Introduction and outline

Diffusion processes and interface motion figure prominently in several chapters of this book. The present chapter is concerned with a class of problems that combines features of both areas, namely the evolution of a solid surface through *surface diffusion*. Compared to collective diffusion in, for example, lattice gas models, surface diffusion entails the additional complication of a dynamically evolving substrate—the surface—whose morphology is determined by the diffusing species itself.

From the point of view of interface motion, the dominance of surface diffusion among the possible relaxation mechanisms implies that the equation of motion for the surface position $h(x,t)$ is of conservation type,

$$\partial_t h + \partial_x J = F. \tag{15.1}$$

Here the current

$$J = J_S + J_R \tag{15.2}$$

contains a systematic part that depends on the *derivatives* of h,

$$J_S = J_S(\partial_x h, \partial_x^2 h, \ldots), \tag{15.3}$$

and a random part J_R with short-range correlations in space and time. In *deposition* processes [1,2] the system is driven by a random external flux F, which, however, does not depend on the height configuration. We shall see below that fully conserved surface diffusion problems ($F = 0$) and deposition problems to a large extent can be treated on the same footing.

The conserved form of the equation of motion does not permit the Kardar-Parisi-Zhang nonlinearity $(\partial_x h)^2$ [3]. Consequently, (15.1) defines a new set of dynamic universality classes (specified by the terms appearing in the systematic current (15.3)), which have attracted much theoretical interest during the past five years. In practical terms, surface diffusion is important because it constitutes the preferred pathway for mass transport on crystal surfaces at moderate temperatures, and thus provides the main smoothing mechanism to counteract the disordering nonequilibrium effects during technologically relevant growth processes such as molecular beam epitaxy (MBE) [4,5].

The goal of the present chapter will be to explain, within the framework of one-dimensional (1D) stochastic models, the appearance of various equilibrium and nonequilibrium contributions to the surface current (15.3), and to outline the consequences for the large-scale behavior of the surface. To provide a unified perspective, much of the discussion will be phrased in terms of the 1D solid-on-solid (SOS) model, defined by the Hamiltonian

$$\mathcal{H} = K \sum_i |h_{i+1} - h_i|, \tag{15.4}$$

for the integer height variables h_i, and equipped with a variety of transition rates.

In Sec. 15.2 the classical Mullins-Herring theory of near-equilibrium morphological evolution is briefly reviewed, and its relation to the *equilibrium* dynamics of the SOS model is sketched. In the following two sections, which constitute the bulk of the chapter, we then demonstrate how nonequilibrium contributions to the surface current (15.3) emerge when detailed balance with respect to the Hamiltonian (15.4) is broken.

Depending on the symmetries that remain in the absence of detailed balance, different functional forms of J_S and different associated dynamical universality classes are possible. It will be emphasized that the precise agent responsible for removing the detailed balance constraint is quite immaterial; while an external deposition flux F may be the most natural mechanism, other external influences such as a transverse electric field or ion impact lead to largely analogous behavior.

We note in passing that equations of the form (15.1) appear also in the theory of *generic scale invariance*, which seeks to identify general principles underlying scale-invariant spatiotemporal correlations in nonequilibrium systems [6,7]. Indeed, (15.1) combines the two most important such principles that have been found: a local conservation law and a continuous symmetry (invariance under shifts $h \rightarrow h + $ constant). A further exploration of this connection is, however, beyond the scope of the present chapter.

15.2 Surface diffusion near equilibrium

15.2.1 The classical theory

The phenomenological theory of capillarity-driven morphological surface evolution was developed in the 1950s by Mullins and Herring [8]. Here we summarize the main points relevant to the case where mass transport occurs through surface diffusion, and the spatial variation of the surface profile is restricted to 1D.

The driving forces for morphological changes near equilibrium are differences in the local chemical potential $\mu(x,t)$. Hence the mass current J_S is proportional to the chemical potential gradient [9-11],

$$J_S = -\sigma(\partial_x h)\,\partial_x \mu, \tag{15.5}$$

where σ is the (generally orientation-dependent) *adatom mobility*, closely related to the *collective* surface diffusion coefficient [12]. The chemical potential, in turn, can be related to the local curvature through a Gibbs-Thomson type equation,

$$\mu = -\hat{\gamma}(\partial_x h)\,\partial_x^2 h. \tag{15.6}$$

The *stiffness* $\hat{\gamma}$ can be expressed in terms of the orientation-dependent surface tension $\gamma(\vartheta)$ as $\hat{\gamma} = (\gamma + \gamma'')\cos\vartheta$, with $\partial_x h = \tan\vartheta$.

Together with the conservation law (15.1) (with $F = 0$), Eq. (15.5) and (15.6) specify a fourth-order partial differential equation for the surface evolution. Due to the orientation dependence of σ and $\hat{\gamma}$ it is generally nonlinear; however, for small height gradients the linearization

$$\partial_t h = -\kappa \partial_x^4 h \tag{15.7}$$

is appropriate, with $\kappa = \sigma(0)\hat{\gamma}(0)$. An immediate consequence of (15.7) is that surface features of wavelength L relax on a time scale of order L^4/κ, as has been verified in numerous experiments [8,12].

15.2.2 Surface diffusion in the SOS model

To set up a relation between the phenomenological theory and the statistical mechanics of the SOS model, we need to identify the two quantities that enter the theory—the mobility σ and the stiffness $\hat{\gamma}$—within the microscopic model. While the mobility is a *transport coefficient* that depends on the microscopic dynamics (to be specified shortly), the stiffness can be computed directly from the Hamiltonian (15.4). In fact the task is trivial because the height differences $u_i = h_{i+1} - h_i$ do not couple in (15.4). The appropriate microscopic expression is [11]

$$\hat{\gamma}(u) = \left[\langle u_i^2 \rangle - \langle u_i \rangle^2\right]^{-1}. \tag{15.8}$$

To include the orientation dependence this quantity has to be evaluated with a 'slope chemical potential' term $\mathcal{H}_s = m \sum_i u_i$, added to (15.4). The calculation is performed at fixed m, and subsequently m is substituted by $u = \langle u_i \rangle(m)$.

To proceed we need to specify microscopic surface diffusion rules that satisfy detailed balance with respect to (15.4). Among the many possibilities, we focus here on two choices, which are both widely used and will serve to illustrate, in Secs. 15.3 and 15.4, an important distinction between transition rates with different degrees of symmetry. For convenience we first define, for the ith surface atom (that is, the topmost atom on column i) the *lateral coordination number* $n_i = \theta(h_{i+1} - h_i) + \theta(h_{i-1} - h_i)$, which can take values $n_i = 0, 1, 2$ ($\theta(x)$ is the Heaviside step function). The energy change associated with moving a surface atom from site i to site $j = i \pm 1$ is then

$$\Delta\mathcal{H}_{ij} = 2K(n_i - n_j). \tag{15.9}$$

Now the two sets of rules can be defined as follows. For *Arrhenius* dynamics [13] the rate for moving an atom from site i to site $j = i \pm 1$ is set to

$$\Gamma_{ij} = \exp[-2(K/T)n_i], \tag{15.10}$$

where T is the temperature and the Boltzmann constant has been set to unity, while for *Metropolis* dynamics,

$$\Gamma_{ij} = \Phi(\Delta\mathcal{H}_{ij}/T), \tag{15.11}$$

where Φ satisfies $\Phi(s) = e^{-s}\Phi(-s)$; popular choices are $\Phi(s) = \min(1, e^{-s})$ and $\Phi(s) = (1 + e^s)^{-1}$. The detailed balance condition for (15.10) is easily checked using (15.9); note that (15.10) is independent of the environment at the final site j.

Given the transition rates (15.10) or (15.11), the adatom mobility σ can be formally expressed in terms of a space-time integral over the microscopic current-current correlation function [14]. The explicit evaluation of such a *Green-Kubo* formula, however, is very difficult even for simple 1D models, and in general only bounds on the transport coefficient can be given [15]. The Arrhenius surface diffusion model is an exception to the rule: it turns out that the nontrivial part of the Green-Kubo formula vanishes identically (in any dimension!), and the mobility reduces to the simple expression [11] $\sigma = \frac{1}{2}e^{-2K/T}$, which is also independent of orientation.

Together with the expression (15.8) for the stiffness, this result completely specifies the large-scale behavior of the SOS model with Arrhenius dynamics, and the nonlinear equation of motion that results from combining (15.1), (15.5), and (15.6) can be used to predict, without adjustable parameters, the evolution of arbitrary macroscopic profiles [11].

If we are interested instead in describing mesoscopic fluctuation phenomena we need to determine the random contribution J_R to the surface current. Since detailed balance has to hold also at the level of the continuum description, no additional information is required; the fluctuation-dissipation relation fixes the covariance of the (Gaussian) random current as

$$\langle J_R(x,t)J_R(x',t')\rangle = 2\sigma\delta(x-x')\delta(t-t'). \tag{15.12}$$

Adding $-\partial_x J_R$ to the r.h.s. of (15.7) we obtain a linear Langevin equation that is easily solved by Fourier transformation. For example, imposing a flat initial condition $h(x, t = 0) \equiv 0$, the buildup of thermal roughness is described by the height-difference correlation function

$$G(r,t) \equiv \langle(h(x+r,t) - h(x,t))^2\rangle = r^{2\zeta} f(r/t^{1/z}), \tag{15.13}$$

where the *roughness exponent* $\zeta = 1/2$, as usual for thermally roughened 1D interfaces, and the *dynamic exponent* $z = 4$ simply reflects the fact that (15.7) is of fourth order in x. More precisely, the limiting behaviors of (15.13) are

$$G(r,t) = \begin{cases} r/\hat{\gamma}, & r \ll \xi(t) = (\kappa t)^{1/4}, \\ \left[2^{5/4}\Gamma(3/4)\pi^{-1}\right]\xi/\hat{\gamma}, & r \gg \xi(t), \end{cases} \tag{15.14}$$

where $\Gamma(3/4) = 1.2254\ldots$ and $\xi(t)$ is a dynamic correlation length that describes the lateral spreading of the roughness. The $t^{1/4}$-scaling in (15.14) can be directly observed in step fluctuations on crystal surfaces—a beautiful example of the application of 1D statistical mechanics to real systems [16].

15.3 The nonequilibrium chemical potential

For reasons that will become clear subsequently, in this section we focus on modifications of the Arrhenius surface diffusion model, and explore the consequences of breaking detailed balance. In Sec. 15.4 the same question will be posed for the Metropolis model. An obvious distinction between different ways of breaking detailed balance concerns whether volume conservation is violated as well. We first turn to the case of *growth*, where the relaxation processes on the surface conserve the volume but the external deposition flux does not.

15.3.1 The Arrhenius growth model

In modeling epitaxial growth the surface diffusion rule (15.10) is supplemented by a random deposition rule, i.e., atoms are *created*, $h_i \rightarrow h_i + 1$, at random positions on the surface, at a constant deposition rate F [13,17,18]. Since evaporation is forbidden, this modification violates detailed balance in an obvious way, which can be incorporated into the continuum description by simply adding a *shot noise term* $\eta(x,t)$ to the r.h.s. of (15.7). The Gaussian noise statistics are specified by

$$\langle \eta(x,t) \rangle = F, \quad \langle \eta(x,t)\eta(x',t') \rangle - F^2 = F\delta(x - x')\delta(t - t'). \quad (15.15)$$

The resulting Langevin equation is still linear and its solution straightforward. The dynamic exponent in (15.13) retains the value $z = 4$ but the roughness exponent is strongly increased, taking the value $\zeta = 3/2$ in 1D [3,17,19]. At the same time the scaling function f develops a short-distance singularity, $f(s) \sim s^{-1}$ for $s \rightarrow 0$, which implies that $G(r,t)$ does not become stationary for $t \rightarrow \infty$ [20]. In a finite system of size L with periodic boundary conditions one obtains the simple result

$$G(r, t \rightarrow \infty) = \frac{1}{24\,\kappa} FLr^2(1 - r/L)^2 \quad (15.16)$$

which is analogous to (15.13) with the dynamic correlation length $\xi(t) \sim t^{1/z}$ replaced by L; (15.16) can be written in a scaling form corresponding to (15.13), but with a singular scaling function $f(s) \sim s^{-1}(1 - s)^2$. This kind of behavior is referred to as anomalous scaling and will concern us further in Sec. 15.3.3.

However, these comparatively trivial considerations do not tell the whole story, since they disregard the possibility that the surface relaxation processes could be affected by the external flux. To gain some insight into this rather subtle effect, we use an expression for the local chemical potential in terms of the coordination numbers n_i, which holds true near equilibrium [11]:

$$\mu_i = 2K + T \ln \langle \exp[-2(K/T)n_i] \rangle. \quad (15.17)$$

The expectation value on the r.h.s. can be rewritten as follows [21]:

$$\langle\exp[-2(K/T)n_i]\rangle = \sum_{n=0}^{2} c_n \exp[-2(K/T)n], \qquad (15.18)$$

where c_n denotes the concentration of surface atoms with n lateral nearest neighbors. A simple computation in thermal equilibrium shows that (15.17) is independent of orientation, as would be expected on the basis of the Gibbs-Thomson relation (15.6)—in equilibrium only *curvature* differences matter. However, in view of the decomposition (15.18) this result is somewhat surprising, since the individual terms in the sum clearly do depend on orientation; for example, for large tilts almost all atoms sit at step edges so that $c_1 \to 1$ and $c_0, c_2 \to 0$ for $u = \langle h_{i+1} - h_i \rangle \to \pm\infty$. Only the constraint of detailed balance ensures that the specific linear combination (15.18) of the c_n remains tilt-independent.

Conversely, we may conclude that any external influence that breaks detailed balance will also upset the delicate relationship between the terms in (15.18) and therefore, via (15.17), give rise to an *orientation-dependent, nonequilibrium contribution* to the local chemical potential. In fact, we can be somewhat more precise: the main effect of deposition will be an increase in the density of isolated adatoms c_0, which gives the dominant contribution to (15.18) (at least at low temperatures). Moreover, this effect is most pronounced at zero tilt, where adatoms cannot be absorbed at steps: for large tilts $c_1 \to 1$, so that $\mu \to 0$ even without detailed balance. Consequently we expect that the nonequilibrium chemical potential is positive, and that it has a maximum at zero tilt. This is summarized in the expansion

$$\mu_{NE}(u) = \mu_0 + (\mu_2/2)u^2 + \cdots \quad \text{with} \quad \mu_0 > 0, \ \mu_2 < 0, \qquad (15.19)$$

which is verified by a direct numerical measurement of (15.17) in the presence of deposition [22].

Inserting (15.19) into (15.5) and adding the shot noise (15.15) we arrive at the following nonlinear Langevin equation for the growing surface,

$$\partial_t h = -\partial_x^2[\kappa\partial_x^2 h + (\lambda/2)(\partial_x h)^2] + \eta, \qquad (15.20)$$

where we have set $\lambda = -\sigma\mu_2$. The terms inside the square bracket are familiar from the Kardar-Parisi-Zhang (KPZ) equation discussed elsewhere in this book (Sec. 4.6); for this reason (15.20) is often referred to as the con-served KPZ equation. In the context of epitaxial growth it was introduced and analyzed independently by Villain [4] and by Lai and Das Sarma [2]. The nonlinearity is relevant in the renormalization group sense. Owing to two exact invariances of the equation, both scaling exponents in (15.13) can be determined, and they take the values $\zeta = 1$, $z = 3$ in 1D [2,23]. This prediction has received some numerical support [18,24-26] (see, however, Sec. 15.3.3).

15.3.2 Impact-enhanced surface diffusion

Next we try to modify the Arrhenius surface diffusion model in such a way that detailed balance is broken but mass conservation is retained. An obvious possibility would be to bias the diffusion moves in one direction (left or right) so that the adatoms are subjected to a transverse 'electric' field, in the spirit of driven diffusive systems [27]. This works for Metropolis dynamics (see Sec. 15.4.3) but not here: the steady state of the Arrhenius model can be shown to be completely invariant under such a bias [11].

We propose instead a modification that is inspired by the phenomenon of surface diffusion by ion impact [28]. In field electron microscopy it has been observed that ions that are accelerated towards the sample can induce mass transport on the surface, thus giving a *nonthermal* contribution to the surface diffusion constant. Imagine, therefore, that our 1D SOS surface is subjected to a beam of ions that knock out surface atoms and move them to neighboring sites at rate p; assume, moreover, that the ions carry enough kinetic energy to move any atom irrespective of its bonding environment. The jump rates (15.10) then become

$$\Gamma_{ij} = p + (1 - p)\exp[-2(K/T)n_i] \qquad (15.21)$$

and detailed balance is broken if $p > 0$.

According to the arguments given in Sec. 15.3.1, the modification should generate an inclination-dependent contribution to the chemical potential, with a functional form given by (15.19). This is borne out by simulations (Fig. 15.1). Moreover, we expect the large-scale behavior of this and similar models to be described by the conserved KPZ equation (15.20) with, however, the shot noise $\eta(x, t)$ replaced by the derivative of a random current of type (15.12). The resulting equation was first proposed and analyzed in [29] on the basis of symmetry considerations. As in the case of (15.20), the equation possesses enough symmetries to allow for the *exact* calculation of the scaling exponents. One obtains $\zeta = 1/3$ and $z = 11/3$ in 1D, in agreement with numerical work [30-32,22].

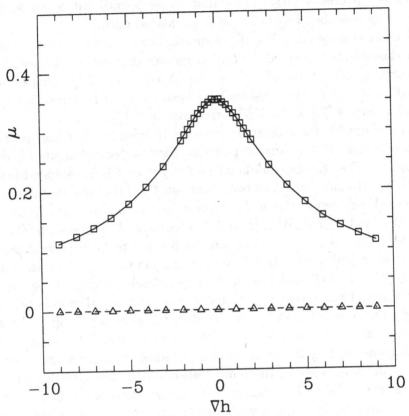

Fig. 15.1. Chemical potential μ (15.17) as a function of inclination, for the model defined by (15.21) with $K/T = 1$, for $p = 0.1$ (squares) and for $p = 0$ (triangles). The system size was $L = 1000$ and each data point was averaged over 5×10^5 Monte Carlo steps per site.

15.3.3 Anomalous scaling

We have seen in Sec. 15.3.1 that the linear fourth-order equation (15.7) subjected to the shot noise (15.15) gives rise to a height-difference correlation function that does not become stationary in time. The essence of this *anomalous scaling* behavior is that *local* roughness measures, such as moments of the height gradient ∇h (or its discrete counterpart $h_{i+1} - h_i$) couple to the long-wavelength fluctuations and diverge as

$$\langle |\nabla h|^q \rangle \sim \xi^{q\alpha_q} \sim t^{q\alpha_q/z}, \quad \alpha_q > 0, \tag{15.22}$$

where $\xi \sim t^{1/z}$ is the usual dynamic correlation length. For the linear fourth-order equation the gradient distribution is Gaussian and therefore

$\alpha_q \equiv \alpha = 1/2$ for all q, but this does not appear to be the typical situation, as will be explained shortly. The stationary long-time limit in a finite system of size L is described by (15.22) with ξ replaced by L.

Several recent numerical studies indicate that anomalous scaling is generic for a class of 1D conserved deposition models that can be defined, for the present purposes, by the absence of the growth-induced currents to be discussed in Sec. 15.4. Most results have been obtained for simplified *limited mobility* models [22] of the kind introduced in [3,17].

In these models the relaxation of atoms to favorable bonding sites is allowed only immediately after deposition. First, a deposition site i is chosen at random. Then the neighborhood $i - \ell \le j \le i + \ell$ is investigated (usually $\ell = 1$), and the site j at which some function $\mathcal{K}_j$ of the local height variables is minimal is chosen as the final incorporation site of the atom. In the models of Wolf and Villain (WV) [3] and Das Sarma and Tamborenea (DT) [17], $\mathcal{K}_j$ is identified with the coordination number at site j; the general case was considered in [25]. Owing to subtle differences in the implementation of the rule the WV and DT models actually have different asymptotic behaviors— the WV model has a weak growth-induced current and thus does not really belong to the class of models considered in this section [22,33]; however, on moderate time and length scales the two models are indistinguishable [34].

Anomalous scaling in limited-mobility models was investigated in [34-36], and its persistence in the more realistic Arrhenius growth model was established in [37-39]. The phenomenon has two remarkable features. First, as was pointed out in [34], it furnishes an example of *multiscaling* in the sense that the exponents α_q in (15.22) generally show a nontrivial q-dependence. In a detailed study of the DT model with $\ell = 1$ it was found that the underlying distribution of height differences has the form of a stretched exponential, with a stretching exponent that decreases with increasing ξ [34].

Second, the exponents α_q appear to be distinctly *nonuniversal*. A dramatic decrease of the exponent α_2/z of the WV model with increasing incorporation length ℓ, from $\alpha_2/z = 0.19 \pm 0.01$ at $\ell = 1$ to $\alpha_2/z \approx 0.027$ at $\ell = 8$, has been observed [40]. Similarly, a strong temperature dependence of α_2/z in the Arrhenius growth model has been reported [37]. In fact, it seems to be possible to suppress anomalous scaling completely, by an appropriate restriction on nearest-neighbor height differences, without otherwise changing the large-scale properties [26].

An understanding of these key features—multiscaling with nonuniversal exponents—is so far lacking. No multiscaling occurs [34] if the conserved KPZ nonlinearity $\partial_x^2(\partial_x h)^2$ in (15.20) is suppressed by imposing tilt invariance of the dynamics; this suggests that the phenomenon might be linked to finite-time divergences of (15.20) that have been observed numerically [24]. However, this scenario seems unlikely in view of recent rigorous results, which establish that the solutions of the deterministic, conserved KPZ equation ((15.20) with $\eta = 0$) remain bounded under quite general conditions [41].

A different explanation has been proposed in [42]. These authors point out that $d = 1$ is the lower critical dimensionality of (15.20), in the sense that all terms in the gradient expansion of the nonequilibrium chemical potential (15.19) have the same naive scaling dimension and therefore are equally relevant by power counting. In a perturbative calculation each of the higher-order terms $\lambda_{2n} \partial_x^2 (\partial_x h)^{2n}$, $n = 2, 3, 4, \ldots$, appearing on the r.h.s. of (15.20), gives rise to a logarithmic correction to the roughness exponent $\zeta = 1$. Exponentiating the series of logarithms one obtains a correction to the exponent itself, which depends explicitly on the coefficients λ_{2n} and hence is nonuniversal. Extending the argument to higher moments of height fluctuations one obtains q-dependent exponents in qualitative agreement with the numerical evidence. It remains to be seen whether these ideas can be turned into a quantitative theory.

15.4 Induced surface currents

The Edwards-Wilkinson (EW) equation [43],

$$\partial_t h = \nu \partial_x^2 h + \eta \tag{15.23}$$

is the simplest representative of the general class of equations specified by (15.1) and (15.3). It is also maximally robust, in the sense that all higher-order terms generated by the gradient expansion of the systematic current (15.3) are irrelevant by power counting when compared to $\nu \partial_x^2 h$ [23].

The reason why (15.23) has not appeared so far in this chapter is that the microscopic origins of the second-order derivative term are somewhat subtle. Within the classical theory of Sec. 15.2.1, a current $J_S \sim \partial_x h$ clearly requires a contribution to the local chemical potential that is proportional to the height h. In the context of the sedimentation of macroscopic particles, which was the concern in [43], such a contribution is naturally provided by gravity. On the atomic level, gravitational effects obviously play no role; however, as was first pointed out in [4], other mechanisms exist, on crystal surfaces, which can give rise to inclination-dependent mass currents and thus to a term $\nu \partial_x^2 h$ in the large-scale description.

We explain the basic idea for the 1D SOS model. Consider a flat surface with some fixed average slope $u = \langle h_{i+1} - h_i \rangle$. In the absence of spatial variations of the inclination the chemical potential is constant, and the classical contribution (15.5) to the surface current vanishes. We ask now whether the expectation

$$J_{NE}(u) = \langle \Gamma_{i\,i+1} - \Gamma_{i+1\,i} \rangle_u \, , \qquad (15.24)$$

evaluated relative to the (equilibrium or nonequilibrium) stationary state at fixed u, is nonzero. Clearly $J_{NE}(u) = -J_{NE}(-u)$ and $J_{NE}(0) = 0$ by reflection symmetry. In equilibrium the detailed balance condition ensures that $J_{NE} = 0$ for all u. Moreover, if the diffusion dynamics is governed by the Arrhenius rates (15.10), we have

$$J_{NE} = \langle e^{-2(K/T)n_i} - e^{-2(K/T)n_{i+1}} \rangle = 0 \qquad (15.25)$$

by translational invariance, irrespective of whether detailed balance holds. This justifies our neglect of the EW term in Sec. 15.3; for Arrhenius surface dynamics the dominant nonequilibrium effect arises from the nonequilibrium chemical potential (15.19) and the associated nonlinearities in (15.20). A similar symmetry argument rules out the EW term in the DT limited-mobility model [22].

For *generic* surface diffusion rates, such as the Metropolis rates (15.11), symmetry arguments to show that $J_{NE} \equiv 0$ are not available, and we have to conclude that, in the absence of detailed balance, (15.24) is some nontrivial, odd function of the inclination u [33]. The total systematic surface current is then

$$J_S = J_{NE}(\partial_x h) - \sigma \partial_x \mu(\partial_x h, \partial_x^2 h) \qquad (15.26)$$

with $\mu = -\hat{\gamma}\partial_x^2 h + \mu_{NE}$. To leading order in a gradient expansion, $J_S = -\nu \partial_x h$, with

$$\nu = -J'_{NE}(u) \qquad (15.27)$$

defining precisely the coefficient in the EW equation (15.23). Note, however, that the argument says nothing about the sign of ν! EW scaling with the familiar exponents $\zeta = 1/2$ and $z = 2$ follows only if $\nu > 0$. For $\nu < 0$ the surface develops an instability that initiates a kind of phase separation process, to be further explored in Sec. 15.4.5.

To determine the sign and magnitude of ν one needs some insight into the microscopic processes that contribute to (15.24). This question will be pursued in the following four sections. Following the pattern of Sec. 15.3, we now focus on nonequilibrium variants of *Metropolis* surface diffusion dynamics with (Sec. 15.4.1) and without (Sec. 15.4.3) deposition. A class of exactly solvable deposition models is described in Sec. 15.4.2, while Sec. 15.4.4 reviews the restricted SOS models for surface diffusion in which the consequences of induced surface currents were first observed [31].

15.4.1 *Growth-induced currents in the Metropolis model*

Metropolis surface diffusion in the presence of a deposition flux has been investigated in [38,44,45]. Using $\Phi(s) = (1+e^s)^{-1}$ in (15.11) and the standard SOS Hamiltonian (15.4) one finds that the growth-induced surface current is directed downhill, so that $\nu = -J'_{NE}(u) > 0$ for small inclinations [33]. Consequently the scaling behavior is of EW type [45].

Different behavior obtains if the SOS Hamiltonian (15.4) is generalized to

$$\mathcal{H}_q = K \sum_i |h_{i+1} - h_i|^q, \tag{15.28}$$

with $q > 1$. For $q = 2$ (15.28) is known as the discrete Gaussian model, which is invariant under tilt transformations $h_i \to h_i + ui$ for integer u [22,38]. As a consequence the growth-induced current (which is forced to vanish at all integer u) is extremely small and can be neglected for all practical purposes. Thus $\nu \approx 0$ and, since the nonequilibrium chemical potential (15.19) is suppressed by the same mechanism, the model is well described by the linear, fourth-order equation [38].

In contrast, for $q = 4$ the surface current is directed uphill, so that $\nu < 0$ and the surface becomes unstable [33,44]. While an explicit calculation of the current from the microscopic dynamics does not seem to be feasible, the qualitative q-dependence can be linked to the presence of *step-edge barriers*, which provide the most important source of growth-induced currents on real crystal surfaces [4] also.

The origin of the barrier is illustrated in Fig. 15.2. The energy of an atom approaching a step edge from above increases by an amount

$$\Delta \mathcal{H} = (2^q - 2)K > 0, \quad \text{for } q > 1, \tag{15.29}$$

relative to its value on the flat terrace. At low temperatures the atom is therefore repelled and preferentially incorporated at the ascending step edge at the other end of the terrace. This gives rise to an uphill mass current.

The argument obviously neglects many other contributions to the surface current. However, since we know that $J_{NE} \approx 0$ at $q = 2$, it seems reasonable to conclude from (15.29) that the *net* current should be uphill for $q > 2$ and downhill for $q < 2$, in agreement with the simulations.

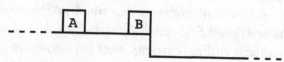

Fig. 15.2. The step-edge barrier (15.29) is the energy difference between atom A and atom B.

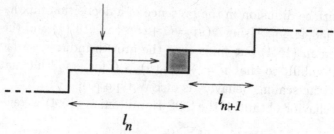

Fig. 15.3. Ideal step flow: each deposited atom is immediately transferred to the ascending step.

15.4.2 Step flow and zero-range processes

The surface current can be easily computed in a simplified model of a 1D vicinal surface, which applies to the limiting case of strong step-edge barriers and low deposition rate [46]. Suppose that (i) the step-edge barrier is infinite, so that each atom is confined to the layer upon which it was deposited, (ii) incorporation at step edges is irreversible, and (iii) the ratio of the deposition rate to the diffusion rate is so small that every atom is incorporated before further atoms appear on the surface. In this limit each deposited atom is certain to attach to the ascending step edge of the terrace on which it has landed, and the surface maintains the shape of an ascending staircase (possibly with multiple steps) for all times (Fig. 15.3). This growth regime is referred to as *step flow*.

Let ℓ denote the average terrace length. Then the net uphill distance traveled by a deposited atom before it reaches its incorporation site is $\ell/2$, and hence the current is [4] $J_{NE} = F\ell/2 = F/2u$, since the surface inclination is $u = 1/\ell$. Using (15.27) it follows that $\nu = F\ell^2/2 > 0$; the vicinal surface is *stabilized* by the step-edge barriers, as was first noted in [47]. Its fluctuations are described by the EW equation, and the height correlation function is $G(r,t) = F|r|/(2\nu) = \ell^{-2}|r|$, for $|r| \ll \xi = (2\nu t)^{1/2} = \ell(Ft)^{1/2}$.

It is instructive to formulate the dynamics in terms of the set of terrace lengths ℓ_n, which can take values $\ell_n = 0, 1, 2, \ldots$ (Fig. 15.3). Depositing an atom on terrace n reduces its length, $\ell_n \to \ell_n - 1$, while increasing the length of terrace $n + 1$, $\ell_{n+1} \to \ell_{n+1} + 1$. This suggests an interpretation of the ℓ_n as 'occupation numbers' of a fictitious particle system, and the deposition event as a directed hop of a particle from site n to site $n + 1$. The probability of deposition on terrace n is proportional to its length ℓ_n, hence the jump rate in the particle model is $c_n = \ell_n$, which implies that the particles are noninteracting. As an immediate corollary one concludes that in the stationary state the terrace lengths have a Poisson distribution [46].

Particle systems with an unlimited number of particles per site and jump rates that depend only on the on-site occupation number are known as zero-range processes [15]. They have the remarkable property that their stationary measure is known for an arbitrary choice of the functional relation between jump rate and occupation number. With the above equivalence in mind, we are therefore led to consider a class of step-flow models in which the probability for a deposited atom to reach the ascending step of the nth terrace is given by some general function $c_n = c(\ell_n)$ of the terrace length. For example, if the adatom has a finite lifetime τ before it redesorbs, one expects [48] $c(k) = 2\ell_D \tanh(k/2\ell_D)$, where $\ell_D = (D\tau)^{1/2}$ is the diffusion length determined by τ and the surface diffusion constant D.

Whatever the function $c(k)$, the stationary distribution of occupation numbers or terrace lengths is [15]

$$P(k) = Z^{-1} e^{\mu k} \prod_{j=1}^{k} c(j)^{-1}, \qquad (15.30)$$

where Z is a normalization constant and the 'chemical potential' μ serves to fix the average terrace length ℓ. It is understood in (15.30) that $P(0) = Z^{-1}$.

Writing (15.30) as a Gibbs distribution, it is seen that the 'energy' of the jth particle at a given site is equal to $\ln c(j)$. Thermodynamic stability requires that $c(j)$ is nondecreasing for large j; otherwise it becomes favorable to place all particles at a single site, and the system collapses. This corresponds precisely to the phenomenon of step bunching, of interest in deterministic theories of step flow; see [47,49] and references therein.

15.4.3 Surface electromigration

We have noted already, in Sec. 15.3.2, that the simplest way of breaking detailed balance without violating mass conservation is to impose a bias in the surface diffusion rates. A physical realization is provided by the effect of *surface electromigration*, which refers to the drift of adatoms under the influence of a DC electric current in the bulk [50]. From a theoretical point of view, the main advantage of this setup is that the resulting nonequilibrium current can be expressed in terms of the (equilibrium) adatom mobility,

$$J_{NE}(u) = \sigma(u)E, \qquad (15.31)$$

at least for small driving fields E; in fact (15.31) *defines* the mobility in the limit $E \to 0$ [14].

As before, the stability of the surface is determined by the sign of $J'_{NE}(u)$, and thus by the inclination dependence of the mobility. Since the mobility is an even function function of u, any unstable orientation can be stabilized by reversing the field direction. This makes it possible to determine numerically the mobility for any orientation by measuring the current at (small) finite fields [11].

In terms of analytic results, the situation is less favorable than it might seem; the direct analytic evaluation of the mobility is possible only when σ is independent of the inclination u, and thus the nonequilibrium current (15.31) has no effect [11]. However, an upper bound that seems to be generally representative of the true mobility is always provided by the averaged jump rate, $\sigma(u) \le \sigma^+(u) \equiv \frac{1}{2}\langle \Gamma_{i\,i+1}\rangle_u$, which can easily be computed.

Biased surface diffusion with Metropolis dynamics was considered in [10]. In the 1D model, jumps are attempted at rate p to the right and rate $1 - p$ to the left, and they are accepted with a probability given by (15.11) with $\Phi(s) = \min(1, e^{-s})$. In terms of an 'electric field' contribution, $\mathcal{H}_E = -E\sum_i ih_i$, to the Hamiltonian, the bias parameter p can be identified as $p = e^{E/T}/(1 + e^{E/T})$.

For Metropolis dynamics the mobility bound $\sigma^+(u)$ has a single minimum at $u = 0$, and tends asymptotically to $1/2$ for $u \to \pm\infty$; the full mobility has the same qualitative behavior [11]. Consequently for $E > 0$ all positive orientations, $u > 0$, are linearly unstable. For a surface that is initially flat ($u = 0$ on average) this implies a 'sawtooth' morphology (Fig. 15.4): Since no stable positive orientations exist, the positively sloped portions of the surface coalesce into 'macrosteps', i.e., segments of infinite slope; by contrast, the negatively sloped parts form facets [51] of a well-defined selected slope $u^* < 0$, which can be computed from continuum theory using the appropriate functions $\hat{\gamma}(u)$ and $\sigma(u)$ [10,52].

A second case of interest is the Gaussian model, defined by (15.28) with $q = 2$, with biased Metropolis rates. Here the discrete tilt symmetry alluded to in Sec. 15.4.1 implies that both the stiffness and the mobility are periodic functions of u. Instead of macrosteps and negatively tilted facets the surface breaks up into two kinds of facet, of finite slopes $u_1 > 0$ and $u_2 < 0$ [52].

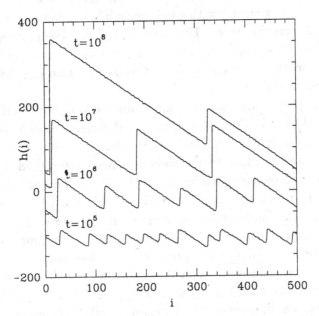

Fig. 15.4. Morphological evolution for the biased Metropolis model with $p = 0.7$ and $K/T = 2$. Subsequent configurations are displaced vertically by 100 lattice units. Time is measured in MC steps per site; from [10].

Apart from the question of slope selection, the most important aspect of these models lies in the coarsening dynamics evident from Fig. 15.4. This will be addressed in Sec. 15.4.5.

15.4.4 Restricted SOS models

Current-induced instabilities due to a negative coefficient ν in (15.23) were first found in a detailed study [31] of a class of restricted solid-on-solid (RSOS) models for surface diffusion introduced in [29]. In the present context these models are of interest because they show how a nonequilibrium surface current can be induced without violating either mass conservation (as in deposition models) or reflection symmetry (as in biased diffusion).

The central feature of RSOS models is a restriction on nearest-neighbor height differences

$$|h_{i+1} - h_i| \leq H, \tag{15.32}$$

which is maintained at all times. Surface diffusion is modeled as usual by attempting to move particles to nearest-neighbor sites, $h_i \to h_i - 1$ and $h_j \to h_j + 1$, where $j = i \pm 1$ with equal probabilities, but moves are accepted only if (15.32) remains true. The same model has been considered in [62] with a different interpretation: the h_i are slope variables and (15.32) defines a restriction on the local curvatures; interface motion can be included by biasing moves to the left.

Rácz et al. [31] made the remarkable observation that surface diffusion in RSOS models may or may not satisfy detailed balance, depending on precisely how the moves are carried out. Suppose that a site i has been chosen at random, a move to the right has been attempted, but the move has been rejected owing to violation of (15.32). There are then two ways to proceed. In procedure (1), a new site is selected and a move is again attempted in a randomly chosen direction. In procedure (2), a move to the left is attempted at the previously selected site i. It turns out that procedure (2) violates detailed balance but procedure (1) does not.

Incidentally, procedure (1) coincides with the model used in the original work [29]. The fact that this dynamics satisfies detailed balance (relative to the distribution that gives equal weight to all configurations allowed by (15.32)) rules out the nonequilibrium term in (15.20) that was postulated in [29]. Indeed, the simulations of [31] confirm that the model with procedure (1) is well described by the equilibrium dynamical equation, (15.7) subject to volume-conserving noise.

The genericity arguments given at the beginning of Sec. 15.4 suggest that procedure (2) should give rise to a nonequilibrium surface current that, because of the reflection symmetry of the dynamics, would be an odd function of inclination. This was confirmed by direct numerical measurement [33]. Moreover, it was found that the coefficient ν defined by (15.27) at $u = 0$ depends on the restriction parameter H in (15.32) and changes sign at $H = 3$, in accordance with the transition from stable to unstable behavior observed in [31].

15.4.5 Coarsening dynamics

The morphological evolution of current-destabilized surfaces beyond the linear regime can be profitably analyzed by exploiting the analogy with phase-ordering kinetics [22,46,53-55]. The analogy is particularly straightforward in the 1D case of interest here. We express the general equation of motion (15.1) in terms of the local slope $u(x,t) = \partial_x h(x,t)$, using the expression (15.26) for the systematic current but neglecting the nonequilibrium contribution $\mu_{NE}(u)$ to the chemical potential [56] as well as the slope dependences of the stiffness $\hat{\gamma}$ and the mobility σ. The result takes the form of a noisy Cahn-Hilliard equation [57],

$$\partial_t u = \partial_x^2[\partial\mathcal{V}(u)/\partial u - \kappa\partial_x^2 u] + \partial_x\eta - \partial_x^2 J_R , \qquad (15.33)$$

with the 'thermodynamic potential'

$$\mathcal{V}(u) = -\int_0^u dv J_{NE}(v). \qquad (15.34)$$

The analogy suggests a simple picture for the surface evolution beyond the linear instability: selected orientations, corresponding to the minima of $\mathcal{V}(u)$, i.e., zeros of J_{NE} with $J'_{NE}(u) < 0$ [33,54], are rapidly established in local domains, and the subsequent evolution can be described as a coarsening process in which the domain size $\xi(t)$ increases in a universal fashion [58]. A complication is that many deposition models give rise to current functions J_{NE} that do not possess any stable zeros [33], and consequently the coarsening process is accompanied by a steepening in which the slopes of the domains increase without bound [38,44,59]. We will return to this effect at the end of the section, but focus presently on the simpler case where the potential (15.34) is of the conventional ϕ^4-form with two degenerate minima.

Coarsening in 1D differs from the higher-dimensional situation in that fluctuations play a crucial role. The reason is that the driving force in 1D is given by the interaction between domain walls, which usually decays exponentially with distance [60]. Therefore, at zero temperature $\xi(t) \sim \ln t$ [57]. In the presence of fluctuations the coarsening is dominated by random encounters between the domain walls. For nonconserved dynamics the domain walls perform independent random walks, leading to $\xi(t) \sim t^{1/2}$, while the conservation of the order parameter enforces a correlated motion that slows down the coarsening, so that $\xi(t) \sim t^{1/3}$ [61,63].

Comparing to (15.33) we see that the case of a conserved order parameter at finite temperature corresponds precisely to a deposition process, in which the (derivative of the) shot noise plays the role of thermal fluctuations. 1D surfaces destabilized by growth-induced currents with stable zeros should therefore coarsen as $t^{1/3}$; the corresponding behavior in two dimensions is found numerically to be $t^{1/4}$, for reasons that are not understood [54,55]. On the other hand, if volume conservation is maintained, as in the systems described in Secs. 15.4.3 and 15.4.4, the only source of fluctuations in (15.33) is the 'doubly conserved' noise term $-\partial_x^2 J_R$, which might be expected to lead to a different coarsening law. In the following we present a simple derivation,

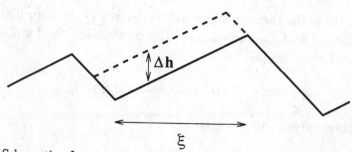

Fig. 15.5. Schematic of coarsening dynamics. The vertical shift of segments, indicated by the broken line, induces a correlated motion of corners (= domain walls).

which also reproduces the result of [61,63] and allows us to understand why coarsening is slowed down in the presence of steepening.

Consider an idealized surface configuration composed of straight segments of slopes $u_1 > 0$ and $u_2 < 0$, with a typical segment length $\xi(t)$ (Fig. 15.5). The 'domain walls' of u correspond to the corners at which two segments meet. Owing to the conservation of u, the corners are not free to move independently. However, the *vertical* motions of the segments can be assumed to be essentially uncorrelated, since they realize the fundamental vertical translation symmetry, $h \rightarrow h +$ constant, of the surface (the 'Goldstone mode'). In the presence of complete volume conservation the height of a segment fluctuates owing to the difference between the influx and the outflux of mass at its boundaries. Assuming that the fluxes are uncorrelated, the mass M below the segment performs a random walk with a diffusion constant of order unity. Denoting by Δh the height fluctuation of the segment, we have

$$\Delta M \sim \Delta h\,\xi \sim t^{1/2}. \tag{15.35}$$

A segment can disappear (two domain walls can collide) when

$$\Delta h \sim \bar{u}\xi, \tag{15.36}$$

where $\bar{u} = \min(|u_1|, |u_2|)$ is the smaller of the two segment slopes (this formulation includes also situations such as in Fig. 15.4, where one slope is infinite). From (15.35) the corresponding time scale is of order ξ^4, and hence the coarsening law is $\xi(t) \sim t^{1/4}$. This has been confirmed in simulations of the biased Metropolis model [10,52].

We can repeat the argument for the deposition case. The height fluctuation of a segment is then given by the shot noise averaged over the domain size ξ, and (15.35) is replaced by $\Delta h \sim (t/\xi)^{1/2}$. The coarsening time scale is again set by the condition (15.36), which yields $\xi \sim t^{1/3}$ in agreement with the detailed analysis [61,63].

Let us finally assume that the segments steepen while they coarsen, according to the power law $\bar{u} \sim \xi^\delta$, $\delta > 0$. Such a behavior has been observed numerically [44] and can be derived from simple models of the current J_{NE} [22,59]. This modifies the condition (15.36) and leads to the coarsening law $\xi \sim t^{1/(n+2\delta)}$, with $n = 4$ ($n = 3$) with (without) strict volume conservation. In either case, steepening slows down coarsening. For a quantitative comparison we consider the simulations [44] for the Metropolis deposition model using the Hamiltonian (15.28) with $q = 4$. The estimate $\delta \approx 2.6$ [44] would imply a coarsening exponent of $1/8.2 \approx 0.12$, not too far from the numerical value of 0.17 [38].

The work described here was performed in collaboration with Harvey Dobbs, Mike Plischke, Martin Schimschak, and Martin Siegert. The author is grateful to Herbert Spohn, Dietrich Wolf, and Andy Zangwill for discussions, and to Sankar Das Sarma and Andy Zangwill for providing him with unpublished manuscripts.

References

[1] In this chapter *deposition* refers to *ideal MBE* processes in the sense of [2], in which desorption and vacancy formation are absent and surface diffusion therefore constitutes the only surface relaxation mechanism.

[2] Z.-W. Lai and S. Das Sarma, *Phys. Rev. Lett.* **66**, 2348 (1991).

[3] D. E. Wolf and J. Villain, *Europhys. Lett.* **13**, 389 (1990).

[4] J. Villain, *J. Physique* I1, 19 (1991).

[5] A. Zangwill, Theory of growth-induced surface roughness, in *Microstructural Evolution of Thin Films*, H. A. Atwater and C. V. Thompson, eds. (Academic, New York, 1995).

[6] G. Grinstein, Generic scale invariance and self-organized criticality, in *Scale Invariance, Interfaces and Non-Equilibrium Dynamics*, p. 261, A. J. McKane, M. Droz, J. Vannimenus and D. E. Wolf, eds. (Plenum Press, New York, 1995).

[7] T. Hwa and M. Kardar, *Phys. Rev.* **A45**, 7002 (1992).

[8] W. W. Mullins, Solid surface morphologies governed by capillarity, in *Metal Surfaces: Structure, Energetics and Kinetics*, N. A. Gjostein and W. D. Robertson, eds. (American Society of Metals, Metals Park, 1963).

[9] Throughout the chapter we write the relations of the continuum theory in a manner appropriate for the description of SOS models, in which only *horizontal* chemical potential differences exist. Otherwise an additional geometric factor $(1 + (\partial_x h)^2)^{-1/2}$ appears in (15.5) and (15.31); see [10,11].

[10] J. Krug and H. T. Dobbs, *Phys. Rev. Lett.* **73**, 1947 (1994).

[11] J. Krug, H. T. Dobbs and S. Majaniemi, *Z. Physik* **B97**, 281 (1995).

[12] H. P. Bonzel, *CRC Crit. Rev. Solid State Sci.* **6**, 171 (1976).

[13] D. D. Vvedensky, A. Zangwill, C. N. Luse and M. R. Wilby, *Phys. Rev.* **48**, 852 (1993).

[14] H. Spohn, J. Stat. Phys. **71**, 1081 (1993).

[15] H. Spohn, Large Scale Dynamics of Interacting Particles (Springer, Heidelberg, 1991).

[16] M. Giesen-Seibert, R. Jentjens, M. Poensgen and H. Ibach, Phys. Rev. Lett. **71**, 3521 (1993).

[17] S. Das Sarma and P. Tamborenea, Phys. Rev. Lett. **66**, 325 (1991).

[18] M. R. Wilby, D. D. Vvedensky and A. Zangwill, Phys. Rev. B**46**, 12896 (1992); B**47**, 16068 (1993).

[19] L. Golubović and R. Bruinsma, Phys. Rev. Lett. **66**, 321 (1991); **67**, 2747.

[20] J. G. Amar, P.-M. Lam and F. Family, Phys. Rev. E**47**, 3242 (1993).

[21] C. N. Luse, A. Zangwill, D. D. Vvedensky and M. R. Wilby, Surf. Sci. Lett. **274**, L535 (1992).

[22] J. Krug, Origins of scale invariance in growth processes, Adv. Phys., to be published.

[23] L.-H. Tang and T. Nattermann, Phys. Rev. Lett. **66**, 2899 (1991).

[24] Y. Tu, Phys. Rev. A**46**, R729 (1992).

[25] J. M. Kim and S. Das Sarma, Phys. Rev. Lett. **72**, 2903 (1994).

[26] Y. Kim, D. K. Park and J. M. Kim, J. Phys. A**27**, L533 (1994).

[27] B. Schmittmann and R. K. P. Zia, Statistical Mechanics of Driven Diffusive Systems. Phase Transitions and Critical Phenomena, Vol. **17**, C. Domb and J. L. Lebowitz, eds. (Academic Press, New York, 1995).

[28] J. Y. Cavaillé and M. Drechsler, Surf. Sci. **75**, 342 (1978).

[29] T. Sun, H. Guo and M. Grant, Phys. Rev. A**40**, 6763 (1989).

[30] Note, however, that the stochastic model proposed in the original work [29] does not belong to the universality class of the conserved KPZ equation, as was subsequently pointed out in [31]. We address this point in Sec. 15.4.4.

[31] Z. Rácz, M. Siegert, D. Liu and M. Plischke, Phys. Rev. A**43**, 5275 (1991).

[32] A. Chakrabarti, J. Phys. A**23**, L919 (1990).

[33] J. Krug, M. Plischke and M. Siegert, Phys. Rev. Lett. **70**, 3271 (1993).

[34] J. Krug, Phys. Rev. Lett. **72**, 2907 (1994).

[35] M. Schroeder, M. Siegert, D. E. Wolf, J. D. Shore and M. Plischke, Europhys. Lett. **24**, 563 (1993).

[36] S. Das Sarma, S. V. Ghaisas and J. M. Kim, Phys. Rev. E**49**, 122 (1994).

[37] C. J. Lanczycki and S. Das Sarma, Phys. Rev. E**50**, 213 (1994).

[38] M. Siegert and M. Plischke, Phys. Rev. E**50**, 917 (1994).

[39] S. Das Sarma, C. J. Lanczycki, R. Kotlyar and S. V. Ghaisas, Phys. Rev. E**53**, 359 (1996).

[40] M. Schroeder, Modelle mit Bezug zum epitaktischen Kristallwachstum, Diploma thesis (University of Duisburg, 1993).

[41] V. Putkaradze, T. Bohr and J. Krug, Global estimates for the solutions of the noiseless conserved Kardar-Parisi-Zhang equation, NBI report 95-04, to be published.

[42] J. K. Bhattacharjee, S. Das Sarma and R. Kotlyar, Phys. Rev. E**53**, R1313 (1996).

[43] S. F. Edwards and D. R. Wilkinson, Proc. Roy. Soc. London A**381**, 17 (1982).

[44] M. Siegert and M. Plischke, Phys. Rev. Lett. **68**, 2035 (1992).

[45] M. Siegert and M. Plischke, J. Physique I**3**, 1371 (1993).

[46] J. Krug and M. Schimschak, J. Physique I**5**, 1065 (1995).

[47] R. L. Schwoebel and E. J. Shipsey, J. Appl. Phys. **37**, 3682 (1966).

[48] R. Ghez and S. S. Iyer, IBM J. Res. Dev. **32**, 804 (1988).

[49] D. Kandel and J. D. Weeks, *Physica* **D66**, 78 (1993).

[50] P. J. Rous, T. L. Einstein and E. D. Williams, *Surf. Sci.* **315**, L995 (1994).

[51] The facets are in fact smooth, since the current suppresses [10] the thermal roughness.

[52] H. T. Dobbs and J. Krug, *J. Physique* I6, 413 (1996).

[53] L. Golubović and R. P. U. Karunasiri, *Phys. Rev. Lett.* **66**, 3156 (1991).

[54] M. Siegert and M. Plischke, *Phys. Rev. Lett.* **73**, 1517 (1994).

[55] M. Siegert, Non-equilibrium ordering dynamics and pattern formation, in *Scale Invariance, Interfaces and Non-Equilibrium Dynamics*, p. 165, A. J. McKane, M. Droz, J. Vannimenus and D. E. Wolf, eds. (Plenum Press, New York, 1995).

[56] The influence of the nonequilibrium chemical potential on the coarsening phenomena described in this section is an interesting problem; however, there are no results to report.

[57] J. S. Langer, *Ann. Phys. (NY)* **65**, 53 (1971).

[58] A. J. Bray, *Adv. Phys.* **43**, 357 (1994).

[59] A. W. Hunt, C. Orme, D. R. M. Williams, B. J. Orr and L. M. Sander, *Europhys. Lett.* **27**, 611 (1994).

[60] An exception is the Metropolis model for surface electromigration, where the effective interaction between macrosteps can be shown to decay as a power law [52].

[61] T. Kawakatsu and T. Munakata, *Prog. Theor. Phys.* **74**, 11 (1985).

[62] J. M. Kim and S. Das Sarma, *Phys. Rev.* **E48**, 2599(1993).

[63] S. J. Cornell, K. Kaski and R. B. Stinchcombe, *Phys. Rev.* **B44**, 12263 (1991).

16

Directed-walk models of polymers and wetting

Julia M. Yeomans

The aim of this chapter is to summarize briefly recent results on directed walks and provide a guide to the literature. We shall restrict consideration to the equilibrium properties of directed interfaces and polymers, focusing particularly on their collapse and binding transitions. The walks will lie in a nonrandom environment.

16.1 Directed walks and polymers

A clear introduction to the physics of directed walks is given by Privman and Švrakić in a book published in 1989 [1]. This summarizes the work up to that time and therefore here we shall aim to describe more recent progress after a brief description of the relevant models.

Many of the interesting results for nonrandom systems have been obtained for walks that should strictly be labeled partially directed. In these movement is allowed along either the positive or negative x-direction but only along the positive t-direction, as shown in Fig. 16.1. Hence the position n_i of the walk in column $t = i$ is unique.

Also shown in Fig. 16.1 for comparison is a fully directed walk, each step of which must have a nonzero component in the positive t-direction. This is a simpler model, which has been very useful in studying the behavior of interfaces in a random environment (not reviewed here; see [34]). The partially directed walk reduces to the fully directed one if the constraint $|n_i - n_{i+1}| = 1$ is imposed.

Regarding the partially directed walk as an interface separating two phases leads to the well-known solid-on-solid model of an interface with no overhangs, which has proved a useful testing ground for theories of wetting and interface fluctuations [2]. The Hamiltonian of a solid-on-solid model on a

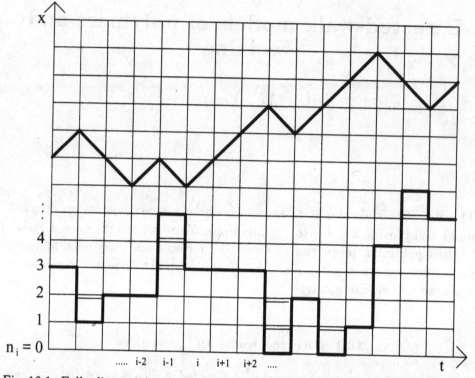

Fig. 16.1. Fully directed (upper solid line) and partially directed (lower solid line) random walks. The interactions introduced in the polymer model (16.3) are shown by double lines.

semi-infinite square lattice with a pinning potential U acting at the surface is

$$\mathcal{H}/(kT) = \sum_i L|n_i - n_{i+1} + 1| - U\delta_{n_i,1}. \qquad (16.1)$$

This Hamiltonian lends itself to solution using a transfer matrix M with elements

$$M_{i,i+1} = \gamma^{|n_i - n_{i+1} + 1|}\kappa^{\delta_{n_i,1}}, \qquad (16.2)$$

where $\gamma = e^{-L}$ and $\kappa = e^U$. The position of the wetting or unbinding transition κ_c follows from an analysis of the spectrum of M. For $\kappa > \kappa_c$ the thermodynamics is dominated by a bound state corresponding to the largest eigenvalue. For $\kappa < \kappa_c$ the bound state does not exist and the continuous spectrum describes a free interface.

The unbinding of a directed walk from a wall is a problem equivalent to the unbinding of two walks. An interesting recent development is the increased understanding of the unbinding transition in a system of N directed walks. Work here has concentrated on continuum theories. Following numerical investigations [3,4] for $N = 3$, it was realized that the unbinding of N identical intersecting or nonintersecting walks with short-range, two-body interactions could be solved exactly using a Bethe ansatz [5,6]. A single transition with exponents independent of N was found. Unbinding of the bundle of strings from a rigid wall could proceed either via a single transition or by a sequence of transitions, each walk unbinding in turn [5]. With the addition of higher-order interactions the problem is no longer solvable but a renormalization group treatment due to Lässig [7], see also [8], has shown that the inclusion of these terms leads to new universality classes.

Results for continuum or restricted ($|n_i - n_{i+1}| = 0, 1$) versions of the Hamiltonian (16.1) with various pinning potentials or bulk fields are summarized in [1]. A recent advance is the solution of the unrestricted solid-on-solid model (16.1) with the addition of a bulk field $-H \sum_i n_i$ [9]. The phase diagram contains the expected wetting transition at a temperature T_ω for $H = 0$ and complete wetting occurs as $H \to 0$ for $T \geq T_\omega$. This is in agreement with the solution of the model for continuous x [10].

Several authors have considered the effect of replacing the surface tension term in the Hamiltonian (16.1) by a bending energy. An exact solution of a restricted solid-on-solid model gave only quantitative differences between the two cases [11] whereas continuum models predict the possibility of a first-order phase transition for both short- and long-range surface interactions [12,13].

Much recent work has viewed the partially directed walk as a polymer rather than an interface. This is equivalent to working within a grand-canonical rather than a canonical ensemble. However, a new feature can be more physically introduced: nearest-neighbor interactions between monomers modeled by an attractive energy J between sites that occupy the same row in adjacent columns (see Fig. 16.1). This leads to a collapse transition with interesting physical and mathematical properties. (We note in passing that the fully directed walk does not have a similar transition [14].)

The grand canonical partition function of the partially directed polymer is

$$Z = \sum_{\text{walks}} \omega^n \kappa^\ell \tau^m, \tag{16.3}$$

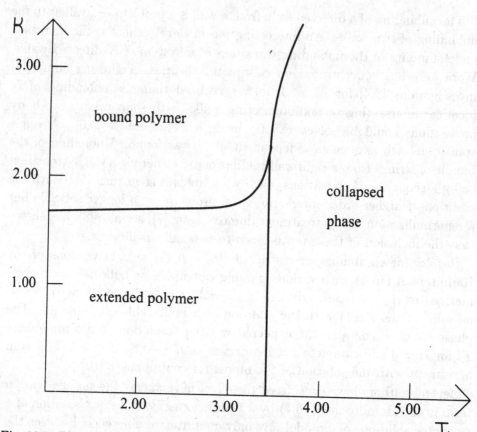

Fig. 16.2. Phase diagram of the partially directed polymer with a binding fugacity κ and a fugacity τ for nearest-neighbor interactions.

where ω is the step fugacity and $\tau = e^{-J/kT}$, n is the number of monomers in the walk, ℓ the number of visits to the wall, and m the number of monomer-monomer interactions. The partition function Z can be written in terms of a generating function or a transfer matrix analogous to (16.2). An infinite polymer limit corresponds to a singularity in the generating function when the largest eigenvalue of the transfer matrix becomes equal to unity.

The phase diagram of the partially directed interacting polymer is shown in Fig. 16.2. Consider first the case $\tau = 0$ where the interesting physical questions concern the binding transition. For the polymer (16.3) there is a continuous binding transition analogous to that of the solid-on-solid model for which κ_c and the critical exponents are well known [1]. More recently, exact results have been derived for a directed polymer inside a parabola [15] and for long-range potentials with power-law decay [16]. In both cases careful tuning of the perturbations results in nonuniversal critical behavior.

For $\tau \neq 0$ regularities in numerical transfer-matrix calculations suggested that an exact solution was likely for the boundary of the collapsed phase [17]. This was found to follow from the structure of the transfer matrix for $\kappa = 0$ [18] and for general κ [19,20]. Generating function techniques have provided a more rigorous derivation of the position of the collapse transition in the former case [23]. A collapse transition in a similar model, but with the walk constrained to fold back on itself at each horizontal step, had been solved much earlier by Zwanzig and Lauritzen [24,25]. The relation between the generating function of this model and that of the partially directed walk is given in [26].

The collapsed phase corresponds to a polymer that completely fills the lattice. The extended-to-collapse transition is continuous and occurs at a value of κ that is independent of τ. The transition between the bound and collapsed phases is first order; its boundary is not known exactly but is thought to remain continuous (as at $\kappa = 1$). Similar results follow for a polymer that interacts with a second polymer rather than a surface [27].

Because analytic progress is possible the collapse transition of the partially directed polymer model has now been well studied. Careful work on the exponents and the structure of the scaling functions is summarized by Owczarek and Prellberg [26]; see also [28-31]. More recent mathematical explanations of the structure of the generating function can be found in references [32,33].

References

[1] V. Privman and N. M. Švrakić, *Directed Models of Polymers, Interfaces and Clusters: Scaling and Finite-Size Properties. Lecture Notes in Physics*, Vol. **338** (Springer, Berlin, 1989).

[2] G. Forgacs, R. Lipowsky and Th. M. Nieuwenhuizen, in *Phase Transitions and Critical Phenomena*, Vol. **14**, p. 135, C. Domb and J. L. Lebowitz, eds. (Academic Press, New York, 1991).

[3] R. R. Netz and R. Lipowsky, *Phys. Rev.* E**47**, 3039 (1993).

[4] R. R. Netz and R. Lipowsky, *J. Physique* I**4**, 47 (1994).

[5] C. Hiergeist, M. Lässig and R. Lipowsky, *Europhys. Lett.* **28**, 103 (1994).

[6] T. W. Burkhardt and P. Schlottmann, *J. Phys.* A**26**, L501 (1993).

[7] M. Lässig, *Phys. Rev. Lett.* **73**, 561 (1994).

[8] T. W. Burkhardt, *Z. Physik* B**97**, 247 (1995).

[9] A. L. Owczarek, and T. Prellberg, *J. Stat. Phys.* **70**, 1175 (1993).

[10] D. B. Abraham and E. R. Smith, *J. Stat. Phys.* **43**, 621 (1986).

[11] G. Forgacs, *J. Phys.* A**24**, L1099 (1991).

[12] G. Gompper and T. W. Burkhardt, *Phys. Rev.* A**40**, 6124 (1989).

[13] T. W. Burkhardt, *J. Phys.* A**26**, L1157 (1993).

[14] G. Forgacs and M. Semak, *J. Phys.* A**24**, L779 (1991).

[15] F. Igloi, *Phys. Rev.* **A45**, 7024 (1992).

[16] F. Igloi, *Europhys. Lett.* **19**, 305 (1992).

[17] A. R. Veal, J. M. Yeomans and G. Jug, *J. Phys.* **A23**, L109 (1990).

[18] P.-M. Binder, A. L. Owczarek, A. R. Veal and J. M. Yeomans, *J. Phys.* **A23**, L975 (1990).

[19] D. P. Foster, *J. Phys.* **A23**, L1135 (1990).

[20] F. Igloi, *Phys. Rev.* **A43**, 3194 (1991).

[21] D. P. Foster and J. M. Yeomans, *Physica* **A17**, 443 (1991).

[22] D. P. Foster, *J. Stat. Phys.* **70**, 1029 (1993).

[23] R. Brak, A. J. Guttmann and S. G. Whittington, *J. Phys.* **A25**, 2437 (1992).

[24] R. Zwanzig and J. I. Lauritzen, *J. Chem. Phys.* **48**, 3351 (1968).

[25] J. I. Lauritzen and R. Zwanzig, *J. Chem. Phys.* **52**, 3740 (1970).

[26] A. L. Owczarek and T. Prellberg, *Physica* **A205**, 203 (1994).

[27] F. Igloi, *Europhys. Lett.* **16**, 171 (1991).

[28] T. Prellberg, A. L. Owczarek, R. Brak and A. J. Guttmann, *Phys. Rev.* **E48**, 2386 (1993).

[29] A. L. Owczarek, T. Prellberg and R. Brak, *J. Stat. Phys.* **72**, 737 (1993).

[30] A. L. Owczarek, *J. Phys.* **A26**, L647 (1993).

[31] D. P. Foster, *Phys. Rev.* **E47**, 1441 (1993) .

[32] T. Prellberg and R. Brak, *J. Stat. Phys.* **78**, 701 (1995).

[33] R. Brak and A. L. Owczarek, *J. Phys.* **A28**, 4709 (1995).

[34] M. Kardar, *J. Appl. Phys.* **61**, 3601 (1987); in *Disorder and Fracture*, J. C. Charmet, S. Roux and E. Guyon, eds. (Plenum, New York, 1990).

Part VI: Diffusion and Transport in One Dimension

Editor's note

In the field of diffusional transport [1-5], problems that are overtly one-dimensional, or reduce to 1D by symmetry, form the core of many modern expositions. The contents in this Part have been limited, therefore, to a selection of *recent* 'hot topics'. Chapter 17 describes exact results for certain 1D diffusion-in-potential models as well as for the Fokker-Planck equation with a simple choice of the potential. The emphasis is on the Laplace transform method.

Chapter 18 reviews resonant activation and ratchets. We note that 'classical' 1D diffusion finds numerous practical applications [1-5]. Chapter 18 also describes new applications that offer novel experimental connections in biology and chemistry. Finally, Ch. 19 is devoted to random walks in a random environment.

[1] N. G. van Kampen, *Stochastic Processes in Physics and Chemistry* (North-Holland, Amsterdam, 1981).
[2] C. Gardiner, *Handbook of Stochastic Methods* (Springer, Berlin, 1983).
[3] H. Risken, *The Fokker-Planck Equation* (Springer, Berlin, 1984).
[4] H. S. Carslaw and J. C. Jaeger, *Conduction of Heat in Solids* (Oxford University Press, 1988).
[5] B. D. Hughes, *Random Walks and Random Environments*, Vol. 1, *Random Walks* (Oxford University Press, 1995).

17

Some recent exact solutions of the Fokker-Planck equation

Harry L. Frisch

17.1 Introduction

This chapter deals with special, exact solutions of the Fokker-Planck equation (FPE) *obtained using Laplace transform methods*. Most of these will deal with the FPE in one-dimensional (1D) configuration space, sometimes called the Smoluchowski equation. Later we shall also be concerned with the case of the FPE of particles in phase space in which the initial temperature of the particles is the same as the medium, this equality being maintained for all times; the so-called 'thermalized' FPE.

Before we do this we shall study diffusion over barriers, which in its simplest form can be treated as overdamped motion in a 1D potential, $U(x)$. In Brownian-motion type notation, the FPE for the probability density $P(x, t)$ of the particle reads

$$\Gamma \frac{\partial P}{\partial t} = kT \frac{\partial^2 P}{\partial x^2} + \frac{\partial}{\partial x} \left(P \frac{\partial U}{\partial x} \right), \qquad (17.1)$$

where $U(x)$ is the potential and Γ equals the particle mass times the viscous friction coefficient. These coefficients are usually combined into the 'noise strength' parameter

$$D = kT/\Gamma \qquad (17.2)$$

and the dimensionless potential U/kT. Together with appropriate boundary conditions this determines P. Using this formulation one can study fluctuation-induced transitions between simultaneously stable states, separated by a barrier, which are of great importance in many fields [1-16] such as chemical kinetics [1,2], electric circuits [3], condensed matter [4,5], laser physics [6] and geophysics [7-9].

337

The steady state of the FPE (17.1) can be solved exactly, but not the time-dependent behavior. Kramers' seminal work [10] led to a frequently used approximation for the latter, valid particularly if two stable states are separated by a large barrier whose height ΔU satisfies $\Delta U \gg D$. Then the process of jumping from one basin of attraction to another stable state is described by an exponential time dependence with a characteristic Kramers time

$$\tau \sim \exp(\Delta U/D), \tag{17.3}$$

with prefactors determined by the shape of the potential function at the initial basin of the minimum and the peak of the barrier. A detailed survey of the Kramers problem can be found in reference [16].

In addition to asymptotic results, the validity of the Kramers approximation has been tested by analytical (in some cases, exact closed-form) solutions for model potentials [12,17-22]. These potentials include the double square-well, the double parabolic, and certain other shapes. Most of these studies use the transformation of the FPE to Schrödinger form [23]. The Kramers time can then be related to the inverse of the leading spectral gap of the Fokker-Planck operator. This asymptotic Kramers regime within the Schrödinger-form approach has also been studied for the W-shaped potential [19] with which we will be concerned in some detail below.

An alternative view of the Kramers theory can be obtained if one focuses, following Gardiner [13], on the total (normalized) probability mass of a given basin of attraction, $p_j(t)$, where $\sum_j p_j(t) = 1$. In this view Kramers' theory is an adiabatic approximation, where $p_j(t)$ varies slowly as compared to diffusion time scales within a basin of attraction, and its time evolution can be described by a master rate equation

$$\frac{dp_j}{dt} = \sum_{i \neq j} \left(\frac{p_i}{\tau_{i \to j}} - \frac{p_j}{\tau_{j \to i}} \right), \tag{17.4}$$

where the $\tau_{i \to j}$ are the appropriate Kramers times.

In some instances the Schrödinger (quantum mechanical) representation leads to difficulties, e.g., when corrections to the Kramers asymptotic limit are considered [22,24-26], involving a continuum of excited states [22,27]. For regular potentials $U(x)$ diverging faster than $\sim |x|$ at both $x = \pm\infty$ these corrections are controlled by the next-to-leading spectral gap (provided $\Delta U \gg D$ holds). But for 'soft potentials' [27] diverging in at least one of the limits $x = \pm\infty$ as $|x|^\alpha$, with $\alpha \leq 1$, which include the bistable piecewise linear W-shaped potential, the Schrödinger formulation involves the calculation of contributions due to a continuum of eigenstates [27], while

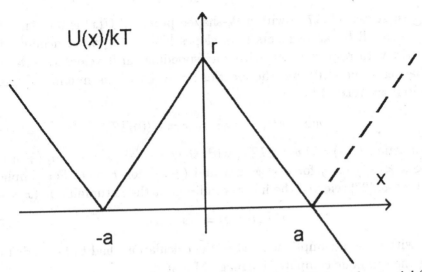

Fig. 17.1. The horizontal-S potential (solid line) and the W-shaped potential (solid line for $x \leq a$, broken line for $x > a$).

the leading Kramers times are associated with the discrete spectrum. We will illustrate this in the Laplace transform nomenclature by explicit calculations for the W-potential [22] (shown in Fig. 17.1) in Sec. 17.2.

Another class of potentials for which the transformation to the quantum mechanical representation of the FPE is difficult to implement, because of the unusual boundary conditions that one must impose, are unstable potentials such as the horizontal-S potential shown in Fig. 17.1. Very few 'quantum mechanical' exact analytical results have been reported [19] for such 'metastable' potentials. For multiple-barrier potentials (see references in Ch. 18) there has been recent interest [28] in various survival-probability type quantities in connection with the studies of general models of metastability [29-31]. Laplace transform methods could also be applied to these multibarrier cases. Single-barrier potentials are of importance in applications such as the surface adsorption of atoms and molecules on metals [32], colloid particle adhesion at substrates of various composition [33,34], etc. Exact analytical results for single-barrier potentials are scarce [19] hence we will review in some detail in Sec. 17.3 exact results for the time variation of some quantities for the horizontal-S potential of Fig. 17.1. In Sec. 17.4 we review the effective-width concept of a barrier and in Sec. 17.5 we will discuss some results for the thermalized FPE and follow that by a short concluding section.

17.2 The W potential

The easiest case of (17.1) with a W-shaped potential $U(x)$ is the symmetric case where all linear segments have slopes $|U'| = w$ (primes denote differentiation with respect to x), with corresponding 'drift velocities' $\pm w$. The tip barrier value $U(0)$ and the locations $x = \pm a$ of the minima of this W potential are related by

$$wa = rD, \qquad \text{where } r = U(0)/kT. \qquad (17.5)$$

The function $U(x) = \pm wx \pm kTr$, with signs $(-,-)$ for $x < -a$, $(+,+)$ for $-a < x < 0$, $(-,+)$ for $0 < x < a$ and $(+,-)$ for $x > a$. For simplicity Frisch *et al.* [22] selected the initial condition at the left minimum $(x = -a)$

$$P(x, t = 0) = \delta(x + a). \qquad (17.6)$$

Even with these assumptions most of the calculations had to be carried out using the symbolic computer language MACSYMA.

In terms of the Laplace transform [22]

$$F(x, s) = \int_0^\infty e^{-st} P(x, t) dt,$$

(17.1) becomes (noting (17.2) and (17.6))

$$sF = DF'' + \left(\frac{D}{kT}\right)(U'F)' + \delta(x + a). \qquad (17.7)$$

Without loss of generality we will use in the rest of this section the units in which $\Gamma = 1$ and the distinction between D and kT can be neglected [cf. (17.2)]. In the four regions between the points $x = -\infty, -a, 0, a, +\infty$, (17.7) reduces to $DF'' \pm wF' - sF = 0$, where the sign is given by the slope of $U(x)$ in the appropriate region.

Then the general solutions in the four regimes of interest are given by

$$F = A_0(s)e^{-(q+r)x/a}, \quad x > a, \qquad (17.8)$$

$$F = B_1(s)e^{(q+r)x/a} + B_2(s)e^{-qx/a}, \quad 0 < x < a, \qquad (17.9)$$

$$F = C_1(s)e^{qx/a} + C_2(s)e^{-(q+r)x/a}, \quad -a < x < 0, \qquad (17.10)$$

$$F = E_0(s)e^{(q+r)x/a}, \quad x < -a, \qquad (17.11)$$

where $q(s)$ is suggested by the appropriate characteristic equation,

$$q \equiv \frac{a}{2D}\left(\sqrt{w^2 + 4Ds} - w\right). \qquad (17.12)$$

The coefficients $A_0, B_{1,2}, C_{1,2}, E_0$ are determined by the continuity of $P(x,t)$ and of the probability current at $x = 0, \pm a$. The continuity of $P(x,t)$ implies the continuity of $F(x,s)$ and yields three equations. The continuity of the current is enforced by integrating (17.7) over the x-ranges $-a \pm \epsilon$, $0 \pm \epsilon$, and $a \pm \epsilon$, with infinitesimal ϵ. This yields the remaining three equations for the six unknown coefficients. For example, integration near $x = -a$ gives $(DF' + wF)_{x=-a_+} - (DF' - wF)_{x=-a_-} = -1$, where the $-a_+$ expression is calculated by putting $x = -a$ in (17.10), while the $-a_-$ expression is calculated by using (17.11). Thus

$$A_0 = r(r + 2q)^2 e^{2r+3q}/H, \qquad B_1 = r^2(r + 2q)e^q/H,$$

$$B_2 = 2qr(r + 2q)e^{r+3q}/H, \qquad C_1 = -2qr^2 e^q \left(e^{r+2q} - 1\right)/H,$$

$$C_2 = re^q \left[r^2 + 4q(r + q)e^{r+2q}\right]/H,$$

$$E_0 = re^{r+q} \left[2qr + r(r - 2q)e^{r+2q} + 4q(r + q)e^{2(r+2q)}\right]/H, \qquad (17.13)$$

where

$$H \equiv 2qw \left[r + 2qe^{r+2q}\right] \left[2(q + r)e^{r+2q} - r\right].$$

These expressions yield a complete solution of the time-dependent problem in the Laplace-transformed form. Since the consideration of the t-dependence, obtained by the inverse Laplace transform, is in itself nontrivial, the authors of [22] focus on one quantity: the probability mass $N_+(t)$ that the diffusing particle is found in the basin of attraction of the minimum at $x = a$ (while the initial conditions were $x = -a$), which is the inverse Laplace transform of

$$N_+(s) = \int_0^\infty dx F(x,s) = \frac{ar(r + 2q)e^q}{2qw(r + q)(r + 2qe^{r+2q})}. \qquad (17.14)$$

The inverse Laplace transform of $N_+(s)$, with the s-dependence entering via (17.12), involves a complex-plane integration over a contour $s = \text{Re}(s) + i\sigma$, with fixed $\text{Re}(s) > 0$, while σ is varied in $(-\infty, +\infty)$. As usual, we shift the contour to run counterclockwise around the singularities of $N_+(s)$ on the relevant Riemann sheet, which are all at $\text{Re}(s) \leq 0$. The most obvious singularity is the branch point at $s_B = -w^2/(4D)$, which suggests that the appropriate Riemann sheet is $\text{Re}\left[(w^2 + 4Ds)^{1/2}\right] \geq 0$, provided the branch-cut is selected along the negative real axis, at $-\infty < \text{Re}(s) < s_B$. Another obvious singularity is the simple pole at $s = 0$ due to the factor q

in the denominator of (17.14), the residue of which can be calculated and, as expected, is just $N_+(\infty) \equiv 1/2$.

Possible singularities due to other factors in the denominator arise. First, the term $r + q$ yields a simple pole at $s = 0$ on the second Riemann sheet, which is of no consequence in the calculations. Secondly, considering the equation $r + 2qe^{r+2q} = 0$, the root corresponding to $2q = -r$ (coinciding in fact with $s = s_B$), is canceled by the factor $r + 2q$ in the numerator. In addition, however, this equation has one real root and an infinite number of complex roots. One can establish by graphical analysis of the corresponding equations for $\mathrm{Re}(q)$ and $\mathrm{Im}(q)$ that the complex roots always lie in the second Riemann sheet of the complex-s plane. (This is in fact a nontrivial analysis [22], details of which are not given here.)

The remaining real root $s_P(r)$ is on the correct Riemann sheet only for $r > 1$. For large r, one has

$$s_P(r) \simeq -\frac{w^2}{2D}e^{-r}, \qquad r \gg 1. \tag{17.15}$$

The residue of the corresponding pole in $N_+(s)$ is given in this limit by $-1/2 + o(e^{-r})$. However, as r is decreased towards $r = 1$, the function $s_P(r)$ approaches the value s_B. This function cannot be obtained in closed form. Its general trend for $r \geq 1$ is similar to the approximate expression (17.15) except that near $r = 1^+$ the full function has zero slope. The corresponding pole in the complex-s plane approaches and enters the tip of the branch-cut as $r \to 1^+$, and emerges on the second Riemann sheet for $0 < r < 1$.

The above analysis suggests that for $r > 1$ one can generally write $N_+(t) - 1/2 = N_P(t) + N_B(t)$, where the notation is self-explanatory. The pole contribution, which has an exponential time dependence, dominates for long times when $r \gg 1$, corresponding to $U(0) \gg D$. In this limit it is given by

$$N_P(t) = \mathrm{residue}(r) \times \exp\left[-|s_P(r)|t\right] \simeq -\frac{1}{2}\exp\left(-\frac{w^2e^{-r}}{2D}t\right). \tag{17.16}$$

The contribution due to the branch-cut, obtained by integrating the discontinuity along it in a standard fashion, can be evaluated for long times. The result of this lengthy calculation is

$$N_B \approx -\frac{a\sqrt{D}e^{-r/2}}{\sqrt{\pi}w^2(r-1)^2}t^{-3/2}\exp\left(-\frac{w^2}{2D}t\right), \qquad r \neq 1, \; t \to \infty. \tag{17.17}$$

Note that N_B becomes comparable to N_P when $r > 1$ is $O(1)$.

For $r \leq 1$, there is no pole contribution, so that formally one can put $N_P \equiv 0$. The branch-cut contribution is then the only time dependence present. It turns out that the long-time limiting form for $r < 1$ is still given by (17.17). However, at $r = 1$ a special behavior is found,

$$N_B \approx -\frac{2\sqrt{a}}{\sqrt{\pi w e}} t^{-1/2} \exp\left(-\frac{w^2}{2D}t\right), \qquad r = 1, \; t \to \infty. \tag{17.18}$$

Finally, one notes that the short-time behavior of $N_+(t)$ can be calculated by a different approach, not detailed here. The leading-order result is

$$N_+(t \to 0) \approx \frac{e^{-r}\sqrt{t}}{\sqrt{\pi a D^{3/2}}} \exp\left(-\frac{a^2 D}{2t}\right), \tag{17.19}$$

illustrating the expected essential singularity at $t = 0$.

The authors of [22] compare the explicit expressions with the results available in the literature. First, they note that the complex-plane pattern of singularities in the Laplace transform formulation is reminiscent of the eigenvalue spectrum in the Schrödinger form [19]. Indeed, $|s_P(r)|$ corresponds to the gap due to the discrete excited state, while $|s_B|$ equals the gap value of the edge of the continuous spectrum.

Next they consider the case of large r. Turning to (17.4) one notes that in the present case, due to the symmetry of the potential, the two time parameters τ are equal. By using the relation $N_- = 1 - N_+$, one can reduce the two equations ($j \leftrightarrow \pm$) to the form

$$\frac{d}{dt}N_+(t) = -\frac{2}{\tau}\left[N_+(t) - \frac{1}{2}\right].$$

Thus in the symmetric case there is an extra factor of 2 relating the Kramers time and the decay rate in, e.g., (17.16). Calculations thus suggest $\tau = (4D/w^2)e^r$.

The original Kramers formulation [10], when put in the notation of [22], suggests the relation

$$\tau = D^{-1}\left[\int_{-a}^{a} dx \exp\left(\frac{\Delta U(x)}{D}\right)\right]\left[\int_{-\infty}^{\infty} dx \exp\left(-\frac{\Delta U_-(x)}{D}\right)\right], \tag{17.20}$$

where $\Delta U(x) \equiv U(x) - U(-a) = U(x)$; see also [26] for a discussion of relations of this kind. The first integral here is over the barrier region and near the tip is usually approximated by the quadratic expansion. In this case the barrier is not quadratic. However, the potential is simple enough to calculate the exact value, $(2D/w)e^r + O(1)$. The second integral is over the region around the minimum at $x = -a$, and $\Delta U_-(x)$ is a *single-minimum potential* that describes the shape of $\Delta U(x)$ at $-a$. Again, the usual quadratic approximation does not apply here. One takes instead $\Delta U_-(x) = w|x + a|$.

The second integral in (17.20) then yields $2D/w$. Collecting the results, a τ value is obtained consistent with the exact solution and with the Markovian character of the process governing the transitions between the two attractors. When on the other hand r is of the order of unity, the dynamics is non-Markovian.

In summary, the work [22] illustrated by exact calculations for the W-shaped potential how the Kramers regime emerges as the leading order in some approximate limit. It also showed how the corrections to this result can be associated with the complex-plane singularities in the Laplace transform formulation—which is particularly useful in cases when the Schrödinger formulation involves a continuous eigenspectrum—and can even dominate the long-time behavior under certain conditions.

17.3 The horizontal-S potential

Avoiding unilluminating mathematical complications [36] the horizontal-S potential, shown in Fig. 17.1, can be taken with the same drift velocity magnitude $w = |U'|/\Gamma = rD/a$ as for the W potential case. Thus the magnitude of the slope is the same in the three regions $x < -a$, $-a < x < 0$, $x > 0$, where $-a$ is the location of the minimum of the potential. The particle is again taken to be located initially at the minimum of the well, cf. (17.6). The Laplace transform of the probability distribution, $F(x, s)$, satisfies again [36] equation (17.7), which, except at $x = 0, -a$, reduces as before to $DF'' \pm wF' - sF = 0$. The plus sign holds in $-a < x < 0$ and the minus sign in the other two regions. Note that we have not assumed $\Gamma = 1$ here. The solution of the above differential equation (or (17.7)) is easily written, in terms of dimensionless variables stemming from the characteristic equation,

$$z = \frac{4a^2}{r^2 D} s, \qquad q = \frac{r}{2}\left(\sqrt{1+z} - 1\right), \tag{17.21}$$

in three regions of the variable, as

$$F = E(s)e^{(q+r)x/a} \quad \text{for} \quad x < -a,$$

$$F = C_1(s)e^{qx/a} + C_2(s)e^{-(q+r)x/a} \quad \text{for} \quad -a < x < 0,$$

$$F = B(s)e^{-qx/a} \quad \text{for} \quad x > 0. \tag{17.22}$$

These expressions contain four coefficients. Two conditions for the coefficients are imposed by the continuity of the probability distribution, and thus of F, at $x = 0$, $-a$. The remaining two conditions are provided by the current conservation at these x values. In the Laplace-transformed form, this amounts to integrating (17.7) over the infinitesimal ranges $0 \pm \epsilon$ and $-a \pm \epsilon$ to derive the appropriate matching conditions.

Using MACSYMA the four coefficients B, E, and $C_{1,2}$, are explicitly obtained, using

$$\Omega = 4q(r+q)e^{r+2q} + r^2,$$

as

$$(\Omega D/a)B = (r+2q)e^q, \quad (\Omega D/a)C_1 = re^q,$$

$$(\Omega D/a)C_2 = 2(r+q)e^q, \quad (\Omega D/a)E = e^{r+q}\left[2(r+q)e^{r+2q} - r\right]. \quad (17.23)$$

These results provide the solution of the diffusion problem in Laplace-transformed form.

The principal interest of the authors studying the horizontal-S potential [36] was in the large-time behavior of the probability $p(t)$ for a particle initially in the $x < 0$ well to remain there up to time t (cf. Fig. 17.1), the survival probability. Asymptotically, for large barriers $U(0) \gg kT$, this probability should obey the rate equation

$$dp/dt \approx -p/\tau, \quad (17.24)$$

where τ is the Kramers time parameter $\tau \sim \exp[U(0)/kT]$. Thus using (17.22), (17.23) the Laplace transform of $p(t)$ is found:

$$\rho(s) = \int_0^\infty e^{-st}p(t)dt = \int_{-\infty}^0 F(x,s)dx$$

$$= \frac{a^2}{(r+q)D\Omega}\left[r(r+q)e^{r+2q} + r^2\frac{1-e^q}{q} - (3r+2q)e^q\right]. \quad (17.25)$$

The complex-plane integration involved in the inverse Laplace transform from $\rho(s)$ to $p(t)$ is over a contour that can be shifted to run counterclockwise around the singularities of $\rho(s)$ on the relevant Riemann sheet. It is convenient to formulate most of the results in terms of the dimensionless variable $z \propto s$, defined in (17.21). The form of this equation suggests a branch-cut at $z_B = -1$ or $s_B = -r^2D/(4a^2)$. One can identify the Riemann sheet $\text{Re}[(1+z)^{1/2}] \geq 0$ as appropriate for the inverse-transform integration, assuming that the branch-cut is selected along the negative real axis, at $-\infty < \text{Re}(z) < -1$. The term $r+q$ in the denominator of ρ (see (17.25)) vanishes at $z = 0$, but on the second Riemann sheet. Thus besides the branch-cut just identified, the only other source of singularities of $\rho(z)$ is zeros of $\Omega(z)$ in the denominator.

Consider first the real-z roots of $\Omega(z)$. One of them is obviously at $z = -1$, just at the tip of the branch-cut. However, this divergence is canceled by a zero of the numerator of (17.25) so that the branch-cut singularity is, in fact, of the form

$$\rho(s) = -\frac{2a^2}{r^3D}\left[\left(2r + e^{-r/2}\right) + \left(\frac{r+1}{r}e^{-r/2}\right)\epsilon + O(\epsilon^2)\right], \qquad (17.26)$$

where $\epsilon \equiv (1+z)^{1/2}$.

Another real root is at $z = Z(r)$, where the numerical values of the function Z, which only depends on the dimensionless barrier-tip value r, are shown in Fig. 17.2. This function cannot be obtained in a closed analytical form. For small r, one has $Z(r) \approx -1 + r^2$, while for large r values, $Z(r) = -e^{-r} + O\left(re^{-2r}\right)$. The appropriate singularity in $\rho(s)$ is of simple-pole type,

$$\rho(s) \approx R(r)/(s - s_P), \qquad (17.27)$$

where

$$s_P = r^2 DZ(r)/(4a^2). \qquad (17.28)$$

The residue function $R(r)$ is shown in Fig. 17.3. For small r, it increases linearly, $R(r) \approx r$. For large r, $R(r) = 1 + (r-3)e^{-r}/4 + O\left(r^2e^{-2r}\right)$. It is interesting to note that the residue function has a broad, not pronounced, maximum near $r \simeq 3.6$, with $R \simeq 1.006$.

With regard to the complex roots of $\Omega(z)$, they all turn out to lie on the second Riemann sheet. This conclusion can be reached [36] by a graphical analysis of the equations for $\mathrm{Re}(q)$ and $\mathrm{Im}(q)$, which is rather cumbersome and is not reproduced here. These roots are all of the form $q = (a_k - r + ib_k)/2$, where $k = 1, 2, \ldots$, and $(2k-1)\pi < b_k(r) < 2k\pi$, $a_k(r) < 0$, for $0 < r < \infty$. These complex-q values correspond to the z (or s) values on the second Riemann sheet.

Thus the only 'physical' singularities are at s_P and s_B. The pole contribution dominates the inverse Laplace transform for large times t, $p(t) \approx R(r)\exp\left[s_P(r)t\right]$. The correction is from the branch-cut integration, and is of order $t^{-3/2}\exp\left(s_Bt\right)$ for large t. From Fig. 17.2, it is clear that the larger is r, the more dominant will the pole contribution be. Thus the approximate rate equation (17.1) becomes accurate for large r.

In the large-r limit, we can use (17.27) and (17.28) to write the dominant contribution explicitly,

$$p(t) \approx \exp\left(-\frac{r^2De^{-r}}{4a^2}t\right). \qquad (17.29)$$

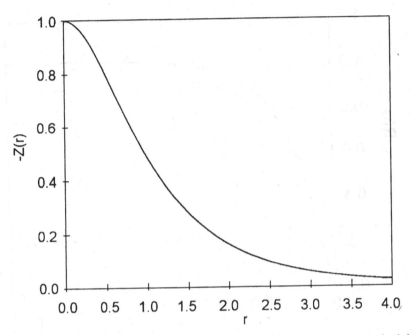

Fig. 17.2. Numerical values $Z(r)$ of the variable z at the simple pole of $\rho(z)$ as a function of the dimensionless parameter $r = U(0)/kT$.

However, this is precisely the limit of applicability of the Kramers theory [10]. Indeed, the Kramers time parameter τ entering (17.1) has been calculated for a piecewise linear barrier [22] and the result in our notation is $\tau = 4a^2 e^r/(r^2 D)$. Thus the exact asymptotic result (17.29) and the Kramers theory yield the same τ value, up to corrections of relative magnitude of order τ^{-1}.

Finally, another interesting observation follows from the exact results of [36]. Indeed, if one considers the full probability distribution $P(x, t)$, then for metastable potentials there is no stationary state. Much discussion has been devoted in the literature (for a review see [35]) to the issue of definition of a 'restricted' probability distribution for the metastable state, i.e., for particles that have not yet escaped over the barrier at time $t > 0$. An attempt to define such a distribution by isolating the part of the probability density that has a simple exponential time dependence such as in (17.24) fails because the resulting space variation cannot be normalized. Indeed,

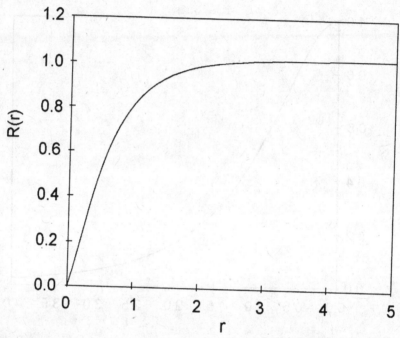

Fig. 17.3. Numerical values of the residue $R(r)$ of $\rho(s)$ at the real-s pole as a function of r.

in our case one can see explicitly that $q(s_P)$ is negative and therefore the resulting x-dependence (see (17.22)) for $x > 0$ is exponentially divergent in the limit $x \to \infty$.

17.4 The effective barrier width for a Brownian particle

Shlyakhtov *et al.* [37] have defined the effective width of a barrier, L, i.e., the barrier penetration coefficient; this appears in the long-time $(t \to \infty)$ probability, $w(t)$, that a single Brownian particle diffusing on an infinite line (with a position-dependent diffusion coefficient $D(x)$), having started from a point $-x_0 < 0$ to the left of a barrier ends up to the right of the barrier. Thus asymptotically

$$w(t) = \int_h^\infty F(x,t)dx \simeq \frac{1}{2} - \frac{x_0 + L}{\sqrt{4\pi D_0 t}} + O\left(t^{-3/2}\right), \tag{17.30}$$

as $t \to \infty$. In (17.30), $F(x,t)$ is the solution of the FPE in configuration space and D_0 is the constant value of the diffusion coefficient to the left of the barrier. The barrier is described by the potential energy (measured in units of kT) $U(x)$, with U vanishing for $x < 0$ and $x > h$.

For the simple rectangular barrier $U(x) = U_0$ in $0 \le x \le h$ and $D(x) = D_0$ one can easily obtain the exact solution of the FPE (17.1) by using either the method of reflections [37] or a Laplace transform. One finds [37], with $\alpha = U_0/2$ the half-barrier height,

$$
F(x < 0, t) = \frac{1}{\sqrt{4\pi D_0 t}} \left\{ \exp\left[-\frac{(x + x_0)^2}{4 D_0 t} \right] + \tanh \alpha \exp\left[-\frac{(x - x_0)^2}{4 D_0 t} \right] \right.
$$
$$
\left. + \sum_{n=1}^{\infty} \frac{\tanh^{2n-1} \alpha}{\cosh^2 \alpha} \exp\left[-\frac{(x + x_0 + 2nh)^2}{4 D_0 t} \right] \right\}; \qquad (17.31)
$$

$$
F(0 < x < h, t) = \frac{1 - \tanh \alpha}{\sqrt{4\pi D_0 t}} \sum_{n=1}^{\infty} \left\{ \tanh^{2n} \alpha \exp\left[-\frac{(x + x_0 + 2nh)^2}{4 D_0 t} \right] \right.
$$
$$
\left. - \tanh^{2n+1} \alpha \exp\left[-\frac{(x - x_0 - 2(n+1)h)^2}{4 D_0 t} \right] \right\}; \quad (17.32)
$$

$$
F(x > h, t) = \frac{\cosh^{-2} \alpha}{\sqrt{4\pi D_0 t}} \sum_{n=1}^{\infty} \tanh^{2n} \alpha \exp\left[-\frac{(x + x_0 + 2nh)^2}{4 D_0 t} \right]. \qquad (17.33)
$$

Note that this solution is not changed by changing the sign of U_0. Using (17.31)-(17.33) in (17.30) one finds an exact expression, $L = h \cosh U_0$. In [37] it is shown that asymptotically for any barrier $U(x)$ and diffusion coefficient $D(x)$,

$$
L = \frac{1}{2} \left[\int_0^h \frac{D_0}{D(x)} e^{U(x)} dx + \int_0^h e^{U(x)} dx \right], \qquad (17.34)
$$

and this figure of merit is comparable in usefulness as a penetration coefficient with the Daynes time lag [38].

17.5 Some exact solutions of the thermalized FPE in phase space

The Fokker-Planck equation for the time-dependent distribution $F(r, u, t')$ of the positions r and velocities u of Brownian particles at time t', moving through a medium at temperature T and in an external potential $V(r)$, has the form

$$
\frac{\partial F}{\partial t'} + u \frac{\partial F}{\partial r} - \frac{\partial V}{\partial r} \frac{\partial F}{\partial u} = \xi \frac{\partial}{\partial u} \left(u + \frac{kT}{m} \frac{\partial}{\partial u} \right) F, \qquad (17.35)
$$

where m denotes the mass of a particle, ξ denotes the friction coefficient, and k is Boltzmann's constant. If the potential depends only on one coordinate, $V(r) = V(x_1)$, and the system is bounded by the impenetrable plane wall $x_1 = 0$, one imposes the following boundary condition for mixed specular and diffusive reflection [39]:

$$F(x_1 = 0, x_2, x_3, u_1, u_2, u_3, t') = \qquad (17.36)$$

$$aF(x_1 = 0, x_2, x_3, -u_1, u_2, u_3, t') + b\left(\frac{m}{2\pi kT}\right)^{3/2} \exp\left(-\frac{mu^2}{2kT}\right),$$

where $u_1 > 0$, the constant a is the probability of specular reflection from the wall and the constant b is determined from the condition of vanishing flux perpendicular to the wall.

The following initial condition was used [39],

$$F(r, u, t' = 0) = \rho\delta(x_1 - x_0)\left(\frac{m}{2\pi kT}\right)^{3/2} \exp\left(-\frac{mu^2}{2kT}\right). \qquad (17.37)$$

Thus initially particles are uniformly distributed on the plane $x_1 = x_0$ with surface density ρ. The initial velocity distribution is the equilibrium Maxwellian at the temperature T of the medium. Under these conditions, the solution of (17.35) is given by the simple product form

$$F(r, u, t') = \rho f(x_1, u_1, t')\left(\frac{m}{2\pi kT}\right)^{1/2} \exp\left(-\frac{mu_2^2}{2kT}\right)$$

$$\times \left(\frac{m}{2\pi kT}\right)^{1/2} \exp\left(-\frac{mu_3^2}{2kT}\right), \qquad (17.38)$$

where $f(x_1, u_1, t')$ satisfies the 1D FPE.

It is convenient to write this FPE in a dimensionless form using the mean thermal velocity $\bar{u} = (kT/m)^{1/2}$ as the characteristic scale for velocity and $\tau = \xi^{-1}$ for the characteristic time scale:

$$\frac{\partial f}{\partial t} + v\frac{\partial f}{\partial x} - \frac{d\Phi}{dx}\frac{\partial f}{\partial v} = \frac{\partial}{\partial v}\left(v + \frac{\partial}{\partial v}\right)f, \qquad (17.39)$$

where $v = u_1/\bar{u}$, $x = x_1/(\bar{u}\tau)$, $t = t'/\tau$, and $\Phi(x) = V(x)/(kT)$ are the dimensionless velocity, position, time, and potential, respectively. Frisch and Nowakowski [39] restricted themselves to a system of Brownian (noninteracting) particles, thermalized by the surrounding medium. That means that the initial temperature of the particles is the same as that of the medium, T, cf. (17.37), and this equality is maintained for all time.

The solution of (17.39) is obtained in [39] for a linear and a quadratic potential (leading to a metastable system). Since Hermite polynomials are eigenfunctions of the Fokker-Planck operator [40] on the r.h.s. of (17.39) the authors of [39] expanded f in a series of Hermite polynomials,

$$f(x, v, t) = \sum_{i=0}^{\infty} (2\pi)^{-1/2} a_i(x, t) H_i(v) \exp(-v^2/2). \qquad (17.40)$$

The coefficients of the expansion are the moments of the local velocity distribution and can be easily calculated employing the property that the Hermite polynomials are orthogonal with the weight $\exp(-v^2/2)$. For example, the spatial density is the zeroth moment,

$$a_0(x, t) = n(x, t) = \int_{-\infty}^{\infty} H_0(v) f(x, v, t) dv = \int_{-\infty}^{\infty} f(x, v, t) dv, \qquad (17.41)$$

and the particle flux is the first moment of the expansion given by (17.40),

$$a_1(x, t) = j(x, t) = \int_{-\infty}^{\infty} H_1(v) f(x, v, t) dv = \int_{-\infty}^{\infty} v f(x, v, t) dv. \qquad (17.42)$$

The second moment calculated for the original, unscaled distribution function involves the mean square velocity and provides the local energy density. The velocity scaling absorbs the temperature kT and this quantity is just the density (or the dimensionless pressure $p(x, t)$),

$$\int_{-\infty}^{\infty} v^2 f(x, v, t) dv = p(x, t) = n(x, t). \qquad (17.43)$$

The equations for the moments $a_i(x, t)$ are derived from the FPE (17.39) by multiplying with $H_i(v)$ and integrating over velocity. Using the property that the ith term in the expansion (17.40) is the eigenfunction of the r.h.s. of the FPE with eigenvalue $-i$, the following infinite system of equations was obtained:

$$\frac{\partial a_i}{\partial t} + \frac{\partial}{\partial x} [a_{i-1} + (i+1)a_{i+1}] + \frac{d\Phi}{dx} a_{i-1} = -i a_i. \qquad (17.44)$$

This system can be transformed to a single partial differential equation for the generating function of the moments, $F(x, t, \theta) = \sum_{i=0}^{\infty} a_i(x, t) \theta^i$. Multiplying (17.44) by θ^i and summing yields the equation for the generating function,

$$\frac{\partial F}{\partial t} + \frac{\partial}{\partial x} \left(\theta F + \frac{\partial F}{\partial \theta} \right) + \frac{d\Phi}{dx} \theta F = -\theta \frac{\partial F}{\partial \theta}. \qquad (17.45)$$

The boundary condition (17.36) translates into the distribution of velocity of the particles at the wall,

$$f(x = 0, v, t) = af(x = 0, -v, t) + (1 - a)M \exp(-v^2/2) \qquad (17.46)$$

for $v > 0$. The coefficient M is found from the vanishing of the flux at the wall $j(0, t) = 0$, which yields

$$M = \int_{-\infty}^{\infty} vf(x = 0, v, t)dv. \qquad (17.47)$$

Relation (17.45) provides the boundary conditions for $a_i(x = 0, t)$,

$$\sum_{k=0}^{\infty} a_k \left\{ [1 + (-1)^{i+1}a]\delta_{ik} + \left[1 - (-1)^{i+1}a\right] x[1 - (-1)^{i+k}]C_{ik} \right\} =$$

$$(1 - a)M(2\pi)^{1/2} \left\{ \delta_{0i} + [1 - (-1)^i]C_{0i} \right\}, \quad i = 0, 1, 2, \ldots . \qquad (17.48)$$

For $a \neq 1$, system (17.48) has the unique solution, independent of the coefficients C_{in},

$$a_0(x = 0, t) = n(x = 0, t) = (2\pi)^{1/2}M, \qquad (17.49)$$

$$a_i(x = 0, t) = 0, \qquad \text{for } i \geq 1, \qquad (17.50)$$

which means that at the reflecting wall the distribution function is locally Maxwellian with a still-undetermined value of the density $n(x = 0, t)$. Equation (17.49) is then identical with (17.47).

In view of (17.43) and (17.44) the equations for the first two moments in the density $n(x, t)$ and the particle flux $j(x, t)$ of the thermalized system form a closed system for any Φ, viz.

$$\frac{\partial n}{\partial t} + \frac{\partial j}{\partial x} = 0, \qquad \frac{\partial j}{\partial t} + \frac{\partial n}{\partial x} + \frac{d\Phi}{dx}n = j.$$

The particle flux can be eliminated, resulting in a closed equation for the density,

$$\frac{\partial^2 n}{\partial t^2} + \frac{\partial n}{\partial t} - \frac{\partial}{\partial x}\left(\frac{\partial n}{\partial x} + \frac{d\Phi}{dx}n\right) = 0. \qquad (17.51)$$

This equation differs from the usual FPE in configuration space (the Smoluchowski equation) only by the term in the second derivative in time, making (17.51) a hypergeometric rather than a parabolic partial differential equation.

Equations (17.45) and (17.51) can be solved for a variety of potentials, e.g., the linear potential

$$\Phi = Kx, \qquad (17.52)$$

or the quadratic potential $\Phi = -\frac{1}{2}w^2(x-b)^2$, $b > 0$, or piecewise combinations, among others using Laplace transforms. These solutions of the thermalized FPE in phase space show that f is a linear functional of the density $n(x,t)$. We will illustrate this in some detail only for the potential given by equation (17.52), for which equation (17.45) can be solved by applying a double Laplace transform for position and time. Let the symbol $\tilde{\ }$ denote the Laplace transform for time, with parameter s, and the symbol $\hat{\ }$ denote the transform for position coordinate x, with parameter q. The equation for the double-transformed generating function $\hat{\tilde{F}}(q,s,\theta)$ has the form

$$s\hat{\tilde{F}} + \theta q\hat{\tilde{F}} + (q+\theta)\frac{\partial\hat{\tilde{F}}}{\partial\theta} + K\theta\hat{\tilde{F}} =$$

$$\hat{F}(q,t=0,\theta) + \theta\tilde{F}(x=0,s,\theta) + \frac{\partial\tilde{F}(x=0,s,\theta)}{\partial\theta}. \quad (17.53)$$

Taking into account the boundary conditions, the solution of the above equation can be written as

$$\hat{\tilde{F}}(q,s,\theta) = e^{-G(q,s,\theta)}\left\{ \int_0^\theta e^{G(q,s,\theta')}\left[\hat{F}(q,t=0,\theta') + \theta'\tilde{n}(0,s)\right]\frac{d\theta'}{q+\theta'}\right.$$

$$\left. + e^{G(q,s,0)}\hat{\tilde{n}}(q,s)\right\}, \quad (17.54)$$

where

$$e^{-G(q,s,\theta)} = (q+\theta)^{s-q^2-qK}e^{(q+K)\theta}.$$

It follows from (17.54) that the complete solution for the generating function is expressed in terms of the lowest moment of the expansion (17.40), i.e., the density. Once the solution for the density is known, all higher moments can be found using (17.54).

The initial condition for the solution of (17.51) is that the particle is located at $x_0 > 0$ at $t = 0$ and there is no initial flux $j(x,t=0) = 0$. This translates to the initial conditions on the density, $n(x,t=0) = \delta(x-x_0)$, $\frac{\partial n}{\partial t}(x,t=0) = 0$. The boundary condition on n at $x = 0$ follows from the vanishing of the flux at the reflecting wall $j(x=0,t) = 0$ as

$$\left[\frac{\partial n}{\partial x}\right]_{x=0} + \left[\frac{d\Phi}{dx}n\right]_{x=0} = 0. \quad (17.55)$$

Using these initial and boundary conditions the transformed density $\tilde{n}(x, s)$ for the linear potential (17.52) satisfies

$$\frac{\partial}{\partial x}\left(\frac{\partial \tilde{n}}{\partial x} + K\tilde{n}\right) - s(s+1)\tilde{n} = -(s+1)\delta(x - x_0). \qquad (17.56)$$

The solution of (17.56) subject to the boundary condition (17.55) is given by

$$\tilde{n} = \left[(s+1)\big/\sqrt{K^2 + 4s(s+1)}\right]\left[e^{\mu_1}(x - x_0)H(x_0 - x)\right.$$
$$\left. + e^{\mu_2}(x - x_0)H(x - x_0) - (\mu_2/\mu_1)e^{(\mu_1+\mu_2)x}\right], \qquad (17.57)$$

where the roots of the characteristic equation of (17.30) are

$$\mu_j = \frac{1}{2}\left[\sqrt{K^2 + 4s(s+1)} + (-1)^j K\right], \quad j = 1, 2. \qquad (17.58)$$

The time-dependent $n(x, t)$ is given by the inverse Laplace transform of (17.57), which involves integration in the complex plane along a line parallel to the imaginary axis. The transformed density (17.57) contains a square root which is a double-valued function in the complex domain defined differently on the two Riemann sheets. The roots of the argument of the square root in (17.58),

$$s_{1,2} = \frac{1}{2}\left(\pm\sqrt{1 - K^2} - 1\right), \qquad (17.59)$$

are the branch points, between which lies the branch-cut. For integration, we choose the sheet for which $\mathrm{Re}[(s - s_i)^{1/2}] > 0$ if $\mathrm{Re}(s - s_i) > 0$.

In the standard procedure of calculation, the contour of integration of the inverse Laplace transform is shifted towards negative values on the real axis. This leads to contributions from the counterclockwise integration around the poles and the branch-cuts of the function on the relevant Riemann sheet. The function $\tilde{n}(x, s)$ of (17.57) has a pole at $s = 0$ (the root of μ_1) and a branch-cut between branch points s_1 and s_2 given by (17.59). The contribution from the pole $s = 0$ provides the limit of $n(x, t)$ as $t \to \infty$, which is the time-independent equilibrium distribution $n_{\mathrm{eq}}(x) = Ke^{-Kx}$. The time dependence is provided by the branch-cut contribution $n_B(x, t)$. Its form depends on whether the branch points (17.59) are real or complex. For a strong potential $K > 1$, the roots (17.59) are complex and the branch-cut contribution $n_B(x, t)$ is given by

$$
n_B = \frac{1}{4\pi} \exp\left[-\frac{K(x - x_0) + t}{2}\right] \int_0^1 ds
$$

$$
\times \left\{ \frac{\cos[(r/2)t\sqrt{1 - s^2}]}{\sqrt{1 - s^2}} - \frac{r\sin[(r/2)t\sqrt{1 - s^2}]}{2} \right\}
$$

$$
\times \left\{ 2\cosh\left(r|x - x_0|s/2\right) + \frac{rs + K}{rs - K} \exp[r(x + x_0)s/2] \right\}, \quad (17.60)
$$

where $r = \sqrt{K^2 - 1}$.

The integral in (17.60) can be evaluated asymptotically in the long-time limit. This yields, for $n_B(x, t \gg 1)$,

$$
n_B \simeq \frac{(1 + r)\sqrt{r}}{4\sqrt{\pi}} \left[xx_0 + \frac{4}{K^2} - \frac{2(x + x_0)}{K} \right] t^{-3/2}
$$

$$
\times \exp\left[-\frac{K(x - x_0) + t}{2}\right] \left[\cos\left(\frac{rt}{2}\right) - \sin\left(\frac{rt}{2}\right) \right]. \quad (17.61)
$$

Equation (17.61) involves a wave propagation feature. In the weak potential case, $K < 1$, the roots (17.59) are real and the branch-cut contribution has the form

$$
n_B(x, t) = \frac{1}{2\pi} \exp\left[-\frac{K(x - x_0) + t}{2}\right] \int_{-1}^1 ds \, \frac{4r[(1/2) + rs]e^{rst}\sqrt{1 - s^2}}{4r^2(1 - s^2) + K^2}
$$

$$
\times \left\{ 2r\sqrt{1 - s^2} \cos\left[r\sqrt{1 - s^2}(x - x_0)\right] \right.
$$

$$
\left. - K \sin\left[r\sqrt{1 - s^2}(x - x_0)\right] \right\}, \quad (17.62)
$$

with $r = \sqrt{1 - K^2}$. The asymptotic dependence of (17.62) for the long-time, $t \gg 1$, limit is

$$
n_B(x, t) \simeq \frac{(1 + r)\sqrt{r}}{4\sqrt{\pi}} \left[xx_0 + \frac{4}{K^2} - \frac{2(x + x_0)}{K} \right] t^{-3/2}
$$

$$
\times \exp\left[-\frac{K(x + x_0) + t}{2}\right]. \quad (17.63)
$$

Equation (17.63) exhibits the exponential decay of $n(x, t)$ characteristic of diffusion processes. The wave propagation feature of (17.51) manifests itself in the short-time behavior of $n(x, t)$. Its form is

$$n(x,t) = \frac{1}{2} \exp\left[-\frac{K(x+x_0)+t}{2}\right] \left[\delta(x+t-x_0) + \delta(x-t-x_0)\right.$$

$$+ \frac{1}{4} I_0 \left(\frac{1}{2}\sqrt{t^2 - (x-x_0)^2}\right)$$

$$\left. + \frac{t}{4\sqrt{t^2 - (x-x_0)^2}} I_1 \left(\frac{1}{2}\sqrt{t^2 - (x-x_0)^2}\right)\right], \qquad (17.64)$$

where I_0 and I_1 denote Bessel functions.

The exact solution of the thermalized FPE in phase space has also been obtained for the rectangular barrier described in Sec. 17.4, with $\Phi = \Phi_0$ in $0 < x < h$ and $\Phi = 0$ otherwise. While this solution is changed when Φ_0 is replaced by $-\Phi_0$, the form of L (given by the appropriate analog of (17.30)) is still given by $L = h \cosh \Phi_0$.

17.6 Concluding remarks

The Laplace transform appears as a suitable method of dealing with a variety of diffusion problems described by the FPE in configuration space and, in some cases, phase space. It is noteworthy that, while diffusion within a single well with approach to a stationary state, barrier crossing in a bistable potential, and metastable diffusion all involve, within the Laplace transformed image, a simplified state described by a pole, the approach to this state is mathematically the consequence of a continuous spectrum arising from branch points in the Laplace image. Simplified figures of merit describing barrier crossing, such as the effective barrier width L or the Daynes time lag in approaching a stationary state, fail to reveal the full mathematical intricacy of these phenomena.

The author is indebted to Prof. V. Privman for his critical reading of this chapter and he acknowledges support by the National Science Foundation Grant DMR-9023541.

References

[1] *Noise in Nonlinear Dynamical Systems*, F. Moss and P. McClintock, eds. (Cambridge University Press, 1989).
[2] M. Büttiker, E. Harris and R. Landauer, *Phys. Rev.* **B28**, 1268 (1983).
[3] R. Landauer, *J. Appl. Phys.* **33**, 2209 (1962).
[4] P. Hänggi, *Z. Phys.* **B68**, 181 (1987).
[5] P. Hänggi, *J. Stat. Phys.* **42**, 105 (1986).

[6] H. Haken, *Synergetics* (Springer, Berlin, 1977).

[7] R. Benzi, G. Parisi, A. Sutera and A. Vulpiani, *Tellus* **34**, 10 (1982).

[8] C. Nicolis, *Tellus* **34**, 1 (1982).

[9] C. Nicolis and G. Nicolis, *Tellus* **33**, 225 (1981).

[10] H. Kramers, *Physica* **7**, 284 (1940).

[11] B. Matkowsky, Z. Schuss and C. Tier, *J. Stat. Phys.* **35**, 43 (1984).

[12] H. Risken, *The Fokker-Planck Equation* (Springer, Berlin, 1984).

[13] C. Gardiner, *Handbook of Stochastic Methods* (Springer, Berlin, 1983).

[14] N. G. van Kampen, *Stochastic Processes in Physics and Chemistry* (North-Holland, Amsterdam, 1981).

[15] B. Caroli, C. Caroli and R. Roulet, *J. Stat. Phys.* **21**, 415 (1979).

[16] P. Hänggi, P. Talkner and M. Borkovec, *Rev. Mod. Phys.* **62**, 251 (1990).

[17] K. Banerjee, J. K. Bhattacharjee and H. S. Mani, *Phys. Rev.* **A29**, 393 (1984).

[18] M. Razany, *Phys. Lett.* **A72**, 89 (1979).

[19] M. Mörsch, H. Risken and H. D. Volmer, *Z. Phys.* **B32**, 245 (1979).

[20] H. Brand and A. Schenzle, *Phys. Lett.* **A68**, 427 (1978).

[21] N. G. van Kampen, *J. Stat. Phys.* **17**, 71 (1977).

[22] H. L. Frisch, V. Privman, C. Nicolis and G. Nicolis, *J. Phys.* **A23**, L1147 (1990).

[23] H. Tomita, A. Ito and H. Kidachi, *Prog. Theor. Phys.* **56**, 786 (1976).

[24] W. Bez and P. Talkner, *Phys. Lett.* **A82**, 313 (1981).

[25] O. Edholm and O. Leimar, *Physica* **A98**, 313 (1979).

[26] R. S. Larson and M. D. Kostin, *J. Chem. Phys.* **69**, 4821 (1978).

[27] F. Marchesoni, P. Sodano and M. Zannetti, *Phys. Rev. Lett.* **61**, 1143 (1988).

[28] B. Gaveau and L. S. Schulman, *Lett. Math. Phys.* **18**, 201 (1989).

[29] J. S. Langer, *Ann. Phys. (NY)* **41**, 108 (1967).

[30] J. S. Langer, *Phys. Rev. Lett.* **21**, 973 (1968).

[31] J. S. Langer, *Ann. Phys. (NY)* **54**, 258 (1969).

[32] A. Zangwill, *Physics at Surfaces, Part 2* (Cambridge University Press, 1988).

[33] P. C. Hiemenz, *Principles of Colloid and Surface Chemistry*, 2nd ed. (Dekker, New York, 1986).

[34] V. Privman, H. L. Frisch, N. Ryde and E. Matijević, *J. Chem. Soc. Faraday Trans.* **87**, 1371 (1990).

[35] L. S. Schulman, in *Finite Size Scaling and Numerical Simulation of Statistical Systems*, Chap. XI, p. 489, V. Privman, ed. (World Scientific, Singapore, 1990).

[36] V. Privman and H. L. Frisch, *J. Chem. Phys.* **94**, 8216 (1991).

[37] A. V. Shlyakhtov, A. R. Khokhlov and H. L. Frisch, *Physica* **A198**, 449 (1993).

[38] H. Daynes, *Proc. Roy. Soc. London* **A97**, 286 (1920); see also J. Crank, *The Mathematics of Diffusion*, 2nd ed. (Clarendon, Oxford, 1975).

[39] H. L. Frisch and B. Nowakowski, *J. Chem. Phys.* **98**, 8965 (1993).

[40] B. Nowakowski and H. L. Frisch, *Int. J. Heat and Mass Transfer*, to be published.

18

Random walks, resonance, and ratchets

Charles R. Doering and Timothy C. Elston

Two recent developments involving activation and transport processes in simple stochastic nonlinear systems are reviewed in this chapter. The first is the idea of 'resonant activation' in which the mean first-passage time for escape over a fluctuating barrier passes through a minimum as the characteristic time scale of the fluctuating barrier is varied. The other is the notion of active transport in a fluctuating environment by so-called 'ratchet' mechanisms. Here, nonequilibrium fluctuations combined with spatial anisotropy conspire to generate systematic motion. The fundamental principles of these phenomena are covered, and some motivations for their study are described.

18.1 Introduction

The study of the interplay of noise and nonlinear dynamics presents many challenges, and interesting phenomena and insights appear even in one-dimensional (1D) systems. Examples include Kramers' fundamental theory of the Arrhenius temperature dependence of activated rate processes [1], Landauer's further insights into the role of multiplicative noise [2], and the theory of noise-induced transitions [3]. This chapter reviews more recent developments which go beyond those studies in that the characteristic time scale of the fluctuations plays a major role in the dynamics of the system, whereas the phenomena in [1-4] are fundamentally white-noise effects. Specifically, the two effects to be described in this chapter are the phenomena of 'resonant activation' and transport in 'stochastic ratchets'.

Resonant activation is a generalization of Kramers' model of activation over a potential barrier to the situation where the barrier itself is fluctuating randomly. The effect of the barrier fluctuations is to change the effective barrier height, but in a manner not altogether expected: the mean first-passage time shows a global minimum as a function of the barrier fluctuation rate (all other parameters held fixed), which is reminiscent of the resonant-like behavior identified with the matching of two time scales. This phenomenon has been seen in simple models theoretically [5] and in analog simulations [6], and has been investigated in more general settings in a number of subsequent studies [7-17].

The second effect, transport in ratchet-like potentials, provides a mechanism for the transduction of nonequilibrium fluctuations into directed motion. By 'ratchet-like' we mean a potential that is periodic but asymmetric. Once again, the potential felt by the particle randomly switches between two or more possible configurations. The resulting current generated by these systems depends not only on the spatial asymmetry of the potential but also on the statistical properties of the nonequilibrium fluctuations [18]. Recently, several models of molecular motors [19-27] and novel mass-separation techniques [28,29] based on fluctuating ratchet potentials have been proposed.

18.2 Resonant activation

The term resonant activation was coined in [5] to describe the nonmonotonic dependence of the mean first-passage time on the characteristic time scale of a fluctuating potential barrier. That is, there exists an optimal fluctuation rate that minimizes the mean first-passage time. The stylized model used in [5] to illustrate resonant activation is conceptually simple and analytically tractable. We use it, therefore, as a pedagogical example to illustrate the underlying principles of resonant activation, and then comment briefly on more general models of this phenomenon.

We begin with a very brief review of the 'escape' or first-passage problem in 1D. This problem addresses the statistical properties of the time required for noise-activated escape over a potential barrier. Figure 18.1 illustrates a typical setup for this problem. A particle is started from the local minimum of a potential located at $x = x_0$, and the mean first-passage time is the average amount of time required for the thermal fluctuations to kick the particle to the local maximum at $x = a$. 1D systems subjected to thermal noise are often described by Langevin equations with the general form

$$m\frac{d^2x}{dt^2} = -v'(x) - \beta\frac{dx}{dt} + \sqrt{\beta kT}\xi, \qquad (18.1)$$

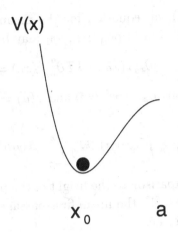

V(x)

x₀   a

Fig. 18.1. Escape over a potential barrier. The mean first-passage time of interest is the average time required for a particle started at x_0 to arrive at a.

where $v(x)$ is the potential felt by the particle, β is the friction coefficient, kT is the Boltzmann constant times the temperature, and ξ is a Gaussian white-noise term of unit strength. The dynamics of many microscopic processes is accurately described by taking the overdamped limit in which inertia can be ignored. In this limit (18.1) reduces to

$$\beta\frac{dx}{dt} = -v'(x) + \sqrt{kT\beta}\,\xi. \tag{18.2}$$

While (18.2) is physically intuitive and useful for producing single realizations of the stochastic process it describes, the statistical properties of the system are more easily obtained via the Fokker-Planck equation associated with (18.2). The Fokker-Planck equation governs the time evolution of the probability density, $\rho(x,t)$, and has the form

$$\partial_t\rho = \partial_x[v'(x)\rho] + kT\partial_x^2\rho, \tag{18.3}$$

where time has been scaled by the friction coefficient. For the problem illustrated in Fig. 18.1 the appropriate boundary conditions are reflecting to the left, $[\{v'(x) + kT\partial_x\}\rho(x,t)]_{x=-\infty} = 0$, and absorbing to the right, $\rho(a,t) = 0$. The initial condition is $\rho(x,0) = \delta(x - x_0)$. The mean first-passage time, τ, is then related to ρ through the expression [30]

$$\tau = \int_{-\infty}^{a} \int_{0}^{\infty} \rho(x,t)\,dt\,dx. \tag{18.4}$$

Using (18.3) and (18.4), an equation for the mean first-passage time as a function of the particle's initial position, x_0, may be derived:

$$-v'(x_0)\partial_{x_0}\tau(x_0) + kT\partial_{x_0}^2\tau(x_0) = -1, \tag{18.5}$$

with boundary conditions $\partial_{x_0}\tau(\infty) = 0$ and $\tau(a) = 0$. The formal solution to (18.5) is

$$\tau = (kT)^{-1}\int_{x_0}^{a} e^{-v(x)/(kT)}\int_{-\infty}^{x} e^{v(y)/(kT)}\,dy\,dx, \tag{18.6}$$

and if kT is small in comparison to the height of the potential barrier then, as first observed by Kramers [1], the mean first-passage time has an Arrhenius temperature dependence, i.e.,

$$\tau \propto \exp\left[\Delta v/(kT)\right], \tag{18.7}$$

where Δv is the potential difference between the bottom of the well and the top of the barrier.

To observe the phenomena of resonant activation, one more level of complexity must be added to the escape problem: the potential barrier felt by the particle is allowed to fluctuate in time. A simple example of this behavior is depicted in Fig. 18.2, in which the slope of a linear potential fluctuates between two values. The particle is started from a reflecting boundary at $x = 0$, and an absorbing boundary is placed at $x = 1$. To make the problem tractable, the fluctuations in the potential are taken to be governed by the master equation

$$\frac{d}{dt}\begin{pmatrix} P_+(t) \\ P_-(t) \end{pmatrix} = \gamma \begin{pmatrix} -1 & 1 \\ 1 & -1 \end{pmatrix}\begin{pmatrix} P_+ \\ P_- \end{pmatrix}, \tag{18.8}$$

where $P_\pm(t)$ is the probability that the potential is in the $v_\pm(x)$ state at time t. Equation (18.8) represents a dichotomous Markov process. The switching between the the two potential configurations, v_+ and v_-, is exponentially correlated and the switching time scale, τ_c, is determined by the rate γ, i.e., $\tau_c = \gamma^{-1}$. This allows the entire process to be described by the following Fokker-Planck equation [3]

$$\frac{\partial}{\partial t}\begin{pmatrix} \rho_+ \\ \rho_- \end{pmatrix} = \begin{pmatrix} L_+(x) - \gamma & \gamma \\ \gamma & L_-(x) - \gamma \end{pmatrix}\begin{pmatrix} \rho_+ \\ \rho_- \end{pmatrix}, \tag{18.9}$$

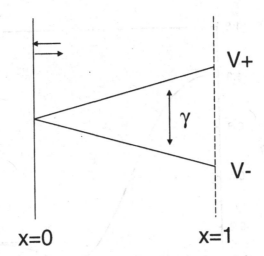

Fig. 18.2. A fluctuating linear potential.

where

$$L_\pm(x) = \partial_x[v'_\pm(x) + kT\partial_x]. \qquad (18.10)$$

Using procedures analogous to the ones described above for a stationary potential, the equations governing the mean first-passage time are found to be [5]

$$-v'_+(x_0)\partial_{x_0}\tau_+ + kT\partial^2_{x_0}\tau_+ - \gamma\tau_+ + \gamma\tau_- = -\tfrac{1}{2},$$

$$-v'_-(x_0)\partial_{x_0}\tau_- + kT\partial^2_{x_0}\tau_- - \gamma\tau_- + \gamma\tau_+ = -\tfrac{1}{2}, \qquad (18.11)$$

where

$$\tau = \tau_+ + \tau_-, \qquad (18.12)$$

and $\tau_\pm$ is the conditional average that the first passage occurs while the potential is in the $\pm$ configuration. Equations (18.11) must be solved subject to the boundary conditions $\tau_+(1) = \tau_-(1) = 0$ and $\partial_{x_0}\tau_+(0) = \partial_{x_0}\tau_-(0) = 0$, and, for the particularly simple case that the potential fluctuates between $v_\pm(x) = \pm Vx$, analytic solutions to these equations can be found [5]. In Fig. 18.3 a log-log plot of the mean first-passage time vs. the switching rate γ is shown. The parameters used to generate this figure were $kT = 1$ and $V = 8$. The two limiting behaviors (shown as broken lines in the figure) are easy to

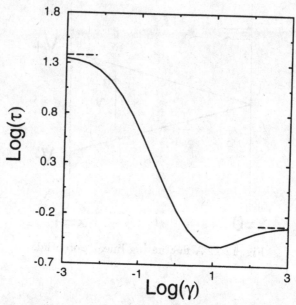

Fig. 18.3. A log-log plot of the mean first-passage time *vs.* switching rate. The broken lines represent the two limiting cases.

understand. For very slow barrier fluctuation rates the mean first-passage time approaches the average of the mean first-passage times for the alternative barrier configurations. On the other hand, for barrier fluctuations fast compared to the typical crossing time, the mean first passage time is that required to cross the average barrier. At intermediate times, however, the mean first-passage time is less than either of these two limits. The minimum in the mean first-passage time occurs when the fluctuations in the potential are roughly equal to the time required for the particle to cross when the barrier is in the 'down' configuration, thus justifying the use of the term resonance.

Certainly the linear potential used in the above example is a great simplification, but potentials that fluctuate between two configurations are ubiquitous in molecular biology. Allosteric transitions in which the binding of a ligand produces a conformational change in the protein help regulate metabolism and generate molecular motion. Resonant activation, therefore, may be useful in understanding the anomalous activation rates often encountered in biochemistry. Many other physical processes, such as molecular dissociation in strongly coupled chemical systems and the dynamics of dye lasers, take place in fluctuating environments. While the precise nature of the fluctuations is not known, the central limit theorem makes it reasonable to model them as Gaussian colored noise. There are now several theoretical treatments of escape over fluctuating barriers driven by Gaussian colored noise, and the interested reader is referred to [11-17] for details.

18.3 Ratchets

A 'stochastic ratchet' is a process in which nonequilibrium fluctuations and a spatially anisotropic periodic potential conspire to produce directed motion. As in the previous section we shall concentrate on a simple model that elucidates the general principles of these ratchets, and refer the interested reader to more biologically and physically realistic models. A major motivation to study these types of system comes from the theoretical modeling of the molecule kinesin. The kinesin molecule belongs to a class of proteins known as motor molecules. These molecules, which also include dyneins and myosin, possess the ability to move unidirectionally along structural filaments such as microtubulin and actin and, among other functions, are used for the transportation of organelles and in muscle contraction. The energy source of these molecules comes from the hydrolysis of ATP. While the simple model presented below may not accurately describe any biological process, we can loosely interpret the fluctuations in the potential as arising from the binding and dissociation of ATP, and the anisotropic periodic potential as representing the electrostatic potential along the long structural filaments.

Consider an overdamped particle that moves in a 1D periodic potential. The Fokker-Planck equation governing this process can be written in the form

$$\partial_t \rho = -\partial_x J, \tag{18.13}$$

where the probability current J is given by

$$J = -v'(x)\rho - kT\partial_x \rho. \tag{18.14}$$

For convenience the period of the potential $v(x)$ is taken to be 1 (i.e., v(0) = v(1)). The time-averaged velocity of the particle is then

$$\left\langle \frac{dx}{dt} \right\rangle = J. \tag{18.15}$$

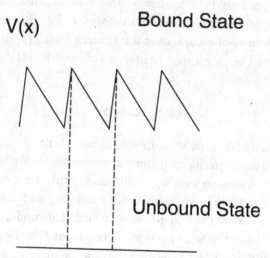

Fig. 18.4. The two possible configurations of the fluctuating potential. The first configuration is termed the bound state as the particle is localized near one of the local minimum. The second state is termed the unbound state, since the particle is free to diffuse.

Because $v'(x)$ is also periodic, we can find the steady-state distribution for this process by setting the l.h.s. of (18.13) equal to zero and solving the resulting ordinary differential equation with the boundary condition $\rho(0) = \rho(1)$ and subject to the normalization condition

$$\int_0^1 \rho(x)dx = 1. \tag{18.16}$$

In doing this we find that J is identically 0, as it must be because the corresponding steady state represents an equilibrium process.

Now consider the slightly more complicated situation in which the potential can fluctuate or 'flash' between the two different configurations shown in Fig. 18.4. The potential configuration labeled as the bound state in this figure is $v_+(x) = u(x)$, a piecewise-linear sawtooth potential. The unbound state corresponds to $v_-(x) = 0$, i.e., free diffusion. Once again the fluctuations in the potential are governed by (18.8) and the Fokker-Planck equation for the process is given by (18.9). Does this system produce a net drift, and if so in which direction? The answer to this question is displayed in Fig. 18.5 where a plot of the current vs. flashing rate is shown. If the whole con-

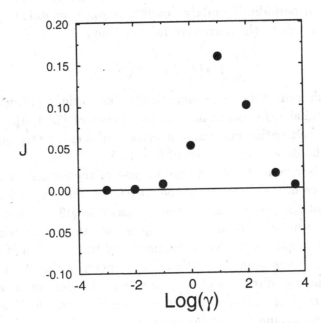

Fig. 18.5. A linear-log plot of the current *vs.* switching rate.

figuration is tilted slightly to the left, then the fluctuations in the potential can actually cause the particle to move uphill!

The mechanism for this process becomes clear when we consider the basin of attraction for one of the fixed points of the bound state. Such a basin of attraction is the region between the broken lines shown in Fig. 18.4. If the thermal fluctuations are small compared to the height of the potential barriers separating the fixed points then, when the system is in the bound state, the particle will remain close to a fixed point. When the system switches to the unbound state the particle begins to diffuse. Owing to the spatial asymmetry of $u(x)$ the particle is closer to the right-hand basin boundary, and is more likely to diffuse across this boundary. Therefore, when the potential switches back to the bound state, the particle is more likely to find itself localized near the fixed point to the right of where it was originally. In this way the particle moves systematically to the right. In the limits of very slow or fast potential fluctuations the net current falls to zero. This results from the fact that for slow fluctuations the current tends to the average of the current produced in each individual potential configuration, and for fast fluctuations the net current is a result of the averaged potential.

To calculate the current requires that the time-independent part of (18.9) be solved with periodic boundary conditions $\rho_+(0) = \rho_+(1)$ and $\rho_-(0) = \rho_-(1)$, and subject to the normalization condition

$$\int_0^1 [\rho_+(x) + \rho_-(x)]\, dx = 1. \tag{18.17}$$

In general this cannot be done analytically, but various perturbative techniques are available for capturing limiting behavior [3,18,31]. An efficient algorithm for calculating currents numerically also exists [31], and was used to generate the data points shown in Fig. 18.5.

In the above example once the mechanism of the ratchet is understood, the dependence of the direction of the current on the properties of the potential fluctuations is obvious. As first discussed in [18], for potentials that fluctuate between more than two configurations the relationship between the statistical properties of the fluctuations and the direction of the current can be subtle. It has been shown that one property of the fluctuations that may determine the direction of the current is the flatness of an additive fluctuating force [18]. The flatness, ϕ, of a random variable X is defined in terms of its second and fourth moments as

$$\phi = \frac{\langle X^4 \rangle}{\langle X^2 \rangle^2}. \tag{18.18}$$

Unexpected current reversals can be observed by varying this parameter. More detailed discussions of the dependence of the currents produced by stochastic ratchets and the statistical properties of the fluctuations can be found in [18,31,32].

There is now an extensive amount of literature devoted to stochastic ratchets [18-29,31-42]. A more complete review of the general properties of these systems can be found in [33], and theoretical articles that discuss stochastic ratchets as the basis for molecular motors include [19-27]. Descriptions of novel mass-separation techniques based on ratchet mechanisms can be found in [28,29], and an experimental (optical) realization of a ratchet is presented in [36].

References

[1] H. A. Kramers, *Physica (Utrecht)* **7**, 284 (1940). For a recent review, see P. Hänggi, P. Talkner and M. Borkovec, *Rev. Mod. Phys.* **62**, 251 (1990).
[2] R. Landauer, *Physics Today* A78, 304 (1980).

[3] W. Horsthemke and R. Lefever, *Noise Induced Transitions* (Springer, New York, 1984).

[4] P. Hänggi, *Phys. Lett.* **A78**, 304 (1980); D. L. Stein, R. G. Palmer, J. L. van Hemmen and C. R. Doering, *Phys. Lett.* **A136**, 353 (1989); D. L. Stein, C. R. Doering, R. G. Palmer, J. L. van Hemmen and R. M. McLaughlin, *J. Phys.* **A23**, L203 (1990).

[5] C. R. Doering and J. C. Gadoua, *Phys. Rev. Lett.* **69**, 2318 (1992).

[6] L. Gammaitoni, F. Marchesoni, E. Manichella-Saetta and S. Santucci, *Phys. Rev.* **E49**, 4878 (1994).

[7] J. Maddox, *Nature* **359**, 771 (1992).

[8] U. Zürcher and C. R. Doering, *Phys. Rev.* **E47**, 3862 (1993).

[9] C. Van den Broeck, *Phys. Rev.* **E47**, 4579 (1993).

[10] M. Bier and R. D. Astumian, *Phys. Rev. Lett.* **71**, 1649 (1993).

[11] P. Hänggi, *Chem. Phys.* **180**, 157 (1994).

[12] P. Pechukas and P. Hänggi, *Phys. Rev. Lett.* **73**, 2772 (1994).

[13] P. Reimann, *Phys. Rev.* **E49**, 4938 (1994).

[14] J. J. Brey and J. Casado-Pascual, *Phys. Rev.* **E50**, 116 (1994).

[15] A. J. R. Madureira, P. Hänggi, V. Buonomano and W. A. Rodrigues, Jr., *Phys. Rev.* **E51**, 3849 (1995).

[16] P. Reimann, *Phys. Rev. Lett.* **74**, 4576 (1995).

[17] P. Reimann, *Phys. Rev.* **E52**, 1579 (1995).

[18] C. Doering, W. Horsthemke, and J. Riordan, *Phys. Rev. Lett.* **72**, 2984 (1994).

[19] M. Magnasco, *Phys. Rev. Lett.* **71**, 1477 (1993).

[20] J. Maddox, *Nature* **365**, 203 (1993).

[21] C. Peskin, B. Ermentrout and G. Oster, *Cell Mechanics and Cellular Engineering* (Springer, Berlin, 1994).

[22] J. Maddox, *Nature* **368**, 287 (1994).

[23] J. Maddox, *Nature* **369**, 181 (1994).

[24] S. Leibler, *Nature* **370**, 412 (1994).

[25] C. Peskin and G. Oster, *Biophys. J.*, to be published.

[26] R. D. Astumian and M. Bier, *Phys. Rev. Lett.* **72**, 1766 (1994).

[27] C. R. Doering, B. Ermentrout and G. Oster, *Biophys. J.* **69**, 2256 (1995).

[28] J. Rousselet, L. Salome, A. Ajdari and J. Prost, *Nature* **370**, (1994).

[29] J. Prost, J. Chauwin, L. Peliti and A. Ajdari, *Phys. Rev. Lett.* **72**, 2652 (1994); M. Bier and R. D. Astumian, *Phys. Rev. Lett.* **76**, 4277 (1996).

[30] C. W. Gardiner, *Handbook of Stochastic Processes* (Springer, Berlin 1983).

[31] T. C. Elston and C. R. Doering, *J. Stat. Phys.* **83**, 359 (1996).

[32] M. Bier, *Phys. Lett.* **A211**, 12 (1996).

[33] C. R. Doering, *Nuovo Cimento* **17D**, 685 (1995).

[34] M. Millonas and M. Dykman, *Phys. Lett.* **A185**, 65 (1994).

[35] R. Bartussek, P. Hänggi, and J. G. Kissner, *Europhys. Lett.* **28**, 459 (1994).

[36] L. P. Faucheux, L. S. Bourdieu, P. D. Kaplan, and A. J. Libchaber, *Phys. Rev. Lett.* **74**, 1504 (1995).

[37] J. Travis, *Science* **267**, 1594 (1995).

[38] B. G. Levi, *Phys. Today* **48**, 17 (1995).

[39] A. Ajdari, D. Mukamel, L. Peliti and J. Prost, *J. Physique* I4, 1551 (1994).

[40] A. Ajdari, *J. Physique* I4, 1577 (1994).

[41] R. Bartussek, P. Reimann and P. Hänggi, *Phys. Rev. Lett.* **76**, 1166 (1996).

[42] A. Mielke, *Ann. Physik (Leipzig)* **4**, 476 (1995).

19

One-dimensional interacting random walks in a random environment

Klaus Ziegler

The dynamics of a grand-canonical ensemble of hard-core particles in a one-dimensional (1D) random environment is considered. Two types of randomness are studied: static and dynamic. The equivalence of a grand-canonical ensemble of hard-core particles and a system of noninteracting fermions is used to evaluate the average number of particles per site and the density of creation and annihilation processes. Exact solutions are obtained for Cauchy distributions of the random environment. It is shown that a new physical state is spontaneously created by dynamic randomness.

19.1 Introduction

The Brownian motion of a particle in a realistic system may be affected by fluctuations of the environment. One can distinguish these fluctuations according to their time scales. There are fluctuations with time scales large compared to the Brownian motion of the particle. Those are usually considered as impurities and can be described by *static* randomness. On the other hand, there are also dynamic stochastic processes that occur on time scales equal to or even shorter than the time scale of the Brownian particle. They can be described by *dynamic* randomness. Mainly for technical reasons it will be assumed that both types of randomness are statistically independent with respect to space and, for the dynamic randomness, also with respect to time.

The purpose of this chapter is to discuss methods for analysis of the dynamics of a 1D ensemble of hard-core particles in a static or dynamic random environment. The emphasis is more on the presentation of various methods and less on explicit calculations for specific models. Calculations can be quite difficult and are often not very instructive. Therefore, a few examples only of solvable problems will be given for demonstration.

The physics of 1D (spatial) random systems is a field that was established some time ago (for reviews, see [1-4]). It has been concerned mainly with the localization of a classical or quantum particle in a random potential. The restriction to 1D allows a recursive treatment. This leads to a dramatic simplification of the evaluation of interesting quantities like the distribution of eigenvectors, Green's functions, etc. In a stationary situation the calculation is reduced to the solution of an integral equation. However, exact solutions are rare and only known for special distributions of randomness. Therefore, numerical calculations also play an important role in obtaining specific results for random 1D systems.

Another interesting aspect of one-dimensionality is the fact that a test particle in an ensemble of hard-core particles behaves in 1D as a single diffusing particle, provided the boundary conditions are periodic. The reason is that the center of mass of the ensemble can diffuse on a circle. (However, there is no diffusion for nonperiodic boundary conditions [5,6].) This is an explicit manifestation of the mean-field idea in a system of interacting particles in 1D. It reflects the fact that the test particle interacts directly only with its neighboring particles. Formally, it is a simple consequence of the fact that hard-core particles are equivalent to noninteracting fermions in 1D, at least in the large-scale limit (cf. the discussion in Sec. 19.3). Therefore, the investigation of hard-core particles is reduced to the analysis of a single particle. It will be shown that this is also true if the spontaneous creation and annihilation of pairs of particles is included.

There are many results in the case of static randomness in 1D, from numerical simulations and from calculations, based on approximate treatments or on exact solutions. The main observation is that static randomness localizes the particle, i.e., diffusion is destroyed. Not much is known for the more complex situation of dynamic randomness. It will be shown in the following that this type of randomness leads to a new physical state characterized by an order parameter. The latter indicates that in the new state creation and annihilation processes (CAPs) play a central role, even when the rate of these processes goes to zero.

The chapter is organized as follows. A brief review of the treatment of static randomness in the dynamics of a single particle is given in Sec. 19.2. The dynamics of a grand-canonical ensemble of hard-core particles, based on a transfer matrix representation, is introduced in Sec. 19.3. It includes the functional representation of the transfer matrix by a functional integral with Grassmann variables. Section 19.3.1 presents the construction of the

Green's function for a test particle in the grand-canonical ensemble. The Green's function is used to define the average number of particles per site and the density of CAPs. These quantities are evaluated at the end of Sec. 19.3 for a system in a homogeneous environment. Moreover, the density of CAPs is calculated for static randomness (Sec. 19.4) and for dynamic randomness (Sec. 19.5). Spontaneous symmetry breaking is discussed in the case of dynamic randomness (Sec. 19.5).

19.2 Dynamics of a single particle in 1D

The random (Brownian) motion of a particle on a 1D lattice (i.e., a chain) with spacing Δ is described by a master equation:

$$\frac{\partial P(r,t)}{\partial t} = W(r,r+\Delta)P(r+\Delta,t) + W(r,r-\Delta)P(r-\Delta,t)$$
$$-[W(r+\Delta,r) + W(r-\Delta,r)]P(r,t) \equiv (\hat{W}P)(r,t). \quad (19.1)$$

The temporal change of $P(r,t)$, which is the probability of finding the particle at time t on site r, depends only on the probability of finding the particle on one of the neighboring sites $r \pm \Delta$. In general, the transition rates W can also depend on time t. A solution for time-independent transition rates reads, with initial probability $P(r,0)$,

$$P(r,t) = \sum_{r'} \exp(t\hat{W})_{r,r'} P(r',0). \quad (19.2)$$

The Laplace transformation of the probability,

$$\tilde{P}(r,E) = \int_0^\infty P(r,t)e^{-Et}dt, \quad (19.3)$$

introduces the Green's function $(E - \hat{W})^{-1}$ at energy E,

$$\tilde{P}(r,E) = \sum_{r'}(E - \hat{W})_{r,r'}^{-1}P(r',0). \quad (19.4)$$

Randomness appears in the transition rates W. Models are separated into different classes depending on the correlations between the transition rates (e.g., trapping, random-barriers, random-forces, etc.). A broad discussion of the various models can be found in [2,4]. One special case will be discussed in this section to illustrate the techniques for calculating characteristic quantities of the single-particle behavior in a random environment.

An example of the operator $\hat{W}$ is one with a random energy $U(r)$, $\hat{W} = \nabla_1^2 + U(r)$, where ∇_1 is the space difference operator with respect to r, i.e., $\nabla_1^2 P(r) = [P(r+1) + P(r-1) - 2P(r)]/4$. The master equation now reads

$$\frac{\partial P(r,t)}{\partial t} = \left[\nabla_1^2 + U(r)\right] P(r,t). \tag{19.5}$$

This equation is motivated by the lattice diffusion equation, which is obtained for $U = 0$; $U \neq 0$ describes the situation where the particle prefers to jump to some site r $(U(r) > 0)$ whereas another site r' is avoided $(U(r') < 0)$. Recursive methods can be applied for the evaluation of eigenvectors and their distributions [1,3,7]. There is an exact solution of (19.5) in the continuum limit $\Delta \to 0$ with a white-noise potential [8,9] $\langle U(r) \rangle = 0$, $\langle U(r)U(r') \rangle = 2\delta(r-r')$. The discrete model is solved, after Fourier transformation in time, $t \to n$, by a recursive procedure along the chain $r = 1, 2, \ldots$ [7], with

$$\tilde{P}_n(r+1) + \tilde{P}_n(r-1) - [2 + E_n - U(r)]\, \tilde{P}_n(r) = 0. \tag{19.6}$$

Division by $\tilde{P}_n(r)$ leads to a first-order recurrence for $z(r) = \tilde{P}_n(r)/\tilde{P}_n(r-1)$, namely, $z(r+1) = -1/z(r) + 2 + E_n - U(r)$. In general, with a first-order recurrence relation for the random variable Y_ℓ of type $Y_{\ell+1} = F(Y_\ell, U_{\ell+1})$, $\ell \geq 0$, where $U_{\ell+1}$ is the random potential on site $\ell + 1$, the distribution of the random variable $Y_{\ell+1}$ satisfies

$$P_{\ell+1}(Y_{\ell+1}) = \iint \delta(Y_{\ell+1} - F(Y_\ell, U_{\ell+1}))P_\ell(Y_\ell)P_0(U_{\ell+1})dY_\ell dU_{\ell+1}, \tag{19.7}$$

where $P_0(U_{\ell+1})dU_{\ell+1}$ is the distribution of the random potential. For a stationary distribution density $\bar{P}(Y)$ the following integral equation holds:

$$\bar{P}(Y) = \iint \delta(Y - F(Y', U))\bar{P}(Y')P_0(U)dY'dU. \tag{19.8}$$

Exact solutions are known, for instance, for the Cauchy distribution (strong randomness) [10],

$$P_0(U) = (\tau/\pi)[\tau^2 + (U - U_0)^2]^{-1}, \tag{19.9}$$

and for the gamma distribution (weak randomness),

$$P_0(U) = [n^n/\Gamma(n)]\, U^{n-1}e^{-nU}, \tag{19.10}$$

in the limit $n \to \infty$ [11].

The behavior of the eigenvectors in space is measured by the Lyapunov exponent $\gamma(E)$. It is defined as the growth rate of $\tilde{P}_n(j)$, (19.6),

$$\gamma = \lim_{j \to \infty} \frac{1}{j}\langle\log[r(j)]\rangle, \tag{19.11}$$

where $r(j) = \sqrt{\tilde{P}_n(j)^2 + \tilde{P}_n(j+1)^2}$. Equation (19.6) implies the following recurrence relation for $r(j)$,

$$\frac{r(j+1)^2}{r(j)^2} = z(j)^2 \frac{1 + z(j+1)^2}{1 + z(j)^2}. \tag{19.12}$$

The iteration of this equation yields

$$\frac{r(j+1)^2}{r(0)^2} = \frac{r(j+1)^2}{r(j)^2} \frac{r(j)^2}{r(j-1)^2} \cdots \frac{r(1)^2}{r(0)^2} = z(j)^2 \ldots z(0)^2 \frac{1 + z(j+1)^2}{1 + z(0)^2}. \tag{19.13}$$

Therefore, the Lyapunov exponent reads

$$\gamma = \lim_{j \to \infty} \frac{1}{j} \sum_{k=1}^{j} \langle \log|z(k)| \rangle = \langle \log|z(j)| \rangle_{\text{stat}}, \tag{19.14}$$

where the stationary distribution is evaluated from (19.8). The exact solution for the Cauchy distribution is a Cauchy distribution itself [10]. This leads to

$$\gamma = \cosh^{-1}\left[\frac{1}{4}\left(\sqrt{(2+U_0)^2 + \tau^2} + \sqrt{(2-U_0)^2 + \tau^2}\right)\right]; \tag{19.15}$$

note that the Lyapunov exponent is always positive for $\tau > 0$, reflecting localized behavior of $\tilde{P}(j)$, and vanishes for $\tau = 0$. Thus the static randomness destroys the diffusion of the particle.

The results depend very much on the specific type of randomness. Nevertheless, static randomness always leads, for symmetric processes (i.e., processes with no external force), to localization of the particle except for the singular points in E of some special models. An example of the latter is a model with random transition rates and with $W(r, r + \Delta) = W(r + \Delta, r)$, where one finds a vanishing Lyapunov exponent at $E = 0$ [12].

19.3 Dynamics of interacting particles

An ensemble of particles is described by the state $\Phi(\{n_r\}; t)$ at time t; n_r is the number of particles (occupation number) on the site r. The discrete time evolution of this state for $t = 0, 1, \ldots$ is defined by a generalization of the master equation of a single particle (19.1) to an ensemble of particles. The discrete dynamics is based on the evolution equation

$$\Phi(\{n_r\}; t+1) = \sum_{\{n'_r\}} \hat{T}(\{n_r\}|\{n'_r\}) \Phi(\{n'_r\}; t), \tag{19.16}$$

using the transfer matrix $\hat{T}$. Each site r is occupied by at most one particle of type a or b, and $n_r \in \{0, 1_a, 1_b, 2\}$ denotes the four possible states of a site: empty, occupied by either particle a or particle b, or doubly occupied, by a and b. The transfer matrix describes a jump process of a particle α on site r to one of the neighboring sites r' with a jump rate $H_{r,r'}$.

The iteration of (19.16) leads to

$$\Phi(\{n_r\}; t) = \sum_{\{n'_r\}} \hat{T}^t(\{n_r\}|\{n'_r\})\Phi(\{n'_r\}; 0). \qquad (19.17)$$

The transfer matrix can be linked to the distribution of particles, or, in other words, to the distribution of the random variables n_r, the occupation state of the site r. The hard-core property of the particles implies a strict constraint for the random variables that corresponds with an algebra of nilpotent operators. In general, such an algebra should be commutative, since classical particles are considered here. Fermions in quantum statistics obey the hard-core (or exclusion) rule (i.e., the Pauli principle) but also anticommutation rules. Therefore, they are not suitable as a description of classical hard-core particles. In general one must couple a statistical gauge field to the fermions to change the anticommutation rule into a commutation rule [13,14]. However, an exceptional case occurs in 1D. Then the particles cannot change their order on the lattice if the crossing of particles during discrete time steps is prohibited; see Fig. 19.1(a). Thus the anticommutation rule does not apply. The noncrossing condition can be achieved simply by assuming an initial state where only every second site of the lattice is allowed to be occupied by a particle; see Fig. 19.1(b). It was shown for a similar model that the exclusion of crossings has an effect on correlation functions but not on local quantities [15]. Since only local quantities will be discussed subsequently, the prohibition of crossing by choosing a special initial state is not necessary here.

The transfer matrix $\hat{T}$ for hard-core particles with Fermi statistics can be constructed with Grassmann variables [16]. Functions and linear mappings (e.g., integration) on the algebra of Grassmann variables are defined analogously to real or complex variables. The transfer matrix requires a set of Grassmann variables $\{\Psi_\alpha(r, t), \bar{\Psi}_\alpha(r, t)\}$ depending on the sites of the chain r, on time t, and on particle type α. At each lattice site r at a given time t a polynomial p can be defined with these Grassmann variables, e.g., $p(\Psi_\alpha(r, t)) = p_0 + p_a\Psi_a(r, t) + p_b\Psi_b(r, t) + p_{ab}\Psi_a(r, t)\Psi_b(r, t)$. The coefficients can also contain Grassmann variables from other lattice sites or from another time. The polynomials are used to present analytic functions of the Grassmann variables such as the exponential function. Generators (or basis

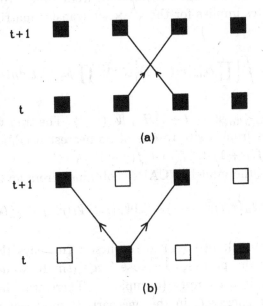

Fig. 19.1. (a) Crossing of particles, and (b) an initial state that avoids crossings.

vectors) of a general polynomial and its conjugate are $\phi_0(r,t) = \bar{\phi}_0(r,t) = 1$, $\phi_{1_{a,b}}(r,t) = \Psi_{a,b}(r,t)$, $\bar{\phi}_{1_{a,b}}(r,t) = \bar{\Psi}_{a,b}(r,t)$, $\phi_2(r,t) = \Psi_a(r,t)\Psi_b(r,t)$, $\bar{\phi}_2(r,t) = \bar{\Psi}_b(r,t)\bar{\Psi}_a(r,t)$. The measure for the integration over the Grassmann variables is

$$d\mu(r,t) = \prod_{\alpha=a,b} \exp[-\Psi_\alpha(r,t)\bar{\Psi}_\alpha(r,t)]d\Psi_\alpha(r,t)d\bar{\Psi}_\alpha(r,t), \qquad (19.18)$$

where the integration is nonzero only for a complete product of Grassmann variables [16],

$$\int \prod_{\alpha=a,b} \bar{\Psi}_\alpha(r,t)\Psi_\alpha(r,t)d\Psi_\alpha(r,t)d\bar{\Psi}_\alpha(r,t) = 1. \qquad (19.19)$$

This integral can be used to express the transfer matrix of two independent types of particle (a and b), which interact among themselves via a hard-core repulsion.

A time-shift operation $t \to t'$ is defined as a mapping between polynomials:

$$p'(\Psi_\alpha(r,t')) = \sum_{n,n'} C_{n,n'} \int \phi_n(r,t')\bar{\phi}_{n'}(r,t)p(\Psi(r,t))d\mu(r,t). \qquad (19.20)$$

The unit operation (time-shift only) is obtained for $C_{n,n'} = \delta_{n,n'}$ where $p' = p$.

The motion of the Grassmann variables from time t to time $t+1$ on a lattice with N sites implies for the $4^N \times 4^N$ transfer matrix (up to normalization),

$$\hat{T}(\{n_r\}|\{n_r'\}) = \int \left[\prod_r \bar{\phi}_{n_r}(r, t+1)\right] e^{-S(t)} \prod_r \phi_{n_r'}(r, t) d\mu(r, t+1) d\mu(r, t),$$

(19.21)

where $S(t) = -\sum_{r,r',\alpha} \Psi_\alpha(r, t+1) H_{r,r'} \bar{\Psi}_\alpha(r', t)$. The transfer matrix allows jumps of particles from a site to one of its nearest neighbors owing to the choice $Hf(r) = [f(r+1) + 2f(r) + f(r-1)]/4$.

An extension of the model to CAPs is obtained from an additional factor

$$\exp\left\{\mu(\alpha, r, t)\Psi_\alpha(r, t)\bar{\Psi}_\alpha(r, t) + i\epsilon[\Psi_a(r, t)\Psi_b(r, t) + \bar{\Psi}_a(r, t)\bar{\Psi}_b(r, t)]\right\}$$

(19.22)

in the measure (19.18); $\mu(\alpha, r, t)$ in the first term plays the role of an inverse fugacity for the particles because large (small) values tend to keep the site r at the time t empty (occupied). Therefore, this quantity corresponds with the energy U in the one-particle model of Sec. 19.2. As a generalization of the one-particle system with random potential energy, the static or dynamic randomness of $\mu(\alpha, r, t)$ describes a random environment for the grand-canonical system of particles. Elementary CAPs, created by the second term, are shown in Fig. 19.2. The integration gives a negative contribution for the elementary CAPs, owing to the anticommutation rule of the Grassmann variables for real $i\epsilon$. Since $(i\epsilon)^2$ always contributes to an elementary closed loop, the imaginary $i\epsilon$ (i.e., ϵ real) guarantees a positive contribution. A correct description of the statistics of CAPs requires distinction of the left- and right-moving particles. This is an elementary extension of the expressions in (19.22) and (19.21). However, in order to keep the calculation short, only a simplified version is used here. It is designed so that it leads to the same large-scale behavior as the correct model. The disadvantage of the simplified version is the negative weights of some processes. An example of the latter is shown in Fig. 19.2(b). Since only the large-scale asymptotics are considered subsequently, the simplified model is sufficient for the following discussion. On shorter scales the negative weights lead to a renormalization of the walks, at least for small ϵ.

Remark. The statistics of interacting closed loops is a well-known problem in equilibrium statistical mechanics. A prominent example is the Ising model on a square lattice [17]. This is equivalent to elements of closed random walks that jump clockwise and counterclockwise around corners on a square lattice. Therefore, the Ising model represents only the statistics of CAPs

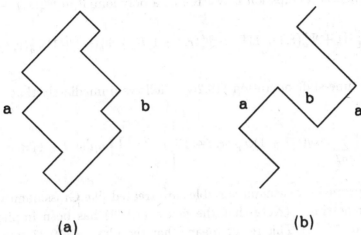

(a)   **(b)**

Fig. 19.2. (a), (b) Elementary processes that involve the creation and annihilation of pairs of particles a and b. Process (b) has rate proportional to ϵ and contributes negative weight in the transfer matrix if the left- and right-moving particles are not distinguished (see text). This leads to a renormalization of the random walk.

and not the mixture of walks of two particle types, where pairs of the two types can be created and annihilated at a tunable rate.

The product of two transfer matrices can be written as a convolution in terms of the integration $d\mu(r,t)$,

$$\hat{T}^2(\{n_r\}|\{n_r'\}) = \sum_{\{n_r''\}} \hat{T}(\{n_r\}|\{n_r''\})\hat{T}(\{n_r''\}|\{n_r'\}) \tag{19.23}$$

$$= \sum_{\{n_r''\}} \int \left[\prod_r \bar{\phi}_{n_r}(r,t+2)\right] e^{-S(t+1)}$$

$$\times \prod_r \phi_{n_r''}(r,t+1)d\mu(r,t+2)d\mu(r,t+1)$$

$$\times \int \left[\prod_r \bar{\phi}_{n_r''}(r,t)\right] e^{-S(t)} \prod_r \phi_{n_r'}(r,t)d\mu(r,t+1)d\mu(r,t).$$

One of the two independent integrations, $\int \cdots d\mu(r,t+1)$, will be performed as an application of (19.20). For this purpose a prime is introduced to distinguish the two independent Grassmann integrations in (19.23) at $t+1$,

$$d\mu'(r,t+1) = \prod_{\alpha=a,b} \exp[-\Psi_\alpha'(r,t+1)\bar{\Psi}_\alpha'(r,t+1)]d\Psi_\alpha'(r,t+1)d\bar{\Psi}_\alpha'(r,t+1).$$

$$\tag{19.24}$$

The exponential expression is written as a polynomial in $\Psi'_\alpha(r, t+1)$,

$$e^{-S'(t)} = \prod_r [1 + \Psi'_a(r, t+1)F_a + \Psi'_b(r, t+1)F_b + \Psi'_b(r, t+1)\Psi'_a(r, t+1)F_{ab}].$$

$$(19.25)$$

From the time-shift operation (19.20) it follows immediately that

$$\int \left[\prod_r \sum_{n''_r} \phi_{n''_r}(r, t+1)\bar{\phi}'_{n''_r}(r, t+1) \right] e^{-S'(t)} \prod_r d\mu'(r, t+1) = e^{-S(t)},$$

$$(19.26)$$

since the primed Grassmann variables are treated like Grassmann variables at a different time. (Note that the factor (19.22) has been implicitly absorbed into $e^{-S(t)}$.) This result means that the l.h.s. of (19.23) can also be written as

$$\hat{T}^2(\{n_r\}|\{n'_r\}) = \iint \left[\prod_r \phi_{n_r}(r, t+2) \right] e^{-S(t+1)-S(t)} \prod_r \bar{\phi}_{n'_r}(r, t)$$
$$\times d\mu(r, t+2)d\mu(r, t+1)d\mu(r, t). \qquad (19.27)$$

By iteration this is generalized to an arbitrary power of the transfer matrix

$$\hat{T}^l(\{n_r\}|\{n'_r\}) = \int \cdots \int \left[\prod_r \phi_{n_r}(r, t+l+1) \right] e^{-S(t+l)-\cdots-S(t)}$$
$$\times \prod_r \bar{\phi}_{n'_r}(r, t)d\mu(r, t+l+1)\cdots d\mu(r, t) \qquad (19.28)$$

which is a functional integral representation of the transfer matrix. It is based on the quadratic form of the Grassmann variables in $\sum_t S(t)$. A renaming of the Grassmann variables of the particle b, $\Psi_b \to \bar{\Psi}_b$, $\bar{\Psi}_b \to -\Psi_b$, leads to the same type of quadratic form in the exponential expression (19.22) as appears in $S(t)$. Therefore, $\sum_t S(t)$ and the arguments of the exponential expressions in the integration-measure (19.18) and in (19.22) can be combined to $\sum_t \bar{S}(t) = -\sum_{t,\ldots} \Psi_\alpha(r, t)\Gamma_{\alpha,r,t;\alpha',r',t'}\bar{\Psi}_{\alpha'}(r', t')$, which implies for the transfer matrix,

$$\hat{T}^l(\{n_r\}|\{n'_r\}) = \int \cdots \int \left[\prod_r \phi_{n_r}(r, t+l+1) \right] e^{-\bar{S}(t+l)-\cdots-\bar{S}(t)}$$
$$\times \prod_r \phi_{n'_r}(r, t) \prod_{t,\alpha=a,b} d\Psi_\alpha(r, t)d\bar{\Psi}_\alpha(r, t). \qquad (19.29)$$

The time-structure of the matrix Γ for $t = 1, \dots, \ell$ reads

$$
\Gamma = \begin{pmatrix}
z_1 & \hat{H} & 0 & \cdots & 0 & \hat{H}_b' \\
\hat{H} & \ddots & \ddots & 0 & \cdots & 0 \\
0 & \ddots & \ddots & \ddots & \ddots & \vdots \\
\vdots & \ddots & \ddots & \ddots & \ddots & 0 \\
0 & \cdots & 0 & \ddots & \ddots & \hat{H} \\
\hat{H}_a' & 0 & \cdots & 0 & \hat{H} & z_\ell
\end{pmatrix},
\tag{19.30}
$$

with appropriate boundary terms $\hat{H}_{a,b}'$. The elements of this matrix depend on the lattice sites $r = 1, \dots, N$, and the particle types a and b. Therefore, these matrix elements are $2N \times 2N$ matrices with

$$
(z_t)_{r,r'} = \begin{pmatrix} \mu(a, r, t) & i\epsilon \\ i\epsilon & \mu(b, r, t) \end{pmatrix} \delta_{r,r'}, \qquad \hat{H} = \begin{pmatrix} H & 0 \\ 0 & H \end{pmatrix}.
\tag{19.31}
$$

It is assumed for further calculations that the particles a and b have the same fugacity $\mu(a, r, t) = \mu(b, r, t) \equiv \mu(r, t)$.

19.3.1 The Green's function and densities

The dynamics of a test particle in an ensemble of hard-core particles can be described by a Green's function. The latter measures the 'probability' for a test particle of type α, which was put at time t_0 onto site r, to walk to site r' at time t_1:

$$
G_\alpha(r, t_0; r', t_1) = \frac{\bar{\Phi}'(t) \hat{T}^{t-t_1-t_0} C_{\alpha-}(r') \hat{T}^{t_1} C_{\alpha+}(r) \hat{T}^{t_0} \Phi(0)}{\bar{\Phi}'(t) \hat{T}^t \Phi(0)}
\tag{19.32}
$$

with $C_{\alpha\pm}(r')(\{n_r\}|\{n_r'\}) = \delta_{n_{r'},n_{r'}'\pm 1_\alpha} \prod_{r \neq r'} \delta_{n_r,n_r'}$. The denominator measures the 'overlap' of a given state $\Phi'(t)$ with a state that is a result of the evolution of the state $\Phi(0)$. Thus the choice of $\Phi'(t)$ determines the boundary conditions in time. The numerator, on the other hand, measures this overlap by including the test particle. The behavior of the Green's function should not depend on the specific boundary conditions, provided t_0 and t_1 are far enough from the boundaries 0 and t. Therefore, periodic boundaries in time will be used for simplicity, and the condition $0 \ll t_0 < t_1 \ll t$. The Green's function then reads

$$
G_\alpha(r, t_0; r', t_1) = \frac{\sum_{\{n_r\}} \hat{T}^{t-t_1-t_0} C_{\alpha-}(r') \hat{T}^{t_1} C_{\alpha+}(r) \hat{T}^{t_0} (\{n_r\}|\{n_r\})}{\sum_{\{n_r\}} \hat{T}^t (\{n_r\}|\{n_r\})}
$$

$$
= \mathrm{Tr} \left[\hat{T}^{t-t_1-t_0} C_{\alpha-}(r') \hat{T}^{t_1} C_{\alpha+}(r) \hat{T}^{t_0} \right] \Big/ \mathrm{Tr} \left(\hat{T}^t \right).
\tag{19.33}
$$

The diagonal element $G_\alpha(r, t_0; r, 0)$ measures the absence of the particle α from the site r at time t_0. This implies, for the average number of particles per site,

$$\bar{n} = 1 - \tfrac{1}{2} \left[G_a(r, t_0; r, 0) + G_b(r, t_0; r, 0) \right]. \qquad (19.34)$$

Another interesting quantity is related to the creation or annihilation of a pair of particles a and b. The probability for the contribution of such events is measured by the following expression,

$$G_\pm(r, t_0) = i \operatorname{Tr} \left[\hat{T}^t C_{a\pm}(r) C_{b\pm}(r) \right] \Big/ \operatorname{Tr} \left(\hat{T}^t \right). \qquad (19.35)$$

Integration over the Grassmann variables can be performed in (19.32) and (19.33). This follows from the general result for Grassmann integration [16],

$$\int \exp \left(\sum_{j,j'} \Psi_j \Gamma_{j,j'} \bar{\Psi}_{j'} \right) \prod_j d\Psi_j d\bar{\Psi}_j = \det \Gamma. \qquad (19.36)$$

For the Green's function one gets Γ^{-1}.

In general, densities can also be identified as response functions to the variation of the corresponding chemical potentials. Since μ is the fugacity of empty sites, the number of particles per site (a and b together) is

$$\bar{n} = 1 - \frac{1}{2Nt} \frac{\partial \log (\det \Gamma)}{\partial \mu} = 1 - \frac{1}{2Nt} \operatorname{Tr} \left(\Gamma^{-1} \right). \qquad (19.37)$$

The density of CAPs, which follows from (19.35), reads

$$\rho = \frac{1}{Nt} \frac{\partial \log (\det \Gamma)}{\partial \epsilon} = \frac{i}{Nt} \operatorname{Tr} \left(\Gamma^{-1} \sigma_1 \right) \qquad (19.38)$$

with the Pauli matrix σ_1; the latter connects the particle types, a and b. Both quantities can be calculated easily in the absence of randomness using the Fourier representation of the Green's function G in time and space. The Fourier transformation is $(r, t) \rightarrow (k, \omega)$. To simplify calculation it is useful to apply a large-scale approximation to the inverse Green's function

$$\Gamma = \begin{pmatrix} \mu - 1 + H\nabla_+ & i\epsilon \\ i\epsilon & \mu - 1 + H\nabla_- \end{pmatrix}, \qquad (19.39)$$

where $\nabla_\pm$ is the time-shift operator $(\nabla_\pm f)(t) = f(t \pm 1)$. Such an approximation is defined for functions of ω and k: a function $g(s)$ ($s = \omega, k$) is a large-scale approximation of a function $f(s)$ if the first two terms of their Taylor expansions agree. In particular, we use the approximation $\nabla_\pm \approx 1 \pm \tfrac{1}{2}(\nabla_+ - \nabla_-)$ and, including the large-scale approximation in space, one has $H\nabla_\pm \approx -1 + H + \nabla_\pm \approx H \pm \frac{1}{2}(\nabla_+ - \nabla_-)$. This gives

$$\Gamma \approx \begin{pmatrix} \mu - 1 + H + \frac{1}{2}(\nabla_+ - \nabla_-) & i\epsilon \\ i\epsilon & \mu - 1 + H - \frac{1}{2}(\nabla_+ - \nabla_-) \end{pmatrix}. \quad (19.40)$$

After Fourier transformation of Γ the average particle number and the density of CAPs are given by

$$\bar{n} = 1 - \int \int \frac{\mu - k^2}{\epsilon^2 + \omega^2 + (\mu - k^2)^2} \frac{d\omega}{2\pi} \frac{dk}{2\pi},$$

$$\rho = \epsilon \int \int [\epsilon^2 + \omega^2 + (\mu - k^2)^2]^{-1} \frac{d\omega}{2\pi} \frac{dk}{2\pi}. \quad (19.41)$$

The ω-integration can be performed,

$$\bar{n} = 1 - \int \frac{\mu - k^2}{\sqrt{\epsilon^2 + (\mu - k^2)^2}} \frac{dk}{2\pi}, \qquad \rho = \epsilon \int [\epsilon^2 + (\mu - k^2)^2]^{-1/2} \frac{dk}{2\pi}. \quad (19.42)$$

The first integral requires a cutoff π for k. For $\epsilon > 0$ both densities are positive and smooth with respect to μ. In the limit $\epsilon \to 0$ the number of particles per site reads

$$\bar{n} \sim \begin{cases} 0, & \mu > \pi^2 \\ 2\sqrt{\mu}/\pi, & 0 \le \mu \le \pi^2 \\ 2, & \mu < 0 \end{cases}. \quad (19.43)$$

The density of CAPs vanishes according to the 'scaling form' if the rate ϵ for the CAPs goes to zero: $\rho \sim \sqrt{\epsilon}\bar{\rho}(\mu/\epsilon)$, with the scaling function $\bar{\rho}(y) = \int [1 + (y - x^2)^2]^{-1/2} dx$. The purpose of the subsequent calculations is to study the behavior of $\rho(\epsilon)$ for $\epsilon \sim 0$ in the presence of static and dynamic randomness.

19.4 Static randomness

After Fourier transformation with respect to time, $t \to \omega$, the matrix Γ in (19.40) reads

$$\begin{pmatrix} i\omega + H + \mu - 1 & i\epsilon \\ i\epsilon & -i\omega + H + \mu - 1 \end{pmatrix}. \quad (19.44)$$

The decay of the Green's function for a test particle is strongly affected by the CAPs. The existence of the latter, which create a finite scale due to the finite lifetime of a-b pairs, leads to an exponential decay of the Green's function even in the pure system. However, in the limit $\epsilon \to 0$ the CAPs disappear in the pure system. As a consequence, the decay of the Green's function becomes a power law, as one can easily calculate from the corresponding expression in (19.33). For a system with static randomness the decay of the Green's function is exponential also for $\epsilon = 0$, owing to localization. This is implied by the exponential behavior of the eigenvectors of the Green's function at $\epsilon = 0$. Since the eigenvectors satisfy the equation $(H + \mu - 1)\varphi_n = E_n\varphi_n$, one can apply the recursive procedure of Sec. 19.2 to evaluate the Lyapunov exponent γ. The result for a Cauchy distribution, given in (19.15), is $\gamma > 0$ for nonvanishing randomness.

The number of particles per site should not change away from $\mu \sim 0$ in a qualitative way as a result of static randomness. The density of CAPs, calculated from the expression (19.38), reads

$$\rho = \frac{\epsilon}{N} \int \mathrm{Tr}_r\left[\{\epsilon^2 + \omega^2 + (H + \mu - 1)^2\}^{-1}\right]\frac{d\omega}{2\pi}$$

$$= -\frac{\epsilon}{N} \int \frac{1}{2i\sqrt{\epsilon^2 + \omega^2}}\mathrm{Tr}_r\left[\{i\sqrt{\epsilon^2 + \omega^2} + H + \mu - 1\}^{-1}\right.$$

$$\left. - \{-i\sqrt{\epsilon^2 + \omega^2} + H + \mu - 1\}^{-1}\right]\frac{d\omega}{2\pi}. \quad (19.45)$$

The averaging of the density ρ with respect to the distribution of μ can be interchanged with the ω-integration. Then it is possible to perform the averaging for a Cauchy distribution (19.9). The reason is that the Cauchy distribution for $\mu(r)$ has the two poles $\mu(r) = \mu_0 \pm i\tau$. The matrices $A_\pm = [\pm i\sqrt{\epsilon^2 + \omega^2} + H + \mu - 1]^{-1}$, as functions of $\mu(r)$, have a pole in only one of the complex half-planes, the upper or the lower half-plane, respectively. Therefore, the path of integration of $\mu(r)$ can be closed in that half-plane where the matrix is analytic. In this case the Cauchy integration contributes only the pole $\mu(r) = \mu_0 - i\tau$ of the distribution to A_+ and $\mu(r) = \mu_0 + i\tau$ to A_-, respectively. The result of the integration for all lattice sites r is $\langle A_\pm \rangle = [\pm i(\sqrt{\epsilon^2 + \omega^2} + \tau) + H + \mu_0 - 1]^{-1}$. This implies for the density

$$\langle \rho \rangle = \epsilon \int \frac{\sqrt{\epsilon^2 + \omega^2} + \tau}{\sqrt{\epsilon^2 + \omega^2}} \int \left[(\sqrt{\epsilon^2 + \omega^2} + \tau)^2 + (\mu_0 - k^2)^2\right]^{-1}\frac{dk}{2\pi}\frac{d\omega}{2\pi}, \quad (19.46)$$

which has again a scaling form: $\sqrt{\epsilon}\bar{\rho}(\mu_0/\epsilon, \tau/\epsilon)$. To evaluate the scaling form the integral can be approximated by pulling $(\sqrt{\epsilon^2 + \omega^2} + \tau)/\sqrt{\epsilon^2 + \omega^2}$ out of the integral as a constant $C > 1$. The resulting expression agrees with (19.41), except for the constant C and the replacements $\epsilon^2 + \omega^2 \to (\sqrt{\epsilon^2 + \omega^2} + \tau)^2$, $\mu \to \mu_0$. On the other hand, the k-integral can be approximated by a constant $C < \tau^{-2}$. The remaining ω-integral gives the asymptotic behavior of the pure system.

Thus $\langle \rho \rangle$ vanishes as $\sqrt{\epsilon}$ with $\epsilon \to 0$ in agreement with the corresponding expression of the pure system; i.e., the localization of the random walks due to static randomness does not change qualitatively the density of CAPs in the asymptotic regime $\epsilon \sim 0$.

19.5 Dynamic randomness

The simple recursive methods of Sec. 19.2 cannot be applied if randomness depends on space and time because this means effectively two-dimensional randomness. Nevertheless, some qualitative and quantitative results are available and presented here. The discussion will be restricted to the density of CAPs because this quantity characterizes new features created by dynamic randomness. It allows us to distinguish the latter from the properties of the pure system, or the system with static randomness, in a qualitative manner.

The matrix Γ of (19.40) appears in the expression of the density of CAPs (19.38), multiplied by the Pauli matrix σ_1,

$$\Gamma\sigma_1 = i\epsilon\sigma_0 + h_1\sigma_1 + h_2\sigma_2, \tag{19.47}$$

with $h_1 = \mu - 1 + H$, $h_2 = i(\nabla_+ - \nabla_-)/2$ and the 2×2 unit matrix σ_0. The Pauli matrices refer to the particles a and b as described above. The matrix $\sigma_1\Gamma^{-1}$ as a function of $\mu(r,t)$ has poles in the upper and in the lower complex half-plane. This becomes obvious when the term with $\mu(r,t)\sigma_1$ is diagonalized by a global orthogonal transformation:

$$\Gamma\sigma_1 \to \tfrac{1}{2}(\sigma_1 + \sigma_3)\Gamma\sigma_1(\sigma_1 + \sigma_3) = i\epsilon\sigma_0 + h_1\sigma_3 - h_2\sigma_2. \tag{19.48}$$

The transformed matrix now depends on $i\epsilon \pm \mu(r,t)$. Therefore, it is not possible to continue analytically $\mu(r,t)$ into the half-planes. However, a transformation leads to a matrix Γ'' that is analytic in the upper half-plane (for $\epsilon > 0$) and which can be used to represent ρ. The first step in its derivation is based on the multiplication of the r.h.s. of (19.48) by $D\sigma_3$, where D is a diagonal matrix that is staggered in time, $D_{t,t'} = (-1)^t \delta_{t,t'}$:

$$\Gamma' = \tfrac{1}{2}(\sigma_1 + \sigma_3)\Gamma\sigma_1(\sigma_1 + \sigma_3)D\sigma_3 = i\epsilon D\sigma_3 + h_1 D\sigma_0 - ih_2 D\sigma_1; \tag{19.49}$$

D commutes with h_1 and anticommutes with h_2. Hermitian conjugation gives $\Gamma'^\dagger = -i\epsilon D\sigma_3 + h_1 D\sigma_0 - ih_2 D\sigma_1$, and the product $\Gamma'^\dagger \Gamma'$ is

$$\tfrac{1}{2}D\sigma_3(\sigma_1 + \sigma_3)\sigma_1\Gamma^\dagger\Gamma\sigma_1(\sigma_1 + \sigma_3)\sigma_3 D$$
$$= (h_1 D\sigma_0 - ih_2 D\sigma_1)^2 + \epsilon^2\sigma_0 = \Gamma''^\dagger\Gamma''. \tag{19.50}$$

Here

$$\Gamma'' = (i\epsilon + h_1 D)\sigma_0 - ih_2 D\sigma_1 ; \qquad (19.51)$$

$(\Gamma'')^{-1}$ is analytic in the upper half-plane of $\mu(r,t)(-1)^t$ as required above, since this matrix depends only on $i\epsilon + \mu(r,t)(-1)^t$. For the discussion it is convenient to rename the random variable $\mu(r,t)(-1)^t$ as $\mu(r,t)$. Therefore, the last matrix is particularly simple for the calculation of the density of CAPs. Moreover, the differentiation with respect to ϵ in (19.38) can now be replaced by differentiation with respect to μ,

$$
\begin{aligned}
Nt\rho &= \epsilon \operatorname{Tr}\left[(\Gamma^\dagger\Gamma)^{-1}\right] = \frac{i}{2}\operatorname{Tr}\left[\Gamma''^{-1} - (\Gamma''^\dagger)^{-1}\right] \\
&= \frac{1}{2}\frac{\partial}{\partial\epsilon}\left[\log\left(\det\Gamma''\right) - \log\left(\det\Gamma''^\dagger\right)\right] \\
&= \frac{i}{2}\sum_x \frac{\partial}{\partial\mu(x)}\left[\log\left(\det\Gamma''\right) - \log\left(\det\Gamma''^\dagger\right)\right], \qquad (19.52)
\end{aligned}
$$

where $x = (r,t)$ denotes space-time coordinates here. The average for a continuous distribution, defined by an integral $\langle\cdots\rangle = \int\cdots\prod_x P(\mu(x))d\mu(x)$, will compensate the differential operator in $[\partial/\partial\mu(x_0)]\log\left(\det\Gamma''\right)$ in a convenient way. The integration is restricted, on a set of space-time points $\{(r,t)\} = S$ with $x_0 \in S$ and with the 'surface' ∂S, to a sufficiently large interval. The restriction is $-\zeta \le \mu(x_0) \le \zeta$ at x_0 and $-\zeta \le \mu(x) \le \zeta + \delta$ on all other points of S. The constant ζ must be inside the interval of the random variable μ and large enough that $[-\zeta, \zeta]$ covers the whole spectrum of $\Pi_S\{\Gamma'' - [\mu - \mu(x_0)]\sigma_0\}\Pi_S$, where Π_S is the projector onto S. If P_S is

$$P_S = \min_{-\zeta \le \mu(x_0) \le \zeta} P(\mu(x_0)) \prod_{x_0 \ne x \in S} \int_{\mu(x_0)}^{\mu(x_0)+\delta} P(\mu(x))d\mu(x), \qquad (19.53)$$

then there is a lower bound [18], $\langle\rho\rangle \ge 2\pi P_S(1 - |\partial S|/|S|)$; $|S|$ and $|\partial S|$ are the numbers of points of the corresponding sets.

Since the matrix $(\Gamma'')^{-1}$ is analytic in the upper half-plane, the integration of the Cauchy distribution can also be carried out for dynamic randomness. The evaluation of the integrals follows the same arguments as discussed in Sec. 19.4 for static randomness, because only the pole of the distribution $\mu(r,t) = \mu_0 - i\tau$ contributes to the average matrix $\langle(\Gamma'')^{-1}\rangle$; i.e., the average over the randomness is equivalent to the replacements $\mu \to \mu_0$ and $\epsilon \to \epsilon' \equiv \epsilon + \tau$: $\langle[\Gamma''(\epsilon,\mu)]^{-1}\rangle = [\Gamma''(\epsilon',\mu_0)]^{-1}$. This implies, for the density of CAPs given by (19.42),

$$\langle\rho\rangle = \epsilon' \int \left[\epsilon'^2 + (\mu_0 - k^2)^2\right]^{-1/2}\frac{dk}{2\pi}. \qquad (19.54)$$

This expression is nonzero in the limit $\epsilon \to 0$, since the original role of ϵ in the pure system is now played by $\tau \neq 0$.

The result of a nonzero density of CAPs in the limit $\epsilon \to 0$ suggests a spontaneous breaking of the continuous symmetry of $[\Gamma\sigma_1]_{\epsilon=0}$,

$$[\Gamma\sigma_1]_{\epsilon=0} \to (c\sigma_0 + s\sigma_3) [\Gamma\sigma_1]_{\epsilon=0} (c\sigma_0 + s\sigma_3) = (c^2 - s^2) [\Gamma\sigma_1]_{\epsilon=0} \quad (19.55)$$

for $c^2 - s^2 = 1$. This condition describes a 1D manifold of the transformation parameter $c(s) = \sqrt{1 + s^2}$. The term with $\epsilon \neq 0$ breaks the symmetry. It corresponds to a nonzero density of CAPs, $\rho \neq 0$, even in the pure system. In fact, the CAPs dominate the dynamics in this case. As a consequence, all correlations decay exponentially with typical scales depending on the rate ϵ for CAPs. In the limit $\epsilon \to 0$ the pure system allows only random walks with conservation of particles. As one can see from (19.46) this remains true in the presence of static randomness, where the density of CAPs vanishes with $\epsilon \to 0$. This shows that static randomness mainly affects by localization only the range of the motion in space.

The situation is more complicated for dynamic randomness. The density of CAPs goes to a nonzero constant for arbitrarily small rates ϵ, i.e., the system is unstable against the creation of CAPs. This reflects a complicated behavior different from diffusion (pure system) and localization (static randomness). Although ρ is an important indicator for spontaneous symmetry breaking (like the order parameter in an equilibrium phase diagram), a better understanding of the physical state requires the analysis of correlation functions. The invariant manifold $c(s)$ of the symmetry suggests long-range correlations in space-time [19]. It would be interesting if this were a kind of a 'glassy' state formed by the CAPs.

The extension of the dynamics of hard-core particles to dimension $D > 1$ is possible. However, the realistic model cannot be constructed with non-interacting fermions (i.e., with a noninteracting Grassmann field) but from hard-core bosons. Then one must apply approximations to analyze the hard-core interaction [13,14]. Nevertheless, a qualitative difference between static and dynamic randomness, related to a picture similar to the 1D case, is expected also in higher dimensions [20].

References

[1] E. H. Lieb and D. C. Mattis, *Mathematical Physics in One Dimension* (Academic Press, New York and London, 1966).
[2] S. Alexander, J. Bernasconi, W. R. Schneider and R. Orbach, *Rev. Mod. Phys.* **53**, 175 (1981).

[3] I. M. Lifshitz, S. A. Gredeskul and L. A. Pastur, *Introduction to the Theory of Disordered Systems* (Wiley, New York, 1988).

[4] J.-P. Bouchaud and A. Georges, *Phys. Rep.* **195**, 127 (1990).

[5] T. E. Harris, *J. Appl. Prob.* **2**, 323 (1965).

[6] H. van Beijeren, K. W. Kehr and R. Kutner, *Phys. Rev.* **B28**, 5711 (1983).

[7] H. Schmidt, *Phys. Rev.* **105**, 425 (1957).

[8] H. L. Frisch and S. R. Lloyd, *Phys. Rev.* **120**, 1179 (1960).

[9] B. I. Halperin, *Phys. Rev.* **139**, A104 (1965).

[10] K. Ishii, *Prog. Theor. Phys. Suppl.* **53**, 77 (1973).

[11] F. J. Dyson, *Phys. Rev.* **92**, 1331 (1953).

[12] R. L. Bush, *J. Phys.* C8, L547 (1975); G. Theodoru and M. H. Cohen, *Phys. Rev.* B13, 4597 (1976).

[13] K. Ziegler, *Europhys. Lett.* **9**, 277 (1989).

[14] K. Ziegler, *J. Stat. Phys.* **64**, 277 (1991).

[15] G. Forgacs and K. Ziegler, *Europhys. Lett.* **29**, 705 (1995).

[16] F. A. Berezin, *Method of Second Quantization* (Academic Press, New York 1966).

[17] T. Schultz, D. C. Mattis and E. H. Lieb, *Rev. Mod. Phys.* **36**, 856 (1964).

[18] K. Ziegler, *Nucl. Phys.* **B285** [FS19], 606 (1987).

[19] K. Ziegler and A. M. M. Pruisken, *Phys. Rev.* **E51**, 3359 (1995).

[20] K. Ziegler, *Z. Phys.* **B84**, 17 (1991).

Part VII: Experimental Results

Editor's note

Experimental systems described by the 1D theories reviewed in the earlier Parts of the book are much less numerous than two- and three-dimensional systems. Nevertheless a surprising variety of experimental situations are effectively one-dimensional.

Reaction-diffusion systems in 1D are notable for experimental observation of fluctuation-dominated behavior, and for the availability of quality data appropriate for comparison to modern theories. Chapters 20 and 21 review reaction-diffusion systems.

Chapter 22 describes deposition on DNA. For the latter system, the quality of the experimental data only allows comparison with theory at the mean-field level (within a Langmuir adsorption type model); deviations from mean-field have been observed as well. Certain other '1D' experimental applications of theories reviewed in the earlier Parts were cited in Chs. 10, 15, and 18 (and of course the 1D diffusion-like equations, Ch. 17, have practical significance in engineering and science).

20

Diffusion-limited exciton kinetics in one-dimensional systems

Ron Kroon and Rudolf Sprik

The experimental verification of models for one-dimensional (1D) reaction kinetics requires well-defined systems obeying pure 1D behavior. There is a number of such systems that can be interpreted in terms of 1D reaction kinetics. Many of them are based on the diffusive or coherent motion of excitons along well-defined chains or channels in the material. In this chapter they will be briefly reviewed.

We also present results of an experimental study on the reaction kinetics of a 1D diffusion-reaction system, on a picosecond-to-millisecond time scale [1]. Tetramethylammonium manganese trichloride (TMMC) is a perfect model system for the study of this problem. The time-resolved luminescence of TMMC has been measured over nine decades in time. The nonexponential shape of the luminescence decay curves depends strongly on the exciting laser's power. This is shown to result from a fusion reaction $(A+A \rightarrow A)$ between photogenerated excitons, which for initial exciton densities $\lesssim 2 \times 10^{-4}$ as a fraction of the number of sites is very well described by the diffusion-limited single-species fusion model. At higher initial exciton densities the diffusion process, and thus the reaction rate, is significantly influenced by the heat produced in the fusion reaction. This is supported by Monte Carlo simulations.

20.1 Introduction

Reactions between (quasi-)particles in low-dimensional systems is an important topic in such diverse fields as heterogeneous catalysis, solid state physics, and biochemistry.

Recent theoretical studies [2-4] have demonstrated the kinetics of diffusion-limited reactions to be determined by the ability of the reaction system to eliminate spatial fluctuations in the concentrations of reactants. Diffusive transport of reactants in three dimensions is well able to eliminate these fluctuations, i.e., to create a spatially uniform concentration of reactants, leading to the well-known classical rate equations to describe the reaction kinetics. However, in systems where diffusive transport is limited to *less* than three dimensions, the classical rate equations are no longer valid. In this case, the diffusing reactants do not explore enough space to eliminate the spatial concentration fluctuations. In fact, the fluctuations grow in time as a consequence of the reactions. This process has been described as a dynamic *self-ordering* of the spatial distribution of reactants [5,6]. The influence of the developing inhomogeneous spatial distribution of reactants on the reaction kinetics is most pronounced in 1D systems.

Many of the theoretical results and insights are addressed in the previous parts of this book. Here we will address the realizations of such low-dimensional systems in experimentally accessible systems.

20.2 Experimental studies of reactions in 1D

An overview of experimental studies on 1D transport leading to reactions is given in Tables 20.1 and 20.2. The migrating species in all systems, apart from cadmium, are quasi-particles: laser-induced electronic excitations (excitons) situated on organic and/or inorganic chromophores. In cadmium, the migrating species are interstitials, i.e., real particles, created by irradiating the Cd crystal with thermal neutrons.

The following properties determine the characteristics of an experimental system:

(1) *The mode of transport,* coherent or diffusive: in coherent transport the wavefunction of the migrating excitation may be viewed as delocalized over the whole chain. Static disorder, weak coupling between sites or strong coupling of the excitation to, e.g., phonons (dynamic disorder) can limit the coherence length of the wavefunction and reduce the mode of transport to the diffusive limit where the migrating species travels like a random walker.

(2) *The anisotropy ratio*: the ratio of the migration rate along the preferred direction of transport to the migration rate in the perpendicular directions.

(3) *The chain length*: the length over which a migrating species can travel without being stopped.

Concerning item (1), diffusive transport is interesting since in this case diffusion-limited reaction rate 'constants' are dependent on time and on the

Table 20.1.

Overview of experimental work in systems showing 1D transport, resulting in reactions. The migrating species in all systems, apart from cadmium, are quasiparticles: electronic excitations. In cadmium, the migrating species are interstitials. AFM: antiferromagnet, Q1D: quasi-1D, FM: ferromagnet, AR: anisotropy ratio, LC: liquid crystals, CL: chain length (expressed as the number of sites). The observed reaction types are reviewed in Table 20.2

System	Transport	Remarks	References
manganese salts			
$(CH_3)_4NMnCl_3$ (TMMC)	diffusive	AFM, $AR \approx 10^8$, $CL \approx 10^6$	[1,8-12]
$(CH_3)_4NMnBr_3$ (TMMB)	diffusive	AFM, $AR \approx 10^8$	[11-13]
$CsMnBr_3$ (CMB)	diffusive	AFM, $AR \approx 10^4$ (Q1D)	[12,14-18]
$CsMnI_3$, $RbMnBr_3$	diffusive	AFM, Q1D	[16]
$CsMnCl_3 \cdot 2H_2O$	diffusive	AFM, Q1D	[19]
$(CH_3)_3NHMnCl_3 \cdot 2H_2O$	diffusive	AFM, Q1D	[20]
other inorganics			
$CsNiF_3$	?	FM, Q1D	[21]
$GdCl_3$	?	FM, Q1D, $CL \approx 2500$	[22]
polymers			
trans-polyacetylene	diffusive	Q1D, $CL \approx 300$	[23-25]
plvn/pmma films	diffusive	see Ch. 21	[26]
polyquanylic acid	diffusive	$CL \leq 95$	[27]
columnar LC			
phtalocyanines	diffusive	$CL \approx 2500$	[28]
colloids			
phtalocyanine particles	diffusive	$CL \approx 50$	[29]
J-aggregates			
pseudoisocyanide (PIC)	coherent		[30]
tdbc	coherent		[31]
other organics			
Tetrachlorobenzene (TCB)	coherent	$AR \approx 10^7$-10^8	[32]
ntcpa	coherent	Q1D	[33]
naphtalene	diffusive	in polycarbonate pores	[5]
metals			
cadmium	·diffusive	see caption; $AR \approx 10^5$	[34]

dimensionality of transport (in 1D, $k \propto t^{-1/2}$; see the next section). Coherent transport is invariably characterized by a time-independent reaction rate constant [5]. Concerning item (2), diffusive transport retains its 1D character only as long as, on average, a 3D hop has not occurred [34]: after a few such hops the characteristics of a purely 1D reaction are completely destroyed, and a quasi-3D diffusion begins to govern the reaction. Item (3) speaks for itself. In the following, we will address these items for the classes

Table 20.2.

Overview of reactions observed in systems showing 1D transport (cf. Table 20.1)

Reaction	References
annihilation and fusion	
$A + B \rightarrow$ heat	[34]
$A + A \rightarrow A +$ heat	[1,12,17,30,31]
$A + A \rightarrow$ not specified in the paper	[28,29]
$A + A \rightarrow$ delayed fluorescence	[5,22,26]
$A + A \rightarrow$ electron + hole	[24,25]
trapping	[8-11,13-16,18-25,27,32,33]
reflecting barriers	[9,18]

of systems listed in Table 20.1. As the experimental systems are described we will mention the types of reactions observed in these systems, as listed in Table 20.2.

In an experiment, exciton annihilation ($A + A \rightarrow 0$), fusion ($A + A \rightarrow A$), and trapping ($A + T \rightarrow T$) manifest themselves as a nonexponential decay in time of the exciton population created by pulsed-laser excitation of the samples (cf. Sec. 20.3). This decay is measured from the time behavior of the exciton luminescence since the intensity of the exciton luminescence is linearly proportional to the exciton population. An experimental setup suitable for this purpose is described in Sec. 20.4.1, where the decay of the exciton population is monitored on a picosecond-to-millisecond time scale [1]. We will now describe the experimental systems listed in Table 20.1.

In the manganese salts the exciton transport is diffusive. This is a result of exciton-phonon coupling-induced lattice distortion around the exciton, which causes a strong narrowing of the exciton band. In this way, coherent transport of the excitons is eliminated. What remains is a thermally activated, diffusive migration. The migration rate and energy barrier for hopping can be measured in studies on trapping of excitons [9,10,13,19]. Typical hopping rates are 10^{11}-10^{12} s^{-1} at room temperature.

In the inorganic systems the (one-)dimensionality of the exciton transport arises from magnetic (exchange) interactions between nearest-neighbor chromophores. This is illustrated in Fig. 20.1, where the structure of antiferromagnetic TMMC, $(CH_3)_4NMnCl_3$, is shown. TMMC consists of linear manganese chloride chains (-Mn-Cl$_3$-Mn-Cl$_3$-), which parallel the c-axis of the crystal [35]. The chains are separated by large tetramethylammonium ions; this diminishes the exchange interactions between the magnetic Mn^{2+} ions ($S = 5/2$) situated on different chains. The difference between the Mn^{2+}-ion separations on the chain ($3.25\,\text{Å}$) and in between chains ($9.15\,\text{Å}$) causes the on-chain exchange interaction to exceed the interchain exchange interaction by a factor of 10^4 [36,37]. As a result, at room temperature the on-chain migration rate of the excitons situated on the Mn^{2+} ions was found to be *eight* orders of magnitude higher than the interchain migration rate [9,10]. TMMB shows a similar result [13]. To our knowledge, this is the highest such anisotropy ratio observed in an experimental system. In other systems, 3D transport is observed to influence the experimental results, e.g., in $CsMnBr_3$ [12,18], which has an anisotropy ratio of $\approx 10^4$. These systems are denoted as quasi-1D.

An instructive demonstration of the influence of magnetic interactions on the dimensionality of exciton transport was given for $GdCl_3$ [22] (hexagonal unit-cell symmetry). In this system, when lowering sample temperature from $4.4\,\text{K}$ to $1.5\,\text{K}$, the 3D exciton transport in the paramagnetic phase was observed to change to (quasi-)1D exciton transport when entering the ferromagnetic phase, which is characterized by a net spin moment parallel to the crystal c-axis. In $GdCl_3$, as well as in $CsNiF_3$, the dimensionality of the exciton transport can be determined but the mode of transport, coherent or diffusive, is very difficult to determine since the trapping rate is, to a high degree, determined by the very low trapping efficiency and not by the actual transport process. In other words, in these systems the trapping rate is not migration limited.

The chain lengths in the inorganic systems are determined by the quality and impurity content of the single-crystal samples. Lattice distortions result in shallow traps, which, however, will not affect exciton transport at room temperature. Impurity ions can act as deep traps, i.e., as chain ends. Cu^{2+}, Co^{2+}, and Ni^{2+} have been observed to act as radiationless traps ('quenchers') [8-10,13,18,19], whereas rare-earth ions like Er^{3+}, Nd^{3+}, Tm^{3+}, and Tb^{3+} have been observed to act as radiating traps [14-16,27]. An impurity ion can act as a trap when its excited-state energy level is below the excited-state energy level of the host ions. The trapping efficiency is found to vary (e.g., Cu^{2+} in $CsMnCl_3 \cdot 2H_2O$ shows an efficiency of $\approx 10^{-4}$ [19]), owing to the existence of an energy barrier for trapping. The physical mechanism may be a lattice distortion around the impurity ion. The energy barrier causes the trap to act also as a reflecting barrier [33]. Cd^{2+} and Mg^{2+} have been observed to act purely as reflecting barriers (or scattering impurities) [9,16,18], on account of their high excited-state energy levels. Thermal activation allows the excitons to cross these barriers. In organic systems,

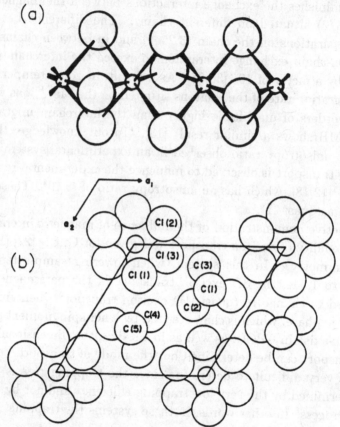

Fig. 20.1. (a) The linear manganese chloride chains (-Mn-Cl₃-Mn-Cl₃-), found in TMMC. (b) Representation of the crystal structure of TMMC, viewed along the manganese chloride chains. The tetramethylammonium ions are located in between the chains, effectively spacing the chains.

isotopic impurities often act as scattering impurities [32,33]. In chemically pure TMMC crystals, chain lengths of 10^5-10^6 sites ($\approx$ 30-300 μm) can be realized, which allows the observation of exciton fusion over a wide range of exciton densities [1].

In polymers the one-dimensionality is a direct reflection of the polymer-chain geometry. However, polymers suffer from many 3D distortions such as kinks in the chains and improper separation of the chains, which render the transport of excitations quasi-1D. Chain imperfections also act both as shallow and deep traps; this was found to reduce chain length strongly [25]. A typical system is trans-polyacetylene, where the highest attainable chain lengths are of the order of 300 carbon atoms (400Å). Transport was observed to be diffusive [23], with a 0.1 ps hopping time [25].

In the columnar liquid crystals of Fig. 20.2, the stacks of disk-like phtalo-cyanine chromophores are well separated by the long alkoxy groups. The intracolumn phtalocyanine separation of ≈ 4 Å, compared to the intercolumnar distance of 30-40 Å, ensures that laser-excited triplet excitons migrate along the columns [28]. In the liquid-crystal phase, transport-limited exciton annihilation has been observed on a time scale of $\leq 5\,\mu$s after pulsed excitation of the sample.

At later times, the decay of the exciton population was governed by exciton trapping. The annihilation and trapping kinetics showed the transport to be diffusive, with a 0.5-10 ps hopping time at around 85 °C. The column length between two traps is of the order of $1\,\mu$m (≈ 2500 sites), considered to correspond to the size of a monodomain [28].

In stacks of aromatic organic molecules such as J-aggregates [30,31], the exciton transport along the stack is coherent at room temperature, a result of the strong wavefunction overlap of adjacent molecules, which leads to the delocalization of the exciton wavefunction over the entire chain. This can also be inferred from the kinetics of the experimentally observed exciton annihilation and trapping: the kinetic equations show a time-independent rate constant, which indicates coherent transport.

At very low temperatures (≤ 4.2 K), coherent transport also takes place in crystals of charge-transfer complexes like naphtalene tetrachlorophtalic anhydride (ntcpa), where the motion of the excitons is along stacks containing alternate donor and acceptor molecules [33]. The presence of scattering impurities on the chain induces a diffusive component in the transport process [32,33].

In the systems described above, the migrating species have all been quasi-particles (electronic excitations). We now turn to real particles. In studies on the radiation damage of cadmium (hexagonal unit-cell symmetry), the recombination of radiation-induced vacancies and self-interstitials at 125 K was shown to be determined by predominantly 1D diffusion of the interstitial Cd atoms along the crystal c-axis, the vacancies being essentially immobile [34]. The diffusion constant showed a fairly high anisotropy ratio: $\mathcal{D}_{\parallel}/\mathcal{D}_{\perp} \approx 10^5$, arising from a channeling effect parallel to the hexagonal axis. At $T = 125$ K, the hopping time of the interstitial Cd atoms along the channel is 460 μs.

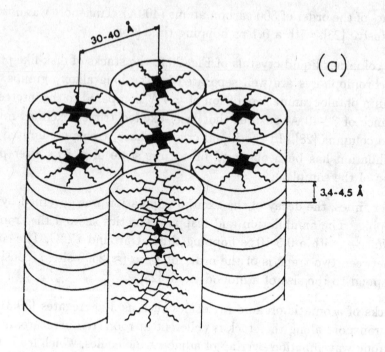

Fig. 20.2. (a) Schematic representation of a columnar liquid-crystalline phase: the alkoxy chains effectively space the stacks of phtalocyanine chromophores. (b) Building block of a column: octasubstituted phtalocyanine $(ROCH_2)_8PcM$, with $M = H_2$ or Zn, and $R = C_{12}H_{25}$ or $C_{18}H_{37}$.

20.3 Diffusion-limited single-species reactions

Diffusion-limited annihilation $(A + A \rightarrow 0)$ and fusion $(A + A \rightarrow A)$ reactions in systems showing three-dimensional transport are characteristic mechanisms for chemical reactions in general. The problem is well understood and many reviews on this topic are available [38]. In the 3D case, with homogeneous starting conditions, the diffusion is efficient enough to maintain the homogeneity in the sample volume during the reaction. As a result, the rate equation for the density $\rho(t)$ in the system obeys the familiar quadratic relation

$$\frac{d\rho(t)}{dt} = -\eta k \rho^2(t), \quad t \rightarrow \infty, \tag{20.1}$$

where k is the reaction rate constant and $\eta = 2$ for the annihilation reaction and $\eta = 1$ for the fusion reaction. After starting the reaction with an initial density $\rho(0)$, the decay in time is described by

$$\rho(t) = \rho(0)/\left[\eta k \rho(0)t + 1\right]. \tag{20.2}$$

On a microscopic level, the spatial fluctuations in the density are important to characterize the reactions. In a system with a spatially uniform density, the statistical properties of the spacings between particles may still be different. This is described by the distribution of spacings between particles, the *interparticle distribution function* (IPDF). A homogeneous density is associated with an equal probability of occupation for each position in the volume. The associated IPDF in the homogeneous situation is exponential and describes the probability of finding the nearest particle at a distance r in any direction:

$$P_E(r) = (1/r_0) \exp\left(-r/r_0\right). \tag{20.3}$$

The exponential IPDF is the starting distribution in experiments where the (quasi-)particles are generated by an instantaneous laser excitation, assuming that the probability of excitation is equal for each site. This is, e.g., the case for laser-excited excitons in the experiments described in Secs. 20.2 and 20.4. In 3D the IPDF remains exponential as the reaction proceeds in time, although the average density gradually decays according to (20.2).

Both the time behavior of the density and of the IPDF are drastically changed for reactions in less than three dimensions [5]. In particular, distributed surfaces with fractal properties or multiply connected chain-like structures, which may in many cases be characterized by an effective dimensionality $1 \leq D < 3$, show distinct deviations from 3D behavior.

Here we limit the discussion to strictly 1D systems and focus on the essential features of the models, in order to compare with the experimental results. The ideal 1D diffusion-limited reactions are influenced by competing decay channels such as those due to finite lifetime of the excitons, interchain transport, and finite chain length.

20.3.1 Ideal diffusion-limited fusion and annihilation

1. Fusion

The diffusion-limited single-species fusion model [5,6] considers the diffusive motion of particles, subject to an irreversible fusion reaction when two of them meet: $A + A \rightarrow A$. In this process, the diffusive transport of the particles constitutes the rate-limiting step. Fusion is assumed to take place instantaneously whenever two particles meet (this is known as the extreme diffusion limit [7]). At long times and fixed temperature the fusion process obeys the rate equation [5,38]

$$\frac{d\rho(t)}{dt} = -k(t)\,\rho^2(t), \quad t \rightarrow \infty, \tag{20.4}$$

in which density $\rho(t)$ is the number of particles per unit length and $k(t)$ is the macroscopic rate 'constant'. The rate constant for a 1D system is different from the 3D case [cf. (20.1)]. The rate constant $k(t)$ is related to the quantity $S(t)$, the random walker's exploration space, i.e., the mean number of distinct lattice sites visited by the random walker at time t [5]: $k(t) = dS(t)/dt$. The expression for $S(t)$ depends strongly on the dimensionality of the lattice [39]. For a 1D lattice $dS(t)/dt \propto t^{-1/2}$. Thus, (20.4) becomes

$$\frac{d\rho(t)}{dt} = -k't^{-1/2}\rho^2(t) = -k'\rho^3(t), \tag{20.5}$$

with $k' = \sqrt{(\Delta x)^2/(\pi \tau_{\text{hop}})} = \sqrt{2D/\pi}$, where Δx is the lattice spacing, τ_{hop} is the average time a particle remains on one site in between two consecutive hops, and D is the diffusion constant [7]. The apparent rate equation (20.5) is cubic in the density instead of quadratic as in the 3D case (20.1).

The general solution of (20.5) is [7,3]

$$\frac{\rho(t)}{\rho(0)} = \frac{1}{1 + \alpha\sqrt{t}} \approx \frac{1}{\alpha\sqrt{t}}, \quad (t \rightarrow \infty), \tag{20.6}$$

in which $\rho(0)$ represents the initial exciton density created under pulsed excitation at $t = 0$. The constant α is given by

$$\alpha = \sqrt{\frac{4}{\pi}}\rho(0)\sqrt{\frac{1}{\tau_{\text{hop}}}}, \tag{20.7}$$

where $\rho(0)$ is expressed as a fraction of the number of chain sites.

The density decay described by (20.6) is usually not directly observable. In the case of systems where optical detection of the luminescence of the excited state is possible (cf. Sec. 20.2), the luminescence intensity $I(t)$ is proportional to the instantaneous exciton density $\rho(t)$, since it results from spontaneous radiative decay of the excitons. This allows us to rewrite (20.6) as

$$\ln\left[\frac{I(0)}{I(t)} - 1\right] = 0.5\ln t + \ln\alpha. \tag{20.8}$$

The rate equation describing the reaction kinetics, (20.4), results from spatial fluctuations in the concentrations of the reactants that are not eliminated by 1D diffusion. This process was described by Kopelman *et al.* [5], and by ben-Avraham *et al.* [6], in terms of the time evolution of the distribution of particles on a chain. The starting point is a completely random distribution of particles, characterized by an exponential IPDF $P_E(x) = \rho(0)\exp[\rho(0)x]$ ((20.3) in 1D). As the reaction proceeds, the ensemble of particles exhibits a dynamic *self-ordering* process, characterized by a change of the exponential IPDF to a skewed Gaussian form, $P_{SG}(x,t) = (\pi/2)\rho^2(t)x\exp[-(\pi/2)\rho^2(t)x^2/2]$. The initial reaction rate depends strongly on the exact shape of the initial IPDF. For particles distributed on a chain according to an exponential IPDF the initial reaction rate is 100 times larger than for a skewed Gaussian initial IPDF (at identical initial particle density).

Starting from an initially random IPDF, one may derive an exact result for the decay of the survival fraction for the $A + A \rightarrow A$ process in a 1D system (after ben-Avraham *et al.* [6]),

$$\frac{\rho(t)}{\rho(0)} = \left[1 - \text{erf}\left(\sqrt{\frac{\rho^2(0)}{\tau_{\text{hop}}}}\,t\right)\right]\exp\left[\frac{\rho^2(0)}{\tau_{\text{hop}}}\,t\right], \tag{20.9}$$

with $\rho(0)$ expressed as a fraction of the number of sites. Comparing the results of (20.6) and (20.9), one finds them identical to within 5% over the time scale of our experiment. This renders (20.6) a very useful approximation for the interpretation of our results.

The effect of the asymptotic skewed Gaussian IPDF and the generating mechanism of the excitations is also apparent when the generation is not instantaneous. Doering and ben-Avraham [40] solved the equilibrium IPDF under a constant and homogeneous feeding, obtaining a different but general result for the IPDF in such a system. The effects of periodic excitation have been studied by Kopelman et al. [41] and may lead to a variety of phenomena in the spatial distribution and the dynamics of the exciton population.

2. Annihilation

The decay of the density for the diffusion-limited 1D annihilation reaction has the same asymptotic behavior as for the fusion case described in the previous section [7,3]. The main difference between annihilation and fusion in 1D reactions is the asymptotic IPDF [5]. The annihilation reaction yields, asymptotically, an IPDF that is a skewed exponential distribution, $P_{SE}(x,t) = (\pi/2)\rho(t)x \exp[-(\pi/2)\rho(t)x/2]$, instead of a skewed Gaussian as is the case for the fusion reaction.

Unfortunately the experiments do not probe the IPDF directly but are only sensitive to the time-dependent density in the system. The asymptotic time dependence for the two reaction models is the same, apart from a trivial rate enhancement by a factor of two. In an experimental situation it is therefore difficult to discriminate between fusion and annihilation if no other information than the asymptotic behavior of the density as a function of time is available.

The influence of simultaneous annihilation and fusion processes has been treated in a lattice representation of the problem [42-44]. This approach maps the master equation describing the reactions onto a 1D Schrödinger-type equation familiar from quantum mechanics. In the limit of infinite systems and pure fusion, the scaling behavior coincides [43] with the continuum equation derived by ben-Avraham et al. [6].

20.3.2 Deviations from ideal

The ideal situation sketched in the previous section is disturbed in experimental systems in many ways. As already mentioned in the discussion of the experimental systems in Sec. 20.2, limited chain length and interchain diffusion limit the applicability of the exact results as described, e.g., by (20.9). A few of the effects influencing the ideal behavior will be mentioned here. The approximate effect on the decay of the density, if the process were the only one contributing to the dynamics, is also noted.

Natural lifetime, in the absence of reactions, $\rho(t) \propto exp(-t/T_1)$: In particular the experimental systems based on laser-generated quasi-particles have an intrinsic decay time T_1. This lifetime may be associated with a luminescent transition, or with any decay mechanism that has an equal transition probability per unit time.

Chain length limitations due to traps or reflecting barriers: For example, for inefficient traps, randomly distributed on a chain, $\rho(t) \propto tK_2(t^{1/2})$, where K_2 is a modified Bessel function of the second kind [21]. In particular, traps may have a drastic influence on exciton reactions. Trapping may compete with the fusion and annihilation reactions and will, at very low density of the reacting species, dominate the reaction kinetics. There is a qualitative difference in the influence of such traps on the reaction rate in three-dimensional systems and lower-dimensional systems.

Interchain diffusion: Diffusive transport retains its 1D character only as long as, on average, a 3D hop has not occurred [34]. After a few such hops the characteristics of a purely 1D reaction are completely destroyed, and a quasi-3D diffusion begins to govern the reaction, ultimately leading to a decay described by (20.1).

Heating: In the next section we will show that the heat produced in reactions can influence the transport properties of the remaining particles.

Interactions and correlations: In Sec. 20.3.1 it was assumed that the reaction and transport properties of the species do not depend on the density. In real systems such a density dependence can be caused by an effective repulsive or attractive interaction potential between the particles (see next section for a discussion of Monte Carlo simulations on such a system).

Non-perfect reactions: In the previous discussion it was assumed that the reaction between two particles is instantaneous and irreversible. Some deviations from this ideal situation can be included in the analytical models (see, e.g., Alcaraz et al. [42]). In experiments on laser-excited systems the reaction time will be of the order of the formation time of the excitons, typically of the order of 100 fs. The reaction probability can be expressed in terms of the ratio of the reaction time to the hopping time, the latter often being strongly dependent on sample temperature.

In Fig. 20.3 the influence of some of the disturbing effects on the ideal behavior are indicated, using the form (20.8) to represent the decay curves. The occurrence of one or more of these disturbing effects usually makes an analytical interpretation of the decay very difficult, and Monte Carlo type simulations are required to obtain the overall density decay.

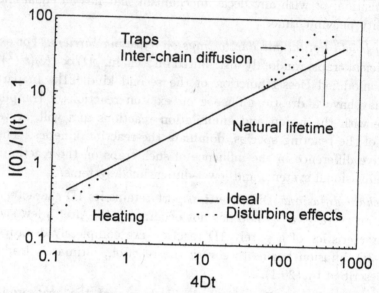

Fig. 20.3. Schematic representation of the influence of various effects on the exciton density $\rho(t)$, as listed in Sec. 20.3.2; $\rho(t)$ is expressed in the form suggested by (20.8).

When studying pure 1D annihilation or fusion, the concentration of (quasi-) particles on a single chain should be high enough to suppress the effects of the reflecting barriers and the traps at the chain ends. This requirement is only fulfilled for systems with sufficiently long chains. At the same time the reaction should be instantaneous and interchain migration should be negligible. As shown in Sec. 20.2, these requirements are beautifully fulfilled by the manganese-based system TMMC, which is therefore an ideal model system to test the theoretically obtained 1D reaction properties.

20.4 Exciton fusion in $(CH_3)_4NMnCl_3$ (TMMC)

In this section, we present the results of a detailed experimental study on the dynamics of a diffusion-limited reaction between photogenerated excitons in 1D TMMC (the properties of TMMC have been described in Sec. 20.2).

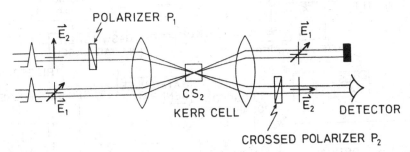

Fig. 20.4. Schematic showing the principle of the optical Kerr gate. Inside the Kerr medium (CS_2), the strong laser field E_1 induces a change in polarization of the weak optical signal E_2, allowing it to be detected. A more detailed explanation is given in the text.

20.4.1 Experimental procedure

1. Picosecond time-resolved luminescence: the optical Kerr gate

To observe the decay in time of the exciton population in TMMC, we have measured the TMMC time-resolved luminescence on a picosecond-to-millisecond time scale. We have applied the optical Kerr effect [45,46] to create a picosecond optical gate, as shown schematically in Fig. 20.4.

When an intense linearly polarized light pulse (E_1) travels through an optically isotropic medium, e.g., liquid CS_2, the material becomes temporarily anisotropic. The laser field E_1 generates a nonlinear polarizability inside the medium that results in an induced optical birefringence, i.e., an induced difference between the indices of refraction parallel and perpendicular to the laser field. In the presence of E_1, the response of the medium to a weak optical signal E_2 will be like that of a uniaxial crystal in which E_1 defines the optic axis. E_2 will experience a change in polarization, which will be maximum when E_2 is polarized at 45° with respect to E_1. E_2 will subsequently pass through polarizer P_2 and be detected. Scanning the position of E_1 in time, with respect to E_2, allows one to measure the transient profile of E_2. In our setup the time resolution is determined by the pulse duration of E_1 (8 ps FWHM), rather than by the damping time of the nonlinear polarization (1.5 ps for CS_2 [47]).

The full setup is shown in Fig. 20.5. The output of a synchronously pumped dye laser is amplified in a three-stage dye amplifier, pumped by the

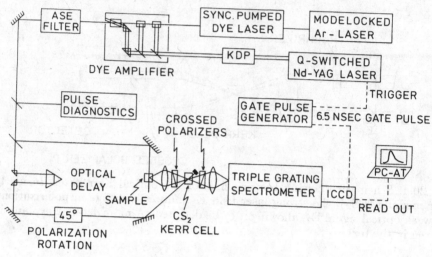

Fig. 20.5. Picosecond time-resolved luminescence setup, employing an optical Kerr gate. An explanation is given in the text.

frequency-doubled output of a Q-switched Nd-YAG laser. The dye-amplifier output consists of 1 mJ pulses of 8 ps duration at 580 nm, at a repetition rate of 20 Hz. The pulse energy is stable within ±8%. A few per cent of the beam is split off to monitor the pulse energy and pulse shape during the experiment. Subsequently, the beam is split into equal parts. One branch is used to excite the TMMC sample ($^6A_{1g} \rightarrow {}^4T_{1g}$ transition, $\mathbf{E} \perp c$-axis [36,48,49]); the other is used to operate the Kerr gate. The TMMC time-resolved luminescence signal is collected at right angles in a 10 nm band-pass around 660 nm ($^4T_{1g} \rightarrow {}^6A_{1g}$ transition) with a triple-grating spectrometer, and detected with an image-intensified CCD. The ICCD is gated with a 6.5 ns electronic gate pulse, synchronized to the laser pulse through use of the Nd-YAG laser Q-switch as a trigger for the ICCD gate. The peak signal is 75 photons per second impinging on the detector at a Kerr-gate efficiency of 25%.

The sensitivity of the setup amounts to 10 signal photons per second impinging on the detector. The main source of noise is spurious photons at the signal frequency, generated inside the Kerr cell under the influence of the strong laser beam by a variety of optical processes such as stimulated Raman scattering, continuum generation, spontaneous Raman scattering, and fluorescence. Integrating over 9000 laser shots renders this background stable within 10 photons per second. The time resolution of the setup is less than 5 ps after deconvolution. The time delay was obtained by scanning an optical delay line over a total of 1.5 ns.

A short comparison of Kerr-gating to other optical spectroscopic techniques that are commonly used to measure time-resolved luminescence reveals the Kerr gate to be superior for this experiment: it shows a good time resolution, very low detection level, and is not sensitive to the signal frequency, which facilitates optical alignment and rapid change in signal frequency. Upconversion [50] does show a better time resolution but is not sensitive to signals below a level of 75 photons per second. Also, alignment problems arise due to phase-matching requirements. Use of a streak camera [51] in pulsed mode, or time-correlated single-photon counting [52], shows a comparable detection level to the Kerr gate but a smaller time resolution (≈ 20 ps), and these systems do not combine very well with a low-repetition-rate laser system. For TMMC, a low repetition rate is required because of saturation of the electronic transition at high repetition rates (a result of the long population relaxation time T_1 of TMMC: $T_1 = 740\,\mu s$ [8]).

Time-resolved luminescence on a nanosecond-to-millisecond time scale was obtained after pulsed excitation of the TMMC samples with the 532 nm frequency-doubled output of the Q-switched Nd-YAG laser used to pump the dye amplifier in Fig. 20.5. The pulse duration was 2 ns. The luminescence signal at 660 nm was detected using the ICCD electronic gate to set the time resolution, which was 3 ns after deconvolution. The time delay was obtained by electronically delaying the trigger pulse from the Nd-YAG laser Q-switch. Signals were reproducible within ±1%.

2. Sample preparation and handling

TMMC single crystals were grown from a slightly acidic (10% HCl [36]) aqueous solution, containing stoichiometric amounts of $(CH_3)_4NCl$ and $MnCl_2 \cdot 4H_2O$, by slow evaporation at room temperature using a dry nitrogen-gas purge. The crystals grow in bright pink hexagonal rods measuring up to 15 mm in length and 5 mm in diameter. The crystallographic c-axis parallels the axis of the rod.

Chemical analysis proved the crystals to be practically free of impurity ions (0.3×10^{-6} mole fraction Cd, 15×10^{-6} mole fraction Cu). The samples were carefully polished using moist lens tissue so as to enhance surface quality, and thus the damage threshold against laser irradiation. We obtained samples measuring typically $1\times3\times10\,mm^3$ with high-quality surfaces. In the experiments, the maximum accessible exciton density did turn out to be limited by laser-irradiation-induced sample damage. The results displayed were obtained below the damage threshold. At incident laser-power densities above the damage threshold the crystal turned grayish over the entire focal volume.

20.4.2 Experimental results

The time-resolved luminescence of TMMC was measured at room temperature as a function of the exciting Nd-YAG laser power, as shown in Fig. 20.6. Pulse energies of 0.74 to 39.9 μJ were focused on a 1-mm-thick sample in a 50 to 70 μm diameter focal spot. The long-time behavior of the luminescence decay curves is identical, as we derived from measurements extending over a time scale of 10 ms (not shown). It corresponds to an exponential decay with a time constant $T_1 = 740\,\mu$s, which agrees well with the literature values on the population relaxation time of TMMC under weak excitation conditions [8]. On a sub-millisecond time scale we observed an additional contribution to the decay of the TMMC luminescence signal, which intensifies with increasing laser power.

Further increase of the laser power incident on the sample results in a drastic change of shape of the luminescence decay curves, depicted in Fig. 20.7. After a fast initial behavior, the decay rate of curves B, C, and D is observed to reduce considerably, leading to a crossing with the 100%-curve of Fig. 20.6, here denoted curve A. At the highest laser powers, the initial decay of the luminescence signal can no longer be resolved by the ICCD gate, as shown in Fig. 20.7 (inset). Use of the picosecond optical Kerr gate indeed reveals a very fast initial luminescence decay behavior. This is shown in Fig. 20.8. The rise time of the luminescence signal is limited by the instrumental function of the Kerr gate, whereas the signal decays on a time scale of a few hundred picoseconds.

On reducing the sample temperature to 80 K, under the excitation conditions of Fig. 20.6, exponential decay on a sub-millisecond time scale is obtained, with a decay time $T_1 = 780\,\mu$s (not shown). Its weak dependence on temperature indicates T_1 to be the radiative lifetime of the $^4T_{1g}$ state [8].

20.4.3 Discussion

Experimental evidence for the occurrence of an exciton annihilation or fusion reaction in TMMC is presented in Fig. 20.9, where we plot the total luminescence signal (i.e., $I(t)$ integrated over time) of Fig. 20.6 vs. Nd-YAG laser pulse energy incident on the sample. In the absence of exciton annihilation or fusion one should observe the experimental data to follow the

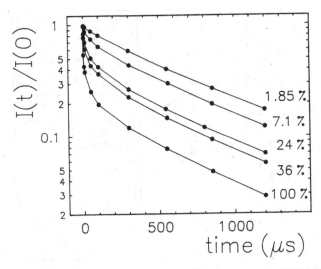

Fig. 20.6. Logarithmic plot of the experimentally observed TMMC luminescence intensity *vs.* time for various excitation energies. Nd-YAG pulse energies of 0.74 to 39.9 μJ were incident on the sample. Measurements were taken at room temperature. The curves have been normalized to their values at zero time delay. The solid lines are guides to the eye.

broken line. We clearly observe a loss of signal at 660 nm (our wavelength of detection).

Exciton annihilation processes have been thoroughly studied in molecular crystals like anthracene [53]. There, two triplet excitons often annihilate to form a singlet exciton at approximately twice the triplet exciton energy. The presence of the singlet exciton can be traced by its luminescence, so-called delayed luminescence [54]. This reaction consumes two excitons: $A+A \rightarrow \hbar\omega$. In TMMC no delayed fluorescence has been observed: the $^4T_{1g} \rightarrow {}^6A_{1g}$ transition is the only emission band of TMMC [36,8,12]. The observed signal loss at 660 nm therefore cannot be ascribed to an annihilation-induced radiative decay mechanism. In likely scenario for the *nonradiative* loss of excitons in TMMC, two excitons meet at a single lattice site, thus exciting the Mn^{2+} ion to twice the exciton energy. Subsequently, emission of phonons causes the ion to relax back to the $^4T_{1g}$ state. This reaction consumes one exciton: $A + A \rightarrow A + \text{heat}$. An exciton-fusion mechanism similar to that described

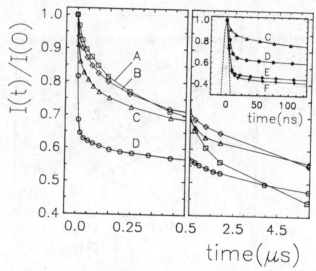

Fig. 20.7. Decay of the experimentally observed normalized luminescence signal of TMMC. Excitation densities (relative to the 100%-curve of Fig. 20.6) are as follows: A, 1.0; B, 2.0; C, 4.1; D, 6.1; E, 8.3; F, 11.2. Measurements were taken at room temperature. The solid lines are guides to the eye. At excitation densities larger than 1.0 the decay curves change shape, with respect to Fig. 20.6. Inset: for high excitation densities the initial decay is unresolved by the ICCD gate. The broken line shows the instrumental function of the ICCD gate.

here has been observed by Diggle *et al.* [55] for excitons in $TbPO_4$, and was used by Wilson *et al.* [56] to explain bi-exciton decay in MnF_2.

Exciton fusion in TMMC is found to cease upon lowering the sample temperature to 80 K under the excitation conditions of Fig. 20.6 (at 80 K we obtain exponential luminescence decay curves). Thus, it is obvious that the transport properties of the excitons play a crucial role in the fusion process.

The nonexponential behavior of the luminescence decay curves, Fig. 20.6, is favorably described by the diffusion-limited single-species fusion model. This is shown in Fig. 20.10(a) where the curves correspond to the form (20.8). The observed linear dependence with slope 0.5 (cf. Table 20.3) accurately identifies the 1D character of the exciton fusion process in TMMC under sufficiently strong laser excitation. For laser pulse energies of 14.2 and 39.9 μJ (the 36% and 100% curves of Fig. 20.6) we are able to follow the exciton (dif)fusion kinetics over four decades in time. At times $\geq 100\mu s$

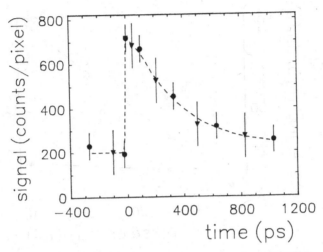

Fig. 20.8. Picosecond time-resolved luminescence signal of TMMC at high initial exciton density (of the order of a per cent of the number of sites, comparable to curve F of Fig. 20.7 (inset)), obtained using an optical Kerr gate. The results of two measurements are shown. The broken line is a guide to the eye.

we observe in the figure the influence of spontaneous emission manifested as the deviation from linear behavior.

The parameters for the best linear least-squares fit to (20.8), represented by the solid lines in Fig. 20.10(a), are listed in Table 20.3. We follow (20.7) to calculate the values for the initial exciton density $\rho(0)$ from the α values obtained from the fits. The values for $\rho(0)$ thus obtained are plotted vs. laser pulse energy incident on the sample in Fig. 20.10(b). The linear scaling of $\rho(0)$ with incident laser energy confirms the adequacy of the theoretical model. For the calculation of $\rho(0)$, in (20.7) a hopping time of 10 ps was used [9-10]. The initial exciton density derived from the experimental excitation conditions (input laser power, interaction volume and TMMC absorbency at 532 nm [8]) is in accordance with the $\rho(0)$ values of Table 20.3.

A drastic change in the shape of the luminescence decay curves results from increasing the laser power on the TMMC samples beyond the 100%-curve of Fig. 20.6. This is shown in Fig. 20.7. In this case, the initial decay rate is considerably higher than predicted by the diffusion-limited single-species fusion model, after which it is observed to reduce substantially. In

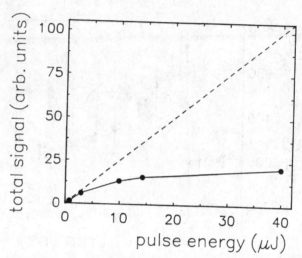

Fig. 20.9. Integral of the experimentally observed TMMC luminescence signal $I(t)$ of Fig. 20.6 over time, plotted *vs.* excitation energy incident on the sample. In the absence of exciton annihilation or fusion one should observe the experimental data to follow the broken line.

Table 20.3.

Initial exciton density $\rho(0)$ *vs.* Nd-YAG laser pulse energy incident on the TMMC sample. The slope and parameter α of the solid lines in Fig. 20.10(a) are derived by fitting the experimentally observed luminescence decay curves of Fig. 20.6 to (20.8). Corresponding values for $\rho(0)$ are derived from (20.7) with $\tau_{\text{hop}} = 10$ ps. $\rho(0)$ is expressed as a fraction of the number of sites [57]

Pulse energy (μJ)	Slope	α	$\rho(0)$
0.74	0.480	2.0	5.66×10^{-6}
2.84	0.475	4.9	1.37×10^{-5}
9.90	0.495	14.5	4.07×10^{-5}
14.23	0.485	20.2	5.66×10^{-5}
39.93	0.460	62.5	1.77×10^{-4}

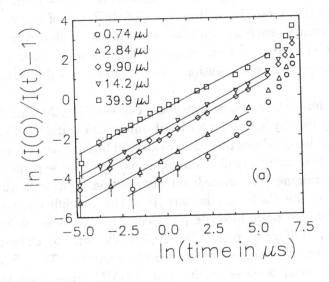

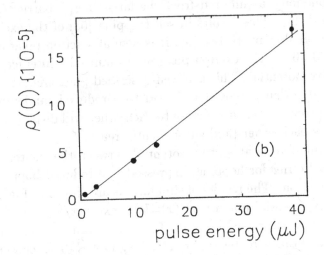

Fig. 20.10. Analysis of the TMMC luminescence decay curves of Fig. 20.6 in terms of the diffusion-limited single-species fusion model. (a) Double logarithmic plot of the luminescence decay curves. Solid lines are best fits to (20.8); the exciting Nd-YAG pulse energies are indicated. (b) Initial exciton density $\rho(0)$ vs. laser pulse energy incident on TMMC, derived from (20.7). The solid line is a guide to the eye. $\rho(0)$ is expressed as a fraction of the number of sites. See also Table 20.3.

[1], we have shown that the observed anomalous enhancement of the initial luminescence decay rate is not the result of the onset of a new population-relaxation channel, such as amplified spontaneous emission in the TMMC sample, but that also in this regime of high incident laser pulse energy the diffusion-limited exciton-fusion process constitutes the dominant decay channel.

The behavior of the luminescence decay curves of Fig. 20.7 implies that the excitons very rapidly attain an 'ordered' interparticle distribution function. As a result, at longer times their decay rate will be considerably reduced. We will now analyze three possible mechanisms that may explain the anomalously fast attainment of an ordered distribution. These mechanisms are analyzed in Monte Carlo simulations [1] of the 1D diffusion-reaction process, which aim to reproduce the picosecond luminescence decay observed with the optical Kerr gate (cf. Fig. 20.8). The first mechanism we consider is an attractive potential between the excitons. The physical origin of such a potential between excitons in TMMC arises from their strong coupling to phonons, as inferred from the large energy barrier for hopping $(800\text{-}1000\,\mathrm{cm}^{-1}$ [9]). This results in self-trapping [58] of the exciton by the lattice distortion that it creates, i.e., it is a small exciton polaron. The deformation field allows the exciton polarons to mutually interact, leading to a net attractive potential. This is a well-described phenomenon for electron polarons [59,60]. The potential will lead to a reduced randomness in the motion of the excitons: excitons close to each other will display a preference to hop towards each other (and subsequently react).

We have simulated an attractive potential between the excitons by reducing the energy barrier for hopping, expressed as a reduced hopping time, in a preferred direction. The results of simulations are shown in Fig. 20.11. We have used an exponential potential $V(x)=V(0)\exp(-x/b)$, with the distance between excitons x expressed in units of lattice parameter. The energy barrier for hopping (taken to be $1000\,\mathrm{cm}^{-1}$, i.e., $1439\,\mathrm{K}$) was chosen to be of the order of magnitude of $V(0)$. We have applied an initial exciton density $\rho(0) = 5\%$, and a hopping time (of a particle alone on a chain) $\tau_{\mathrm{hop}} = 10\,\mathrm{ps}$. The attractive potential induces a strong enhancement of the initial reaction rate, both as a function of potential strength $V(0)$ (Fig. 20.11(a)), and as a function of the range of interaction b (Fig. 20.11(b)). It is seen, however, that the notion of an attractive potential does not explain the results obtained in the picosecond experiment. The shape of the simulated decay curve is actually opposite to the experimental result. The initial decay is too fast, the subsequent decay is too slow. Several other potentials $[V(x) \propto x^{-1},\ x^{-3}]$ yielded similar results.

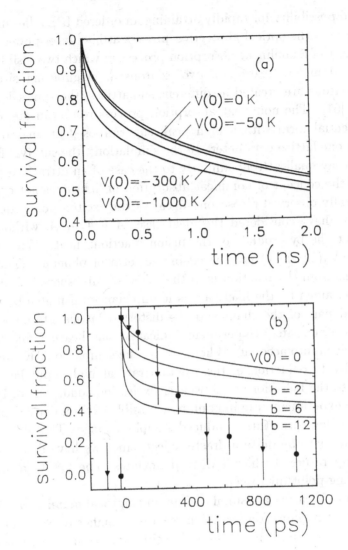

Fig. 20.11. Results of Monte Carlo calculations showing the effect of an attractive potential between the exciton polarons, $V(x) = V(0)\exp(-x/b)$, on the diffusion-limited $A + A \rightarrow A$ reaction in 1D. $\rho(0) = 5\%$, $\tau_{hop} = 10\,ps$. (a) The survival fraction as a function of $V(0)$ with $b = 2$. (b) The survival fraction as a function of b with $V(0) = -1000\,K$. It is seen that the shape of the simulated curves does not match the experimental results.

A second possibility for rapidly attaining an ordered IPDF lies in a change in the nature of the optical-absorption process at high laser-power densities, i.e., the onset of a nonlinear absorption process in which two visible photons are absorbed under creation of two excitons at neighboring sites. Since the two excitons are created at different sublattices, the spin selection rule is fulfilled [61]. The nonlinear absorption process results in an IPDF that has exponential form with a high peak superimposed at an inter-exciton distance of one lattice parameter. In the simulations, the survival fraction is found to decay qualitatively similarly to the case of an attractive potential. Therefore, the concept of nonlinear absorption of light does not explain the experimentally observed picosecond decay of the exciton population either.

So far, we have considered the ideal case, $A + A \rightarrow A$, without taking into account the by-product of the fusion reaction: heat. With every reaction, $1.9\,\mathrm{eV}$ ($22\,000\,\mathrm{K}$) is released in the form of phonons. The exciton that remains from the reaction is in the midst of this 'shower' and will feel its influence, aided by the likely acoustic impedance mismatch between the unperturbed part of the chain and the distorted lattice site on which the remaining exciton sits. Experiment indicates that heating due to fusion should be taken into account. When cooling the sample to $80\,\mathrm{K}$, we observe only a moderate reduction of the reaction rate at high input laser power, in contrast to the behavior at the weaker excitation conditions of Fig. 20.6. Also, in the experiments the maximum accessible exciton density proved to be limited by laser-irradiation-induced sample damage. This shows fusion-induced heating to be quite a drastic effect, since we have determined the (lattice) temperature at which TMMC dissociates to be $368\,^\circ\mathrm{C}$ (as measured with a melting-point microscope).

The effects of heating are simulated, in an empirical manner, by assigning an exciton that remains from a fusion reaction a higher temperature (T^*_{exc}), i.e., a reduced hopping time. The 'hot' excitons are assumed to react before losing their temperature, on account of the high density of excitons and the acoustic impedance mismatch mentioned above. The energy barrier for hopping ($1439\,\mathrm{K}$) was chosen to be of the order of magnitude of T^*_{exc}. Results of the simulations are shown in Fig. 20.12.

In Fig. 20.12(a), the behavior of the simulated survival fraction is seen to approximate the shape of the experimental data. Monte Carlo results for two values of the hopping time are shown, in view of the experimental uncertainty regarding its value [9,10]. Values for $\rho(0)$ are chosen to fit the experimental result. Their order of magnitude is in accordance with the value derived from the experimental excitation conditions. The effect of the temperature that we assign to the hot excitons is found to saturate rapidly;

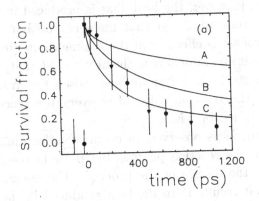

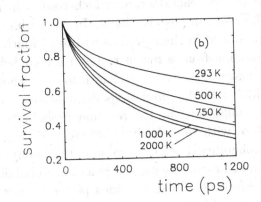

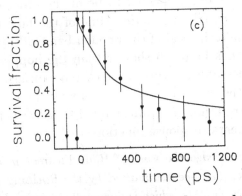

Fig. 20.12. Monte Carlo calculations showing the influence of the heat produced in the fusion reaction on the decay of the exciton population. (a) A, $T^*_{exc} = 293\,K$ (no heating), $\rho(0) = 5\%$, $\tau_{hop} = 10\,ps$; B, $T^*_{exc} = 1000\,K$, $\rho(0) = 5\%$, $\tau_{hop} = 10\,ps$; C, $T^*_{exc} = 1000\,K$, $\rho(0) = 2.5\%$, $\tau_{hop} = 1\,ps$. (b) The effect of T^*_{exc} rapidly saturates. The assigned T^*_{exc} is indicated. $\rho(0) = 5\%$, $\tau_{hop} = 10\,ps$. (c) We have included heating of the lattice proportionally to the reaction rate, as described in the text. $T^*_{exc} = 1000\,K$, $\rho(0) = 5\%$, $\tau_{hop} = 10\,ps$. The shape of the simulated curve is seen to match the experimental result very well.

see Fig. 20.12(b). However, the heat that is produced in the reactions will also influence excitons which, at that moment, are not participating in a reaction. This process is effective in three dimensions (in TMMC at room temperature there is no anisotropy in the heat conductivity [62]). It has been simulated in Fig. 20.12(c), where we have raised the average temperature proportionally to the reaction rate: after every ten reactions the average temperature was raised by 5 K.

The heating process is observed to be most influential, not so much in the initial stage of the reaction process, but more in the subsequent stage. This follows from the nature of the process. The enhanced hopping rate of the excitons that results from the heat produced in the reactions brings about an increase of the reaction rate, which results in more heat being produced, etc. The reaction sustains itself. In the regime of lower exciton density the heat produced in the reactions will be of much less importance, since excitons remaining from a fusion reaction are cooled down to lattice temperature long before they have the chance to meet another exciton.

Results of our Monte Carlo simulations demonstrate the initial reaction rate to be strongly influenced by the reaction probability (when two particles meet they fuse with probability η, where $\eta \leq 1$, and pass through each other with probability $1 - \eta$). In view of the highly efficient reaction process observed in the experiments, the reaction probability was assumed to be unity in all Monte Carlo simulations presented here. The reaction probability is determined by the ratio of the reaction time to the hopping time. The reaction time is expected to be of the order of the time of formation of the excitons (the relaxation time of the lattice to a distortion), which is equal to the reciprocal of the phonon bandwidth, i.e., $\approx 100\,\mathrm{fs}$. The reaction time will thus be much shorter than the hopping time; hence our interpretation of the data in terms of the extreme diffusion limit. The short formation time of the excitons also explains why the observed rise time in the picosecond Kerr-gating experiment (cf. Fig. 20.8) is limited to the rise time of the instrumental response function.

We gratefully acknowledge the work of Hilde Fleurent in the experiments on TMMC. This research has been supported by the Stichting voor Fundamenteel Onderzoek der Materie (FOM), which is financially supported by the Nederlandse Organisatie voor Wetenschappelijk Onderzoek (NWO).

References

[1] R. Kroon, H. Fleurent and R. Sprik, *Phys. Rev.* E47, 2462 (1993).

[2] D. Toussaint and F. Wilczek, *J. Chem. Phys.* 78, 2642 (1983).

[3] G. Zumofen, A. Blumen and J. Klafter, *J. Chem. Phys.* 82, 3198 (1985).

[4] K. Kang and S. Redner, *Phys. Rev.* A32, 435 (1985).

[5] R. Kopelman, *Science* 241, 1620 (1988); *J. Stat. Phys.* 42, 185 (1986); R. Kopelman, S. J. Parus and J. Prasad, *Chem. Phys.* 128, 209 (1988).

[6] D. ben-Avraham, M. A. Burschka and C. R. Doering, *J. Stat. Phys.* 60, 695 (1990); C. R. Doering and D. ben-Avraham, *Phys. Rev.* A38, 3035 (1988).

[7] D. C. Torney and H. M. McConnell, *J. Phys. Chem.* 87, 1941 (1983).

[8] H. Yamamoto, D. S. McClure, C. Marzzacco and M. Waldman, *Chem. Phys.* 22, 79 (1977).

[9] R. A. Auerbach and G. L. McPherson, *Phys. Rev.* B33, 6815 (1986).

[10] R. Knochenmuss and H. U. Güdel, *J. Chem. Phys.* 86, 1104 (1987).

[11] W. J. Rodriguez, M. F. Herman and G. L. McPherson, *Phys. Rev.* B39, 13187 (1989).

[12] R. L. Blakley, C. E. Martinez, M. F. Herman and G. L. McPherson, *Chem. Phys.* 146, 373 (1990).

[13] W. J. Rodriguez, R. A. Auerbach and G. L. McPherson, *J. Chem. Phys.* 85, 6442 (1986).

[14] U. Kambli and H. U. Güdel, *Inorg. Chem.* 23, 3479 (1984).

[15] G. L. McPherson and A. H. Francis, *Phys. Rev. Lett.* 41, 1681 (1978).

[16] K. F. Talluto, V. F. Trautmann and G. L. McPherson, *Chem. Phys.* 88, 299 (1984).

[17] W. J. Rodriguez, R. A. Auerbach and G. L. McPherson, *Chem. Phys. Lett.* 132, 558 (1986).

[18] G. L. McPherson, Y. Y. Waguespack, T. C. Vanoy and W. J. Rodriguez, *J. Chem. Phys.* 92, 1768 (1990).

[19] V. V. Eremenko, V. A. Kartachevtsev and A. R. Kazanchkov, *Phys. Rev.* B49, 11799 (1994).

[20] T. Tsuboi, *J. Phys. Condens. Matter* 5, 1143 (1993).

[21] J. Cibert, M. C. Terrile and Y. Merle d'Aubigné, *J. Phys., Coll.* C7, 135 (1985).

[22] R. Mahiou, B. Jacquier and C. Madej, *J. Chem. Phys.* 89, 5931 (1988).

[23] Z. Vardeny *et al.*, *Phys. Rev. Lett.* 49, 1657 (1982).

[24] C. V. Shank *et al.*, *Phys. Rev. Lett.* 49, 1660 (1982).

[25] S. D. Phillips and A. J. Heeger, *Phys. Rev.* B38, 6211 (1988); A. J. Heeger, S. Kivelson, J. R. Schrieffer and W.-P. Su, *Rev. Mod. Phys.* 60, 781 (1988).

[26] R. Kopelman, C. S. Li and Z.-Y. Shi, *J. Lumin.* 45, 40 (1990); see also Ch. 21.

[27] Yu. P. Blagoy, I. A. Levitsky, Yu. V. Rubin and V. V. Slavin, *Chem. Phys. Lett.* 203, 265 (1993).

[28] D. Markovitsi, I. Lécuyer and J. Simon, *J. Phys. Chem.* 95, 3620 (1991).

[29] V. Gulbinas, M. Chachisvilis and V. Sundström, *J. Phys. Chem.* 98, 8118 (1994).

[30] V. Sundström, T. Gillbro, R. A. Gadonas and A. Piskarskas, *J. Chem. Phys.* 89, 2754 (1988).

[31] M. van Burgel, D. A. Wiersma and K. Duppen, *J. Chem. Phys.* 102, 20 (1995).

[32] D. D. Dlott, M. D. Fayer and R. D. Wieting, *J. Chem. Phys.* 69, 2752 (1978).

[33] A. A. Avdeenko, V. V. Eremenko and V. A. Karachevtsev, *Sov. Phys. JETP* 67, 1677 (1988).

[34] U. Gösele and A. Seeger, *Phil. Mag.* **34**, 177 (1976); A. Seeger and U. Gösele, *Radiat. Effects* **24**, 123 (1975).

[35] B. Morosin and E. J. Graeber, *Acta Cryst.* **A23**, 766 (1967).

[36] R. Dingle, M. E. Lines and S. L. Holt, *Phys. Rev.* **187**, 643 (1969).

[37] M. Steiner, J. Villain and C. G. Windsor, *Adv. Phys.* **25**, 87 (1976).

[38] R. M. Noyes, *Prog. React. Kinet.* **1**, 128 (1961).

[39] E. W. Montroll and G. H. Weiss, *J. Math. Phys.* **6**, 167 (1965).

[40] C. R. Doering and D. ben-Avraham, *Phys. Rev. Lett* **62**, 2563 (1989). See also Part I of this book.

[41] R. Kopelman, Z.-Y. Shi and C. S. Li, *J. Lumin.* **48-49**, 143 (1991); see also Ch. 21.

[42] F. C. Alcaraz, M. Droz, M. Henkel and V. Rittenberg, *Ann. Phys.* **230**, 250 (1994).

[43] K. Krebs, M. Pfannmüller, B. Wehefritz and H. Hinrichsen, *J. Stat. Phys.* **78**, 1429 (1995).

[44] M. Henkel, E. Orlandini and G. M. Schültz, *J. Phys.* **A28**, 6335 (1995).

[45] M. A. Duguay and J. W. Hansen, *Opt. Comm.* **1**, 254 (1969).

[46] P. P. Ho and R. R. Alfano, *Phys. Rev.* **A20**, 2170 (1979); K. Sala and M. C. Richardson, *ibid.* **12**, 1036 (1975).

[47] B. I. Greene and R. C. Farrow, *Chem. Phys. Lett.* **98**, 273 (1983).

[48] K. E. Lawson, *J. Chem. Phys.* **47**, 3627 (1967).

[49] P. Day and L. Dubicki, *J. Chem. Soc. Faraday Trans.* II**69**, 363 (1973).

[50] A. Mokhtari, A. Chebira and J. Chesnoy, *J. Opt. Soc. Am.* B**7**, 1551 (1990).

[51] Y. Tsuchiya and Y. Shimoda, *Proceedings of SPIE*, Vol. **533**, p. 110, M. J. Soileau, ed. (1985).

[52] D. V. O'Connor and D. Phillips, *Time-correlated Single Photon Counting* (Academic Press, London, 1984).

[53] For a comprehensive review see K. C. Kao and W. Hwang, *Electrical Transport in Solids* (Pergamon, Oxford, 1981).

[54] R. G. Kepler, J. C. Caris, P. Avakian and E. Abramson, *Phys. Rev. Lett.* **10**, 400 (1963).

[55] P. C. Diggle, K. A. Gehring and R. M. Mcfarlane, *Solid State Comm.* **18**, 391 (1976).

[56] B. A. Wilson, J. Hegarty and W. M. Yen, *Phys. Rev. Lett.* **41**, 268 (1978).

[57] We thank Prof. V. Rittenberg for pointing out an inconsistency in the table published in Kroon *et al.* [1].

[58] Y. Toyozawa, in *Relaxation of Elementary Excitations, Proc. Taniguchi Internat. Symp., Susono-shi, Japan*, R. Kubo and E. Hanamura, eds. (Springer-Verlag, Berlin, 1980).

[59] H. Hiramoto and Y. Toyozawa, *J. Phys. Soc. Jpn* **54**, 245 (1985).

[60] H. De Raedt and A. Lagendijk, *Z. Phys.* **B65**, 43 (1986).

[61] R. Moncorgé and B. Jacquier, in *Collective Excitations in Solids, Proc. NATO Advanced Studies Inst. Collective Excitations in Solids, Enrice, Italy*, B. Di Bartolo and J. Danko, eds. (Plenum, New York, 1983).

[62] H. Miike and K. Hirakawa, *J. Phys. Soc. Jpn* **39**, 1133 (1975).

21

Experimental investigations of molecular and excitonic elementary reaction kinetics in one-dimensional systems

Raoul Kopelman and Anna L. Lin

It has been well established by theory and simulations that the reaction kinetics of diffusion-limited reactions can be affected by the spatial dimension in which they occur. The types of reactions $A + B \rightarrow C$, $A + A \rightarrow A$, and $A + C \rightarrow C$ have been shown, theoretically and/or by simulation, to exhibit nonclassical reaction kinetics in 1D. We present here experimental results that have been collected for effectively 1D systems.

An $A + B \rightarrow C$ type reaction has been experimentally investigated in a long, thin capillary tube in which the reactants, A and B, are initially segregated. This initial segregation of reactants means that the net diffusion is along the length of the capillary only, making the system effectively 1D and allowing some of the properties of the resulting reaction front to be studied. The reaction rates of molecular coagulation and excitonic fusion reactions, $A + A \rightarrow A$, as well as trapping reactions, $A + C \rightarrow C$, were observed via the phosphorescence (P) and delayed fluorescence (F) of naphthalene within the channels of Nuclepore membranes and Vycor glass and in the isolated chains of dilute polymer blends. In these experiments, the nonclassical kinetics is measured in terms of the heterogeneity exponent, h, from the equation rate $\sim F = kt^{-h}P^n$, which gives the time dependence of the rate coefficient. Classically $h = 0$, while $h = 1/2$ in 1D for $A + A \rightarrow A$ as well as $A + C \rightarrow C$ type reactions. Continuous and pulsed excitation experiments were also performed on naphthalene embedded in Vycor glass; these elucidate the effects of Poissonian *vs.* non-Poissonian distributions of reactants on the reaction kinetics in 1D.

21.1 Introduction

Nonclassical diffusion-limited reaction kinetics are of interest both in their own right and also as a tool in material sciences, catalysis, complex liquids and intracellular biochemistry [1,2]. Theoretically, the simplest cases of nonclassical reactions occur for one-dimensional (1D) topologies [3-25], even though fractal, disjoint-islands, two-dimensional, and even three-dimensional topologies may exhibit anomalous, nonclassical, reaction kinetics [18]. Subsequently, there has been an abundance of theory and of simulation work on reaction kinetics in 1D topologies. Experimentally, 1D systems pose obvious difficulties. However, there are several experimental systems that have been utilized successfully to observe nonclassical reaction kinetics in 1D. Three types of bimolecular elementary reactions are discussed in this chapter: *one-kind* ($A + A \rightarrow$ products), *two-kind* ($A + B \rightarrow$ products), and pseudo-monomolecular ($A + C \rightarrow C +$ products).

The nonclassical behavior of an $A + B \rightarrow C$ type reaction occurring in 1D was studied experimentally using the inorganic complexation reaction of disodium ethyl bis(5-tetrazolylazo)acetate trihydrate ('tetra') with Cu^{2+}. The reaction is carried out in long thin capillary tubes with the reactants A and B initially segregated in space. The reactants meet in the center of the capillary forming a reaction front. The kinetic behavior of this reaction front was predicted, via a scaling ansatz [24], to be nonclassical with respect to the temporal behavior of the position of the center of the reaction front, x_f, as well as the width of the front, w, and the local reaction rate at the center of the front, r_f. These predictions were borne out in an experiment which we discuss in Sec. 21.2.

A number of experiments have been performed that exhibit the nonclassical behavior of reactions of the type $A + A \rightarrow$ products (homofusion) and $A + C \rightarrow C +$ products (heterofusion) in 1D. Both the photochemical dimerization reaction of naphthalene molecules dissolved in a solvent that fills the cylindrical pores of Nuclepore membranes and the photophysical fusion 'reaction' of excitons traveling through crystalline-naphthalene-packed cylindrical channels of Nuclepore membranes or through the pseudo-1D channels of Vycor glass exhibit nonclassical reaction kinetics: the *heterogeneity coefficient*, h, where $k(t) = k_0 t^{-h}$, equals 1/2 instead of the classical value, $h = 0$, in the long-time limit.

The heterogeneity coefficient was chosen as the parameter by which to measure experimentally the nonclassical rates of the $A + A \rightarrow A$ and $A + C \rightarrow C$ elementary reactions because the differential rate equations can be written easily in terms of h for both reactions. For any batch reaction ('batch' = all reactants introduced at time $t = 0$), both $A + A \rightarrow A$ and $A + C \rightarrow C$ reactions exhibit $h = 1/2$ behavior in 1D. On the other hand, the integrated rate equations of the two reactions differ in the asymptotic limit: the $A + A \rightarrow A$ reaction has an algebraic reactant density decay while the $A + C \rightarrow C$ reaction has a stretched-exponential decay. However, even in the integrated cases, the time exponents are both $1 - h = 1/2$.

In experiments in which a continuous or pulsed excitation source was used to create the initial excited state (reactant) population, the ratio of the phosphorescence of excited triplet-state naphthalene and the delayed fluorescence intensity of its fusion product were compared. In 1D, the two decay rates are different as a result of the differing initial spatial distribution of reactive species created by the different excitation sources, a phenomenon resulting in a nonclassical rate law.

Exciton fusion studies were also performed on very dilute blends of P1VN/PMMA, where isolated chains of P1VN are insulated from exciton cross-talk by the exciton-inert environment of the PMMA host [26]. Time-resolved phosphorescence and delayed fluorescence intensities relate to the exciton population and the fusion rate, respectively, as a function of time. This enables the comparison of experiment with theory as well as characterization of the polymer system and the exciton interactions.

We note that the systems discussed here differ significantly from previously studied quasi-1D systems [27-30]. The latter are essentially two- or three-dimensional systems with highly anisotropic exciton-exchange interactions. So, for a short time, the exciton is confined in 1D. There is a finite probability, however, of moving along other directions (interchange hopping), resulting in two- or three-dimensional behavior over longer, asymptotically infinite, times and this usually limits the measurements to ultrashort times. Moreover, the phonons and exciton-phonon interactions in these systems are seldom 1D. In contrast, our systems are truly 1D over long times and there is no reactant (exciton) escape or tunneling out of the thin, 1D systems. As in the cylindrical pores, a system may be 3D at very early times, but becomes 1D at longer times and stays so for asymptotically infinite times.

21.2 $A + B \rightarrow C$ where A and B are segregated throughout the reaction

Analytical work on an $A + B \rightarrow C$ type reaction-diffusion process in an effectively 1D system was done by Galfi and Racz [24]. In their model, the reactants A, with a constant density a_0, and B, with a constant density b_0, are initially separated. They meet at time 0, forming a single reaction boundary, which makes the system effectively 1D. The motion of the reaction front with time is shown in Fig. 21.1. This model is similar to that of Weiss et al. [32] for the reaction $A + C \to C$, where A is a 1D continuous solute and C is a single trap.

The results of a set of reaction-diffusion equations for a and b, which are valid in the long-time limit, show that x_f, the position of the center of the reaction front, scales as $t^{1/2}$ while w, the width of the reaction front, scales as $t^{1/6}$ and the reaction rate at the center of the front, r_f, scales as $t^{-2/3}$. It was also determined that the global reaction rate, R, scales as $t^{-1/2}$ [33,34], i.e.,

$$R = \frac{dC}{dt} \sim t^{-1/2}. \tag{21.1}$$

Such a system as is described above was investigated experimentally using the inorganic complexation reaction of disodium ethyl bis(5-tetrazolylazo)-acetate trihydrate ('tetra') with Cu^{2+} [33]. This reaction produces a 1:1 complex in water. Gelatin solutions of the two reactants are introduced into opposite ends of a long thin capillary tube (see Fig. 21.2). The two reactants meet in the center of the capillary, forming a reaction front. The use of gelatin allows efficient diffusion to occur but little or no convection, and aids in the formation of a sharp reaction boundary. The absorbance of the product, a $1 : 1$ Cu^{2+} : 'tetra' complex, is monitored at fixed time intervals by scanning with a detector and a light source in parallel, along the reaction front domain, while the capillary reactor remains fixed in space.

Using the instantaneous reaction of Cu^{2+} with 'tetra', investigated under initially segregated reactant conditions, allowed the study of the further segregation of the reactants in time, as well as other time-dependent properties of the reaction front. In contrast to classical expectations, the two reactants do not inter-diffuse with time. Rather, a reactant gap forms and grows larger with increasing time. The concentration of A and B within the gap is very low and decreases with time as the gap size increases (see Fig. 21.1). Thus, the initial segregation is maintained throughout the reaction and causes the reaction rate to decrease in time as $t^{-1/2}$ (the reaction rate in the reaction-limited regime increases with time as $t^{1/2}$). The results are summarized in Table 21.1 and agree well with theoretical expectations.

Self-ordering and segregation effects have been predicted repeatedly, by theory and simulations, to affect the reaction rate of diffusion-limited elementary $A + B \to C$ reactions in 1D (as well as in higher dimensions). However, the work discussed here was the first experimental study to show the existence and persistence of reactant segregation in low-dimensional systems.

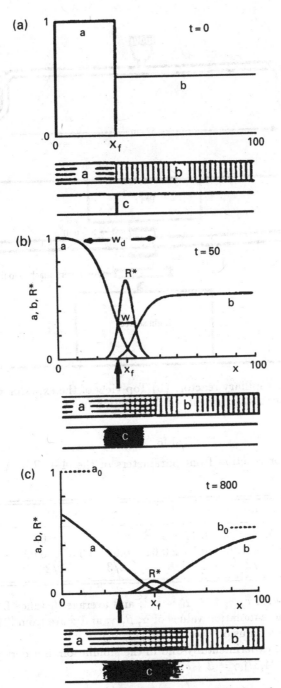

Fig. 21.1. Motion of the reaction front with time. The length x, time t, densities a, b of reagents A, B, and the magnified production rate ($R^* = 100R = 100kab$) of C are all scaled to be dimensionless. Initial conditions are $a = 1$, $b = 0$ for $x < 30$, and $a = 0$, $b = 0.5$ for $x > 30$. The position of the center of the reaction front, x_f, is indicated by the arrow.

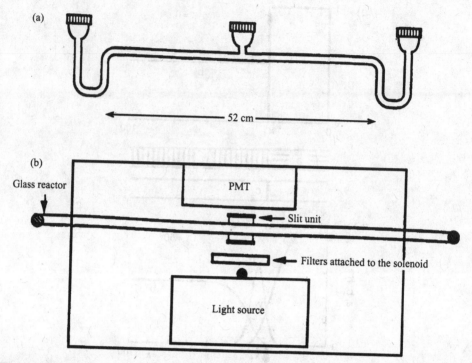

Fig. 21.2. (a) Glass capillary reactor. (b) Top view of the experimental setup used to obtain absorbance measurements of the reaction front.

Table 21.1.

Time exponents[a] for reaction front parameters of the $A + B \rightarrow C$ reaction in a capillary

	α	β	γ	δ	ϵ
experiment	0.51 ± 0.03	0.17 ± 0.03	0.70^{b}	0.53 ± 0.02	0.45 ± 0.01
simulation[a]	0.52 ± 0.03	0.16 ± 0.01	0.68^{b}	0.52 ± 0.02	–
theory[a]	$1/2$	$1/6$	$2/3$	$1/2$	$1/2$

[a]Here $x_f \sim t^{\alpha}$, $w \sim t^{\beta}$, $r_f \sim t^{-\gamma}$, $R \sim t^{-\delta}$, and average distance from the origin $\sim t^{\epsilon}$. The theory and simulation values of α, β, γ, and δ are from [24,37] and that of ϵ is from Einstein's diffusion theory (see [35,36]).

[b]Difficult to measure experimentally and in the simulation, but derivable from the theoretical relation [24,37] $\gamma = \beta + \delta$.

21.3 Homofusion and heterofusion

Several of the experiments discussed in the sections below employ naphthalene as the reactant A undergoing the process $A + A \rightarrow$ products, or $A + C \rightarrow C +$ products, where C are trapped naphthalene species (also designated by A' in this chapter). Depending on the experiment, the naphthalene participates in either molecular (chemical) reactions or exciton (physical) reactions. In both cases, fusion (homofusion) and trapping (heterofusion) reactions occur and exhibit binary and pseudo-monomolecular kinetics, respectively. This is a result of the fact that the reactant, excited state naphthalene, exists in two forms, mobile-species and trapped-species.

In a perfect molecular dimerization reaction in solution or in perfect crystalline samples, the molecules or excitons move freely and at random, resulting in binary exciton-exciton fusion [25,39,40],

$$A + A \rightarrow A + h\nu, \tag{21.2}$$

where $h\nu$ is the delayed fluorescence. The fusion rate R and the fluorescence rate F are given by

$$F \sim R = k\left(\rho_A\right)^2, \tag{21.3}$$

where ρ_A is the excited molecule, or exciton, density and k is the rate constant of the reaction. In most samples, however, a fraction of the free excitons (excited molecules) are quickly trapped, resulting in a roughly constant density of trapped excitons (excited molecules), A', which gives rise to a pseudo-unary (pseudo-monomolecular) fusion reaction and rate

$$A + A' \rightarrow A' + h\nu, \tag{21.4}$$

$$F \sim R = k'\rho_A, \tag{21.5}$$

where $k' = k\rho_{A'}$. In addition, the triplet excitations undergo natural decay, $A \rightarrow h\nu$, with lifetime τ:

$$P \sim \rho_A. \tag{21.6}$$

Therefore, the overall kinetic analysis of the data should be in the form

$$k \sim F/P^n. \tag{21.7}$$

Here $n = 2$ for perfect, trapless samples where only homofusion occurs and $n = 1$ for real samples with traps, where heterofusion prevails. For 3D samples, k, and therefore $\log k$, is expected to be constant in time, while for 1D samples $k \sim t^{-1/2}$ and so a plot of $\log k$ vs. $\log t$ has a slope of $-1/2$. This relationship is tested through naphthalene exciton fusion and trapping in porous glass and porous membranes, by monitoring the delayed fluorescence of the exciton-exciton collision product and the phosphorescence of the reactant triplet excitons (see Sec. 21.4); effectively, one is measuring the long-time-limit kinetics. Similar experimental tests involve excited naphthalene molecules in solution.

21.4 Pulsed or continuous excitation

For the experiments described in Sec. 21.6 and 21.7, two different excitation techniques were employed to create an initial population of excited (reactant) species. The observed reaction rates for the so-called 'pulsed' excitation experiments are the result of the anomalous diffusion (compact random walk) that occurs in 1D subsequent to random creation, while the reaction rates for the so-called 'continuous' excitation experiments are the result of the nonrandom steady-state distributions of the reactive populations in 1D created by a continuous source. This is expatiated upon below.

The standard theoretical derivations [8,13,40] of the nonclassical relations $R = K\rho_A^3$ and $k \sim t^{-1/2}$ in 1D emphasize that these are asymptotic relations and implicitly or explicitly require that the ensemble of reactants has a Poissonian distribution at $t = 0$. One could wonder about the validity of such relations in the 'early-time limit' or whether they would also result from a nonrandom 'initial' distribution. The excitation ensembles for the experiments described in this chapter were prepared either under steady-state excitation, which produces a nonrandom distribution of reactants in 1D at $t = 0$, or with a pulsed excitation, which creates a random distribution of excitons (reactants) at $t = 0$. In 1D, the nonrandom steady-state initial condition results in nonclassical kinetics at early times as well as at late times while the random pulsed exciton source results in nonclassical kinetics in the intermediate and long-time domains only. In 3D, neither excitation source produces nonclassical kinetics for exciton-exciton reactions. Hence, this provides a method with which to test a reaction-diffusion system for nonclassical kinetic behavior in 1D.

The analysis for exciton fusion kinetics is based on the instantaneous reactive collision probability per unit density, the rate constant k. In the experiments presented in this chapter, k is monitored via (21.7), i.e., by measuring the delayed fluorescence intensity generated by the exciton-exciton fusion product (the instantaneous reaction rate) and by measuring the phosphorescence from the first excited triplet-state excitons (the instantaneous reactant concentration).

21.5 Molecular $A + A$ reactions in porous membranes

We describe here an experiment in which naphthalene molecules dissolved in methanol undergo a photodimerization reaction within the confines of the pores of Nuclepore membranes. The simple $A + A$ type photodimerization reaction of naphthalene (N), which has been studied in detail [44-46], is

$$N^\dagger + N^\dagger \rightarrow N_2^{\dagger\dagger} \rightarrow N + N + h\nu. \tag{21.8}$$

Here $N^\dagger$ is a naphthalene molecule in its first *triplet* state, $N^{\dagger\dagger}$ is a transient singlet 'excimer', and $h\nu$ is the ultraviolet fluorescence quantum. The first step of (21.8) is rate determining and diffusion controlled [41,42]. We note that this reaction is analogous to the charge exchange reactions that may occur in biological membranes $(A^+ + A^+ \rightarrow A^{2+} + A$, etc.).

Samples were prepared by dissolving zone-refined naphthalene in methanol. A fluorescence cell was filled with this solution $(c = 10^4\text{-}10^5 \text{ mol})$ and packed tightly with Nuclepore membranes (Fig. 21.3). While a given membrane has uniformly sized pores, membranes with different-size pore diameters are available. Thus, separate experiments were done with pore diameters ranging over 100-20 000 Å to test the effect of increasing pore diameter on the effective dimensionality of the system. Membranes were fully submerged in the solutions throughout the experiment. They are optically transparent and so scatter little light when in solution. The samples were excited at 310 nm with a xenon arc lamp in conjunction with a monochrometer, photomultiplier tube and photon counter. Sequential delayed fluorescence decays were obtained by shuttering the excitation beam and using various filters with a photo tube and signal averager [44,45].

The reactant $(N^\dagger)$ is produced by direct triplet excitation. At a given continuous light intensity a steady state is produced. Both the steady state and the relaxation from it are monitored. The instantaneous reactant concentration is monitored in time via the green phosphorescence decay, which mirrors the natural decay of the first excited triplet state. The instantaneous product concentration is monitored via the ultraviolet fluorescence, which mirrors the natural decay of the first excited singlet state (product). In the relaxation part of the experiment, the product's instantaneous concentration is directly proportional to the reaction rate. The excited singlet decays 10^6 times faster than the excited triplet. Thus, this is a highly controlled and doubly time-monitored photo-reaction. Because the excited-state population results from a steady-state excitation source, one can analyze both the short-time and the long-time kinetics for nonclassical behavior.

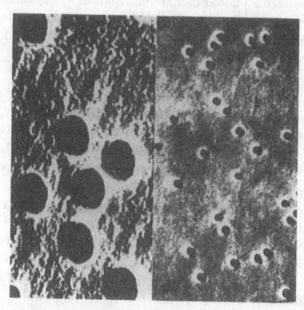

Fig. 21.3. Polycarbonate channel-pore membranes. The magnification is about 10^4; the channel length is 6 μm. The radius of the channels in the membrane on the left is 1 μm while for those in the membrane on the right it is 0.1 μm.

In the 'fractal kinetics' formalism [25], the rate of a dimerization reaction is given by

$$-\frac{d\rho_A}{dt} = kt^{-h}\rho_A^2,\qquad(21.9)$$

where ρ_A is the monomer concentration, t is the time, k is a constant, and h is the heterogeneity exponent. For homogeneous three-dimensional reactions, $h = 0$ and k is the rate constant. For 1D reactions, $h = 1/2$.

As noted in Sec. 21.3, the excited naphthalene molecules exist in two forms: mobile-species and trapped-species. The trapped-species form includes monomers and polymeric aggregates of naphthalene at defect sites near the pore wall. This situation, having both free and trapped excited naphthalene, results in pseudo-first-order reaction kinetics. The following logical chain of events leading to such a result was suggested by Gentry [43]. (1) Mobile excitations are trapped out by the lower-energy levels of the 'traps'. This process depletes some of the mobile excitations. (2) A fusion process occurs. This is mostly heterofusion between a free excitation and a trapped excitation. (3) Heterofusion further depletes the free excitations while the trapped-triplet (A') excitation density remains nearly constant in time: $A + A' \rightarrow A'$, because most of the fusion products are again triplet traps, whether a highly excited triplet or an excited singlet was originally created.

Thus, (21.9) needs to be rewritten

$$-\frac{d\rho_A}{dt} = kt^{-h}\rho_A\rho_{A'} = k't^{-h}\rho_A. \tag{21.10}$$

Here $\rho_{A'}$ is the nearly constant density of excited-state traps and $k' = k\rho_{A'}$; thus (21.10) is a pseudo-first-order reaction. The delayed fluorescence of the product is proportional to the fusion rate, $F \propto |d\rho_A/dt|$. The observed phosphorescence (instantaneous reactant concentration) at longer times is $P(t) \propto \rho_A(t)$. Therefore

$$F = \bar{k}t^{-h}P, \tag{21.11}$$

$$\log(F/P) = -h\log t + \log \bar{k}. \tag{21.12}$$

Employing the kinetic analysis described in Sec. 21.3 for these data by plotting $\log(F/P)$ vs. $\log t$ yields a slope $-h$ (see Fig. 21.4). We assume that the mobile excitation species are the diffusing naphthalene molecules (excited monomers) while the trapped species are excited naphthalene aggregates or monomers adsorbed onto the pore walls. While the overall fraction of trapped species is small, the fraction of excited species that are trapped is large owing to excitation supertrapping [42]. This condition, observed by Gentry in another context, justifies (21.10) and (21.12). The data are inconsistent with the homofusion model (21.9) but consistent with the heterofusion model (21.10) and (21.12).

It was found that the 20 000 Å Nuclepore channels yield an h value of 0.07 ± 0.02 while the 500 Å channels yield an h value of 0.41 ± 0.002. These two limiting values are in good agreement with the theoretically expected values, $h = 0$ in 3D and $h = 1/2$ in 1D. A crossover from 1D to 3D behavior occurs at diameters of about 2000 Å. This gives an upper limit on the diffusion length of the free molecules in solution. For channels with a diameter of 20 000 Å the excited-state molecules hardly feel the channel walls within their lifetimes, and thus exhibit 3D behavior, while in channels with a diameter of 500 Å the diffusion of the free excited molecules is severely confined and therefore exhibits 1D behavior [45].

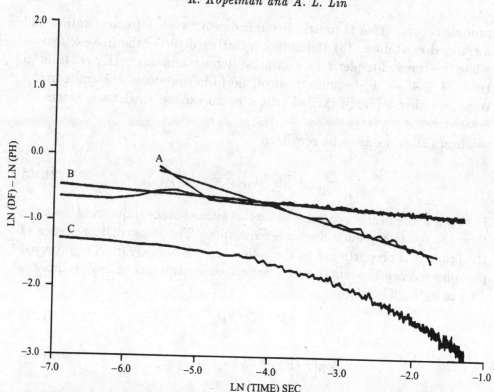

Fig. 21.4. Excited molecular coagulation rate in naphthalene solutions given by the fusion rate coefficient in Nuclepore membranes at $T = 200$ K, $DF/P^n = kt^{-h}$, vs. time on a ln-ln scale; $DF \equiv F =$ delayed fluorescence; $PH \equiv P$; $n = 1$. Curve A, 500 Å channel diameter; curve B, 20 000 Å channel diameter; curve C, 2000 Å channel diameter. Note the crossover of the 2000 Å channels from 3D to 1D behavior.

21.6 Exciton annihilation and trapping in porous glass and membranes

In the experiments discussed in this section, exciton fusion in naphthalene-embedded Vycor glass and Nuclepore membranes was monitored via phosphorescence and delayed fluorescence. Preparation of the naphthalene-embedded samples and the optical arrangement are described elsewhere [46,47]. The Nuclepore membranes described in Sec. 21.5 were used in this experiment also. Vycor glass samples, and Nuclepore membranes, provide effective 1D structures in which to test nonclassical rate laws. Naphthalene embedded in these porous substrates forms crystalline naphthalene 'wires', the diameters of which are that of the pore size in which they are embedded. Typical results are shown in Figs. 21.5, 21.6, and 21.7. The data are analyzed in

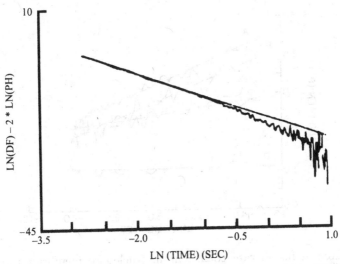

Fig. 21.5. Exciton fusion rate coefficient in porous Vycor glass, $DF/P^n = kt^{-h}$ vs. time on a ln-ln scale; $n = 2$; $T = 6\,K$; $h = 0.44$; $DF \equiv F$; $PH \equiv P$.

the long-time limit. For the Nuclepore membranes, only the pseudo-unary model ($n = 1$ in (21.7)) results in linear slopes. The binary model, where $n = 2$, cannot be fitted with straight lines. Moreover, it results in non-constant-k curves for even the thickest naphthalene wires, which have a diameter of $1.2\,\mu m$. However, for Vycor glass $n = 2$ holds, as can be seen in Fig. 21.5. Thus, heterofusion is the predominant reaction in the Nuclepore membranes and homofusion is the predominant reaction in Vycor glass.

In the Nuclepore membranes, the thinnest naphthalene wires yield the value $h \simeq 0.5$ while the thickest wires yield $h \simeq 0$ at both temperatures; extrapolating the Nuclepore embedded 'wires' to zero diameter gives $h \rightarrow 0.49 \pm 0.02$, while for $1\,\mu m$ diameter wires $h = 0.02 \pm 0.02$. These limiting values are in excellent agreement with the theoretically predicted values of $h = 1/2$ in 1D and $h = 0$ in 3D. The crossover from $h = 0$ to $h = 1/2$ occurs at diameters of approximately 250 and $400\,Å$ at 4 and $77\,K$ respectively; see Figs. 21.6 and 21.7. The crossovers are relatively sharp and the temperature dependence is relatively mild as can be seen from Fig. 21.8.

For the naphthalene wires in Vycor samples, an early-time analysis of the triplet fusion kinetics was also performed. Random populations of excitons were produced by 5 ms pulsed excitations created with a mechanically shuttered xenon arc lamp. Steady-state populations were created by leaving the shutter open for several seconds, well over the time interval required to establish a constant phosphorescent signal. Neutral density filters were used

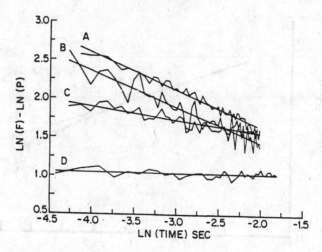

Fig. 21.6. Exciton fusion rate coefficient in Nuclepore membranes at $T = 4\,\mathrm{K}$, $F/P^n = kt^{-h}$ vs. time on a ln-ln scale; $n = 1$. The pore radii are: curve A, $75\,\text{Å}$; curve B, $150\,\text{Å}$; curve C, $250\,\text{Å}$; curve D, $400\,\text{Å}$. The trapped exciton phosphorescence is excluded by an interference filter centered at the free exciton peak.

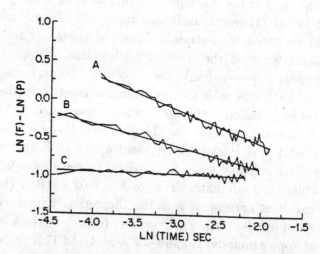

Fig. 21.7. Exciton fusion rate coefficient in Nuclepore membranes at $t = 77\,\mathrm{K}$, $F/P^n = kt^{-h}$ vs. time on a ln-ln scale; $n = 1$. The pore radii are: curve A, $250\,\text{Å}$; curve B, $400\,\text{Å}$; curve C, $1000\,\text{Å}$.

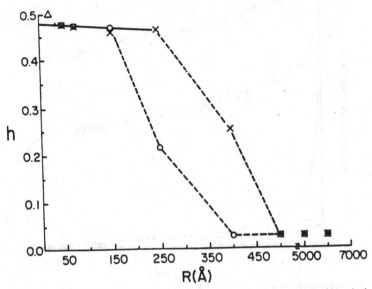

Fig. 21.8. Exciton reaction heterogeneity exponent, h, vs. naphthalene 'wire' radius, r (Å), at 4 K (○) and 77 K (×) in Nuclepore membranes. The 20 Å point at 58 K (△), is for porous Vycor. Note the break in the scale.

to make the phosphorescence intensity of the steady-state excitation at $t = 0$ (when the shutter closes) equal to that produced by the pulsed excitation. This ensures equal initial global exciton densities.

Although the initial global densities are equal, the observed phosphorescence and delayed-fluorescence decays are quite different for the random (pulsed) as compared to the steady-state (continuous) cases. Figures 21.9 and 21.10 show the nonclassical results for naphthalene-embedded Vycor glass. These results are in agreement with simulation results obtained from an initially random source and a steady-state source; see Fig. 21.11. The unequal decay curves illustrate the reaction rate's nonclassical dependence on history, i.e., its dependence on how the initial reactant population was obtained. In contrast, the reaction rate in 3D systems of isotropically mixed crystals displays no dependence on the excitation source [51].

21.7 Exciton annihilation and trapping in isolated polymer chains

In this section we report on studies of exciton fusion in isolated chains of poly(1-vinyl)naphthalene polymer (P1VN), molecular weight $\sim 10^5$, blended

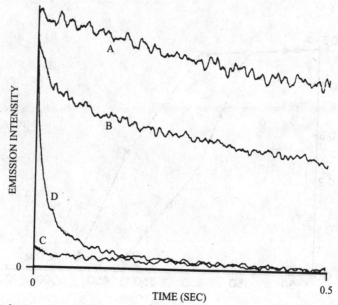

Fig. 21.9. Relative intensities of phosphorescence and delayed-fluorescence decays of naphthalene excitons in porous Vycor glass at $T = 18\,K$ after continuous (11 s duration) or pulsed (20 ms duration) excitation at 319 nm. Curve A: phosphorescence from continuous (steady-state) excitation with a neutral density filter used to equalize the initial intensity to that of curve B. Curve B: phosphorescence from pulsed (random) excitation. Curve C: delayed fluorescence from excitation conditions in curve A. Curve D: delayed fluorescence from excitation conditions in curve B. The intensity scale for C and D is different from that for A and B.

in poly(methylmethacrylate) (PMMA), molecular weight $\sim 1.54 \times 10^5$. Samples of 0.005% and 0.01% (wt/wt) P1VN/PMMA were cast as thin films using the doctor-blade technique [48]. An excimer laser was used as the pulsed excitation source. Decay data were collected about 10 milliseconds after the excitation pulse [26].

The exciton fusion process in P1VN/PMMA samples has been shown to be orders of magnitude more efficient than it is in naphthalene/PMMA samples [26,49], evidence that the triplet excitations migrate efficiently along the entire polymer chain, the average length of which is of the order of 100 monomers. The premise for these experiments is that chain-chain hopping can be neglected for these very dilute blends, but triplet exciton hops between adjacent naphthalene pendants on the same chain are quite efficient at the experimental temperature, 77 K. This hypothesis is corroborated by the rate-law measurements.

As in the previously discussed exciton fusion experiments, two fusion reactions are contributing to the delayed fluorescence decay: homofusion

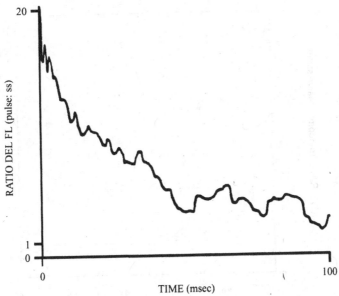

Fig. 21.10. Delayed fluorescence of naphthalene excitons in porous Vycor glass: the ratio of the pulsed (random) and the continuous (steady-state) excitation results with equalized initial phosphorescence intensities.

$(A + A \rightarrow$ products) and heterofusion $(A + C \rightarrow C)$. We hypothesize that the reaction rate observed is that of the trapping, heterofusion reaction. The heterofusion rate law is valid after most of the traps (defects, etc.) have been discovered and occupied by freely hopping excitons. The electronic excitation is confined to the naphthalene moieties of the P1VN. The resulting triplet excitation hopping is limited to a range of approximately 5 Å; thus in the absence of neighboring P1VN chains, exciton hopping is limited to a single chain. These chains, not being tightly coiled, are very close to true 1D systems since they have an approximate 5 Å diameter and the triplet exciton hopping length is of the order of 5 Å as well. In order to ensure 1D topology, the blends are necessarily as dilute as 0.005% and 0.01%. Higher P1VN concentrations, i.e., 1%, allow cross-talk between chains, resulting in a crossover from 1D to 3D kinetics as the concentration of P1VN is increased.

Data obtained from this experiment are consistent with a heterofusion model—the data fit (21.7) when $n = 1$; see Fig. 21.12. The heterogeneity exponents for the 0.005% and 0.01% P1VN/PMMA samples were found to be $h = 0.47 \pm 0.05$ and 0.53 ± 0.03, respectively, in good agreement with the theoretical expectation of $h = 1/2$. On the other hand, the heterogeneity exponent for the pure (100%) P1VN material was found to be $h = 0.02 \pm 0.02$,

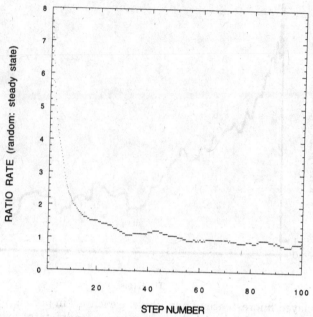

Fig. 21.11. Monte Carlo computer simulation: the ratio R_{ur}/R_{ss} of fusion reaction rates resulting from uniformly random and from steady-state generation of reactants, *vs.* step number for the simulation of $A + A \rightarrow 0$ on a 30 000-site 1D lattice.

in good agreement with the expectation $h = 0$ for 3D (classical) systems [50].

In summary, for the 1D topology of the P1VN chains embedded within a PMMA host, the reaction is dominated by heterofusion on the time scale in which the measurements are made and obeys the diffusion-controlled, nonclassical kinetics for the pseudo-monomolecular reaction $A + C \rightarrow C +$ products, in 1D.

21.8 Conclusion

We have presented the results of several experimental systems that exhibit 1D reaction kinetics in the asymptotic limit. Under some circumstances, i.e., a continuous excitation source, 1D behavior is observed in early-time regimes as well. The bimolecular reaction of Cu^{2+} with 'tetra' was observed in capillary tubes where the reactants are initially segregated in space. The nonclassical behavior of some temporal properties of the reaction front are manifested in anomalous power laws. Molecular coagulation and excitonic

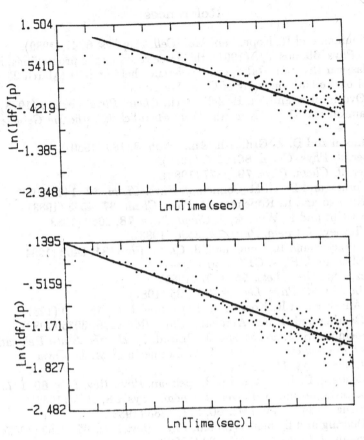

Fig. 21.12. Exciton fusion rate coefficients in 0.05% (top) and 0.01% (bottom) P1VN/PMMA at $T = 77\,\mathrm{K}$, $DF/P^n = kt^{-h}$ vs. time on a ln-ln scale; $n = 1$.

fusion reactions were investigated using naphthalene, either in solution or in crystalline form, in porous membranes and glasses. Fusion reaction kinetics was also observed on isolated P1VN chains in P1VN/PMMA polymer blends. The nonclassical reaction kinetics observed in these systems was measured in terms of the heterogeneity coefficient, h. The classical value is $h = 0$, while the observed value was approximately $1/2$ in these 1D topologies. Of all the exciton fusion experiments discussed here, only those done in Vycor glass displayed homofusion (binary, $n = 2$) kinetics. All the other exciton fusion experiments exhibited heterofusion (pseudo-monomolecular, $n = 1$) kinetics.

We would like to thank Steve Parus, Zhong-You Shi, and Andrew Yen for helpful suggestions. Our work was supported by NSF grant DMR-9410709.

References

[1] E. E. Fauman and R. Kopelman, *Mol. Cell. Biophys.* **6**, 47 (1989).

[2] *J. Stat. Phys.* **65**, nos. 5/6 (1991): this issue contains the proceedings, *Models of Non-Classical Reaction Rates*, of a conference held at NIH (March 25-27, 1991) in honor of the 60th birthday of G. H. Weiss.

[3] A. A. Ovchinnikov and Ya. B. Zel'dovich, *Chem. Phys.* **28**, 215 (1978).

[4] M. Bramson and D. Z. Griffeath, *Wahrscheinlichskeitstheorie Gebiete* **53**, 183 (1980).

[5] M. Bramson and D. Z. Griffeath, *Ann. Prob.* **8**, 183 (1980).

[6] J. Keizer, *J. Phys. Chem.* **86**, 5052 (1982).

[7] J. Keizer, *J. Chem. Phys.* **79**, 4877 (1983).

[8] D. C. Torney and H. M. McConnell, *J. Phys. Chem.* **87**, 1441 (1983).

[9] P. W. Klymko and R. Kopelman, *J. Phys. Chem.* **87**, 4565 (1983).

[10] D. Toussaint and F. Wilczek, *J. Chem. Phys.* **78**, 2642 (1983).

[11] D. C. Torney, *J. Chem. Phys.* **79**, 3606 (1983).

[12] L. W. Anacker and R. Kopelman, *J. Chem. Phys.* **81**, 6402 (1984).

[13] P. V. Elyutin, *J. Phys.* C**17**, 1867 (1984).

[14] Z. Racz, *Phys. Rev. Lett.* **55**, 1707 (1985).

[15] A. A. Lushnikov, *Phys. Lett.* A**120**, 135 (1987).

[16] L. W. Anacker and R. Kopelman, *Phys. Rev. Lett.* **58**, 289 (1987).

[17] C. R. Doering and D. ben-Avraham, *Phys. Rev.* A**38**, 3035 (1988).

[18] R. Kopelman, S. J. Parus and J. Prasad, in *Excited State Relaxation and Transport Phenomena in Solids*, J. L. Skinner and M. D. Fayer, eds., special issue of *Chem. Phys.* **128**, 209 (1988).

[19] K. Lindenberg, B. J. West and R. Kopelman, *Phys. Rev. Lett.* **60**, 1777 (1988).

[20] E. Peacock-López and J. Keizer, *J. Chem. Phys.* **88**, 1997 (1988).

[21] J. L. Spouge, *Phys. Rev. Lett.* **60**, 871, 1885 (1988).

[22] C. R. Doering and D. ben-Avraham, *Phys. Rev. Lett.* **62**, 2563 (1989).

[23] V. Privman, *J. Stat. Phys.* **69**, 629 (1992).

[24] L. Galfi and Z. Racz, *Phys. Rev.*, A**38**, 3151 (1988).

[25] R. Kopelman, *Science* **241**, 1620 (1988).

[26] C. S. Li, Ph.D. thesis (University of Michigan, Ann Arbor, 1988).

[27] R. M. Hochstrasser and J.D. Whiteman, *J. Chem. Phys.* **56**, 5945 (1972).

[28] M. D. Fayer, in *Modern Problems in Solid State Physics*, Vol. **4**, p. 185, V. M. Agranovich and R. M. Hoschstrasser, eds. (North-Holland, Amsterdam, 1983).

[29] R. Kroon and R. Sprik, see Ch. 20.

[30] S. D. D. V. Rughooputh, D. Bloor, D. Phillips and B. Movaghar, *Phys. Rev.* B**35**, 8103 (1987).

[31] W. J. Rodriguez, R. A. Auerbach and G. L. McPherson, *J. Chem. Phys.* **85**, 6442 (1986).

[32] G. Weiss, R. Kopelman and S. Havlin, *Phys. Rev.* A**39**, 466 (1989).

[33] Y. E. Koo and R. Kopelman, *J. Stat. Phys.* **65**, 893 (1991).

[34] Y. E. Koo, L. Li and R. Kopelman, *Mol. Cryst. Liq. Cryst.* **183**, 187 (1990).

[35] M. V. Smoluchowski, *Z. Phys. Chem.* **92**, 129 (1917).

[36] S. Chandrasekhar, *Rev. Mod. Phys.* **15**, 1 (1943).

[37] L. Li, Ph.D. thesis (University of Michigan, Ann Arbor, 1989).

[38] S. Redner, in *Extended Abstracts of the Materials Research Society, Symposium on Dynamics in Small Confining Systems*, p. 109, J. M. Drake, J. Klafter and R. Kopelman, eds. (1990).

[39] R. Kopelman, *J. Stat. Phys.* **42**, 185 (1986).

[40] P. W. Klymko and R. Kopelman, *J. Phys. Chem.* **87**, 4565 (1983).

[41] E. Doller and Th. Forster, *Z. Phys. Chem.* **31** 274 (1962).

[42] T. Gentry and R. Kopelman, *J. Chem. Phys.* **81**, 3014 (1984).

[43] T. Gentry and R. Kopelman, *J. Chem. Phys.* **81**, 3022 (1984).

[44] J. Prasad and R. Kopelman, *J. Phys. Chem.* **91**, 265 (1987).

[45] J. Prasad and R. Kopelman, *Chem. Phys. Lett.* **157**, 535 (1989).

[46] R. Kopelman, S. Parus and J. Prasad, *Phys. Rev. Lett.* **56**, 1742 (1986).

[47] R. Kopelman, S. J. Parus and J. Prasad, *Chem. Phys.* **128**, 209 (1988).

[48] E. O. Allen, in *Plastic Polymer Science and Technology*, p. 600, M. D. Baijal, ed. (Wiley, NY, 1982).

[49] Z.-Y. Shi, C.-S. Li and R. Kopelman, in *Polymer Based Molecular Composites*, *Proc. Materials Research Symp.*, J. E. Mark and D. W. Shaefer, eds. (Pittsburgh, Pennsylvania 1989).

[50] Z.-Y. Shi, Ph.D. thesis (University of Michigan, Ann Arbor, 1990).

[51] S. Parus and R. Kopelman, *J. Mol. Liquid Crystals* **175**, 119 (1989).

22

Luminescence quenching as a probe of particle distribution

Stefan H. Bossmann and Lawrence S. Schulman

When the metal complex $[Ru(phen)_2(dppz)]^{2+}$ is bound to DNA it can luminesce. If the metal complex $[Rh(phi)_2(phen)]^{3+}$ is nearby on the strand, the luminescence is quenched by electron transfer. By varying concentrations and by varying the DNA it is possible to probe the distribution of complexes in this one-dimensional (1D) system, and to gather information about the electron transfer length and interparticle forces. Our model assumes random deposition with allowance for interactions among the complexes. Long strands of calf thymus (CT) DNA and short strands of a synthetic 28-mer were used in the experiments and, for fixed $[Ru(phen)_2(dppz)]^{2+}$ concentration, quenching was measured as a function of $[Rh(phi)_2(phen)]^{3+}$ concentration. In previous work, to be cited later, we reported an electron transfer length based on the CT-DNA data. However, the short-strand (28-mer) experiments show a remarkable difference from the previously analyzed data. In particular, the electron-transfer quenching upon irradiation is enhanced by a factor of approximately four. This requires the consideration of new physical effects on the short strands. Our proposal is to introduce complex-complex repulsion as an additional feature. This allows a reasonable fit within the context of the random-deposition model, although it does not take into account changes in the structure of the 28-mer introduced by the metal complexes during the loading process.

22.1 Introduction: The need for random deposition models in experimental science

In this chapter we show that random deposition models can be applied to the evaluation of physical data in microheterogeneous systems of biochemical, technical, and environmental importance. Many of the commonly applied—

443

and effective—models presuppose particular interaction mechanisms. The random deposition model is less specific. It drops the details of particular models, but it gains the ability to compare data obtained in *different* systems.

It is clear that the development of science has profited from the elaboration of paradigms, leading to ingenious interpretational techniques for experimental data [1]. On the other hand, the existence of these different and specific models sometimes makes it difficult to observe the basic principles, structures, and dynamics of reactions, across a broader range of phenomena. In other words, we sometimes desire a more general mathematical language that permits the description of experimental data beyond these specific paradigms, and therefore enhances interdisciplinary communication. Examples of well-developed, but also specific, models are the method of McGhee and van Hippel [2] for the determination of the binding behavior of ligands or metal complexes on DNA, and also the classic paradigm of steric recognition of metal complex enantiomers by DNA [3]. Whereas Λ-[Ru(phen)$_3$]$^{2+}$ is believed to bind by intercalation to DNA in the region of the major groove, the Δ-enantiomer should be surface-bound at the DNA's supramolecular structure [4]. Despite the authors' opinion that intercalation exists, this paradigm appears to neglect the influence of the complex's distribution along the DNA-strand on the observed luminescence and quenching behavior.

So, a challenging situation arose when the combined application of photophysical investigation methods and the continuous-wave electron spin resonance of nitroxide-labeled ruthenium complexes demonstrated that, indeed, three fractions of ruthenium complexes were found in the presence of DNA, each possessing a different mobility and freedom of motion [5]. The two slow-motion components could be assigned to be the intercalated and the surface-bound species, as predicted by the classic paradigm for the DNA-recognition of ruthenium complexes. But the main component (up to 85%, depending on experimental conditions) was in more rapid motion (along the DNA) than the two other components. Furthermore, the data supported the existence of an exchange mechanism between the ruthenium complexes next to the DNA and those possessing a high mobility.

In addition to these findings, it is widely known that DNA undergoes a remarkable structural change during its interaction with positively charged metal complexes [6]. Typically the effect of unwinding is observed, leading eventually to the loss of clearly defined supramolecular regions such as the minor and the major groove, especially under conditions of high metal complex concentrations. Similar effects can be observed as a function of salt concentration, pH, etc.

Thus, despite significant advances in experimental science, it is not possible to perform titration experiments of DNA with metal complexes under well-defined conditions. For this reason we favor the application of a phenomenological model, such as random deposition on a 1D DNA strand [7]. This avoids excessive interpretation of the experimental data. Furthermore, comparison of results obtained from the same set of experimental data, employing different methods of interpretation, should demonstrate the degree of model dependence of the conclusions drawn from these experiments. Such strategies are of course independent of any actual experimental example.

In this chapter, though, we will apply these ideas to a particular set of experiments, involving the interaction of metal complexes and DNA. This example is 1D, but we believe the random deposition model should apply to all interactions of metals and metal complexes with one-, two-, and three-dimensional surfaces possessing opposite charge and/or hydrophobic features. Examples of this wide and highly interdisciplinary field of applications are the applied science of water-soluble polymers, catalysis employing semiconductors (TiO_2, SiO_2-TiO_2, zeolites, and other xerogels in combination with photoactive metal complexes like $[Ru(bpy)_3]^{2+}$), and the diffusion dynamics in waters containing humic acids as naturally occurring polymers from the degradation of plant tissues and other macromolecules of environmental importance.

22.2 Experimental conditions for the application of random deposition models

In principle the method of investigation should provide time resolution at least an order of magnitude faster than the diffusion dynamics in the system to be observed. In these cases the metal complexes move along the 1D structure and interact with the target and/or with each other (including cases where more than one type of metal complex is employed, for instance an electron transfer photodonor-acceptor pair). With such time resolution, the interactions appear to be 'frozen in time', and time-dependent phenomena can be neglected.

Each experimental investigation method thus provides a typical 'observation window', which then permits the use of the data obtained as input for random deposition models. Photophysical methods such as single-photon-counting or laser-flash-photolysis usually permit the observation of time dependent phenomena by time-resolved emission or absorption spectroscopy in the time domain between 1 nanosecond and 1 millisecond [8]. Note that faster optical measurements are also available today, for instance, picosecond and femtosecond spectroscopy, but since we want to use the experimental data as input for random deposition models there will always be species undergoing diffusion or more general motion in the systems. Diffusion faster than the nanosecond time scale is seldom observed.

In principle the same considerations apply for all other experimental methods permitting quantitative measurements in microheterogeneous systems.

In the case of ESR spectroscopy, usually the 'observation window' is between the sub-nanosecond domain up to a millisecond for continuous-wave-ESR spectroscopy [9] and from several nanoseconds to a millisecond for Fourier-transform-ESR spectroscopy [10]. For nuclear magnetic resonance spectroscopy (NMR) this range should be from a microsecond up to one second, depending on the nucleus observed [11]. The combination of all these experimental techniques permits in principle the application of random deposition methods for the description of many different systems, as indicated above.

Other spectroscopic techniques such as infrared and complementary Raman spectroscopy can hardly be used for the generation of data intended as input for random deposition models, because their 'time window', which corresponds to the correlation time for the occurrence of vibrations in a molecule (10^{-12} to 10^{-10} seconds) is, as in the case of optical picosecond and femtosecond spectroscopy, too short to observe the interactions of different species along a one-, two-, or three-dimensional surface [12].

In cases where the time resolution of spectroscopic methods does not permit the application of 'pure' random deposition models, a combination of a random deposition model with a 'classic' model for the dynamics in microheterogeneous systems can be derived [13]. Such breakdowns can occur when we observe different contributions to experimental results, one which is in the right time domain for applying a random deposition model and another which is so fast that our criterion does not apply, that is, the diffusion processes do *not* appear to be 'frozen'. An indication of such a breakdown would be bi- or multi-exponential fits of the luminescence decay traces or corresponding absorption measurements.

22.3 Does DNA serve as an electron wire?

The question whether DNA can serve as a long-range electron wire on a molecular scale is of great importance for the understanding of the working principles of life [14]. A spin-off from this research is the development of efficient anti-cancer drugs [15]. In recent years, evidence for the occurrence of DNA-mediated long-range electron transfer has been presented [14,16], leading to a lively discussion about the validity of the experimental findings and interpretation. Since these data had been interpreted using the classic paradigm of DNA intercalation, and because of the importance of the conclusions regarding long-range electron transfer, we thought it worthwhile to apply a random deposition model and to compare results.

The experimental probes providing data for our calculations are two positively charged metal complexes: Bis (1,10-phenanthroline) (dipyridophenazine) ruthenium (II), i.e., $[Ru(phen)_2(dppz)]^{2+}$; Bis (9,10-phenanthrenequinone diimine) (1,10-phenanthroline) rhodium (III), i.e., $[Rh(phi)_2(phen)]^{3+}$; see Fig. 22.1. Two types of DNA have been used: calf thymus (CT) DNA, with an average size of 10 000 base pairs, and a synthetic 28-mer [16]. Calf thymus DNA is an example of a DNA double strand occurring in living cells. Note that the DNA-binding proteins are removed during purification of the sample.

In principle, there are the following differences between CT-DNA and small oligomers.

1. The dynamics is different. Generally, the synthetic oligomer (28-mer) behaves like a stiff rod [6]. On the other hand, the macromolecule CT-DNA undergoes a variety of complex, temperature-dependent motions, coiling and stretching as well as pairing and unpairing of base pairs.

2. Since DNA-oligomers are much smaller than CT-DNA double strands, end effects may be significant. The same consideration applies for repulsion effects between metal complexes possessing the same charge, which is the case for $[Ru(phen)_2(dppz)]^{2+}$ and $[Rh(phi)_2(phen)]^{3+}$.

Both metal complexes are avid binders to DNA and possess binding constants $K_b > 10^6 \, M^{-1}$ [17]. $[Ru(phen)_2(dppz)]^{2+}$ serves in this system as photo-electron donor to the electron-acceptor complex $[Rh(phi)_2(phen)]^{3+}$, after excitation with a photon in the range of blue light. Furthermore the chosen ruthenium complex, $[Ru(phen)_2(dppz)]^{2+}$, has a special property: it does not emit photons in water or buffer solution, but it shows emission in alcohols, acetonitrile, micelles, and especially in interaction with DNA [17]. This permits observation of the DNA-bound ruthenium complexes ex-

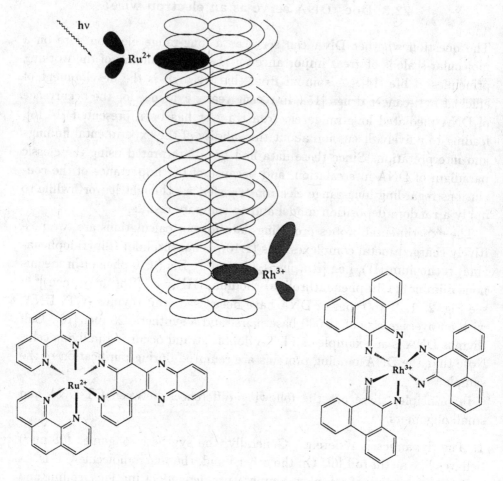

Fig. 22.1. Shown are structures of the complexes [Ru(phen)$_2$(dppz)]$^{2+}$ and [Rh(phi)$_2$(phen)]$^{3+}$ serving as photodonor/electron-acceptor pairs intercalated in calf thymus DNA. The experiments measure the quenching of the luminescence of [Ru(phen)$_2$(dppz)]$^{2+}$ when [Rh(phi)$_2$(phen)]$^{3+}$ is nearby.

clusively. If electron-transfer quenching to [Rh(phi)$_2$(phen)]$^{3+}$ occurs, the luminescence of the ruthenium complexes is quenched and consequently decreased. So we observe the ratio I^0/I (I^0 = luminescence intensity with no quencher added, I = remaining luminescence as a function of quencher concentration) for the quantitative determination of the amount of quenched ruthenium complexes [16]. Time-resolved laser-flash experiments as well as single-photon-counting measurements demonstrated clearly that the electron quenching occurs in a time domain faster than a nanosecond. Consequently the application of a random deposition model is justified.

22.4 Experimental parameters

The experimental conditions [16] during the electron-transfer-quenching experiments were:

$[\text{Ru(phen)}_2(\text{dppz})]^{2+}$	10 μM
$[\text{Rh(phi)}_2(\text{phen})]^{3+}$	0-100 μM
CT-DNA	1000 μM nucleoside
28-mer	1000 μM nucleoside
buffer	5 mM Tris-HCl/50 mM NaCl, pH $= 7.2$
temperature	25°C

22.5 Theoretical model

There are two primary theoretical issues and several secondary ones, the latter representing effects that do not seem to play an important role in the experimental system we study.

First, we need an estimate of the gap distribution function p_n, the probability of finding a gap of length n. (To be precise, by 'a gap of length n' we mean that there are $4(n+1)$ base pairs between the intercalation sites. We use this convention because the effective region covered by both complexes on the DNA strand is four base pairs.) Several such functions must be considered, for donor-donor, acceptor-acceptor, donor-acceptor, and anything-anything. In general these are interrelated. The simplest form for the gap distribution function is the Poisson distribution, namely, if we denote a per-site density of substance by ρ, then the probability that an occupied site has a gap of length exactly k to its right is $\rho(\rho^k/k!)\exp(-\rho)$. Unfortunately, naive application of this well-known formula [19] proved inadequate for the description of our data. For the work in [7], the main problem was the determination of the correct ρ to use; this is discussed in detail below. Correlations that could invalidate the mean-field assumption did not prove to be significant. Of course it was necessary to study variants of this formula before one had confidence that a data fit could *not* be improved by allowing deviations from Poisson. For example, one such variant involves the correlation parameter R mentioned below (and described in detail in [7]). As we shall see in the present chapter, there are significant correlations for the 28-mer data, so that from the mathematical point of view things begin to get interesting.

The other essential function connecting the experimentalist's data to the theoretical model is the relation between the amount of compound in solution and the amount actually on the DNA strand. Under some circumstances this is well understood theoretically, for example, in equilibrium and with a limited variety of particle-particle interaction [2]. Pure 'random sequential absorption' (RSA) [18] represents an opposite extreme in which equilibrium is never attained because a particle that lands always sticks, never shifting to configurations of lower free energy. The evidence for the experiment we study—and very likely for many others like it—points to something in between: a state that is relatively predictable during the typical data-taking epochs, but is described by neither extreme model. (For these experiments, the binding constants are not so large as to imply RSA, while if one were to wait for 'equilibrium' the DNA would be destroyed by the materials intended to probe it.)

For this reason we advocate a phenomenological approach. The simplest relation would be $\tilde{\rho} = \rho$, where $\tilde{\rho}$ is the density of the material actually on the strand, and ρ is the density of it in solution. Our ansatz, however, represents a simple functional form that has the same functional dependence for small ρ but for large ρ saturates, reaches a limiting density. The functional form we use is

$$\tilde{\rho} = \frac{\rho}{1 + (\rho/\rho_\infty)}. \tag{22.1}$$

This introduces a single fitting parameter ρ_∞. Other functional forms can provide the same asymptotics, so that one should not think that all possible physical richness is consigned to this one parameter. Nevertheless, we have found that this form fits the data well, where others (such as an exponential in which the approach to ρ_∞ is more precipitate) are less effective. Furthermore, although our approach remains phenomenological, the form (22.1) is close to that derived from certain theoretical models such as the Langmuir adsorption isotherm (which gives the same function) or the McGhee-von Hippel [2] model, whose behavior bears a qualitative resemblance to (22.1) after appropriate manipulation of the parameters.

Other physical questions enter before the phenomenology is complete. We need a *form* for the dropoff of quenching as the donor and quencher separate. We use the following exponential form: quenching $\sim \exp(-n/L)$, where L is a characteristic length, later interpreted as the electron transfer length, and the distance (in our parametrization) n, measured in units of four base pairs (which is the distance taken up by either donors or acceptors on the DNA strands studied). This form has analytic advantages and one can muster physical arguments in its favor. However, we are wary of such arguments, and the actual use was checked by trying other forms, such as a sharp cutoff or a power law. No essential difference was noted, although this question was not investigated to the point of comparing goodness of fit of the optimized forms.

This introduces an interesting observation. Our goal was to fit about 10 data points, and we used at least two, perhaps five, parameters (if you count attempts to vary the phenomenology by introducing new parameters and then rejecting them when the fit was not improved). It is remarkable that these rather bland-looking data were difficult to fit with any functional form that arose from even our nonspecific model. If you overshoot on the range it goes badly wrong at one end, and no amount of fiddling with the other parameter can repair it.

Other physical and chemical issues that need to be addressed are, firstly, the matter of blocking: if something lands between a donor and a acceptor, does it block their interaction? Secondly, competing binding: do the two species have the same—or sufficiently similar—affinity for DNA? What about supramolecular species? Our course of action on these and other questions is reported in detail in [7].

One matter discussed extensively in [7] was the possibility of attraction or repulsion between members of the same species or of other species. Roughly speaking, this was handled by replacing the Poisson distribution by one in which p_0 was assigned independently, while p_n, for $n > 0$, was given the Poisson form. After normalizing, this left one parameter related to the donor-acceptor interaction, with p_0 larger or smaller than its Poisson value for attraction or repulsion, respectively.

For this particular additional possibility we did optimize for various couplings, since the physical consequences seemed especially significant. (As it turns out, we do find evidence that in other circumstances there is an important effect.) We concluded that within the range of plausible couplings and good data fits, there was little effect on the electron transfer length.

We now consider another data set. This is from experiments on short strands of DNA, specially constructed to be 28 base pairs in length. As we relate below, use of the same parameters that successfully fitted the CT-DNA severely underestimates the quenching. Therefore it is necessary to hypothesize new mechanisms, significant on the 28-mer, but not on the longer strand. The full analysis of this work is not yet complete, but the direction in which it takes us, and some of the physical implications, are sufficiently clear and significant to warrant reporting.

One effect that enhances quenching in a short-strand environment is moderate-range acceptor-acceptor repulsion. On a long strand this would tend to lower the overall occupancy of the strand, but on a short strand it will heavily favor single-acceptor occupancy on each short strand. Such occupancy will in turn greatly improve the quenching effectiveness of the acceptors, since they will be placed for maximum mischief. It may also be hypothesized that the degree of acceptor-acceptor repulsion depends strongly on the environment, differing significantly on short, stiff, rodlike strands from that on longer, spaghetti-like strands. In any case, it will be seen that such repulsion does a reasonable job of explaining the data, whereas no other hypothetical form comes close. The physical explanation should then be guided by these phenomenological results.

The mathematical form of the fitting makes use of random deposition supplemented by an energy term associated with the repulsion. Because the strand is short the quenching calculation must allow finite-size effects; for example a donor at the end of the strand is less likely to be quenched than one in the center or on an 'infinite' CT-DNA strand. Moreover, even homogeneous random deposition does not yield a Poisson distribution, but only a binomial form for which the Poisson distribution represents a limit (as usual). The energy term is a Boltzmann-like factor, of the form $\exp[-\mu(k-1)]$, with μ a dimensionless quantity related to chemical potential and k the number of acceptors on the short strand. This particular form suggests a moderate range, saturated interaction (i.e., one neighbor, at any distance up to the length of the strand, gives the full effect, so that the expression is linear rather than quadratic in k).

22.6 Analysis of data

In Fig. 22.2 we show a combination of CT-DNA and 28-mer data. Note the significantly higher quenching in the 28-mer system (circles in the figure). The solid line through the plus symbols (CT-DNA) represents the fit reported in [7], using $L = 0.7917$ (in units of four base pairs) and $\rho_\infty = 1.7677$. (Recall that $L = 0$ represents direct contact, so to arrive at center-to-center electron transfer distances one should add 1 to L.) The curves in this figure show what adjustment of the acceptor-donor interaction parameter, R, can—or more to the point, cannot—do. In this case, attraction enhances quenching (bringing the complexes together), but even though at the largest loadings the measured level is attained, the rest of the pattern shows a large discrepancy from the data.

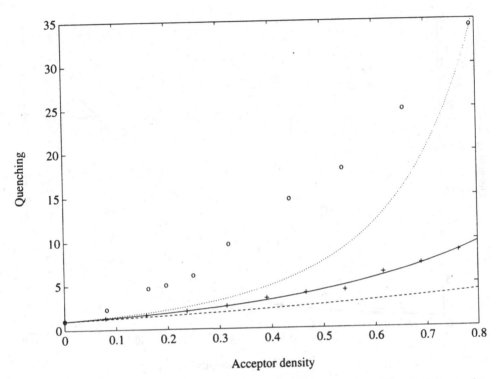

Fig. 22.2. Quenching, I_0/I, as a function of acceptor loading. The circles are the 28-mer data and the plus symbols the CT-DNA data; see Appendix A. The solid curve represents the fit in [7], using values $L = 0.7917$ and $\rho_\infty = 1.7677$. (L is measured in units of four base pairs, and for center-to-center distance, L should be increased by 1.) This fit assumed essentially no attractive or repulsive effects in the donor-acceptor distribution function. The dotted line results from a strong attractive force, while the broken line corresponds to donor-acceptor repulsion.

As indicated above, to fit the data we found it necessary to postulate acceptor-acceptor *repulsion*. Figure 22.3 shows a variety of attempted fits, the one we consider to be most acceptable being the upper, solid, line. This uses the same values of L and ρ_∞ that we found best for the CT data, the only difference being a repulsion of strength $\mu = 1.65$. (If the Boltzmann factor picture that motivated us above has quantitative validity, this pure number would be the chemical potential divided by $k_B T$.) The lowest, broken, curve gives the quenching on a 28-mer that would be expected if

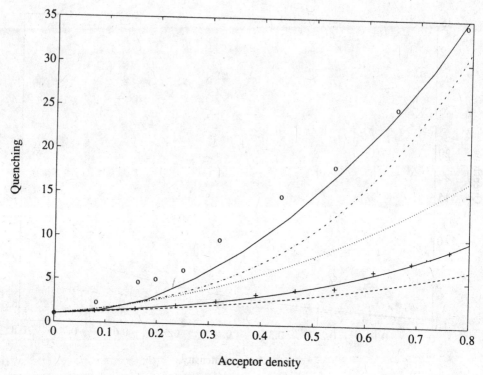

Fig. 22.3. Same experimental data as Fig. 22.2. The curves correspond to the following parameter values and calculation schemes. Lowest, broken, curve: $L = 0.7917$, $\rho_\infty = 1.7677$, $\mu = 0$, 28-mer calculation, with blocking. Next curve, counting upward on the right: solid curve, standard fit to CT data. Next, dotted, curve: $L = 10$, $\rho_\infty = 0.8$, $\mu = 0$, back to the 28-mer calculation (as for all the following curves). Next, broken-and-dotted curve: $L = 10$, $\rho_\infty = 1.7677$, $\mu = 0$. Finally, the uppermost, solid, curve is for $L = 0.7917$, $\rho_\infty = 1.7677$, $\mu = 1.65$.

$\mu = 0$. This is quite reasonable, since breaking a 10 000-base-pair strand into little pieces would create quenching-reducing end effects that our calculation takes into account. The dotted and broken-and-dotted lines between the two solid curves represent examples of parameter fiddling, to see if by hook or by crook a fit could be attained. The broken-and-dotted curve has $L = 10$, $\rho_\infty = 1.7677$, $\mu = 0$ while the dotted curve also uses $L = 10$ but tries a smaller value of ρ_∞ (= 0.8). These curves are intended to show that within

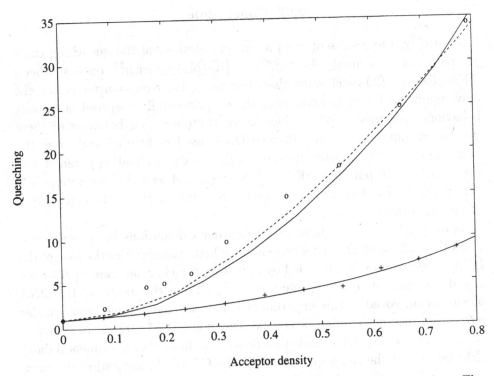

Fig. 22.4. Same experimental data as Fig. 22.2. Comparison of two μ values. The solid lines are the same as in the previous figure, but for the broken line we took $L = 10$, $\rho_\infty = 0.8$ and $\mu = 1.5$.

the overall context of the models we have developed, there seems to be no way to account for the data without including repulsion.

In an effort to estimate the precision with which the repulsion parameter is known, we varied L and ρ_∞ away from the CT optimized values, since the properties that these parameters represent could certainly change on the short strand (although the $L = 10$ of the last paragraph was extreme, representing essentially perfect quenching along the entire length of the strand). In Fig. 22.4 the broken curve has $\mu = 1.5$, $L = 2$, and $\rho_\infty = 0.8$. This particular choice represents the best of an unsystematic but extensive search in the parameter space. It would be difficult to make a case for one set over the other. Both are less than optimum at low loading; moreover, quality of fit is to be balanced by the need to depart significantly from previously optimized parameters.

22.7 Conclusions

Interestingly, the results of our present calculations of the quenching characteristics of $[Ru(phen)_2(dppz)]^{2+}$ and $[Rh(phi)_2(phen)]^{3+}$ on a synthetic DNA-oligomer (28-mer) show that long-range electron transfer is not the only significant factor in explaining the experimentally observed quenching behavior. Our model, [7], developed for the quenching behavior of these two metal complexes on calf thymus DNA, used as fitting parameters the long-range electron transfer distance L, the maximum loading parameter of the quencher $[Rh(phi)_2(phen)]^{3+}$ at DNA, ρ_∞, and an additional interaction parameter R. The latter did not contribute significantly to the explanation of the experimental findings.

As outlined above, we have performed our calculations in several steps. From Figs. 22.2 and 22.3 it follows that the high quenching in the case of the DNA 28-mer cannot be modeled by adjusting the electron transfer distance L and the maximum loading parameter ρ_∞. From the nature of the DNA oligomers employed in this experiment it is clear that the electron transfer interaction distance cannot exceed the size of the DNA strand. In contrast, the repulsion parameter μ leads to a reasonable fit of the experimental data, although not of the same quality as for the CT-DNA, suggesting the presence of other factors. It is of interest that for the 28-mer data there is greater sensitivity to the parameter μ, the measure of repulsion, than to the principal parameters in our previous work, namely L and ρ_∞. A drastic adjustment of the latter nevertheless allowed equivalent fitting of the data with a small change in μ; see Fig. 22.4. (Not just any drastic change was allowed, of course, and arbitrary pairings allowed no reasonable fit.) This finding shows the significance of repulsion between the charged metal complexes in small DNA oligomers. The main difference between B-DNA and DNA oligomers is their shape, motion characteristics, and structure of the electrical (double) layer around the microheterogeneous structure [3,6]. In the latter case, end effects and the much smaller size of the DNA oligomers lead to a lower electric attraction potential between the negative charges embedded in the DNA structure (two negatively charged phosphate groups per DNA base pair). Therefore repulsion between the three-fold positively charged $[Rh(phi)_2(phen)]^{3+}$-quenchers can play a more dominant role in DNA oligomers than in B-DNA. (Whether or not this is the explanation of the importance of repulsion for the 28-mer data, it is clear that there *is* repulsion and that it strongly affects luminescence quenching.)

By the standards of say, Onsager, our calculations are just a step beyond hand-waving. Nevertheless, this allows a reasonable fit to the data, and because of sensitivity to some modeling parameters and not others, a differentiation of chemical mechanisms. This sensitivity may be considered an aspect of good experimental design. A natural step beyond our phenomenological model is to perform Monte Carlo simulations. A large variety of hypotheses can be considered in this way, without the compromises we found necessary in order to arrive at analytical forms. However, this has the disadvantage of requiring extensive programming for each experimental situation. Whether one's intuition is better informed by analytic results or by a collection of simulations may be a question of culture.

We wish to thank J. K. Barton, D. ben-Avraham, R. J. Gooding, V. Privman, C. Turro, N. J. Turro, and H. van Beijeren. Our work was supported in part by the US NSF grant PHY-9316681, the Deutsche Forschungsgemeinschaft, and by a grant from Hewlett-Packard.

Appendix A: Data

The following are the experimental data. Error bars for the quantity I_0/I (the quenching) are approximately 8%.

Calf thymus (10^4-mer)			28-mer		
Rh(uM)	ρ_s	I_0/I	Rh(uM)	ρ_s	I_0/I
0.00	0.0000	1.000	0.00	0.000	1.00
9.95	0.0796	1.288	10.2	0.082	2.2
19.80	0.1584	1.590	20.35	0.163	4.65
29.56	0.2365	2.038	24.6	0.197	5.02
39.22	0.3138	2.574	31.2	0.250	6.10
48.78	0.3902	3.416	39.8	0.318	9.68
58.25	0.4660	3.970	54.5	0.436	14.7
67.63	0.5410	4.246	67.3	0.538	18.1
76.92	0.6154	6.238	82.1	0.657	24.85
86.12	0.6890	7.254	98.7	0.790	34.3
95.24	0.7619	8.669			

References

[1] N. J. Turro, *Angew. Chem. Int. Ed. Engl.* **25**, 882 (1986).

[2] J. D. McGhee and P. H. v. Hippel, *Mol. Biol.* **86**, 469 (1974); Y. Dong, W. Shao, W. Tang and A. Dai, *Huaxue Xuebao* **49**, 1478 (1991); S. Maiti and K. L. Bhattacharya, *Indian J. Biochem. Biophys.* **23**, 197 (1986); B. Gaugain, J. Barbet, N. Capelle and B. Roques, *Biochemistry* **14**, 5078 (1978).

[3] N. J. Turro, J. K. Barton and D. A. Tomalia, *Acc. Chem. Res.* **24**, 332 (1991).

[4] M. D. Purugganan, C. V. Kumar, N. J. Turro and J. K. Barton, *Science* **24**, 1, 1645 (1988); G. Orellana, A. K.-D. Mesmaeker, J. K. Barton and N. J. Turro, *Photochem. Photobiol.* **54**, 499 (1991).

[5] M. F. Ottaviani, N. D. Ghatlia, S. H. Bossmann, N. J. Turro, J. K. Barton and H. J. Dürr, *J. Am. Chem. Soc.* **114**, 8946 (1992).

[6] J. C. Wang and G. N. Giaver, *Science* **240**, 300 (1988); D. Porschke, *Biophys. Chem.* **40**, 169 (1991); D. M. Crothers, *J. Mol. Biol.* **172**, 263 (1984); W. Saenger, *Principles of Nucleic Acid Structure*, p. 556 (Springer-Verlag, New York, 1988).

[7] L. S. Schulman, S. H. Bossmann and N. J. Turro, *J. Phys. Chem.* **99**, 9283 (1995).

[8] H. G. O. Becker, H. Büttcher, F. Dietz, D. Rehorek, G. Roewer, K. Schiller and H.-J. Timpe, *Einführung in die Photochemie*, p. 503 (Verlag der Wissenschaften, Berlin, 1991).

[9] G. Martini, *Colloids and Surfaces* **45**, 83 (1990) and references cited therein; L. J. Berliner, ed., *Spin Labeling II: Theory and Applications* (Academic Press, New York, 1979) and references cited therein.

[10] A. Schweiger, *Angew. Chem. Int. Ed. Engl.* **103**, 223 (1991); A. Hudson, in *Electron Spin Resonance*, Vol. **11**B, p. 9, M. C. R. Symons, ed. (The Royal Society of Chemistry, Thomas Graham House, Science Park, Cambridge, UK, 1989); C. P. Keijzers, E. J. Reijerse and J. Schmidt, eds., *Pulsed EPR: A New Field of Applications*, p. 238 (Koninklijke Nederlandse Akademie van Wetenschappen, Amsterdam, 1989); L. Kevan, *Modern Pulsed and Continuous Wave Electron Spin Resonance* (Wiley, New York, 1990).

[11] A. E. Derome, *Modern NMR Techniques for Chemistry Research* (Pergamon Press, 1987); R. R. Ernst, G. Bodenhausen and A. Nokaun, *Principles of Nuclear Magnetic Resonance in One and Two Dimensions* (Oxford University Press, 1987); H. Günther, *NMR-Spektroskopie* (Thieme, Stuttgart, 1992).

[12] H. Naumer and W. Heller, *Untersuchungsmethoden in der Chemie*, p. 386 (Thieme, Stuttgart, 1986); H. Galla, *Spektroskopische Methoden in der Biochemie*, p. 147 (Thieme, Stuttgart, 1988).

[13] S. Reekmans and F. C. De Schryver, in *Frontiers in supramolecular Organic Chemistry and Photochemistry*, p. 287, H.-J. Schneider and H. Dürr, eds. (VCH, Weinheim, 1991).

[14] C. J. Murphy, M. R. Arkin, Y. Jenkins, N. D. Ghatlia, S. H. Bossmann, N. J. Turro and J. K. Barton, *Science* **262**, 1025 (1993).

[15] R. A. Marcus and N. Sutin, *Biochim. Biophys. Acta* **811**, 265 (1985); B. E. Bowler, A. L. Raphael and H. B. Gray, *Prog. Inorg. Chem.* **38**, 259 (1990).

[16] C. J. Murphy, M. R. Arkin, N. D. Ghatlia, S. H. Bossmann, N. J. Turro and J. K. Barton, *Proc. Natl. Acad. Sci. USA* **91**, 5315 (1994).

[17] A. E. Friedman, J.-C. Chambron, J.-P. Sauvage, N. J. Turro and J. K. Barton, *J. Am. Chem. Soc.* **112**, 4960 (1990); A. E. Friedman, C. V. Kumar, N. J. Turro and J. K. Barton, *Nucleic Acids Res.* **19**, 2595 (1991); R. M. Hartshorn and J. K.

Barton, *J. Am. Chem. Soc.* **114**, 5919 (1992); Y. Jenkins, A. E. Friedman, N. J. Turro and J. K. Barton, *Biochemistry* **31**, 10809 (1992); A. Sitlani, E. C. Long, A. M. Pyle and J. K. J. Barton, *Am. Chem. Soc.* **114**, 2303 (1992); K. Uchida, A. M. Pyle, T. Morii and J. K. Barton, *Nucleic Acids Res.* **17**, 10259 (1989); N. Gupta, N. Grover, G. A. Neyhart, W. Liang, P. Singh and H. H. Thorp, *Angew. Chem.* **104**, 1058 (1992).

[18] J. J. Gonzales, P. C. Hemmer and J. S. Høye, *Chem. Phys.* **3**, 228 (1974); P. Nielaba and V. Privman, *Mod. Phys. Lett.* **B6**, 533 (1992); V. Privman, J. S. Wang and P. Nielaba, *Phys. Rev.* **B43**, 3366 (1991).

[19] T. Liggett, *Interacting Particle Systems* (Springer, New York, 1985); R. F. Pasternack, M. Caccam, B. Keogh, T. A. Stephenson, A. P. Williams and E. J. Gibbs, *J. Am. Chem. Soc.* **113**, 6835 (1991); V. Privman and M. Barma, *J. Chem. Phys.* **97**, 6714 (1992).

Index

Abbreviations

1D	one dimension, one-dimensional
2D	two dimensions, two-dimensional
3D	three dimensions, three-dimensional
AFM	antiferromagnet
AR	anisotropy ratio
ASEP	asymmetric simple exclusion process
ATP	adenosine triphosphate
BAW	branching annihilating random walk
CAP	creation and annihilation process
CCD	charge-coupled device
CL	chain length
CMB	$CsMnBr_3$
CP	contact process
CPU	central processing unit
CSA	cooperative sequential adsorption
CT	calf thymus (DNA)
DC	direct current
DE	deposition-evaporation (dynamics)
DF	delayed fluorescence (F)
DLA	diffusion-limited aggregation
DNA	deoxyribo(se)nucleic acid
DP	directed percolation
DT	Das Sarma-Tamborenea (model)
ESR	electron spin resonance
EW	Edwards-Wilkinson (equation, model)
FM	ferromagnet
FPE	Fokker-Planck equation
FWHM	full width at half maximum

HB	heat-bath
HCRWCS	hard-core random walkers with conserved spin
HTE	high-temperature expansion
ICCD	image-intensified CCD
IDL	interval distribution of level sets (analysis)
IMD	interacting monomer dimer (model)
IPDF	interparticle distribution function
IS	irreducible string
KCA	kink cellular automaton
KPZ	Kardar-Parisi-Zhang (equation, theory)
LC	liquid crystal
MBE	molecular beam epitaxy
MC	Monte Carlo (method, simulation)
MSD	many-sector decomposable (system)
NE	nonequilibrium (quantities)
NER	nonequilibrium relaxation
NMR	nuclear magnetic resonance
NN	nearest-neighbor (sites, spins)
ntcpa	naphtalene tetrachlorophtalic anhydride
P1VN	poly(1-vinyl)naphthalene
PCP	pair contact process
PH	phosphorescence (P)
PIC	pseudoisocyanide
PMMA	poly(methylmethacrylate)
Q1D	quasi-1D
Q2D	quasi-2D
YAG	yttrium aluminum garnet
RG	renormalization group
RSA	random sequential adsorption
RSOS	restricted solid-on-solid (model)
SDT	statistical dependence time
SE	skewed exponential (distribution)
SG	skewed Gaussian (distribution)
SOS	solid-on-solid (model)
TCB	tetrachlorobenzene
TDGL	time-dependent Ginzburg-Landau (dynamics, equation)
TMMB	$(CH_3)_4NMnBr_3$
TMMC	$(CH_3)_4NMnCl_3$
WV	Wolf-Villain (model)
ZGB	Ziff-Gulari-Barshad (model)